Additive Manufacturing

Additive Manufacturing

Alloy Design and Process Innovations

Special Issue Editors

Prashanth Konda Gokuldoss
Zhi Wang

MDPI • Basel • Beijing • Wuhan • Barcelona • Belgrade • Manchester • Tokyo • Cluj • Tianjin

Special Issue Editors

Prashanth Konda Gokuldoss
Head of Additive Manufacturing
Laboratory, Department of
Mechanical and Industrial
Engineering, Tallinn University
of Technology
Estonia

Zhi Wang
National Engineering Research
Center of Near-net-shape
Forming for Metallic Materials,
South China University of
Technology
China

Editorial Office
MDPI
St. Alban-Anlage 66
4052 Basel, Switzerland

This is a reprint of articles from the Special Issue published online in the open access journal *Materials* (ISSN 1996-1944) (available at: https://www.mdpi.com/journal/materials/special_issues/ AMADPI).

For citation purposes, cite each article independently as indicated on the article page online and as indicated below:

LastName, A.A.; LastName, B.B.; LastName, C.C. Article Title. *Journal Name* **Year**, *Article Number*, Page Range.

Volume 2
ISBN 978-3-03928-414-6 (Pbk)
ISBN 978-3-03928-415-3 (PDF)

Volume 1-2
ISBN 978-3-03928-416-0 (Pbk)
ISBN 978-3-03928-417-7 (PDF)

Cover image courtesy of Prasanth Konda Gokuldoss

Contents

About the Special Issue Editors

Prashanth Konda Gokuldoss (Prof.) is Head of the Additive Manufacturing Laboratory and Professor of Additive Manufacturing in the Department of Mechanical and Industrial Engineering, Tallinn University of Technology, Tallinn, Estonia. In addition, he is a guest scientist at the Erich Schmid Institute of Materials Science, Austrian Academy of Science, Leoben, Austria, and an Adjunct Professor at the department of CBCMT, School of Engineering, Vellore Institute of Technology, Vellore, India. was a awarded a PhD by the Technical University Dresden, Germany (2014), and conducted postdoctoral research at the Leibniz Institute of Solid State and Materials Research (IFW) Dresden, Germany. He also worked as an R&D Engineer (Sandvik, Sweden), Senior Scientist (Erich Schmid Institute of Materials Science, Austrian Academy of Science, Leoben, Austria), and Associate Professor (Norwegian University of Science and Technology, Gjøvik, Norway) before taking a Full Professorship at the Tallinn University of Technology, Tallinn, Estonia. His present research is focused on but not limited to additive manufacturing (alloy, process and product development), fabrication of meta-stable materials, powder metallurgy, light materials, solidification and biomaterials. He has published over 125 peer-reviewed journal papers and has an h-index of 32 (google scholar). A multiple award winner, he actively collaborates with/visits China, India, the USA, Austria, Poland, Norway, Germany, Spain, Taiwan, South Korea and Iran.

Zhi Wang (Prof.) is a Full Professor at the National Engineering Research Center of Near-net-shape Forming for Metallic Materials, School of Mechanical and Automotive Engineering, South China University of Technology, Guangzhou, China. He received his PhD from Technical University Dresden, Germany (2014), and did his postdoctoral research at the WPI-Advanced Institute for Materials Research, Tohoku University, Japan. He research revolved around powder metallurgy, additive manufacturing, Al and its alloys, metal matrix composites, solidification, and meta-stable materials. He is an active reviewer for more than 20 Elsevier journals and has published more than 65 articles in international peer reviewed journals. He is actively collaborating with groups around the world, including in Estonia, China, India, Germany, Austria, Australia, Poland and the USA.

Preface to "Additive Manufacturing"

Additive manufacturing (AM) is revolutionizing the manufacturing sector, particularly in terms of the production of metallic components with added functionality, including complex or intricate geometries, conformal cooling channels, and highly customized parts with small production cycles. In addition, material saving and weight reduction of the components in automobile and aerospace sectors, helping reduce fuel consumption, and promoting a green environment fall into this category. The AM field is undergoing rapid development with improvements and innovations taking place at all times. However, the AM field will face several challenges before it can be adapted completely to an industrial environment. The challenges include the process capabilities and material aspects, including the microstructure formation and properties, and the process cycle. It is, therefore, necessary to devote attention to the research and development activities in the field of AM, and to promote the industrialization process. This includes the production of powder, the properties of powder, AM process developments, alloy systems used for the AM process, and post-processing of the components. In view of the growing importance of additive manufacturing, this book addresses key issues related to emerging science and technology in this area (especially alloy design for additive manufacturing and process innovations). Detailed and informative articles are presented in this book, related to different additive manufacturing processes like selective laser melting (laser-based powder bed fusion process), electron beam melting, direct metal laser sintering, laser classing, ultrasonic consolidation, and wire arc additive manufacturing. Of key importance in the area of materials science are the end properties and the correlation with their microstructure. Accordingly, the articles presented critically discuss the effects of microstructural features, such as porosity, forming defects, and heat treatment induced effects on mechanical properties, with a focus on applications in the aerospace, automobile, defense and aerospace sectors. Overall, the cutting edge information presented in this book is of significant importance to researchers in the field of additive manufacturing.

Prashanth Konda Gokuldoss, Zhi Wang
Special Issue Editors

Article

Simulation of Stress Field during the Selective Laser Melting Process of the Nickel-Based Superalloy, GH4169

Zhanyong Zhao, Liang Li, Le Tan, Peikang Bai *, Jing Li, Liyun Wu, Haihong Liao and Yahui Cheng

School of Materials Science and Engineering, North University of China, Taiyuan 030051, China; syuzzy@126.com (Z.Z.); nucliliang@126.com (L.L.); hltanle@foxmail.com (L.T.); jing.li3d@hotmail.com (J.L.); wuliyunnuc@126.com (L.W.); scizhao@foxmail.com (H.L.); nuccyh@foxmail.com (Y.C.)
* Correspondence: baipeikang@nuc.edu.cn

Received: 1 August 2018; Accepted: 22 August 2018; Published: 24 August 2018

Abstract: In this paper, GH4169 alloy's distributions of temperature and stress during the selective laser melting (SLM) process were studied. The SLM process is a dynamic process of rapid melting and solidification, and we found there were larger temperature gradients near the turning of scan direction and at the overlap of the scanning line, which produced thermal strain and stress concentration and gave rise to warping deformations. The stresses increased as the distance became further away from the melt pool. There was tensile stress in the most-forming zones, but compressive stress occurred near the melt pool area. When the parts were cooled to room temperature after the SLM process, tensile stress was concentrated around the parts' boundaries. Residual stress along the z direction caused the warping deformations, and although there was tensile stress in the parts' surfaces, but there was compressive stress near the substrate.

Keywords: selective laser melting; GH4169; temperature and stress fields; simulation; model

---

1. Introduction

As a precipitation aging-enhanced nickel-based superalloy, GH4169 alloys are used extensively in the important high-temperature parts of the aerospace and nuclear industry due to its good corrosion resistance, anti-radiation, and excellent mechanical properties. Traditional manufacturing methods of GH4169 nickel-based superalloy may not have been the best choice, due to processing problems such as tool consumption, processing complex, high costs, and complex processes. Maybe this is why the application of selective laser melting (SLM) technology to form GH4169 nickel-based superalloy has caused widespread interest, as it is low-cost, has a simple process, and uses direct molding parts [1].

SLM technology can fabricate metal parts with complex shape, good mechanical properties, high precision, and high density [2], which cannot be produced using traditional methods [3,4]. SLM has been widely used in the fields of medical, military, aerospace, and automobile manufacturing because it is now able to process a variety of metals, including, but not limited to, chemical elements such as Al, Cu, Ti, etc. [5–7]. Up till now, many researchers have also carried out research on this technique: Michael et al. investigated the SLM process of the nickel-based superalloy IN738LC and the cobalt based alloy Mar-M509, where results showed that the microstructural and mechanical characteristics were attributed to the recovery and recrystallization behavior of IN738LC and Mar-M509 [8]; Vilaro et al. was able to prepare the Nimonic 263 with good microstructure and mechanical properties by using SLM [9]; Xia et al. established the mesoscopic model to investigate the thermodynamic mechanisms and densification behavior of nickel-based superalloy during additive manufacturing/three-dimensional (3D) printing (AM/3DP) processes [10]; Pröbstle et al. investigated creep properties of the polycrystalline nickel-based superalloy prepared by SLM, and found that the parts prepared by SLM had better creep strength than that of conventional casted parts [11]; Fabian et al. studied correlating laser-scanning strategies with the resulting textures and corresponding anisotropy of the elastic behavior of bulk materials [12]; and Carter et al. developed a processing route for the SLM powder-bed fabrication of the nickel superalloy CM247LC, and found that the island scan strategy strongly influenced the grain structure of the material [13].

The SLM process is a dynamic one of rapid melting and solidification with severe temperature gradients producing large thermal strain and stress concentration, resulting in warping deformations and cracks in the parts [14]. The distributions of temperature and stress in the alloy were difficult to measure by using the traditional method, prompting the widespread use of the finite element simulation method instead for the analysis of the temperature and stress field distributions [15,16]. Gu et al. established a three-dimensional, transient, finite element method (FEM) model to predict the stress distribution of parts shaped during the SLM process. By simulating the laser-beam scanning process, the peak values of the thermal stresses were first recorded at the onset of the first track where the first heating-cooling cycle occurred. After the whole part was cooled down, the largest residual stresses were found at the end of the first and last tracks. The simulation results were then verified by conducting the experimental investigation with the same parameters [17]. Hodge et al. discussed various perturbations of the process parameters and modeling strategies and compared the model-generated solid mechanics results [18]. Hussein et al. used three-dimensional finite element simulation to investigate the temperature and stress fields in single 316L stainless steel layers built on the powder bed without support in SLM [19]. Wu et al. established proper numerical models to investigate the residual stress evolution of AlSi10Mg alloy in a point exposure SLM process [20].

Although finite element simulation for the temperature and stress of the parts prepared by SLM has been carried out, due to differences in the physical properties of the materials, the distributions in temperature and stress of GH4169 alloys during the SLM process are difficult to predict. Furthermore, residual stress distribution in the parts prepared by SLM, which cool to room temperature, needs further research as temperature distribution and stress fields of GH4169 alloys' parts during the SLM process are the main subjects of study. Our aim in this study was to disclose the distributions of temperature and stress during the SLM process by establishing proper three-dimensional finite element models and providing theoretical guidance for the SLM formation of GH4169 alloys.

2. Experiment and Simulation

2.1. Experiment

The experimental equipment used was the Nd:YAG laser (LWY400P, Huagong Ltd., Wuhan, China), and the substrate plate was Q235. The experimental material used was GH4169 alloy powder, where its size range was 20–50 μm. The effects of the scanning process on distributions of temperature and stress during the SLM process were studied.

2.2. The Calculation Model

2.2.1. Finite Element Method for Transient Heat Conduction

The SLM is a rapid and intense forming process that uses the interaction of a laser beam. The material undergoes complex processes, such as thermal conduction, heat loss due to convection and radiation, phase transformation, and melting and cooling solidification. Therefore, a three-dimensional transient temperature field model was first established. During the heat transfer process, we assumed that the powder bed and its surroundings would constitute a closed and thermally insulated system. Energy balance follows the first law of thermodynamics, and in the Cartesian coordinate system, the temperature field of Ω can be expressed as a three-dimensional heat transfer differential equation, such as the following [21]:

$$\rho c \frac{\partial T}{\partial t} = Q + k_x \frac{\partial^2 T}{\partial x^2} + k_y \frac{\partial^2 T}{\partial y^2} + k_z \frac{\partial^2 T}{\partial z^2} \quad (x, y, z \in \Omega) \tag{1}$$

where ρ is material density (kg/m^3); c is the specific heat capacity (J/(kg·°C)); T is temperature (°C); t is interaction time (s); k_x, k_y, k_z are the effective thermal conductivity of the powder bed in the x, y, and z directions, respectively (W/(m·K)); and Q is the heat generated per volume within the component (W/m^3), which is described more specifically in the following sections.

In order to solve the heat transfer differential equation, the initial and boundary conditions need to be determined. Before the forming process begins, we assumed that the initial temperature was T_0, which can be expressed by:

$$T(x, y, z, 0) = T_0 \quad (x, y, z \in S) \tag{2}$$

In the SLM process, the surface of the powder bed interacting as a laser heat source is simplified to the heat flux input, which belongs to the second type of boundary conditions. This can be defined by:

$$k_x \frac{\partial T}{\partial x} n_x + k_y \frac{\partial T}{\partial y} n_y + k_z \frac{\partial T}{\partial z} n_z = Q \tag{3}$$

The convective heat dissipation process with air (or protective atmosphere) as the medium on the surface of the powder bed belongs to the third type of boundary condition, which can be defined by:

$$k_x \frac{\partial T}{\partial x} n_x + k_y \frac{\partial T}{\partial y} n_y + k_z \frac{\partial T}{\partial z} n_z = h(T - T_0) \tag{4}$$

The surface of the powder bed radiates thermal energy to the surrounding environment and belongs to the fourth type of boundary condition, which can be defined by:

$$k_x \frac{\partial T}{\partial x} n_x + k_y \frac{\partial T}{\partial y} n_y + k_z \frac{\partial T}{\partial z} n_z = \sigma \varepsilon \left(T^4 - T_0^4 \right) \tag{5}$$

In summary, the temperature field boundary condition can be defined as:

$$k \frac{\partial T}{\partial n} = Q + h(T - T_0) + \sigma \varepsilon \left(T^4 - T_0^4 \right) \quad (x, y, z \in S) \tag{6}$$

In Equations (3)–(6), n is the normal vector of the top surface S; h is the heat transfer coefficient of the natural thermal convection; T_0 is the ambient temperature, considered to be 25 °C; σ is the emissivity; and ε is the Stefan-Boltzmann constant, which has the value of 5.67×10^8 W/m^2 K^4.

2.2.2. Basic Theory of Stress Field Simulation

The stress perpendicular to the scanning surface is called the normal stress σ, and the normal stress along the x, y, and z axes are expressed as σ_x, σ_y and σ_z. Stress tangent to the scanning surface is called shear stress, and the shear stress along the x, y, and z axes are expressed as τ_x, τ_y, and τ_z. As there will be some residual stress formed after the cooling solidification stage, if the residual stress was greater than the yield strength of the material, it will produce local deformation. Thus, the equivalent stress of Mises is the stress at yield [22].

During the SLM process, the material strain rate $\dot{\varepsilon}$ influenced by external force and temperature includes the elastic strain rate $\dot{\varepsilon}^e$, plastic strain rate $\dot{\varepsilon}^p$, creep strain rate $\dot{\varepsilon}^c$, and the strain rate caused by temperature change $\dot{\varepsilon}^T$. Their relationship can be described as [23]:

$$\dot{\varepsilon} = \dot{\varepsilon}^e + \dot{\varepsilon}^p + \dot{\varepsilon}^c + \dot{\varepsilon}^T. \tag{7}$$

The elastic constant varying with temperature can be defined by:

$$\dot{\varepsilon}^e = \frac{d\left(D_e^{-1}\sigma\right)}{dt} = D_e^{-1}\dot{\sigma} + \frac{d}{dt}\left(D_e^{-1}\right)\sigma \tag{8}$$

where D_e is the elastic matrix, which is the time derivative of stress.

Based on the flow theory, plastic strain rate can be determined by:

$$\dot{\varepsilon}^p = \dot{\lambda}\frac{\partial F}{\partial \sigma} \tag{9}$$

where F is the yield function, $\dot{\lambda}$ is the plastic growth factor.

Based on the creep theory, the creep strain rate can be described as:

$$\dot{\varepsilon}^c = \frac{3}{2}\frac{\bar{\dot{\varepsilon}}^c}{\bar{\sigma}}\sigma' \tag{10}$$

where $\bar{\sigma}$ is equivalent stress, $\bar{\dot{\varepsilon}}^c$ is equivalent creep strain rate, and $\bar{\dot{\varepsilon}}^c$ can be described as:

$$\bar{\dot{\varepsilon}}^c = \frac{d\bar{\varepsilon}^c}{dt} = \frac{\sqrt{2}}{3}\left[\left(\dot{\varepsilon}_{11}^c - \dot{\varepsilon}_{22}^c\right)^2 + \left(\dot{\varepsilon}_{22}^c - \dot{\varepsilon}_{23}^c\right)^2 + \left(\dot{\varepsilon}_{33}^c - \dot{\varepsilon}_{11}^c\right)^2 + 6\left(\dot{\varepsilon}_{12}^{c\,2} + \dot{\varepsilon}_{23}^{c\,2} + \dot{\varepsilon}_{31}^{c\,2}\right)\right]^{\frac{1}{2}} \tag{11}$$

The temperature strain rate can be defined by:

$$\dot{\varepsilon}^T = \dot{T}A_1 \tag{12}$$

where A_1 can be described by $1 = a\{1, 1, 1, 0, 0, 0\}^T$, a is the linear expansion coefficient, and $\dot{T}$ is the change rate of temperature with time.

Therefore,

$$\dot{\varepsilon} = D_e^{-1}\dot{\sigma} + \left(\frac{d}{dt}D_e^{-1}\right)\sigma + \dot{\lambda}\frac{\partial F}{\partial \sigma} + \dot{\varepsilon}^c + \dot{\varepsilon}^T \tag{13}$$

The type (14) multiplied by the elastic coefficient matrix can be calculated as:

$$\dot{\sigma} = D_e\dot{\varepsilon} - \dot{\lambda}D_e\frac{\partial F}{\partial \sigma} - D_e^{-1}\left(\dot{\varepsilon}^c + \dot{\varepsilon}^T\right) + \frac{dD_e}{dt}\dot{\varepsilon}^c \tag{14}$$

When the yield condition is:

$$F\left(\sigma_{ij}, \varepsilon_{ij}^p, T\right) = 0 \tag{15}$$

It can be calculated:

$$\dot{\lambda} = \frac{q^T D_e\dot{\varepsilon} - q^T D_e\left(\dot{\varepsilon}^c + \dot{\varepsilon}^T\right) + q^T\frac{dD_e}{dt}\dot{\varepsilon}^c + \frac{\partial F}{\partial T}\dot{T}}{p^T q + q^T D_e q} \tag{16}$$

$$q = \frac{\partial F}{\partial \sigma} \tag{17}$$

Where p^T and q^T are the corresponding matrices, the incremental elastic–plastic strain can be described as [24].

$$\sigma = \left[D_e - \frac{D_e q(D_e q)^T}{W}\right]\left(\dot{\varepsilon} - \dot{\varepsilon}^c - \dot{\varepsilon}^T + \dot{\varepsilon}^e\right) - D_e\frac{\partial F\dot{T}}{\partial TW} \tag{18}$$

$$W = p^T q + q^T D_e q \tag{19}$$

2.2.3. Mechanical Properties of Materials

GH4169 was used as the melting material, and its melting temperature is 1260–1320 °C. The mechanical parameters and coefficient of the thermal expansion of GH4169 alloys are listed in Tables 1 and 2. The process parameters in the simulation and experiments are shown in Table 3, which, through the experiments and simulations, have been proven to result in high-quality products.

Table 1. Mechanical properties of GH4169 alloy.

Temperature/°C	20	100	200	300	400	500	600	700
Elastic Modulus/GPa	205	201	196	189	183	176	169	164
Shear Modulus/GPa	79	77	70	73	70	67	64	61
Poisson Ratio	0.3	0.3	0.3	0.3	0.31	0.31	0.32	0.34

Table 2. Coefficient of thermal expansion of GH4169 alloy.

Temperature/°C	100	200	300	400	500	600	700	800	900	1000
Thermal Expansion Coefficient/10^{-6}·°C^{-1}	13.2	13.3	13.8	14	14.6	15	15.8	17	18.4	18.7

Table 3. The process parameters in the simulation and experiments.

Parameter	100
Laser Power	150 W
Scanning Speed	150 mm/min
Hatching Space	100 μm
Spot Diameter	150 μm
Layer Thickness	50 μm

2.2.4. The Moving Heat Source Model

The accuracy of the temperature and stress field simulations were affected by the heat source model. LWY400P-type YAG-pulsed laser equipment was used, and its laser heat source was loaded into the powder bed in the form of a heat flux, which obeyed the Gauss distribution [25]:

$$Q = \frac{2AP}{\pi\omega^2}exp\left(-\frac{2r^2}{\omega^2}\right) \tag{20}$$

where q is laser power density, P is laser power, and A_2 is the heat absorption rate of materials. In the study, A_2 was optimized as 0.38, the wavelength of the laser beam was 1.06 μm, ω is the laser spot radius, and r is the distance from a point on the surface of the powder bed to the center of the laser beam. This can be determined by:

$$r = (x - x_0)^2 + (y - y_0)^2 \tag{21}$$

In Figure 1a, a simplified model of the Gauss heat source is described. The scope of the laser beam was approximated as being circular, which accounted for a cell size of 3 × 3. The heat flow density in the middle shaded area was 1, and in the four blank corners is 0.5. The scanning strategy is shown in Figure 1b [21].

The three-dimensional finite element model was divided into two components: powder bed and substrate. The size of the three-layer model was 0.6 mm × 0.6 mm × 0.45 mm. An eight-node hexahedron SOLID70 3D solid element was adopted as the mesh type in the model, and its dimension was 0.033 mm × 0.033 mm × 0.033 mm. Meanwhile, the mesh size of Q235 was 4 mm × 4 mm × 0.9 mm, and a three-dimensional, twenty-node thermal solid element, SOLID90, was adopted as the mesh type. The effect of the substrate temperature on the powder bed could be minimized by using the partition of different units with the powder bed mode. The PCG solver was mainly used for this simulation and the birth–death element method was used to load the heat source onto different units at different times. The finite element model is shown in Figure 2.

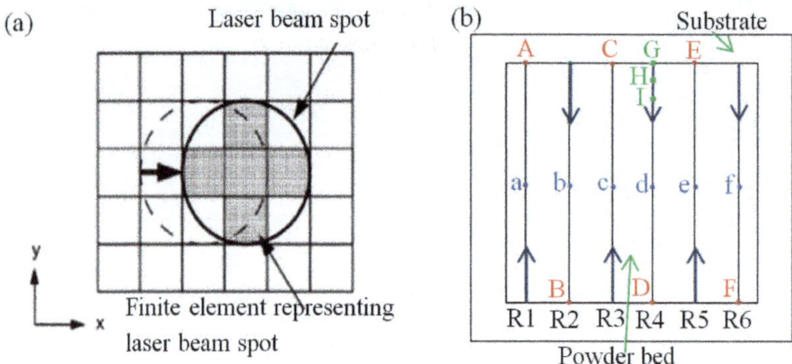

Figure 1. Movement of the laser beam spot and schematic diagram of powder bed. (**a**) Movement of laser beam spot by five elements; (**b**) Scanning strategy.

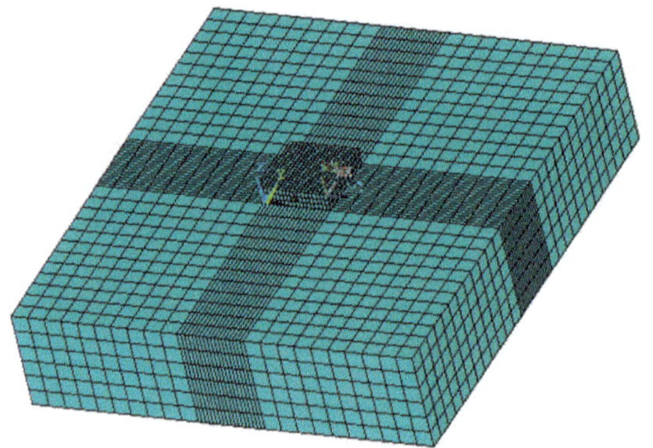

Figure 2. The single-layer finite element model.

3. Results and Discussions

3.1. Temperature Distribution during the SLM Process

Figure 3 shows the temperature curves of the points a, b, c, d, e, and f in Figure 1, with the changes in interaction times. Figure 4 is the corresponding node temperature change-rate curve. We will be discussing the two curves by taking point a as an example, where the curves of the other points can then be deduced by analogy.

When the laser beam closes into midpoint a of the scan line, the temperature gradually increases while the heating rate increases rapidly. The maximum values of the temperature and heating rates can be obtained when the laser beam reaches the midpoint, as shown by t = 0.17 s in Figures 3 and 4, respectively. When the laser beam leaves the point, the temperature of point *a* begins to decrease, while the cooling rate begins to increase. The laser beam continues to scan toward point *b* for point *a*, during which the laser beam separation process and, afterwards, the laser beam approach is experienced. Accordingly, the temperature at point *a* decreases at first and then rises slowly. As shown in Figure 3, the time at which the second peak of point *a* and the highest temperature of point *b* occur coincide exactly.

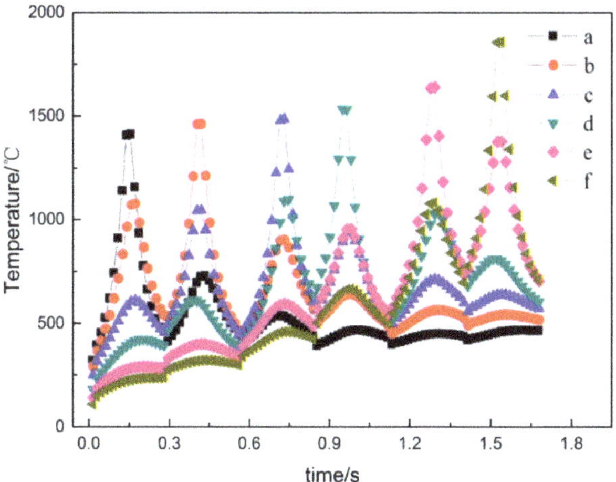

Figure 3. The temperature curves of points a, b, c, d, e, and f.

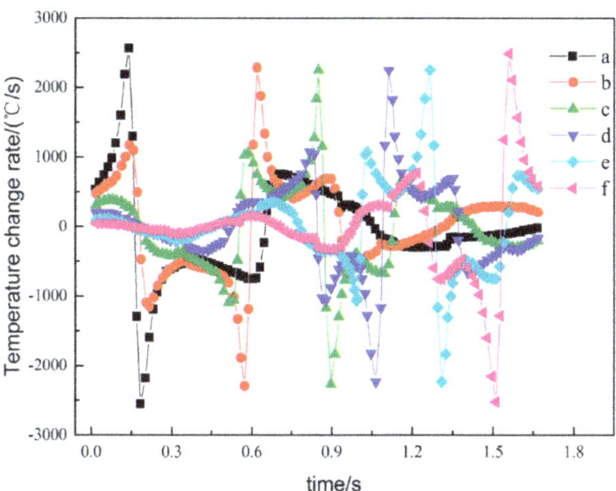

Figure 4. The temperature change rate curves of points a, b, c, d, e, and f.

Based on the above discussion, the temperature change trend of each scan line has six peaks, and the curve change trend is basically the same. The maximum temperature was obtained when the laser beam scanned the point. When the overlapping area between two adjacent tracks was formed, the temperature increased; however, the heat effect of the other scanning line on the point was less. As shown in Figure 3, because of the heat accumulation effect, the maximum temperature from point *a* to point *f* slightly increased. The first formed area has a preheating effect on the post-forming area, and the post-formed area re-melts the previously formed area to achieve metallurgical bonding, thereby ensuring the quality of the formed part. The heating rate was described as being greater than zero, and the cooling rate was described as being less than zero, as shown in Figure 4. The SLM forming process had high heating and cooling rates, thus causing higher temperature gradients and related thermal stresses in the parts.

Figure 5 shows the isothermal diagrams at the endpoint of each scan line in the single layer scanning. There was a certain preheating function in the vicinity of the powder bed under the action of the laser heat source. The densest temperature isotherm around the laser heat sources caused the steep temperature gradient. Simulation results show that the maximum temperatures from point A to point F at the end of each scan line are 1829, 1524, 1494, 1602, 1708, and 1988 °C, respectively, which is different to the temperature change of the midpoint of the adjacent scanning line. It is shown that at first, the temperature decreased due to the good absorption of powders to laser radiation. Afterwards, the cumulative heat of the powder bed made the temperature increase. The temperature reached a maximum point at the end of the scan line, which had a greater effect at the beginning of the next scan line due to the complexities in thermal convection and heat radiation. It can thus be said that it is easy to cause thermal stresses and deformations with the turning of scan direction.

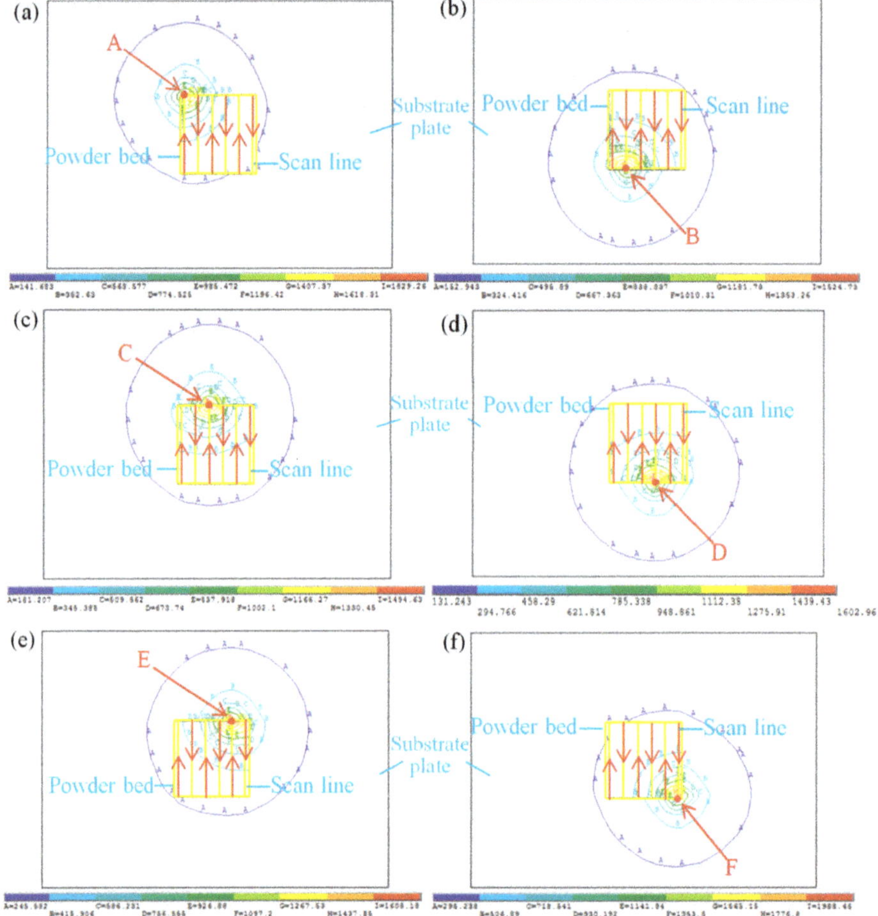

Figure 5. The isothermal diagrams of the endpoint of the scan line. (**a**) Point A; (**b**) Point B; (**c**) Point C; (**d**) Point D; (**e**) Point E; (**f**) Point F.

3.2. Stress Analysis

3.2.1. Stress Field Distribution during the SLM Process

The isothermal temperature and stress fields during the SLM process are shown in Figure 6. In the center of the laser heat source, the temperature gradient was larger (Figure 6a), which resulted in compressive stress. In order to maintain the balance, the low-temperature region also had a corresponding tensile force (Figure 6b), and the large tensile stress caused warping deformations or cracks at the boundary of the forming part. As shown in Figure 6c, whereas the alloy powder's lower-temperature zone created obvious tensile stress, compressive stress was formed in the higher-temperature zone, although the stress concentration and stress value were small. Stress was formed in the forming part due to the temperature gradient, resulting in the deformations. As shown in Figure 6d, the Von Mises stress equivalent was concentrated at the overlap of the scanning line, where the stress value was 326 MPa, which closed to the yield strength of the material. That was because the temperature variation of the SLM process was complex and severe, as shown in Figures 3 and 4. When the adjacent scanning lines overlapped, the temperature difference between the melting liquid metal and solidification parts was bigger, and thus caused higher temperature gradients, leading to instantaneous stress close to or exceeding the yield strength, and thus easily causing cracks and deformations.

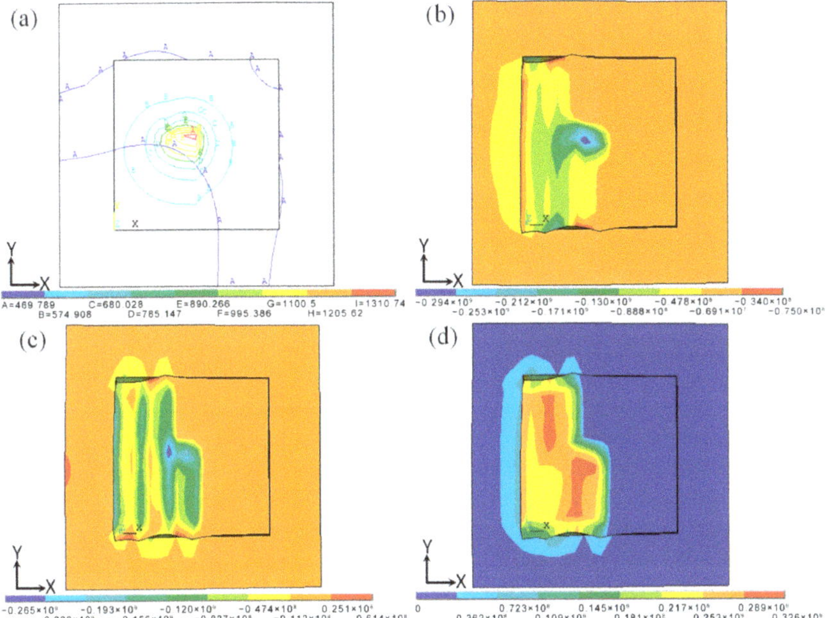

Figure 6. The isotherm temperature and stress fields during the selective laser melting (SLM) process. (**a**) The isotherm temperature field; (**b**) Tensile stress distribution; (**c**) Compressive stress distribution; (**d**) Equivalent stress distribution.

As shown in Figures 3–5, thermal stress and deformation occurred easily due to complex temperature variation near the turning of scan direction. The corresponding stress field is shown in Figure 7. There was a large amount of stress concentration at point C, while the stress value decreased when the laser beam scanned point G—however, the stress concentration did not decrease due to the large temperature gradient. The range of stress concentration was greatly reduced when the laser beam moved to point I, reason being because the appropriate forming lap ratio made a good lap between adjacent scan lines and thus caused re-melting, which made the residual stress in the interior release and the stress concentration reduce.

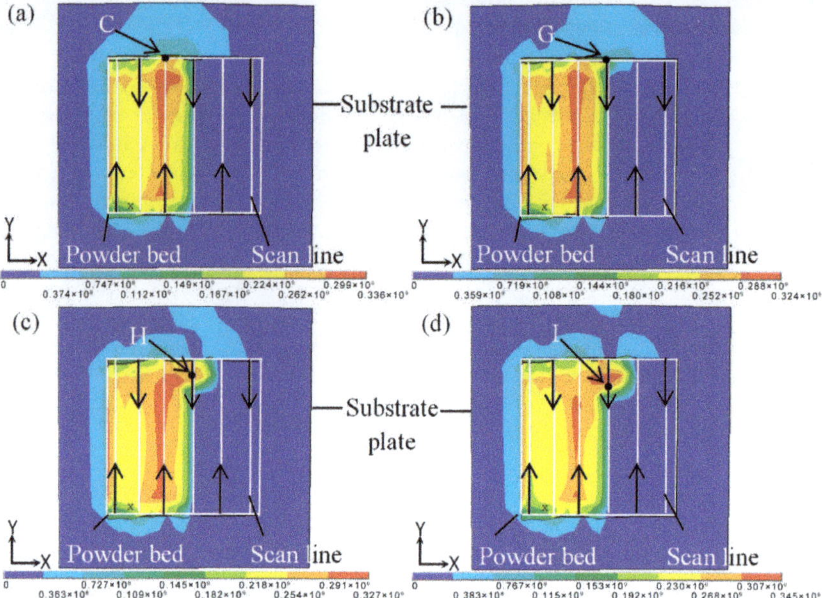

Figure 7. The stress distribution fields near the turning of scan direction. (**a**) Point C; (**b**) Point G; (**c**) Point H; (**d**) Point I.

3.2.2. Residual Stress Distribution of the Product at Room Temperature

Figure 8 shows the residual stress filed when the parts cooled to room temperature. The residual stress σ_x of the forming area was mainly tensile stress (Figure 8a). The maximum value of tensile stress was obtained near the boundary of the parts, which was because the boundary region was not hindered by the surrounding powder. This resulted in greater tensile stress during the thermal expansion process of the materials, ultimately causing cracks and warping deformations. Figure 8b shows that larger tensile stress was the key component of residual stress σ_y, which occurred in the turning of scan direction. The compressive stress was concentrated in the overlapping area of the scanning line, which was because the scanning direction of single-layer scanning was along the y axis, and changed direction at the x axis. The longer cooling time resulted in a lower temperature gradient and stress concentration than when it had a longer scanning distance along the y axis. In the scan direction turning process, the short scanning distance and large temperature gradient, was what caused stress concentration. The residual stress σ_y was the main reason for the formation of cracks and deformations in the turning of scan direction. The residual stress σ_z along the z direction was the main cause of warpage, as shown in Figure 8c. This is because the directions of the compressive stress caused by the lower temperature in the substrate and the tensile stress caused by thermal expansion in the powder bed were contrary to one another, making it more likely to cause warping deformations. This defect could be reduced by preheating the substrate. The value of the equivalent stress of Von Mises did not exceed the material yield strength, which could ensure the quality of the forming parts (Figure 8d).

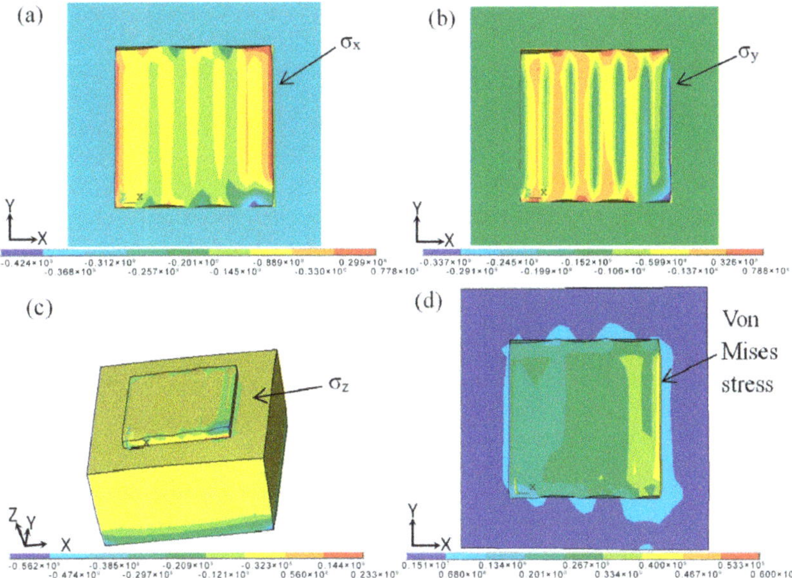

Figure 8. The residual stress fields cooling to room temperature. (**a**) Residual stress σ_x distribution; (**b**) Residual stress σ_y distribution; (**c**) Residual stress σ_z distribution; (**d**) Equivalent stress distribution.

In order to observe the change in internal stress during the SLM process, the curve of the residual stress σ_x at the midpoint of each scan line was plotted and the change trend of each point was roughly the same, as shown in Figure 9. The corresponding temperature distributions and temperature change rates of the midpoints of each scan line are shown in Figures 3 and 4. When laser beam scanning reached the midpoint, the value of residual stress σ_x decreased accordingly due to the heat concentration and rise in temperature. As the laser beam moved away, the value of residual stress σ_x increased accordingly. During the cooling process after processing was completed, the overall trend in the changes in residual stress σ_x was generally stable at the beginning, and then increased rapidly. This occurred because there was a certain temperature when the forming process had just finished. In order to balance out the temperature gradient, the internal stress of the forming parts changed gradually and finally became stable after the cooling of the forming parts, which formed the residual stress σ_x.

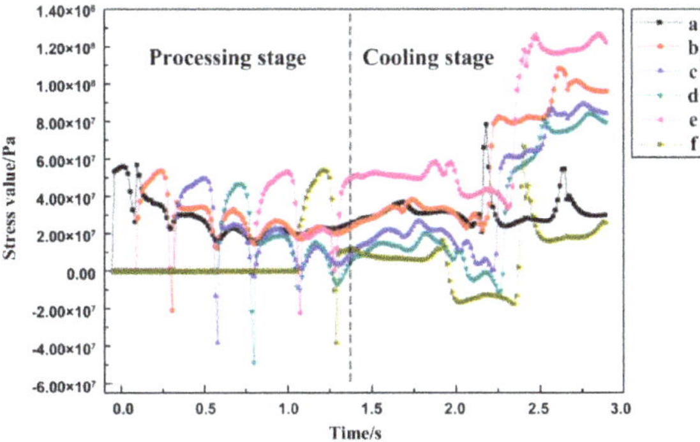

Figure 9. The variation of residual stress σ_x with time at the midpoint of each scan line.

Warping and cracking in the formed parts have always been a problem in the SLM-forming process. The most fundamental reason for this phenomenon is that the interaction with the laser heat source unbalances the powder bed and forms a steep temperature gradient, resulting in inconsistent shrinkage of the material system. Figure 10a shows a typical warping deformation sample. It is clear that there is obvious crack at the bottom of the formed part, and the experimental results are consistent with the simulation analysis. During the molding process, as the scanning area and layer thickness increase, warping first occurs at the boundary region of a single layer caused by the inappropriate process parameters. Contact with the simulation analysis knows that because of the excessive stress in σ_x, σ_y and σ_z at the junction of the formed part and the substrate, cracks and warpage occur. Under the guidance of the simulated results, we were able to obtain high-quality SLM parts (Figure 10c) using optimized process parameters. Furthermore, common methods for reducing defects include preheating the substrate and using it while having the thermal expansion coefficient slightly larger than the SLM-formed part, which can effectively reduce the temperature gradient.

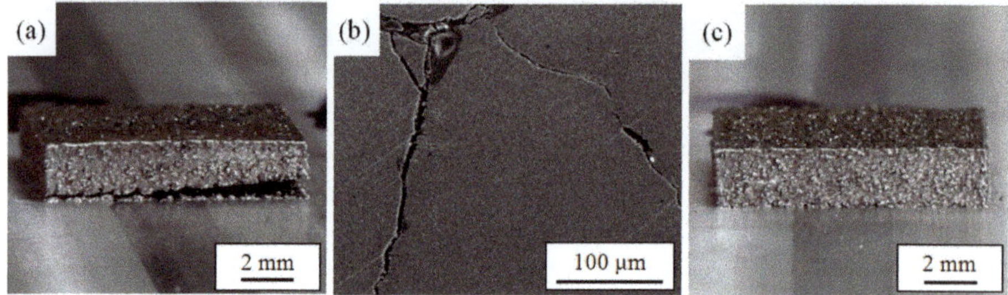

Figure 10. Experimental SLM-forming samples. (**a**) Warping deformations in the sample; (**b**) Cracks in the sample; (**c**) A high-quality sample.

4. Conclusions

(1) The SLM process was a dynamic process of rapid melting and solidification, and there were larger temperature gradients near the scan direction turning point and at the overlap of the scanning line, which produced thermal strain and concentrations of stress, and also gave rise to warping deformations.

(2) The stresses increased as the distance away from the melt pool also increased. There was tensile stress in zones with the highest levels of concentration, but compressive stress occurred near the melt pool area.

(3) After the SLM process, when the parts were cooled to room temperature, the tensile stress concentrated on the boundary of the parts. Residual stress along the z direction caused the warping deformations. Although there was tensile stress in the parts' surfaces, there was compressive stress near the substrate.

Author Contributions: Conceptualization, Z.Z.; Data Curation, L.L.; Investigation, L.L., L.W. and Y.C.; Methodology, Z.Z. and H.L.; Project Administration, P.B.; Supervision, Z.Z.; Validation, J.L.; Writing-original Draft, L.L. and Y.C.; Writing-review & Editing, Z.Z., L.L. and L.T.

Funding: This research was funded by the National Natural Science Foundation of China (Grant No. 51604246 and Grant No. 51775521) and the North University of China for Young Academic Leaders.

Conflicts of Interest: The authors declare no conflict of interest.

References

1. Heckl, A.; Neumeier, S.; Göken, M.; Singer, R.F. The effect of Re and Ru on γ/γ' microstructure, γ-solid solution strengthening and creep strength in nickel-base superalloys. *Mater. Sci. Eng. A* **2011**, *528*, 3435–3444. [CrossRef]

2. Abe, F.; Osakada, K.; Shiomi, M.; Uematsu, K.; Matsumoto, M. The manufacturing of hard tools from metallic powders by selective laser melting. *J. Mater. Process Technol.* **2001**, *111*, 210–213. [CrossRef]

3. Zhao, Z.Y.; Bai, P.K.; Guan, R.G.; Murugadoss, V.; Liu, H.; Wang, X.J.; Guo, Z.H. Microstructural evolution and mechanical strengthening mechanism of Mg-3Sn-1Mn-1La alloy after heat treatments. *Mater. Sci. Eng. A* **2018**, *734*, 200–209. [CrossRef]

4. Zhao, Z.Y.; Guan, R.G.; Zhang, J.H.; Zhao, Z.Y.; Bai, P.K. Effects of process parameters of semisolid stirringon microstructure of Mg-3Sn-1Mn-3SiC (wt%) strip processed by rheo-rolling. *Acta Metall. Sin.* **2017**, *30*, 66–72. [CrossRef]

5. Prashanth, K.G.; Shahabi, H.S.; Attar, H.; Srivastava, V.C.; Ellendt, N.; Uhlenwinkel, V.; Eckert, J.; Scudino, S. Production of high strength Al85Nd8Ni5Co2 alloy by selective laser melting. *Addit. Maunf.* **2015**, *6*, 1–5. [CrossRef]

6. Scudino, S.; Unterdörfer, C.; Prashanth, K.G.; Attar, H.; Ellendt, N.; Uhlenwinkel, V.; Eckert, J. Additive manufacturing of Cu-10Sn bronze. *Mater. Lett.* **2015**, *156*, 202–204. [CrossRef]

7. Attar, H.; Ehtemam-Haghighi, S.; Kent, D.; Dargusch, M.S. Recent developments and opportunities in additive manufacturing of titanium-based matrix composites: A review. *Int. J. Mach. Tool Manuf.* **2018**, *133*, 85–102. [CrossRef]

8. Cloots, M.; Kunze, K.; Uggowitzer, P.J.; Wegener, K. Microstructural characteristics of the nickel-based alloy IN738LC and the cobalt-based alloy Mar-M509 produced by selective laser melting. *Mater. Sci. Eng. A* **2016**, *658*, 68–76. [CrossRef]

9. Vilaro, T.; Colin, C.; Bartout, J.D.; Nazé, L.; Sennour, M. Microstructural and mechanical approaches of the selective laser melting process applied to a nickel-base superalloy. *Mater. Sci. Eng. A* **2012**, *534*, 446–451. [CrossRef]

10. Xia, M.; Gu, D.; Yu, G.; Dai, D.; Chen, H.; Shi, Q. Selective laser melting 3D printing of Ni-based superalloy: understanding thermodynamic mechanisms. *Sci. Bull.* **2016**, *61*, 1013–1022. [CrossRef]

11. Pröbstle, M.; Neumeier, S.; Hopfenmüller, J.; Freund, L.P.; Niendorf, T.; Schwarze, D.; Göken, M. Superior creep strength of a nickel-based superalloy produced by selective laser melting. *Mater. Sci. Eng. A* **2016**, *674*, 299–307. [CrossRef]

12. Geiger, F.; Kunze, K.; Etter, T. Tailoring the texture of IN738LC processed by selective laser melting (SLM) by specific scanning strategies. *Mater. Sci. Eng. A* **2016**, *661*, 240–246. [CrossRef]

13. Carter, L.N.; Martin, C.; Withers, P.J.; Attallah, M.M. The influence of the laser scan strategy on grain structure and cracking behaviour in SLM powder-bed fabricated nickel superalloy. *J. Alloy Compd.* **2014**, *615*, 338–347. [CrossRef]

14. Zhao, Z.Y.; Li, L.; Bai, P.K.; Jin, Y.; Wu, L.Y.; Li, J.; Guan, R.G.; Qu, H.Q. The Heat treatment influence on the microstructure and hardness of TC4 titanium alloy manufactured via selective laser melting. *Materials* **2018**, *11*, 1318. [CrossRef] [PubMed]

15. Li, Z.H.; Xu, R.J.; Zhang, Z.W.; Kucukkoc, I. The influence of scan length on fabricating thin-walled components in selective laser melting. *Int. J. Mach. Tool Manuf.* **2017**, *126*, 1–12. [CrossRef]

16. Dai, K.; Shaw, L. Thermal and mechanical finite element modeling of laser forming from metal and ceramic powders. *Acta Mater.* **2004**, *52*, 69–80. [CrossRef]

17. Gu, D.; He, B. Finite element simulation and experimental investigation of residual stresses in selective laser melted Ti-Ni shape memory alloy. *Comp. Mater. Sci.* **2016**, *117*, 221–232. [CrossRef]

18. Hodge, N.E.; Ferencz, R.M.; Vignes, R.M. Experimental comparison of residual stresses for a thermomechanical model for the simulation of selective laser melting. *Addit. Manuf.* **2016**, *12*, 159–168. [CrossRef]

19. Hussein, A.; Hao, L.; Yan, C.; Everson, R. Finite element simulation of the temperature and stress fields in single layers built without-support in selective laser melting. *Mater. Des.* **2013**, *52*, 638–647. [CrossRef]

20. Wu, J.; Wang, L.; An, X. Numerical analysis of residual stress evolution of AlSi10Mg manufactured by selective laser melting. *Optik-Int. J. Light Electron Opt.* **2017**, *137*, 65–78. [CrossRef]

21. Matsumoto, M.; Shiomi, M.; Osakada, K.; Abe, F. Finite element analysis of single layer forming on metallic powder bed in rapid prototyping by selective laser processing. *Int. J. Mach. Tool Manuf.* **2002**, *42*, 61–67. [CrossRef]

22. Li, C.; Liu, J.F.; Guo, Y.B. Prediction of residual stress and part distortion in selective laser melting. *Procedia CIRP* **2016**, *45*, 171–174. [CrossRef]

23. Parry, L.; Ashcroft, I.A.; Wildman, R.D. Understanding the effect of laser scan strategy on residual stress in selective laser melting through thermo-mechanical simulation. *Addit. Manuf.* **2016**, *12*, 1–15. [CrossRef]
24. Xia, Z.; Okabe, T.; Curtin, W.A. Shear-lag versus finite element models for stress transfer in fiber-reinforced composites. *Compos. Sci. Technol.* **2002**, *62*, 1141–1149. [CrossRef]
25. Zhao, X.; Iyer, A.; Promoppatum, P.; Yao, S.C. Numerical modeling of the thermal behavior and residual stress in the direct metal laser sintering process of titanium alloy products. *Addit. Manuf.* **2017**, *14*, 126–136. [CrossRef]

 *materials*

Article

Analytical Modeling of In-Process Temperature in Powder Bed Additive Manufacturing Considering Laser Power Absorption, Latent Heat, Scanning Strategy, and Powder Packing

Jinqiang Ning [1,*], Daniel E. Sievers [2], Hamid Garmestani [3] and Steven Y. Liang [1,*]

1 Georgia Institute of Technology, George W. Woodruff School of Mechanical Engineering, 801 Ferst Drive, Atlanta, GA 30332-0405, USA
2 The Boeing Company, 499 Boeing Boulevard, Huntsville, AL 35824, USA; daniel.e.sievers@boeing.com
3 Georgia Institute of Technology, School of Materials Science and Engineering, 771 Ferst Drive NW, Atlanta, GA 30332, USA; hamid.garmestani@mse.gatech.edu
* Correspondence: jinqiangning@gatech.edu (J.N.); steven.liang@me.gatech.edu (S.Y.L.)

Received: 5 February 2019; Accepted: 5 March 2019; Published: 8 March 2019

Abstract: Temperature distribution gradient in metal powder bed additive manufacturing (MPBAM) directly controls the mechanical properties and dimensional accuracy of the build part. Experimental approach and numerical modeling approach for temperature in MPBAM are limited by the restricted accessibility and high computational cost, respectively. Analytical models were reported with high computational efficiency, but the developed models employed a moving coordinate and semi-infinite medium assumption, which neglected the part dimensions, and thus reduced their usefulness in real applications. This paper investigates the in-process temperature in MPBAM through analytical modeling using a stationary coordinate with an origin at the part boundary (absolute coordinate). Analytical solutions are developed for temperature prediction of single-track scan and multi-track scans considering scanning strategy. Inconel 625 is chosen to test the proposed model. Laser power absorption is inversely identified with the prediction of molten pool dimensions. Latent heat is considered using the heat integration method. The molten pool evolution is investigated with respect to scanning time. The stabilized temperatures in the single-track scan and bidirectional scans are predicted under various process conditions. Close agreements are observed upon validation to the experimental values in the literature. Furthermore, a positive relationship between molten pool dimensions and powder packing porosity was observed through sensitivity analysis. With benefits of the absolute coordinate, and high computational efficiency, the presented model can predict the temperature for a dimensional part during MPBAM, which can be used to further investigate residual stress and distortion in real applications.

Keywords: in-process temperature in MPBAM; analytical modeling; high computational efficiency; molten pool evolution; laser power absorption; latent heat; scanning strategy; powder packing

---

1. Introduction

Metal powder bed additive manufacturing (MPBAM), alternatively named powder bed fusion (PBF), is one of the widely used additive manufacturing processes, in which high-density laser power is used to selectively melt and fuse powders to the build part in a layer by layer manner. MPBAM is capable of producing geometrically complex parts with effective cost [1]. The large temperature gradient in MPBAM is frequently observed due to the repeatedly rapid heat and solidification, which are detrimental to the quality of the produced part by causing defects, such as cracking [2], undesired residual stress [3], and part distortion [4,5], and thus alternating the part's mechanical properties and functionality [6–8]. Therefore, the monitor or prediction of the temperature distribution in MPBAM is needed.

In situ temperature measurement is difficult and inconvenient due to the restricted accessibility under elevated temperature conditions [9,10]. Non-contact and contact techniques are employed for temperature measurements in the additive manufacturing (AM) process. Non-contact thermal photographic techniques, such as an infrared (IR) pyrometer and an IR camera, can only measure temperatures on the exposed surfaces [11,12]. Contact techniques, such as an embedded thermocouple, can only measure temperatures inside the substrate rather than the build [13,14]. In fact, the temperatures inside the build have a direct influence on the quality of the produced part. More in situ measurement techniques in the AM process can be found in the review literature [15,16]. In addition, molten pool measurement using an optical microscope based on the solidified microstructure is a post-processing technique for thermal analysis, which requires extensive experimental work for sample preparation, such as cutting, polishing, and etching [17,18].

Numerical models based on finite element method (FEM) were developed to address the difficulty and inconvenience in monitoring the additive manufacturing (AM) process. Roberts developed a FEM model to predict the temperature distribution in selective laser melting (SLM) of Ti-6Al-4V involving multiple layers with an element birth and death technique [19]. Similar numerical models were developed with different types of heat sources considering the absorptivity and the different shape of the heat source [20,21]. Fu et al. developed a FEM model to predict the temperature distribution in the SLM of Ti-6Al-4V using powder material properties and bulk material properties. Improved prediction accuracy was reported using powder material properties upon validation to experimental molten pool dimensions [22]. FEM models were also developed for temperature prediction in SLM of various materials, including aluminum, titanium, stainless steel, and Inconel alloy [23–26]. Criales et al. investigated the sensitivity of material properties and process parameters in the prediction of SLM of Inconel 625 [27]. Papadakis et al. developed a computational reduced model for prediction in SLM with improved computational efficiency, in which the heat in each scanning vector was quantified as an input [28]. The recent numerical models have considered the influence of powder packing and molten pool dynamics in the temperature prediction, which allowed the investigation of defects in the produced parts [29–31]. Qi et al. and Kolossov et al. developed FEM models to predict the temperature distribution in a coaxial laser cladding process and selective laser sintering (SLS) process, respectively [32,33]. Residual stress and part distortion in the AM process were also investigated through numerical modeling [34–36]. The detailed discussion of FEM models is out of the scope of this work. More details can be found in the review literature [37–39]. Although FEM models have made considerable progress in predicting the AM processes, the high computational cost is still the major drawback.

Analytical models have demonstrated their high computational efficiency, high prediction accuracy, and broad applicability in predicting the manufacturing processes, and in the inverse determination of the material constants [40,41]. To overcome the drawback in computational efficiency, analytical models were also developed to predict the AM process. Peyre et al. developed a semi-analytical model to predict the temperature in direct metal deposition (DMD), in which an analytical model and a FEM model were used to characterize the deposition geometry and temperature distribution, respectively [42]. Yang et al. developed another semi-analytical model to predict the temperature in SLM, in which an analytical model and a FEM model were used to characterize the moving heat source and impose heat transfer boundary condition, respectively [43]. Van Elsen et al. summarized three moving heat source solutions based on the coordinate with a moving origin at the heat source location, namely moving point heat source, moving semi-ellipsoidal heat source, and moving uniform heat source [44]. Isotropic and homogeneous material and semi-infinite medium were assumed in those models. The aforementioned heat source solutions were originally developed by Carslaw and Jaeger [45]. The moving point heat source was further developed with consideration of heat source shape for temperature prediction in SLM [46]. Rosenthal developed a moving line heat source solution to predict the temperature in welding for an infinite thin plate [47]. The line heat source solution was adopted to predict the temperature in coaxial laser cladding, in which single-track scan prediction and semi-infinite medium assumption were enforced [48]. Tan et al. further developed the line heat source solution by transforming the moving coordinate to the absolute coordinate for consideration of the part dimensions. The final solution was constructed from the superposition of the actual heat source and two image heat sources [49]. This solution is limited only for continuous single-track scans. However, the FEM models used in semi-analytical models prevent optimized computational efficiency; most developed analytical models employed a moving coordinate and assumed a semi-infinite medium and steady-state condition through the process. Those assumptions reduced the usefulness of the developed model in real applications because of the lack of time-dependence and location-dependence related to part dimensions. Moreover, Green's Function has been widely used for temperature and stress predictions of the bounded medium due to thermal and mechanical loads, but the high mathematical complexity resulted in an unoptimized computational efficiency [45,50,51]. The temperatures in a dimensional part cannot be accurately and efficiently predicted with the developed analytical models for multi-track scans in MPBAM.

This paper presents an analytical model to predict in-process temperatures in MPBAM based on a stationary coordinate, whose origin is located at the part boundary (absolute coordinate), with consideration of the laser power absorption, latent heat, scanning strategy, and powder packing porosity. Analytical solutions are developed to predict temperatures in single-track scans and multi-track scans under various process conditions. Molten pool dimensions are then obtained by comparing the predicted temperatures to the material melting temperature. Inconel 625 was chosen to test the presented models with validation to the molten pool dimensions in the literature, which were measured based on the solidification microstructure [18]. The following tasks were performed in this study: (1) to inversely determine the laser power absorption with comparison between predicted molten pool dimensions and experimental values using trial-and-error method; (2) to investigate the molten pool evolution with respect to scanning time; (3) to predict the stabilized molten pool dimensions in single-track scans and bidirectional scans under various process conditions, and then perform validations against experimental values; (4) to record and present the computation time; (5) to investigate the influence of powder packing porosity on the predicted temperatures through sensitivity analyses.

The employed absolute coordinate and time consideration allows the temperature prediction for a dimensional part, which significantly improves the usefulness of the developed model in real applications. For comparison, other analytical models assume steady-state condition, and thus they are not applicable for temperature prediction at the beginning of the scan, where the part boundary is located at. The in-process temperature analysis allows the investigation of molten pool evolution, specifically the growth and stabilization of the molten pool. The high computational efficiency allows the inverse determination of laser powder absorption based on the experimental measurements.

2. Methodology

In this work, the in-process temperature in MPBAM is predicted through analytical modeling based on the absolute coordinate. The heat balance governing equation is expressed as

$$\frac{\partial \rho u}{\partial t} + \frac{\partial \rho H V}{\partial x} = \nabla \cdot (k \nabla T) + \dot{q} \tag{1}$$

where u is internal energy, H is enthalpy, ρ is density, k is conductivity, and $\dot{q}$ is a volumetric heat source, t is time, x is distance, V is heat source moving speed, and T is temperature. With $V = 0$, and $du = C_p dT$, where C_p is specific heat, the heat balance equation becomes the heat conduction equation expressed as

$$\frac{\partial^2 T}{\partial x^2} + \frac{\partial^2 T}{\partial y^2} + \frac{\partial^2 T}{\partial z^2} = \frac{1}{\kappa}\frac{\partial T}{\partial t} + \dot{q} \tag{2}$$

where κ is thermal diffusivity ($\kappa = k/\rho C_p$), x, y, z denote three mutually perpendicular directions in the absolute coordinate.

A point heat source solution is developed by Carslaw and Jaeger [45] with satisfaction of the heat conduction equation as the following.

$$\theta(x, y, z) = \frac{Q}{8(\pi \kappa t)^{\frac{3}{2}}} \exp\left[-\frac{x^2 + y^2 + z^2}{4\kappa t}\right] \tag{3}$$

where Q is the amount of heat, θ is temperature change ($\theta = T - T_0$).

The moving point heat source solution at the current time (t) and location $(X - V(t - t\prime), Y, Z)$ due to the heat input at the previous time (t') was derived from the point heat source solution as the following.

$$\theta(x, y, z, t) = \frac{Pdt\prime}{8\rho c[\pi \kappa(t - t\prime)]^{\frac{3}{2}}} \exp\left[-\frac{\{x - V(t - t')\}^2 + y^2 + z^2}{4\kappa(t - t')}\right] \tag{4}$$

where P, V are laser power and laser scanning velocity, respectively, x-direction is assumed to be laser scanning direction, y-direction is assumed to the hatch direction, as illustrated in Figure 1a.

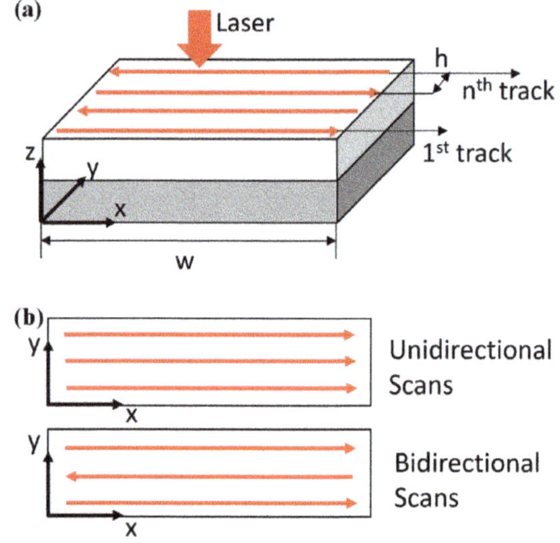

Figure 1. (**a**) Schematic view of the additive manufacturing process, where w, h denote track length and hatch space, respectively; (**b**) schematic view of unidirectional scans and bidirectional scans.

The 3D in-process temperature can be calculated by integrating the temperature solution with a time range from 0 to t. With the consideration of laser power absorption (η), the temperature solution becomes the following.

$$\theta(x, y, z, t) = \frac{P\eta}{8\rho C_p(\pi\kappa)^{\frac{3}{2}}} \int_0^t \frac{\exp\left[-\frac{(x - V(t - t'))^2 + y^2 + z^2}{4\kappa(t - t')}\right]}{(t - t')^{\frac{3}{2}}} dt' \tag{5}$$

The temperature solution can be further derived by integrating t' from 0 to t as the following.

$$\theta(x,\,y,z,\,t) = \frac{P\eta}{2Rk_t\pi^{\frac{3}{2}}}exp\left(\frac{Vx}{2\kappa}\right)\int_{\frac{R}{2\sqrt{\pi t}}}^{\infty} exp\left[-\xi^2 - \left(\frac{V^2R^2}{16\kappa^2\xi^2}\right)\right]d\xi \tag{6}$$

where $R^2 = x^2 + y^2 + z^2$, t is the current time, $t\prime$ is previous time, x, y, z are the corresponding distances from the laser source, ξ is an integration variable which leads to the concise expression.

The 2D in-process temperatures in the single-track scan can be calculated by setting y-direction distance as zero.

$$\theta(x,0,z,t) = \frac{P\eta}{8\rho C_p(\pi\kappa)^{\frac{3}{2}}}\int_0^t \frac{exp\left[-\frac{(x-V(t-t'))^2+z^2}{4\kappa(t-t')}\right]}{(t-t')^{\frac{3}{2}}}dt' \tag{7}$$

The 2D in-process temperatures in the multi-track scan with hatch space (h) is calculated by combining temperature at the current track location due to the energy input from the previous track (θ_1) and the temperature due to the added energy input from the current track (θ_2). The in-process temperatures can be expressed as the following.

$$\theta(x,\,0,z,\,t) = \theta_1 + \theta_2 \tag{8}$$

$$\theta_1(x,h,z,t_1) = \frac{P\eta}{2Rk_t\pi^{\frac{3}{2}}}exp\left(\frac{Vx}{2\kappa}\right)\int_{\frac{R}{2\sqrt{\pi t}}}^{\infty} exp\left[-\xi^2 - \left(\frac{V^2R^2}{16\kappa^2\xi^2}\right)\right]d\xi \tag{9}$$

$$\theta_2(x,0,z,t) = \frac{P}{8\rho C_p(\pi\kappa)^{\frac{3}{2}}}\int_{t1}^t \frac{exp\left[-\frac{(x-V(t-t'))^2+y^2+z^2}{4\kappa(t-t')}\right]}{(t-t')^{\frac{3}{2}}}dt' \tag{10}$$

$$\theta_2(x,0,z,t) = \frac{P}{8\rho C_p(\pi\kappa)^{\frac{3}{2}}}\int_{t1}^t \frac{exp\left[-\frac{(x+V(t-t'))^2+y^2+z^2}{4\kappa(t-t')}\right]}{(t-t')^{\frac{3}{2}}}dt' \tag{11}$$

where a positive scanning velocity is used for unidirectional scans, as in Equation (9); a negative scanning velocity is used for bidirectional scans, as in Equation (10). The unidirectional scans and bidirectional scans are illustrated in Figure 1b. Additionally, t_1 is the required time for completing previous track in two consecutive tracks and t is the current time in two consecutive tracks. Then, the transient temperature in multi-track scans can be calculated in the same manner.

In addition, the latent heat is considered using the heat integration method, in which the temperature of the molten pool material is lowered by an amount as the following, due to the phase transformation [44].

$$\Delta T = H_f/C_p \tag{12}$$

The influence of powder packing porosity on the predicted temperatures is investigated using effective material properties as the following.

$$\rho_e = (1-\tau)^{\gamma}\rho \tag{13}$$

$$k_e = (1-\tau)^{\beta}k \tag{14}$$

where subscript e denotes effective values, τ is powder packing porosity, γ, β are coefficients that can be taken as 1, as suggested by Criales et al. [27].

3. Results and Discussion

To investigate the accuracy and effectiveness of the presented model, temperatures in MPBAM of Inconel 625 were predicted under different process conditions in single-track scans and bidirectional scans. Six different process conditions were used for single-track scans and bidirectional scans separately, as given in Table 1. Molten pool dimensions were then obtained from the comparison between the predicted temperatures and material melting temperature, as illustrated in Figure 2. The determined molten pool dimensions were validated by the experimental measurements in the literature, in which an EOS M2070 machine (ASTM International, West Conshohocken, PA, USA) was used in the MPBAM process with gas atomized powders, and a digital optical microscope was used to measure the molten pool dimensions based on the solidification microstructure. The machining uses a single-mode, continuous wave ytterbium fiber laser under nitrogen gas ambiance. The gas automized powder has an average size of 35 μm with powder size distribution of D60% = 29.4 μm, D10% = 13.5 μm, D90% = 43 μm (D denotes powder diameter). More details of experimental design can be found in the reference [18].

Table 1. Process conditions for single-track scans and bidirectional scans in SLM of Inconel 625 [18]. The layer thickness is 20 μm.

Test	Laser Powder P (W)	Scanning Velocity V (mm/s)	Hatch Space h (mm)
1	169	875	0.1
2	195	875	0.1
3	182	800	0.1
4	195	725	0.1
5	169	725	0.1
6	195	800	0.1

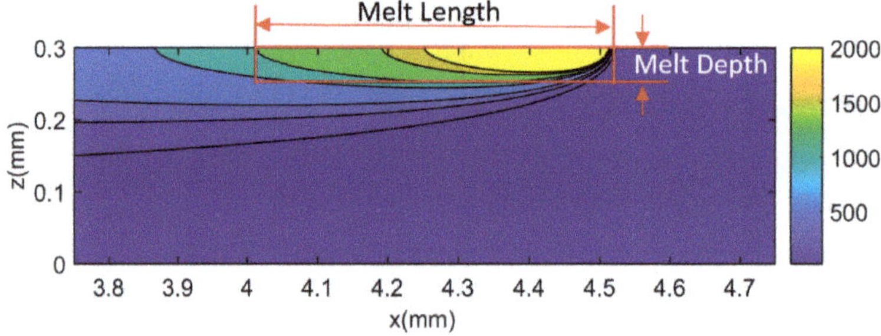

Figure 2. Predicted temperature distribution and molten pool geometry under test 6 process conditions, using 40% absorption. It should be noted that the color bar values correspond to the temperature contours.

The thermophysical material properties of Inconel 625 alloy are given in Table 2. Laser power absorption is affected by laser and powder materials, such as laser wavelength, powder material properties, powder packing-related surface roughness, laser-workpiece standoff distance, etc. [52,53]. Therefore, a given laser power absorption should be valid only for a specific experimental setup. Since the laser power absorption has not been measured and reported in the literature, the absorption was inversely determined using the presented model by minimizing the difference between predicted molten pool depth and experimental molten pool depth under test 6 process conditions. Inverse analysis has been widely used in the determination of materials constants and physical parameters [54]. The trial-and-error method was employed with varying absorption values. The laser absorption was determined as 40% for the current study, as shown in Figure 3. A linear relationship was observed with underestimated absorptions. A non-linear relationship was observed with overestimated absorptions, which might be caused by the more significant influence of latent heat on the overestimated temperature zone. The determined laser absorption (40%) was used in the following studies in this work. The simple calculations in the presented solutions without mesh and iteration, which FEM models rely on, lead to the high computational efficiency of the developed model. Computational time and implementation details were provided for the investigation of the computational efficiency of the presented model. A MATLAB program was used to implement the predictions using the presented model on a personal computer running at 2.8 GHz. The computation time under each process condition was 27 s less for an area near a heat source, which has a length (x = 1 mm) with 5 µm increments and a height (z = 0.3 mm) with 1 µm increments. Other related variables of molten pool length, molten pool depth, and corresponding prediction error are given in Table 3.

Table 2. Material properties of Inconel 625 and inversely determined laser absorption (T_0 = 20 °C) [14,27].

Density ρ (kg/m^3)	Thermal Conductivity k (W/m–°C)	Specific Heat C_p (J/kg–°C)	Solidus Temperature T_s (°C)	Liquidus Temperature T_L (°C)	Latent Heat H_f (J/kg)	Absorption η (%)
8840	9.8	410	1290	1350	227,000	40

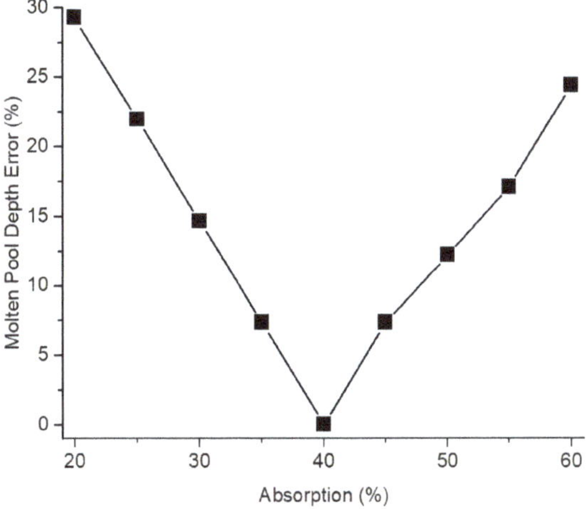

Figure 3. Inverse determination of laser absorption based on melt depth prediction using the trial-and-error method under test 6 process condition.

Table 3. Inverse determination of laser power absorption in this study.

Absorption η (%)	Molten Pool Length L (µm)	Molten Pool Depth D (µm)	Molten Pool Depth Error $Error$ (%)	Computation Time t_c (s)
20	180	29	29.27	26.11
25	225	32	21.95	24.46
30	270	35	14.63	23.06
35	310	38	7.32	22.95
40	355	41	0.00	23.04
45	395	44	7.32	23.23
50	440	46	12.20	22.95
55	480	48	17.07	23.10
60	525	51	24.39	23.20

To investigate the evolution of molten pool during the MPBAM, the in-process temperatures were predicted in a single-track scan under test 6 process conditions. Predicted temperature profiles on the top view and the cross-sectional view along the laser scanning direction were illustrated in Figure 4.

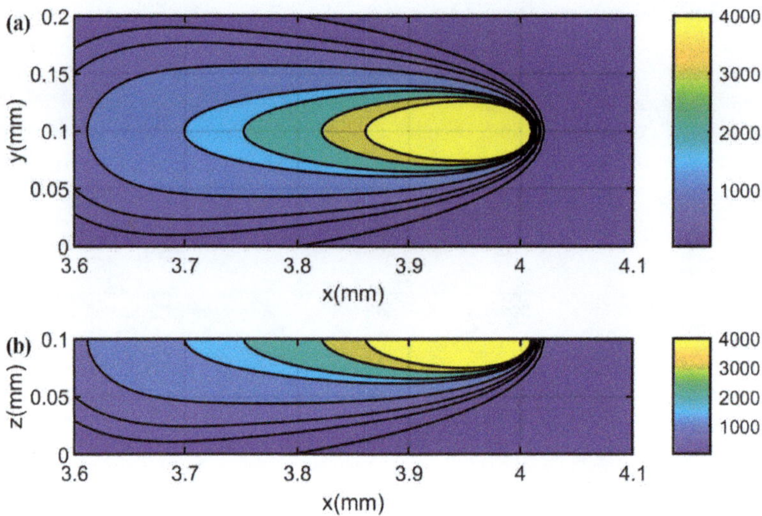

Figure 4. Predicted temperature distribution at t = 5 ms (laser location x = 4 mm, y = 0.1 mm) in single track under test 6 process conditions using 40% absorption. (**a**) Top view; (**b**) cross-sectional view at laser scanning location. It should be noted that the color bar values are corresponding to the temperature contours.

The molten pool growth for the stabilization process was investigated with a time interval from 0.001 ms to 10 ms, which corresponded to a distance interval from 0.0008 mm to 8 mm. The trends of molten pool length, molten pool width, and molten pool depth over time were predicted using the presented model, as shown in Figure 5a. The molten pool became stable after 1 ms. The trend of molten pool volume over time is shown in Figure 5b, which was calculated as the following [22]. All associated data is given in Table A1 in the Appendix A.

$$Vol = \frac{\pi DLW}{6} \tag{15}$$

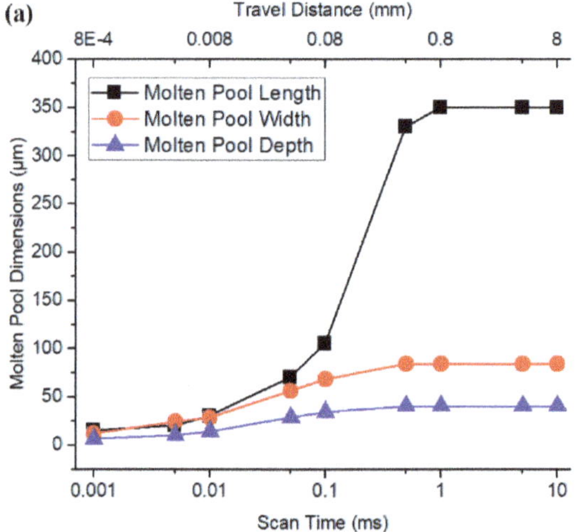

Figure 5. *Cont.*

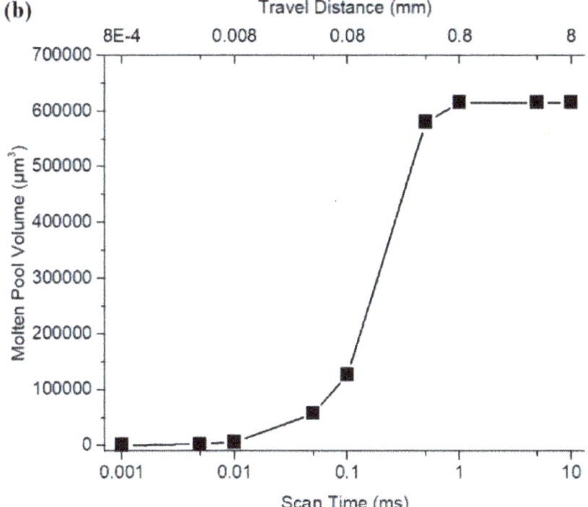

Figure 5. The growth and stabilization of the predicted molten pool in single track scan using the determined absorption of 40% in terms of (**a**) molten pool length, molten pool width, and molten pool depth, and (**b**) molten pool volume.

Moreover, the molten pool growth and stabilization process in the second track of two consecutive bidirectional scans was predicted, as shown in Figure 6. The molten pool in the second track of bidirectional scans became stable after 0.5 ms, which was faster than that in the single track scan (1 ms). In addition, the larger molten pool length and width were observed in the bidirectional scan after stabilization because of the heat affected zone due to the previous track.

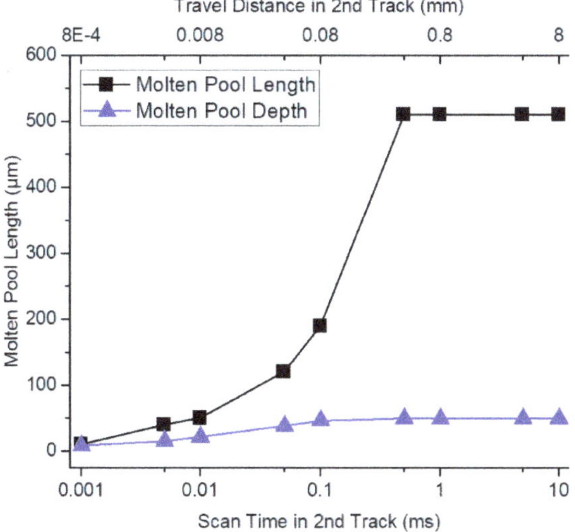

Figure 6. The growth and stabilization of predicted molten pool in the second track of bidirectional scans using the determined absorption of 40% in terms of molten pool length and molten pool depth.

The stabilized melting depth values under various process conditions were predicted and validated to the experimental values in the literature [18]. The temperature distributions along the depth (z-direction) in a single-track scan and bidirectional scans under test 6 process conditions were predicted, as shown in Figure 7. The maximum temperature in a bidirectional scan is higher than that in the single-track scan. The constant temperature region in both scans is due to the consideration of latent heat, in which phase transformation takes place rather than temperature increase with continuous heat input. A similar trend was also reported in the literature [19,32,44]. Close agreements were observed upon validation to experimental values under various process conditions for single-track scans and bidirectional scans, as shown in Figure 8a,b, respectively. The deviations between predictions and experiments might be caused by the molten pool shrinkage [55], which was assumed to be negligible. The predicted melt length, melt depth, and computation time are given in Table 4. The average computation time for 2D temperature prediction in single-track scans was 19.44 s; the average computation time for 2D temperature prediction in bidirectional scans was 88.17 s. However, a 3D temperature prediction requires much longer computational time, especially in bidirectional scans, because a large number of 2D temperature profiles need to be calculated iteratively with consideration of the influence of the heat affected zone produced by the previous scans.

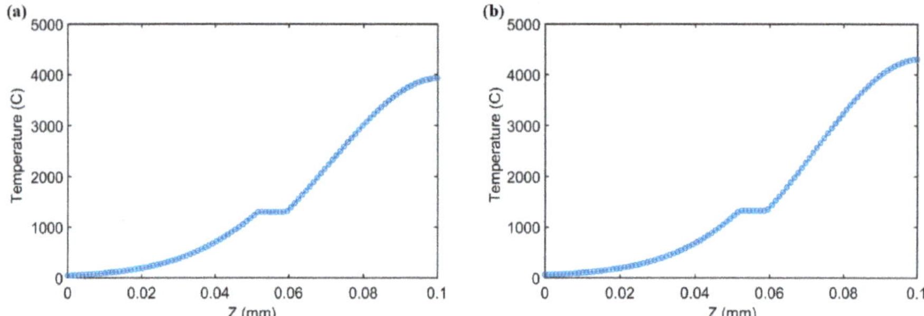

Figure 7. Stabilized temperature distribution from prediction along depths (z-direction) at laser scan locations (x = 4 mm, y = 0.1 mm) for (**a**) single-track scan and (**b**) bidirectional scan under test 6 conditions. The minimum temperatures are at room temperature (20 °C). The determined absorption of 40% was used in the predictions.

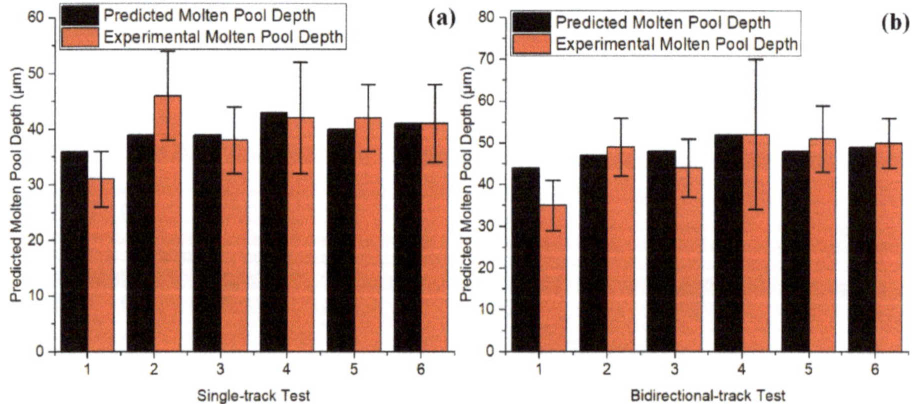

Figure 8. Validation of predicted molten pool depth in (**a**) a single track scan and (**b**) a bidirectional scan after stabilization. Black color represents predicted melt depth. Red color represents experimental melt depth based on solidified microstructure. The determined absorption of 40% was used in the predictions.

Table 4. Predicted molten pool dimensions for single-track scans and bidirectional scan under steady-state conditions at $x = 4$ mm, $y = 0.1$ mm, with 2 µm increments in x and z directions.

Single-Track Test	Molten Pool Length L (µm)	Molten Pool Depth D (µm)	Computation Time t_c (s)	Bidirectional Test	Molten Pool Length L (µm)	Molten Pool Depth D (µm)	Computation Time t_c (s)
1	310	36	19.60	1	450	44	87.78
2	360	39	18.90	2	510	47	88.18
3	330	39	19.72	3	480	48	91.38
4	360	43	19.73	4	510	52	89.68
5	310	40	19.64	5	450	48	85.91
6	360	41	19.04	6	510	49	86.12

Moreover, the influence of powder packing porosity on predicted temperature was investigated by sensitivity analyses for single-track tests and bidirectional tests separately. The packing porosity was deliberately given as 0%, 20%, 40%, and 60% in an increasing trend. The simple fraction models (in Equations (13) and (14)) were employed to correlate between material properties and powder packing. A similar method has been employed for sensitivity analysis, as reported in the literature [27]. It should be noted that the fraction models calculate materials' properties considering the porosity of packed powder rather than the porosity of the build. The build porosity is affected by powder packing and process conditions. Positive correlations were observed between the packing porosity and melt depth, as illustrated in Figure 9. Therefore, the increase in powder packing porosity leads to an increase in molten pool dimensions, and vice versa. This finding confirms the instinctive trend because larger porosity leads to a lower value for thermal conductivity, which prevents energy being dissipated into the build.

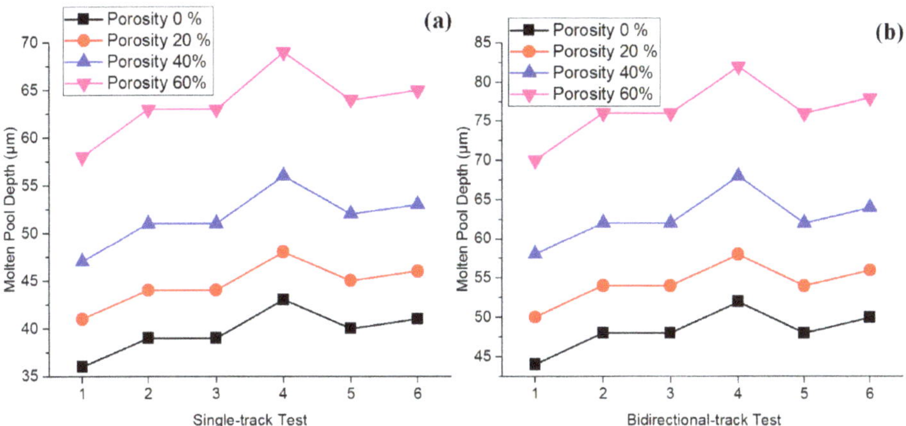

Figure 9. Influence of powder packing porosity on temperature distribution through sensitivity analyses in (**a**) single track scans and (**b**) bidirectional scans. Note: the presented results are obtained from prediction using the simple fraction models with the determined absorption of 40%.

This work investigated the molten pool evolution, including the growth and stabilization of the molten pool using the presented temperature model with analytical solutions. The recorded and presented computation time demonstrated the high computational efficiency of the presented model. With the absolute coordinate and the extended predictive capability for multi-track scans, the temperature distribution can now be predicted for a dimensional part, which significantly improved the usefulness of the presented model in real applications. The high computational efficiency also allows the process-parameters planning for desired temperature conditions. The predicted temperature can be used to further investigate residual stress and part distortion [56,57]. It should be noted that the cooling state after laser turning off and cooling-associated molten pool shrinkage were not considered in this study. The limitations of the presented model are the following: the limited predictive capability for single layer scanning; the assumption of temperature-independent material properties; the assumption of simplified point heat source. The cooling state, extended predictive capability for multiple layer scans, temperature-dependent material properties, and three-dimensional heat source should be considered in future works.

4. Conclusions

This work investigates the in-process temperature in MPBAM by further developing a moving point heat source solution, as originally proposed by Carslaw and Jaeger [45], to extend its applicability from a single-track scan to multi-track scans with consideration of laser power absorption, latent heat, scanning strategy, and powder packing porosity. The absolute coordinate allows the temperature prediction for a dimensional part with consideration of time-dependence and location-dependence, with respect to the beginning time and location of MPBAM. The laser power absorption was inversely determined using trial and error method based on experimental measurement of molten pool dimensions. The latent heat was considered by heat integration method. The extended applicability and the absolute coordinate significantly increase the usefulness of the developed model in real applications. For comparison, other analytical models predicted temperature distribution in single-track scans with the semi-infinite medium assumption, in which a moving coordinate was defined with the origin located at the laser heat source. To test the presented model, Inconel 625 alloy was chosen to predict the temperatures in MPBAM under various process condition. Good agreements were observed upon validation against experimental values in the literature. The molten pool evolution was investigated to demonstrate the molten pool growth and stabilization during the scanning process. It should be noted that the cooling state in the post-process has not been considered in the presented model. A positive correlation between molten pool dimensions (temperature distribution) and powder packing porosity was revealed through sensitivity analyses using the presented model. With the benefits of extended applicability for multi-track scans, the absolute coordinate, and the high computational efficiency, the developed model can be used for temperature investigation in real applications.

Author Contributions: J.N. performed formal analysis and investigation, extracted, analyzed, and validated the data, and wrote the manuscript. S.Y.L. supervised the research project and proofed the manuscript. D.E.S. and H.G. provided resources and critical feedback.

Funding: This research was funded by The Boeing Company grant number [confidential].

Acknowledgments: The authors would like to acknowledge the funding support from the Boeing Company.

Conflicts of Interest: The authors declare no conflict of interest.

Nomenclature

MPBAM	metal powder bed additive manufacturing
PBF	powder bed fusion
SLM	selective laser melting
SLS	selective laser sintering
DMD	direct metal deposition
FEM	finite element method
IR	infrared
u	internal energy
H	enthalpy
ρ	density
ρ_e	effective density
k	thermal conductivity

k_e	effective thermal conductivity
c_p	specific heat
κ	thermal diffusivity
$\dot{q}$	a volumetric heat source
Q	amount of heat
H_f	latent heat
P	laser power
η	absorption
V	laser heat source moving velocity
h	hatch space
w	track length
x, y, z	coordinate
$T,\ T_0, T_m$	temperature, room temperature, material melting temperature
ΔT	temperature change due to the consideration of latent heat
θ	temperature change due to the moving laser heat source
t	time
ζ	time related integration variable
τ, γ, β	powder packing related coefficients
$L,\ D,\ W,\ Vol$	molten pool length, depth, width, volume

Appendix A

Table A1. Variables in studying molten pool growth and stabilization during a single-track scan under test 6 condition.

Scan Time t (ms)	Melt Length L (μm)	Melt Width W (μm)	Melt Depth D (μm)	Melt Volume Vol (μm^3)
0.001	15	12	6	565
0.005	20	24	10	2513
0.01	30	28	14	6158
0.05	70	56	28	57,470
0.1	105	68	34	127,109
0.5	330	84	40	580,566
1	350	84	40	615,752
5	350	84	40	615,752
10	350	84	40	615,752
50	350	84	40	615,752
100	350	84	40	615,752
500	350	84	40	615,752
1000	350	84	40	615,752

References

1. Gu, D.D.; Meiners, W.; Wissenbach, K.; Poprawe, R. Laser additive manufacturing of metallic components: materials, processes and mechanisms. *Int. Mater. Rev.* **2012**, *57*, 133–164. [CrossRef]
2. Gong, H.; Rafi, K.; Gu, H.; Starr, T.; Stucker, B. Analysis of defect generation in Ti–6Al–4V parts made using powder bed fusion additive manufacturing processes. *Addit. Manuf.* **2014**, *1*, 87–98. [CrossRef]
3. Shiomi, M.; Osakada, K.; Nakamura, K.; Yamashita, T.; Abe, F. Residual stress within metallic model made by selective laser melting process. *CIRP Annals* **2004**, *53*, 195–198. [CrossRef]
4. Heigel, J.C.; Michaleris, P.; Palmer, T.A. In situ monitoring and characterization of distortion during laser cladding of Inconel® 625. *J. Mater. Process. Technol.* **2015**, *220*, 135–145. [CrossRef]
5. Buchbinder, D.; Meiners, W.; Pirch, N.; Wissenbach, K.; Schrage, J. Investigation on reducing distortion by preheating during manufacture of aluminum components using selective laser melting. *J. Laser Appl.* **2014**, *26*, 012004. [CrossRef]

6. Wauthle, R.; Vrancken, B.; Beynaerts, B.; Jorissen, K.; Schrooten, J.; Kruth, J.P.; Van Humbeeck, J. Effects of build orientation and heat treatment on the microstructure and mechanical properties of selective laser melted Ti6Al4V lattice structures. *Addit. Manuf.* **2015**, *5*, 77–84. [CrossRef]

7. Yang, Y.; Chen, Y.; Zhang, J.; Gu, X.; Qin, P.; Dai, N.; Li, X.; Kruth, J.P.; Zhang, L.C. Improved corrosion behavior of ultrafine-grained eutectic Al-12Si alloy produced by selective laser melting. *Mater. Des.* **2018**, *146*, 239–248. [CrossRef]

8. Van Hooreweder, B.; Kruth, J.P. Advanced fatigue analysis of metal lattice structures produced by Selective Laser Melting. *CIRP Ann.* **2017**, *66*, 221–224. [CrossRef]

9. Ning, J.; Liang, S. Predictive Modeling of Machining Temperatures with Force–Temperature Correlation Using Cutting Mechanics and Constitutive Relation. *Materials* **2019**, *12*, 284. [CrossRef] [PubMed]

10. Ning, J.; Liang, S. Prediction of Temperature Distribution in Orthogonal Machining Based on the Mechanics of the Cutting Process Using a Constitutive Model. *J. Manuf. Mater. Process.* **2018**, *2*, 37. [CrossRef]

11. Craeghs, T.; Clijsters, S.; Yasa, E.; Bechmann, F.; Berumen, S.; Kruth, J.P. Determination of geometrical factors in Layerwise Laser Melting using optical process monitoring. *Opt. Laser Eng.* **2011**, *49*, 1440–1446. [CrossRef]

12. Clijsters, S.; Craeghs, T.; Buls, S.; Kempen, K.; Kruth, J.P. In situ quality control of the selective laser melting process using a high-speed, real-time melt pool monitoring system. *Int. J. Adv. Manuf. Technol.* **2014**, *75*, 1089–1101. [CrossRef]

13. Denlinger, E.R.; Jagdale, V.; Srinivasan, G.V.; El-Wardany, T.; Michaleris, P. Thermal modeling of Inconel 718 processed with powder bed fusion and experimental validation using in situ measurements. *Addit. Manuf.* **2016**, *11*, 7–15. [CrossRef]

14. Heigel, J.C.; Gouge, M.F.; Michaleris, P.; Palmer, T.A. Selection of powder or wire feedstock material for the laser cladding of Inconel® 625. *J. Mater. Process. Technol.* **2016**, *231*, 357–365. [CrossRef]

15. Everton, S.K.; Hirsch, M.; Stravroulakis, P.; Leach, R.K.; Clare, A.T. Review of in-situ process monitoring and in-situ metrology for metal additive manufacturing. *Mater. Des.* **2016**, *95*, 431–445. [CrossRef]

16. Tapia, G.; Elwany, A. A review on process monitoring and control in metal-based additive manufacturing. *J. Manuf. Sci. Eng.* **2014**, *136*, 060801. [CrossRef]

17. Wang, Y.; Kamath, C.; Voisin, T.; Li, Z. A processing diagram for high-density Ti-6Al-4V by selective laser melting. *Rapid Prototyp. J.* **2018**, *24*, 1469–1478. [CrossRef]

18. Criales, L.E.; Arısoy, Y.M.; Lane, B.; Moylan, S.; Donmez, A.; Özel, T. Laser powder bed fusion of nickel alloy 625: experimental investigations of effects of process parameters on melt pool size and shape with spatter analysis. *Int. J. Mach. Tool. Manuf.* **2017**, *121*, 22–36. [CrossRef]

19. Roberts, I.A.; Wang, C.J.; Esterlein, R.; Stanford, M.; Mynors, D.J. A three-dimensional finite element analysis of the temperature field during laser melting of metal powders in additive layer manufacturing. *Int. J. Mach. Tools Manuf.* **2009**, *49*, 916–923. [CrossRef]

20. Hussein, A.; Hao, L.; Yan, C.; Everson, R. Finite element simulation of the temperature and stress fields in single layers built without-support in selective laser melting. *Mater. Des.* **2013**, *52*, 638–647. [CrossRef]

21. Patil, R.B.; Yadava, V. Finite element analysis of temperature distribution in single metallic powder layer during metal laser sintering. *Int. J. Mach. Tools Manuf.* **2007**, *47*, 1069–1080. [CrossRef]

22. Fu, C.H.; Guo, Y.B. Three-dimensional temperature gradient mechanism in selective laser melting of Ti-6Al-4V. *J. Manuf. Sci. Eng.* **2014**, *136*, 061004. [CrossRef]

23. Loh, L.E.; Chua, C.K.; Yeong, W.Y.; Song, J.; Mapar, M.; Sing, S.L.; Liu, Z.-H.; Zhang, D.Q. Numerical investigation and an effective modelling on the Selective Laser Melting (SLM) process with aluminium alloy 6061. *Int. J. Heat Mass Trans.* **2015**, *80*, 288–300. [CrossRef]

24. Li, Y.; Gu, D. Thermal behavior during selective laser melting of commercially pure titanium powder: Numerical simulation and experimental study. *Addit. Manuf.* **2014**, *1*, 99–109. [CrossRef]

25. Antony, K.; Arivazhagan, N.; Senthilkumaran, K. Numerical and experimental investigations on laser melting of stainless steel 316L metal powders. *J. Manuf. Process.* **2014**, *16*, 345–355. [CrossRef]

26. Andreotta, R.; Ladani, L.; Brindley, W. Finite element simulation of laser additive melting and solidification of Inconel 718 with experimentally tested thermal properties. *Finite Elem. Anal. Des.* **2017**, *135*, 36–43. [CrossRef]

27. Criales, L.E.; Arısoy, Y.M.; Özel, T. Sensitivity analysis of material and process parameters in finite element modeling of selective laser melting of Inconel 625. *Int. J. Adv. Manuf. Technol.* **2016**, *86*, 2653–2666. [CrossRef]

28. Papadakis, L.; Loizou, A.; Risse, J.; Bremen, S.; Schrage, J. A computational reduction model for appraising structural effects in selective laser melting manufacturing: a methodical model reduction proposed for time-efficient finite element analysis of larger components in Selective Laser Melting. *Virtual Phys. Prototyp.* **2014**, *9*, 17–25. [CrossRef]

29. Xia, M.; Gu, D.; Yu, G.; Dai, D.; Chen, H.; Shi, Q. Porosity evolution and its thermodynamic mechanism of randomly packed powder-bed during selective laser melting of Inconel 718 alloy. *Int. J. Mach. Tools Manuf.* **2017**, *116*, 96–106. [CrossRef]

30. Xiang, Y.; Zhang, S.; Wei, Z.; Li, J.; Wei, P.; Chen, Z.; Yang, L.; Jiang, L. Forming and defect analysis for single track scanning in selective laser melting of Ti6Al4V. *Appl. Phys. A* **2018**, *124*, 685. [CrossRef]

31. Wei, P.; Wei, Z.; Chen, Z.; He, Y.; Du, J. Thermal behavior in single track during selective laser melting of AlSi10Mg powder. *Appl. Phys. A* **2017**, *123*, 604. [CrossRef]

32. Qi, H.; Mazumder, J.; Ki, H. Numerical simulation of heat transfer and fluid flow in coaxial laser cladding process for direct metal deposition. *J. Appl. Phys.* **2006**, *100*, 024903. [CrossRef]

33. Kolossov, S.; Boillat, E.; Glardon, R.; Fischer, P.; Locher, M. 3D FE simulation for temperature evolution in the selective laser sintering process. *Int. J. Mach. Tools Manuf.* **2004**, *44*, 117–123. [CrossRef]

34. Wang, Z.; Denlinger, E.; Michaleris, P.; Stoica, A.D.; Ma, D.; Beese, A.M. Residual stress mapping in Inconel 625 fabricated through additive manufacturing: Method for neutron diffraction measurements to validate thermomechanical model predictions. *Mater. Des.* **2017**, *113*, 169–177. [CrossRef]

35. Afazov, S.; Denmark, W.A.; Toralles, B.L.; Holloway, A.; Yaghi, A. Distortion prediction and compensation in selective laser melting. *Addit. Manuf.* **2017**, *17*, 15–22. [CrossRef]

36. Cao, J.; Gharghouri, M.A.; Nash, P. Finite-element analysis and experimental validation of thermal residual stress and distortion in electron beam additive manufactured Ti-6Al-4V build plates. *J. Mater. Process. Technol.* **2016**, *237*, 409–419. [CrossRef]

37. Bikas, H.; Stavropoulos, P.; Chryssolouris, G. Additive manufacturing methods and modelling approaches: a critical review. *Int. J. Adv. Manuf. Technol.* **2016**, *83*, 389–405. [CrossRef]

38. Schoinochoritis, B.; Chantzis, D.; Salonitis, K. Simulation of metallic powder bed additive manufacturing processes with the finite element method: A critical review. *Proc. Inst. Mech. Eng. Part B J. Eng. Manuf.* **2017**, *231*, 96–117. [CrossRef]

39. Pinkerton, A.J. Advances in the modeling of laser direct metal deposition. *J. Laser Appl.* **2015**, *27*, S15001. [CrossRef]

40. Ning, J.; Nguyen, V.; Liang, S.Y. Analytical modeling of machining forces of ultra-fine-grained titanium. *Int. J. Adv. Manuf. Technol.* **2018**, 1–10. [CrossRef]

41. Ning, J.; Nguyen, V.; Huang, Y.; Hartwig, K.T.; Liang, S.Y. Inverse determination of Johnson–Cook model constants of ultra-fine-grained titanium based on chip formation model and iterative gradient search. *Int. J. Adv. Manuf. Technol.* **2018**, *99*, 1131–1140. [CrossRef]

42. Peyre, P.; Aubry, P.; Fabbro, R.; Neveu, R.; Longuet, A. Analytical and numerical modelling of the direct metal deposition laser process. *J. Phys. D Appl. Phys.* **2008**, *41*, 025403. [CrossRef]

43. Yang, Y.; Knol, M.F.; van Keulen, F.; Ayas, C. A semi-analytical thermal modelling approach for selective laser melting. *Addit. Manuf.* **2018**, *21*, 284–297. [CrossRef]

44. Van Elsen, M.; Baelmans, M.; Mercelis, P.; Kruth, J.P. Solutions for modelling moving heat sources in a semi-infinite medium and applications to laser material processing. *Int. J. Heat Mass Transf.* **2007**, *50*, 4872–4882. [CrossRef]

45. Carslaw, H.; Jaeger, J. *Conduction of Heat in Solids*; Oxford Science Publication: Oxford, UK, 1990.

46. De La Batut, B.; Fergani, O.; Brotan, V.; Bambach, M.; El Mansouri, M. Analytical and numerical temperature prediction in direct metal deposition of Ti6Al4V. *J. Manuf. Mater. Process.* **2017**, *1*, 3. [CrossRef]

47. Rosenthal, D. The theory of moving sources of heat and its application of metal treatments. *Trans. ASME* **1946**, *68*, 849–866.

48. Pinkerton, A.J.; Li, L. The significance of deposition point standoff variations in multiple-layer coaxial laser cladding (coaxial cladding standoff effects). *Int. J. Mach. Tools Manuf.* **2004**, *44*, 573–584. [CrossRef]

49. Tan, H.; Chen, J.; Zhang, F.; Lin, X.; Huang, W. Process analysis for laser solid forming of thin-wall structure. *Int. J. Mach. Tools Manuf.* **2010**, *50*, 1–8. [CrossRef]

50. Saif, M.T.A.; Hui, C.Y.; Zehnder, A.T. Interface shear stresses induced by non-uniform heating of a film on a substrate. *Thin Solid Films* **1993**, *224*, 159–167. [CrossRef]

51. Moulik, P.N.; Yang, H.T.Y.; Chandrasekar, S. Simulation of thermal stresses due to grinding. *Int. J. Mech. Sci.* **2001**, *43*, 831–851. [CrossRef]

52. Boley, C.D.; Khairallah, S.A.; Rubenchik, A.M. Calculation of laser absorption by metal powders in additive manufacturing. *Appl. Opt.* **2015**, *54*, 2477–2482. [CrossRef]

53. Li, Y.; Gu, D. Parametric analysis of thermal behavior during selective laser melting additive manufacturing of aluminum alloy powder. *Mater. Des.* **2014**, *63*, 856–867. [CrossRef]

54. Ning, J.; Liang, S.Y. Model-driven determination of Johnson-Cook material constants using temperature and force measurements. *Int. J. Adv. Manuf. Technol.* **2018**, *97*, 1053–1060. [CrossRef]

55. Dai, K.; Shaw, L. Finite element analysis of the effect of volume shrinkage during laser densification. *Acta Mater.* **2005**, *53*, 4743–4754. [CrossRef]

56. Denlinger, E.R.; Heigel, J.C.; Michaleris, P.; Palmer, T.A. Effect of inter-layer dwell time on distortion and residual stress in additive manufacturing of titanium and nickel alloys. *J. Mater. Process. Technol.* **2015**, *215*, 123–131. [CrossRef]

57. Fergani, O.; Berto, F.; Welo, T.; Liang, S.Y. Analytical modelling of residual stress in additive manufacturing. *Fatigue Fract. Eng. Mater. Struct.* **2017**, *40*, 971–978. [CrossRef]

Article

Design, Development and FE Thermal Analysis of a Radially Grooved Brake Disc Developed through Direct Metal Laser Sintering

Gulam Mohammed Sayeed Ahmed * and Salem Algarni

Department of Mechanical Engineering, College of Engineering, King Khalid University, P.O. Box 394, Abha 61421, Saudi Arabia; saalgarni@kku.edu.sa

* Correspondence: gmsa786@kku.edu.sa; Tel.: +966-50-948-2050

Received: 6 May 2018; Accepted: 10 July 2018; Published: 13 July 2018

Abstract: The present research work analyzed the effect of design modification with radial grooves on disc brake performance and its thermal behavior by using additive manufacturing based 3D printed material maraging steel. Temperature distribution across the disc surface was estimated with different boundary conditions such as rotor speed, braking pressure, and braking time. Design modification and number of radial grooves were decided based on existing dimensions. Radial grooves were incorporated on disc surface through Direct Metal Laser Sintering (DMLS) process to increase surface area for maximum heat dissipation and reduce the stresses induced during braking process. The radial grooves act as a cooling channels which provides an effective means of cooling the disc surface which is under severe condition of sudden fall and rise of temperatures during running conditions. ANSYS software is used for transient structural and thermal analysis to investigate the variations in temperatures profile across the disc with induced heat flux. FE based thermo-structural analysis was done to determine thermal strains induced in disc due to sudden temperature fluctuations. The maximum temperature and Von Mises stress in disc brake without grooves on disc surface were observed which can severely affect thermal fatigue and rupture brake disc surface. It was been observed by incorporating the radial grooves that the disc brake surface is thermally stable. Experimental results are in good agreement with FE thermal analysis. DMLS provides easy fabrication of disc brake with radial grooves and enhancement of disc brake performance at higher speeds and temperatures. Therefore, DMLS provides an effective means of implementing product development technology.

Keywords: design; disc brake; 3D metal printing; direct metal laser sintering; thermal stress analysis; radial grooves

---

1. Introduction

One of the remarkable additive manufacturing (AM) processes that allows direct production by means of material layer addition is direct metal laser sintering, which was developed to overcome the disadvantages of traditional processes. DMLS process allows the direct production of components at net shape quality. The major advantage of AM process is fewer post-production processes, no geometrical restrictions, and a smallest possible feature size of about 100 μm. DMLS uses laser based processing techniques with different variety of 3D printed materials such as polymers, metals, ceramics and composites. The main objective of this research work was to fill the observed research gap on avoiding undesirable thermal strains, thermal fatigue and thermal damage on disc surface due to sudden variations in thermal boundary conditions. This research work designed a disc brake with radial grooves on disc surface. These modifications are achieved by using additive manufacturing based 3D printed DMLS processes in fabrication. Murr et al. [1] used a DMLS machine equipped with powder feed, bed, and laser energy sources for sintering and melting. A CAD model of the energy provided to metal powder during DMLS is based on full melting of powdered state of material by a high power laser beam (100–500 W). The starting 3D printed material in the form of powder is deposited on a substrate plate; the deposited layer is some tens of microns (25–50 μm). Different types of metals can be processed through DMLS; most common metals used for this process are aluminum alloys, nickel alloys, tool steels and stainless steels. In this present work, maraging steel was considered to study thermal behavior of 3D printed metal under sudden increase and decrease of temperature conditions. Maraging steel evolves from a group of martensitic steels with less carbon content. These steels are designated from the combination of "martensitic" and "aging", since these steels go through different aging heat treatments to improve strength and hardness. Maraging steel is most suitable for DMLS, since they have good weldability property at micro level due to micron-sized melt-pool in the DMLS process with high cooling rates. The property of these steels is proven to be well-matched with typical heavy duty applications in aerospace, automobile and tooling industries. The most commercially available grades for maraging are 200, 250, 300 and 350, which specify yield stress in kilo-pound per square inch with nominal yield stress for maraging being from 1500 to 2500 MPa. Vickers Hardness value in 10 kgf is HV123 (10 Kg), density is 8.2082 g/cc, temperatures range is from 450 to 650 °C for ageing treatment and proper selection of working temperatures must be made for desirable strength for reasonable time interval. Maraging steels are applied for applications with less cobalt and nickel content and have been produced to decrease costs. The strength of Martensite is controlled by carbon content, while Ni and Al have less effect than other alloying elements. Electron dispersive X-ray spectroscopy (EDX) reveals 3D printed Maraging Steel (MS1) characterization, as shown in Figure 1. Each element has unique emission spectrum. Measuring the spectrum peak intensities after an appropriate calibration, a quantitative estimation of chemical composition can be attained for material composition of maraging steel given in Table 1. The requirement for metallic 3D printing is expected to grow faster than plastic. In the present context, a research gap has been observed to avoid these undesirable thermal strains, thermal fatigue, and thermal damage on disc surface due to sudden variations in thermal boundary conditions. Thus, this research work designed a disc brake with modifications on disc surface with radial grooves and these modifications were achieved using DMLS processes for easy fabrication. Today, automotive manufacturers use the 3D printing technology for prototyping and functional parts manufacturing. In Figure 2, 3D-CAD geometries are shown without and with radial grooves having optimal dimensions of grooves. 3D printing can also improve quality through lighter parts, better ergonomics and more design freedom. Automotive companies also found overall product cycle times decreases by experimenting with 3D printing for assembly fixture, customized fixtures and tooling, making parts cheaper, lighter and faster is often a key goal of automotive industry, indicating opportunities for 3D printing manufacturing.

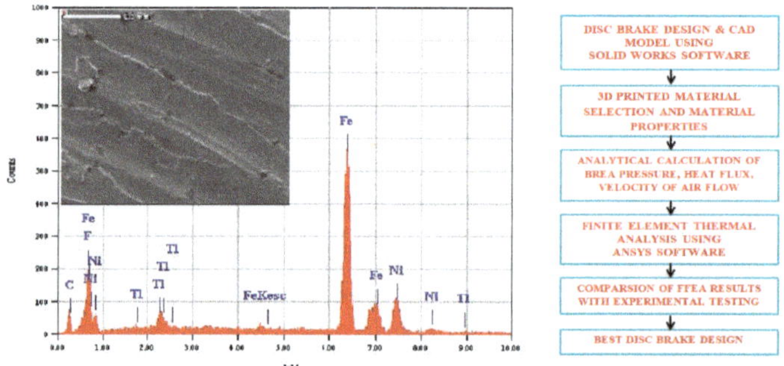

Figure 1. 3D printed maraging steel Characterization, flow chart of Design-Analysis of disc brake.

Table 1. Material Composition, wt % of maraging steel.

Metal	Wt %
Ni	17–19
Co	8.5–9.5
P/S	Max. 0.01
Mo	4.5–5.2
Ti	0.6–0.8
Al	0.052–0.15
C	Max. 0.03
Cr/Cu %	Max. 0.5
Fe	Balance

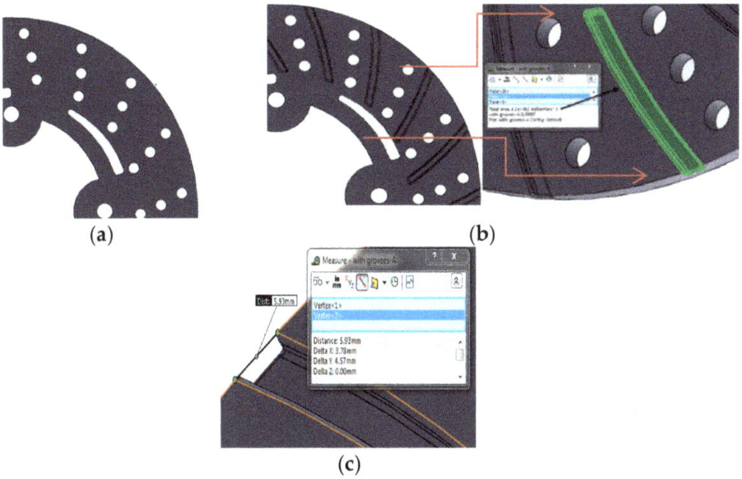

Figure 2. 3D-CAD geometries: (**a**) without Radial grooves; (**b**) with radial grooves; and (**c**) optimal dimensions and areas of grooves.

2. Development of Disc Brake by DMLS

DMLS based additive manufacturing process refers to digital 3D CAD data being used to build up a solid model in layers by depositing molten metal, and helps in developing complex products that can be light and stable [2]. The solid model during laser sintering is prepared by positioning and slicing through 3D printing software, namely Magics RP. The STL format of CAD data is converted into layer data by means of buildup processor available in 3DP machine. DMLS machine has building volume of 250 mm × 250 mm × 325 mm, equipped with fiber laser of 400 W having scanning speed of 7 m/s [3]. During DMLS, metal powder upon exposure to laser power beam melt droplets are created and, due to moving beam, melt pools are created and can be regarded as small castings. 3D printed maraging steel properties are given in Table 2. After getting final 3D printed disc brake of maraging steel, it was heat treated for enhancing hardness and mechanical properties, dimensional stability, reduction in residual stresses, corrosion cracking, and fatigue. Ageing temperature is 490 °C for 5–8 h with air cooling. The 3D printing machine EOS M-290 is a flexible, fast and cost-effective production system for metallic parts. In this work, 3D printed maraging steel (MS1) was considered due to its remarkable properties, such as excellent strength, high toughness, excellent surface finish, and good thermal conductivity. Herzog et al. [4] showed that maraging steel provides applications in critical parts in automotive, aerospace, structural component, tooling, machine tools, fasteners and production sectors.

Table 2. 3D printed maraging steel properties.

Physical Properties	
1. Typical accuracy:	0–20 µm
2. Age hardening shrinkage:	0.08%
3. Smallest wall thickness:	0.3–0.4 mm
4. Relative density	100%
5. Specific density	8–8.1 g/cm^3
6. Surface roughness Ra	4–6.5 µm
7. Surface roughness Rz	20 µm
Mechanical Properties	
1. Ultimate tensile strength:	2050 ± 100 MPa
2. Ultimate tensile strength (y)	1100 ± 100 MPa
3. Hardness	50–56 HRC
4. Tensile—Young's Modulus	180,000 ± 20,000 MPa
Thermal Properties	
1. Thermal conductivity	20 ± 1 W/m °C
2. Specific heat capacity	450 ± 20 J/kg °C
3. Operating temperature:	400 °C

The of brake disc developed by EOS M-290 system and disc brake developed by DMLS process are depicted in Figure 3. Micrographs of disc brake surface, powder morphology analysis of top surfaces of disc brake under heat flux and temperature surface conditions were carried out by Zeiss EVO 50 (ZEISS, Jena, Vienna, Germany) scanning electron microscope (SEM). SEM micrograph of brake disc surface, shown in Figure 4 at radii 50 mm and 100 mm, represents the high temperature region of maraging steel by means of high energy collimated electron beam Leitz Aristomet (Leitz, Hicksville, NY, USA) microscope with high resolution and high depth of focus. Differential scanning calorimetry analyses, shown in Figure 5, were performed using HITACHI (Hitachi Systems, Chiyoda, Tokyo, Japan), Differential Scanning Calorimeter DSC7000 Series, TGDSC-DTA equipment at KELVIN LABS (Kelvin Lab Inc, Hyderabad, Telengana, India), in nitrogen gas atmosphere at three different heating rates (30, 40, and 50 °C/min) between 30 and 1000 °C, with scanning rate 0.01 °C to 150 °C/min, TG measurement range as ±400 mg. Maraging specimens were prepared by cutting small samples having a weight of about 5.120 mg. DSC tests were conducted to assess phase changes and reactions sequence. The DSC testing instrument consists of empty crucible and another one containing the maraging specimen. They are simultaneously heated and kept at the certain temperature. A DSC curve reveals heat flow, i.e. amount of energy exchanged by specimen versus subjected temperature. DSC makes it possible to study phase transformation sequence under precise non-isothermal temperatures. It was observed in DSC curve that MS1 specimen is stable. No phase change was observed during the testing temperature between 300 and 1000 °C. Typical DSC Specifications are given in Table 3.

Figure 3. (**a**) Development of brake disc in EOS M-290 and (**b**) Disc brake developed by DMLS.

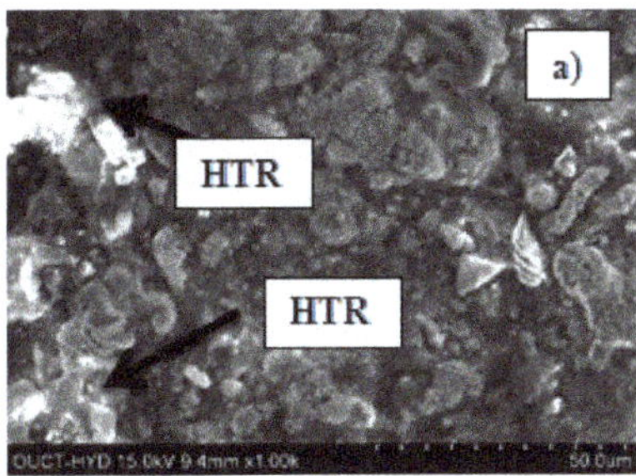

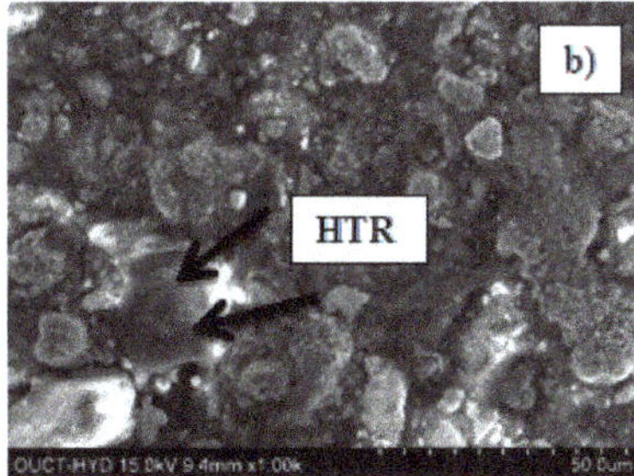

Figure 4. SEM of brake disc surface subjected to High Temperature Region (HTR) at: (**a**) R = 50 mm; and (**b**) R = 100 mm.

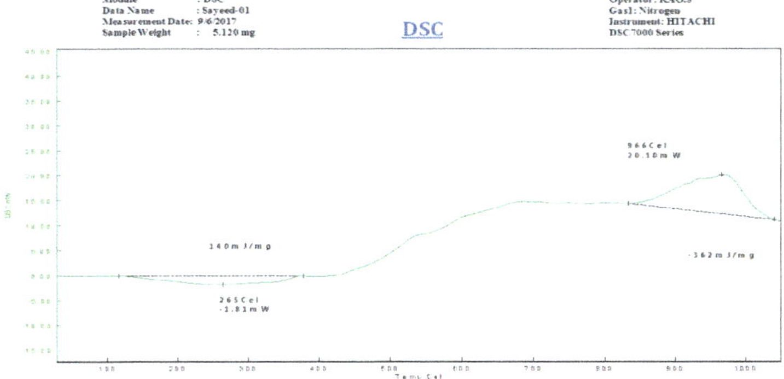

Figure 5. Differential Scanning (DSC) analyses of maraging steel from ambient to 1000 °C.

Table 3. Typical DSC specifications.

1. DSC Test Methods	ASTM E1269-05
2. Temperature Range	Ambient to 1100 °C.
3. Temperature Accuracy	±0.2 K
4. Temperature Precision	±0.02 K
5. Furnace-temperature Resolution	±0.00006 K
6. Heating Rate	0.02 to 300 K/min
7. Cooling Rate	0.02 to 50 K/min
8. Calorimetric Resolution	0.01 μW
9. Measuring Environment	Nitrogen.
10. TG Measurement range	±400 mg
11. Scanning rate	0.01 °C to 150 °C/min
12. Sensitivity	0.2 μg.

3. Previous Studies on Thermal Analysis of Disc Brake

Belhocine et al. [5] presented a simulated thermal behavior in brake disc and determined the initial flux entering the disc to evaluate convective coefficient and visualize disc temperature (3D). We conclude that temperature is influenced by construction and materials and mode. Choi et al. [6] reported that sudden rise and fall of temperature change in metal parts of sliding systems induces uneven thermal stresses due to thermal expansion. This phenomenon is particularly evident in disc brakes under high thermal loads. This paper deals with the finite element modeling of frictional heating process in disc brakes to study the temperature and stress distributions during operation. Andinet et al. [7] presented factors influencing braking performance of train during braking time and found major factors are temperatures and friction coefficient between pad and brake disc. Thermal transient analysis of disc braking system was performed to evaluate nodal temperature under different thermal and operating conditions. Balaji et al. [8] reported thermal degradation is significant to determine thermal stability of product considering the brake application. The present paper deals with the role of various fibers such as aramid, acrylic and cellulose fibers. Thermogravimetric analysis has shown that composite NA03 had minimum weight loss and more thermal stable. Grzes et al. [9] and Marko et al. [10] reported the influence of the pad cover angle on temperature fields on disc brake. A three-dimensional finite element (FE) model of pad–disc system was developed and calculations were carried out for single braking process at constant deceleration with contact pressure corresponds with cover angle of the pad and the evaluated distributions of temperature for both contact surface of pad and disc surface. They developed and evaluated three-dimensional (3D) thermal–structure coupling model and implemented transient thermal analysis of thermo-elastic contact of disc brakes with variation in frictionally generated heat. They found the source of thermal fatigue was the thermo-elastic problem using finite element method. The results demonstrate that the maximum surface equivalent stress may exceed the material yield strength during an emergency braking, which may cause a plastic damage accumulation in a brake disc, while residual tensile hoop stress is incurred on cooling. FE structural and thermal analysis has been carried out with the dimensions and specifications of Corolla car model. The demand of this automobile car is increasing worldwide due to its low cost, high fuel economy and different towing capacity as per the requirement of roads. The maximum speed of this car is 200 kph. This high speed is the reason this car was considered for structural–thermal analysis of disc brake, and maximum speed was used for thermal analysis. Dimensions of disc and pads given in Table 4 were used to develop the 3D model in solid works with and without radial grooves. The average stopping distance with fully loaded disc brake at 25 °C traveling speed varies from 100 to 200 m under the experimental test conditions, requiring an average of 81 m stopping distance with the deceleration rate 8 m/s^2 in 4.5 s. For the analysis, speed of the car reduced from 33.34 to 0 m/s, within 4.5 s. Single stop cycle of braking was used for thermal and structural analysis since material regains its original elastic condition after brake force is removed. Lopez et al. [11] made several assumptions to simplify thermal analysis complexity and output surface temperature was measured experimentally and compared with FE analysis [12,13]. Heat dissipated through brake disc surface during application of brake and heat flux applied to surface was considered with and without radial grooves. Huajiang et al. [14] considered heat transfer convection only after brake application was completed and car accelerated to regain its original speed. These areas include an effective surface area for applying braking pressure with and without radial grooves on disc surface. The remaining surface area of the disc was considered insulated for the purpose of comparison of surface temperatures with and without grooves under the brake pressure of 1 MPa.

Table 4. Properties of pad and disc.

Properties	Disc	Brake Pad
Young's modulus (N/mm^2)	180,000 ± 20,000 MPa	1500
Density (kg/m^3)	8–8.1 g/cm^3	2595
Poisson's ratio	0.3	0.25
Thermal conductivity, W/m °C	20 ± 1 W/m °C	1.212
Ultimate tensile strength, N/mm^2	2050 ± 100 MPa	-
Coefficient of friction	0.35	0.35
Specific heat (J/kg °C)	450 ± 20 J/kg °C	1465

4. Temperature Distributions in Disc Brake

To investigate temperature behavior of brake discs, it is necessary to obtain temperature distributions as a function of braking time, speed, and braking pressure. This research attempted to incorporate radial groove features on disc surface by using DMLS and to predict the temperature response due to these design modifications. Rotary motion of the brake disc causes the sliding surface contact between pad and disc to generates heat. Hudson et al. [15] showed that the surface temperature due to friction generated heat should be considered. The friction force is determined from pressure distribution at contact surfaces of disc and pad. According to Mahmoud et al. [16], the kinetic energy of a car, once the brakes are applied to the pads, which press against brake rotor, is converted to thermal energy [17]. In the case of disc brakes, it is kinetic energy converted to thermal energy. M_v is total mass of vehicle and V_i is initial speed of vehicle and the heat dissipated by each disc is $K_e = 1/2 M_v V_i^2$, i.e., the rate of heat generated due to friction is equal to friction power, and this frictional heat is absorbed by brake disc and pads. If it is assumed that whole friction power is transferred to heat energy, then heat partition γ_p coefficient needs to be considered. Thermal energy generated at friction surface of brake can be transferred to both rotor and pads. This partitioning of energy is dependent on thermal resistances of brake discs and pad that are further dependent on density, material thermal conductivities and heat capacities. The brake pad thermal resistance must be more than rotor thermal resistance to avoid brake fluid from high temperatures. The partitioning coefficient (γ) for thermal input to brake disc and pad was determined from thermal effusivity ξ, given by

$$\xi_{ed} = \sqrt{k_d \rho_d c_d}, \; \xi_{ep} = \sqrt{k_p \rho_p c_p} \tag{1}$$

The friction contact area of pad and disc are determined from equations given as

$$A_{cp} = \varnothing_c \int_{r_1}^{r_2} r\, dr, \; A_{cd} = 2 \times \pi \int_{R1}^{R_2} R\, dr \tag{2}$$

The total heat generated on frictional contact interface Q_{Total} equals the heat flux into the disc Q_{disc} and pad Q_{Pad}, and braking energy, which is termed as heat partition coefficient γ_p, is determined from the following equation:

$$\gamma_p = \frac{\xi_{ed} A_{cd}}{\xi_{ed} A_{cd} + \xi_{ep} A_{cp}} \tag{3}$$

where ξ_{ed} and ξ_{ep} are thermal effusivities of disc and pad; this partitioning of thermal energy is dependent on thermal resistances of pad and brake disc rotor, which are related to their thermal conductivities, materials densities, and heat capacities. Solid works CAD models of disc and pad with 6, 9 and 18 radial grooves on disc surface are shown in Figure 6. The heat flux generated by pressing pad against rubbing surface of rotor is only source of heat input to the disc; magnitude of this heat flux was calculated from basic energy principle and input of energy is in terms of rotor disc speed, radius of rotor, coefficient of friction and pressure distribution [18]. The frictional heat generation due to friction of contact surfaces of two surfaces of brake system, coefficient of friction, speed of vehicle, geometry of the disc rotor and pad, and pressure distribution at the sliding surfaces. In uniform pressure distribution, heat flux Q_{Total} on contact area under the pressure distribution is taken into account in thermal and structural analysis [19].

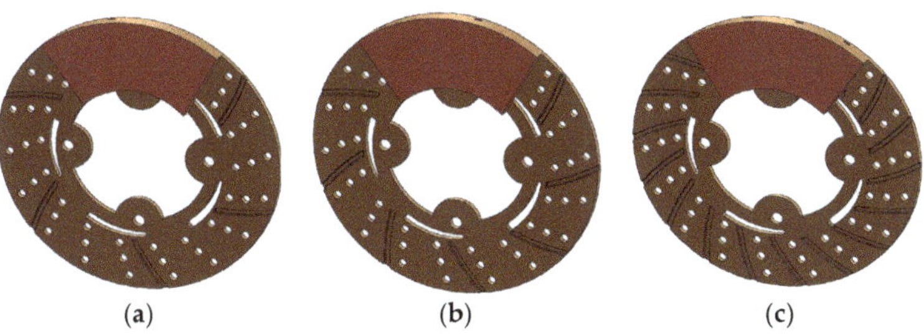

Figure 6. CAD model of brake disc and pad with (**a**) 6, (**b**) 9 and (**c**) 18 radial grooves on disc surface.

In the present thermal analysis, ambient temperature was assumed at 25 °C and brake disc surface temperature was 37.2 °C at 100 kph. It was assumed that heat dissipation from brake disc surface to atmosphere through convection process is governed by $Q_f = h_C A_{Cd}(T_s - T_a)$, where Q_f is in Watts; hc is the convection heat transfer coefficient; Input parameters and dimensions of brake discs are tabulated in Table 5, Acd and Acp are the contact surface area of the disc and pads, respectively, in m^2; Ts is surface temperature of brake disc; and Ta is ambient air temperature in °C. Heat transfer coefficient is applied to brake discs as heat flux boundary condition. Thus, increasing the rate of heat transfer from surface brake discs reduces disc surface temperature on total surface area of the brake discs [20].

Table 5. Input parameters and dimensions of brake discs.

External Brake disc Radius, mm	120
Internal Brake disc Radius, mm	60
Internal radius of pad, mm	60
External radius of pad, mm	120
Brake pad thickness, mm	12
Brake disc Thickness, mm	24
Brake disc Height, mm	49
Initial speed v, Km/h	30
Mass of Vehicle m, Kg.	1385
The cover angle of pad (in degrees), 20%	650
Deceleration ad, m/s^2	8
Vent thickness, mm	6
Brake Disc Effective Radius, R effective, mm	100
Factor of Swept distribution of the disc, ε_p	0.5
Surface disc swept by the pad A cd, mm^2	33,912
Contact Pressure P, MPa	1
Heat partition coefficient γ_p	0.95
Thermal Effusivity for Brake Pad.	2645.7
Thermal Effusivity for Brake Disc.	8971.3

Density of air ρ_{air} (kg/mm^3) is given by = 1.225 Kg/m^3, where m_a is the mass flow rate of air (m^3/s) and V_{avge} is the average air velocity (m/s). Convective heat transfer coefficient at different air velocities are obtained from the formula [21].

$$h_c = 0.70\,K_{air}/d_o(R_e)^{0.55} \tag{4}$$

$$V_{avge} = 0.015\omega \left[\frac{A_{outet} + A_{inlet}}{A_{outet}}\right]\left[d_o^2 - d_i^2\right]^{0.5} m/\sec \tag{5}$$

Radial grooves areas at inlet and outlet are 8.75 mm^2 Air thermal conductivity $K_{air} = 0.024$ W/mK, The contac areas of the pad and disc are 0.0061236 m^2, 0.033912 m^2, respectively. The Convective heat transfer Q_f into disc surface can be calculated using Equation (6) [22].

$$Q_f = \frac{1 - \varnothing}{2}\frac{gmzV_0}{2A_{cd}\varepsilon_p} \tag{6}$$

where $\varnothing$ is rate coverage sector of braking forces between the front and rear axle, $Z = \frac{a_d}{g}$ is braking effectiveness, A_{cd} is disc surface swept by brake pad (m^2), ε_p is factor load distributed on brake disc surface, m is mass of vehicle (kg), g = 9.81 is acceleration of gravity (m/s^2), V_0 is initial speed of vehicle (m/s), and ad is the deceleration of the vehicle (m/s^2).The disc brake groove passage and sector chosen for numerical analysis are shown in Figure 1. The dimensions of grooves on disc surface are 3.5 mm by 2.5 mm each, with outer and inner diameters as 240 mm and 120 mm, respectively. Heat transfer coefficient hc associated with laminar flow for radial and non-radial grooves on brake discs was calculated for Re < 2.4×10^5, where do is outer diameter of discs mm, Re is Reynolds number, and Ka is thermal conductivity of air, W/m °C. Experimental validation was done on modified brake disc with and without radial grooves brake using non-contact thermometer Fluke-561, Infrared thermometer, which can measure contact and ambient temperatures. IR thermometer is used to measure hot moving energized, hard-to-reach objects instantly. Experimental results are given in Table 6 for generation of heat on disc surfaces. Disc surface temperatures increase with increasing braking time for different disc design configurations.

Table 6. Calculation of heat flux at different influential parameters.

Time	a_d	$\varnothing$	m	g	V0	z	A cd	ε_p	$(1 - \varnothing)$	$gmzV_0$	2 A cd ε_p	Heat Flux
0	8	0.2	1385	9.8	33.34	0.8	0.033	0.5	0.8	369,407.2	0.03	4.925
1	8	0.2	1385	9.8	27.7	0.8	0.033	0.5	0.8	307,615.7	0.03	4.101
2	8	0.2	1385	9.8	22.2	0.8	0.033	0.5	0.8	246,269.8	0.03	3.283
3	8	0.2	1385	9.8	16.6	0.8	0.033	0.5	0.8	184,591.6	0.03	2.461
4	8	0.2	1385	9.8	11.1	0.8	0.033	0.5	0.8	123,134.9	0.03	1.641
5	8	0.2	1385	9.8	5.5	0.8	0.033	0.5	0.8	60,903.0	0.03	0.812

5. Experimental Validations

The experimental has been conducted on disc brake with different radial grooves; Temperatures are recorded as shown in Figure 7 with Infra-Red Digital thermometer at different speeds. The heat flux were calculated are tabulated in Table 7 with the effect of radial grooves areas on disc brake.

Figure 7. Temperature recording on modified brake disc surface with FLUKE IR thermometer.

Table 7. Effect of radial grooves area on heat flux of disc brake.

	Disc Brake with 6-Grooves $A_{cd} = 0.03637$ m^2	Ts, °C	Disc Brake with 9-Grooves $A_{cd} = 0.037602$ m^2	Ts, °C	Disc Brake with 18-Grooves $A_{cd} = 0.041292$ m^2	Ts, °C
	Qo, W/mm^2		Qo, W/mm^2		Qo, W/mm^2	
1.	3.383	36.61	3.272	35.87	2.979	34.02
2.	4.060	38.97	3.927	38.05	3.576	35.82
3.	4.736	41.27	4.581	40.22	4.171	37.62
4.	5.413	43.60	5.235	42.40	4.768	39.43
5.	6.089	45.92	5.889	44.57	5.363	41.23
6.	6.766	48.25	6.544	46.75	5.959	43.04
7.	7.443	50.58	7.199	48.93	6.556	44.84
8.	8.119	52.90	7.853	51.10	7.151	46.64
9.	8.797	55.23	8.509	53.28	7.749	48.45
10.	9.472	57.55	9.162	55.45	8.343	50.25
11.	10.14	59.88	9.817	57.63	8.939	52.06
12.	10.82	62.18	10.470	59.80	9.535	53.86
13.	11.50	64.53	11.125	61.98	10.131	55.67
14.	12.17	66.85	11.779	64.15	10.726	57.47
15.	12.85	69.16	12.433	66.33	11.322	59.27
16.	13.53	71.51	13.089	68.51	11.919	61.08

6. Results and Discussions

Belhocine et al. [23–25] used FE software ANSYS for simulation of structural deformation of brake disc and observed induced stress depends on conditions such as speed, contact pressure and coefficient of friction. CAD model was created and FE analysis was performed using ANSYS 15. Tetrahedron Element type was selected for thermal analysis and it is a higher order four-node thermal element. Four-node elements have excellent compatible temperature shapes and are well suited to model curved boundaries and these elements are used for meshing grooved and un-grooved portion of disc brake shown in Figure 8, Transient thermal boundary conditions are introduced into thermal module of ANSYS 15.0, by choosing the first mode of simulation by defining material models and physical properties of materials [26,27]. The transient thermal simulation was carried out using simulation conditions tabulated in Table 8.

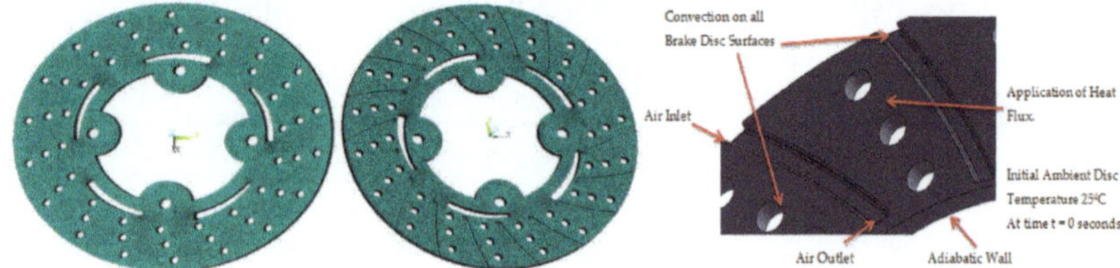

Figure 8. FE meshes model and radially grooved disc brake with boundary conditions.

Table 8. Transient thermal simulation conditions.

3D-Printed Materials	Maraging Steel
Total time of simulation	45 s
Initial temperature of the disc	60 °C
Increment of initial time	0.25 s
Minimal Time Increment	0.5 s
Maximal Time Increment	0.125 s

6.1. Nodal Temperatures and Contact Pressure

In Figures 9 and 10, the temperature distribution of the disc brake is shown. Belhocine et al. [28,29], in their thermal studies on brake disc pad systems, tried to reduce thermal stresses by changing the design factors associated with the temperature distribution in disc brake and raises to squeal phenomenon. In comparison with experimental results, nodal temperatures obtained from FE simulation are in good agreement as surface temperatures Ts are reduced from 77 to 70.0 °C with and without radial grooves. The pressure distribution on disc surface area interface with brake pad taken at various times in FE simulation. For static deformation analysis, the pressure distribution scale varies from 0 to 3.0 MPa. The maximum contact pressure is located on edges of the brake disc and decreases from leading edge towards trailing edge due to friction. The maximum values of the thermal deformation vary from 979.336 to 433.938 µm with and without radial grooves, as shown in Figure 11. This value was obtained for speed of vehicle at 100 kph. The Comparative results of Static Deformation are given in Table 9. The same trend of the stress distribution was observed and highest stress areas are located in the same region with and without radial grooves. This pressure distribution is symmetrical compared with and without radial grooves.

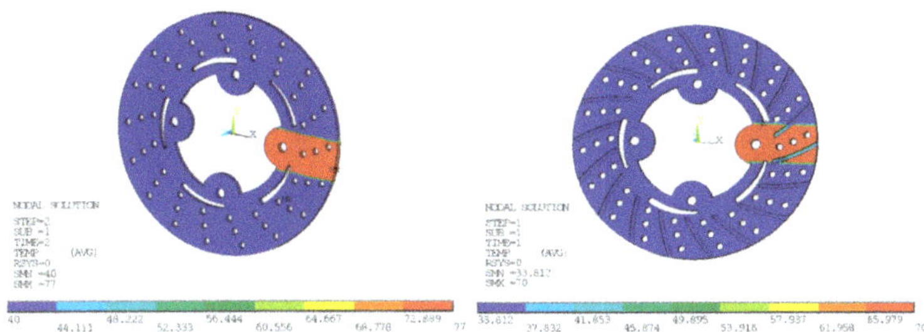

Figure 9. Nodal temperatures at braking pressure of 1 MPa, without and with grooves on disc surfaces.

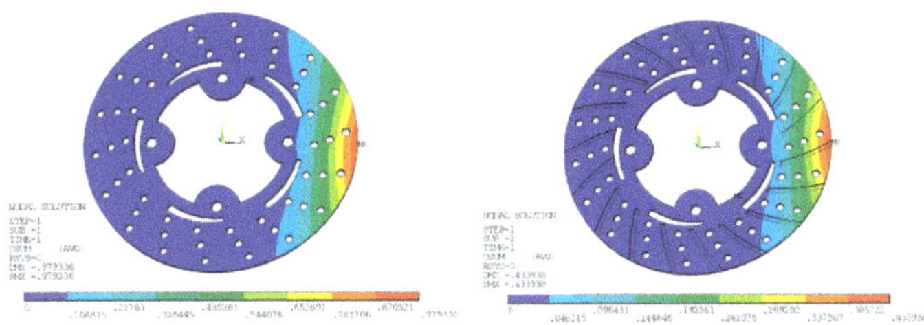

Figure 10. Thermal Deformation at Braking Pressure of 1 MPa, without and with grooves on disc surfaces.

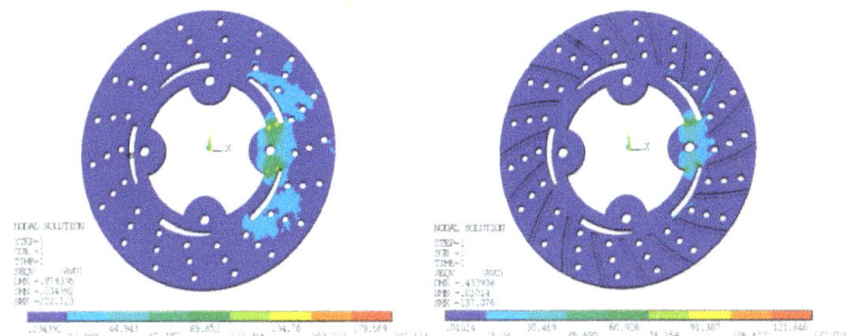

Figure 11. Von Mises and stress for conventional brake disc and radially grooved disc brake.

Table 9. Comparison of Static Deformation results.

Parameters	Without Groove	With Groove
Nodal temperature	77 °C	70 °C
Thermal Deformation	0.979336 mm	0.433938 mm
Vonmises Stress	202.123 N/mm^2	137.076 N/mm^2

6.2. Von Mises and Stress

The maximum value recorded during this FE simulation for von Mises stresses at disc surface temperature 77 °C in the case without grooves is 202.123 N/mm^2 while with grooves is 137.076 N/mm^2. A significant reduction has been observed from ANSYS results. At higher speeds, the von Mises and stress are likely increased because of more frictional heat generated at the time of maximum braking time and pressure applied. The Transient finite element simulation gives variation of temperature distribution with respect to time as shown in Figure 12.

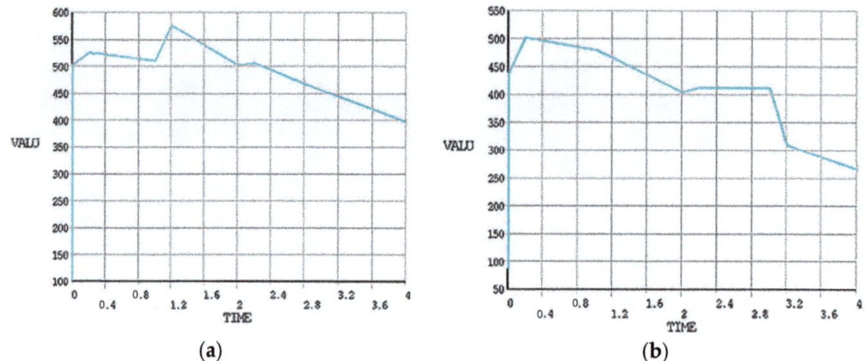

Figure 12. Transient FE analysis of (**a**) conventional and (**b**) grooved brake disc.

In Figure 13, fitted line plot for heat flux and main effect plots for disc temperatures are shown. The different temperature profiles were observed with 6, 9 and 18 radial grooves but these temperatures are lower compared to disc brake without grooves, as shown in Figure 14. Estimation of heat flux is possible with these surface plots with respect to vehicle speeds. Transient FE analysis of conventional brake and grooved disc brake was carried out with simulated conditions given in Table 9. It has been observed that there is a decrease of temperature gradient after moving a certain distance after releasing of brake and with increasing speed of the vehicle as shown in Figure 15. The variation of disc surface temperatures was observed to increase with input heat flux, since, as the speed of the vehicle increases, the heat flux entering in o the disc also increases at the time of friction between two sliding surfaces and thereby sudden rise in temperature and fall at the braking time occurs. Main effect plots for disc surface temperatures represents information about amount of heat flux entered into the discs at respective speeds of the vehicle. Contour plot is typical graphical calculator that gives helpful information on design of disc brake under various thermal loading conditions. It is an additional tool to assess the surface temperatures and compare the nodal temperatures obtained through experimental and FE analysis. Experimentally determined temperatures with grooves and without grooves were 61.08 and 77.25 °C, respectively. It is in good agreement with 95% permissible confidence index. Maximum thermal stress is localized on corner of inner and outer edges of solid disc surfaces. variation in surface temperatures with different vehicle speed are presented in Table 10.

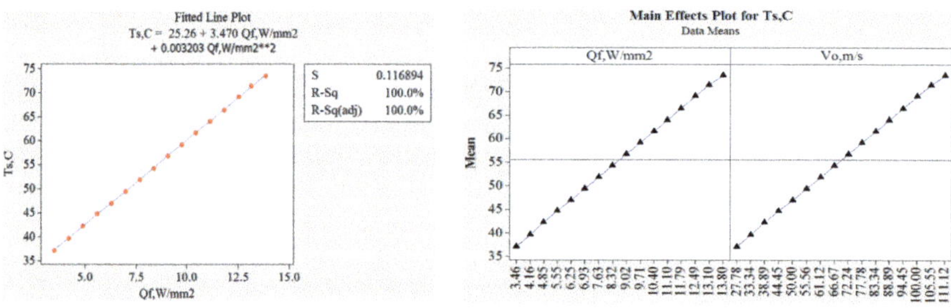

Figure 13. Fitted line plot for Temperature and heat flux and main effect plots for disc temperatures.

The maximum stress is located on inner surface side of holes. For the radially grooved disc configurations, maximum stresses occurred around holes that has been strengthened with more supports near the holes along the circumference of the disc. Hence, the specific regions where maximum stresses are observed have been strengthened to prevent potential crack and fatigue problems. The design modifications with radial grooves on disc surface could be one of the possible solutions for brake disc design for maximum heat dissipation, reduced thermal loading and less thermal fatigue. Maximum thermal stress is localized on the corner of the inner and outer edges of the solid disc surfaces. The maximum stress is located on inner surface and side holes in the disc.

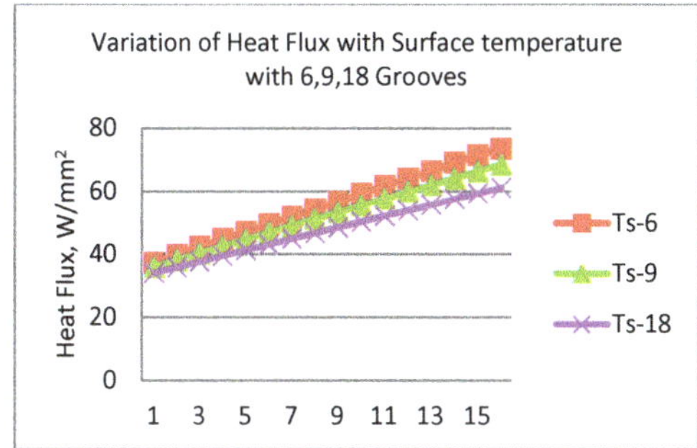

Figure 14. Variation of heat flux with surface temperature with 6, 9, and 18 Radial Grooves.

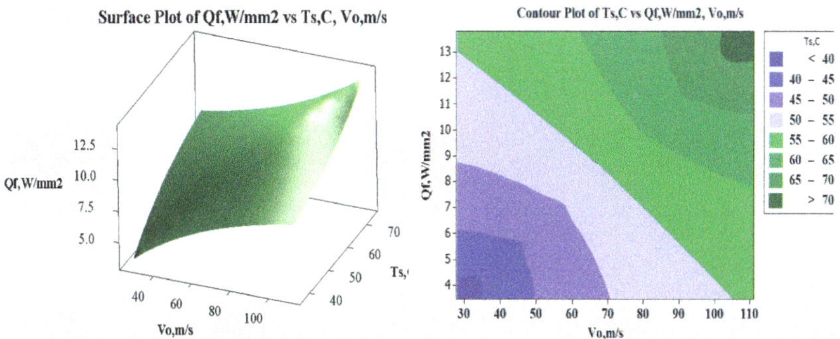

Figure 15. Variation of heat flux with different speed, surface temperatures and contour plots for surface temperature range at different heat flux.

Table 10. Variation in temperatures with vehicle speed.

	With Groove		Without Groove	
Vo, m/s	Experiment, Temp, (°C)	FEA, Temp, (°C)	Experiment, Temp, (°C)	FEA, Temp, (°C)
27.78	34.02	33.5	38.37	41.15
33.34	35.82	33.8	41.05	36.59
38.89	37.62	34.6	43.72	38.24
44.45	39.43	36.2	46.39	41.36
50.00	41.23	38.0	49.07	43.58
55.56	43.04	39.3	51.74	47.21
61.12	44.84	41.5	54.42	48.96
66.67	46.64	50.1	57.09	52.24
72.24	48.45	46.8	59.77	55.23
77.78	50.25	45.4	62.44	58.51
83.34	52.06	50.4	65.12	59.54
88.89	53.86	52.8	67.79	61.27
94.45	55.67	56.2	70.47	66.98
100.0	57.47	58.5	73.14	69.45
105.5	59.27	54.1	75.81	72.12
111.1	61.08	70.0	78.49	77.25

7. Conclusions

In this research work, an attempt was made to study the effect of design modification on disc brake surface incorporated with radial grooves. FE Thermal Analysis was performed for analyzing thermal aspects with geometrical parameters influencing structural and thermal analysis of disc brakes without and with radial grooves. Surface modification on disc brake was achieved by additive manufacturing using DMLS process. Braking pressure, braking time and speed of the vehicle are influential parameters on thermal characterization of brake surface. From experimental results, it was concluded that 3D printed metals such as maraging steel can be used for longer life of disc brake with desired surface quality and it is suitable for high temperatures disc brake applications. Von Mises stresses were observed to be reduced in the case of grooved disc brake by 32% and also nodal temperature variation was around 10%. A transient thermal analysis was carried out using the direct time integration technique for the application of braking force due to friction for time duration of 4 s. The results obtained from this research work has revealed that, to enhance the maximum heat dissipation area on disc brake surface, radial grooves can be incorporated as they increase rate of heat transfer and reduce the disc surface temperature. Variation in heat flux for the 18 radial grooves are less when compare to 6 and 9 grooves. From the Transient temperature distribution, it has been observed that, for more braking time, the heat flux into the disc is increased and surface temperature also increases. It is recommended that using maraging steel disc brake material is safe based on the strength and rigidity criteria. All simulated values obtained from the finite element analysis are permitted values within the design tolerance and hence the brake disc design is safe based on comparative thermal analysis. The effective utilization of DMLS process can be further extended in designing and analyzing thermal deformation of brake disc pads developed through 3D printed materials and may be considered as future scope of present work.

Author Contributions: Conceptualization, G.M.S.A. and S.A.; Methodology, G.M.S.A. and S.A.; Software, G.M.S.A.; Investigation, Writing-Original Draft Preparation, G.M.S.A. and S.A.; Writing-Review & Editing S.A.; Supervision S.A.; Project Administration, S.A.; Funding Acquisition, G.M.S.A.

Funding: This research was funded by King Khalid University, ABHA, Kingdom of Saudi Arabia with grant number G.R.P.-324-38.

Acknowledgments: Authors would like to thank Deanship of Scientific Research at King Khalid University for funding this research work through research groups program under grant number G.R.P.-324-38.

Conflicts of Interest: The authors declare no conflict of interest.

Nomenclature

M_v	is total mass of vehicle
V_i	Initial speed of vehicle
K_e	Kinetic Energy
γ_p	Heat partition coefficient
ξ_{ep}, ξ_{ed}	Thermal Effusivity of pad and Disc.
A_{cp}, A_{cd}	Friction contact area of pad and Disc.
Q_{Total}	Heat flux on contact area.
Q_{disc}, Q_{Pad}	Heat flux into the disc and pad.
Q_f	Convective heat transfer
Ts	Surface temperature of brake disc.
Ta	Ambient air temperature in °C.
Ø	coverage sector of braking forces
Z	Braking Effectiveness.
g	Acceleration of gravity
m_a	Mass flow rate of air
V_{avge}	Average air velocity

References

1. Murr, L.E.; Johnson, W.L. 3D metal droplet printing development and advanced materials additive manufacturing. *J. Mater. Res. Technol.* **2017**, *6*, 77–89. [CrossRef]
2. Duda, T.; Raghavan, L.V. 3D Metal Printing Technology. *IFAC* **2016**, *49*, 103–110. [CrossRef]
3. Naiju, C.D.; Adithan, M.; Radhakrishnan, P. Statistical analysis of compressive strength for the reliability of parts produced by direct metal laser sintering (DMLS). *Int. J. Mater. Eng. Innov.* **2012**, *3*, 282–294. [CrossRef]
4. Herzog, D.; Seyda, V.; Wycisk, E. Claus Emmelmann, Additive manufacturing of metals. *Acta Mater.* **2016**, *117*, 371–392. [CrossRef]
5. Belhocine, A.; Bouchetara, M. Thermal analysis of a solid brake disc. *Appl. Therm. Eng.* **2012**, *32*, 59–67. [CrossRef]
6. Choi, J.-H.; Lee, I. Finite element analysis of transient thermoelastic behaviors in disc brakes. *Wear* **2004**, *257*, 47–58. [CrossRef]
7. Eticha, A.K. Analysis of the Performance of Disc Brake System of Addis Ababa Light Rail Transit Using Temperature and Coefficient of Friction as a Parameter. *Int. J. Mech. Eng. Appl.* **2016**, *4*, 205–211.
8. Sai Balaji, M.A.; Kalaichelvan, K. Experimental studies of various reinforcing fibres in automotive disc brake pad on friction stability, thermal stability and wear. *Int. J. Mater. Prod. Technol.* **2012**, *45*, 132–144. [CrossRef]
9. Grzes, P. Finite element analysis of temperature distribution in axisymmetric model of disc brake. *Acta Mechan. Autom.* **2010**, *4*, 23–28.
10. Tirovic, M. Development of a lightweight wheel carrier for commercial vehicles. *Int. J. Veh. Des.* **2012**, *60*, 138–154. [CrossRef]
11. Galindo-Lopez, C.H.; Tirovic, M. Understanding and improving the convective cooling of brake disc with radial vanes. *J. Automob. Eng.* **2008**, *222*, 1211–1229. [CrossRef]
12. Gao, C.H.; Huang, J.M.; Lin, X.Z.; Tang, X.S. Stress analysis of thermal fatigue fracture of brake discs based on thermomechanical coupling. *ASME J. Tribol.* **2007**, *129*, 536–543. [CrossRef]
13. Gao, C.H.; Lin, X.Z. Transient temperature field analysis of a brake in a non-axisymmetric three-dimensional model. *J. Mater. Process. Technol.* **2002**, *129*, 513–517. [CrossRef]
14. Ouyang, H.; Abu-Bakar, A.R.; Li, L. A combined analysis of heat conduction, contact pressure and transient vibration of a disc brake. *Int. J. Veh. Des.* **2009**, *51*, 190–206. [CrossRef]
15. Hudson, M.; Ruhl, R. Ventilated Brake Rotor Air Flow Investigation. SAE Technical Paper 971033. 1997. Available online: https://www.sae.org/publications/technical-papers/content/971033/ (accessed on 15 August 2017).
16. Mahmoud, K.R.M.; Mourad, M. Parameters affecting wedge disc brake performance. *Int. J. Veh. Perform.* **2014**, *1*, 254–263. [CrossRef]
17. Mackin, T.J.; Noe, S.C.; Ball, K.J. Thermal cracking in dick brakes. *Eng. Fail. Anal.* **2002**, *9*, 63–76. [CrossRef]
18. Duzgun, M. Investigation of thermo-structural behaviors of different ventilation applications on brake discs. *J. Mech. Sci. Technol.* **2012**, *26*, 235–240. [CrossRef]
19. Milošević, M. Modeling Thermal Effects in Braking Systems of Railway Vehicles. *Therm. Sci.* **2012**, *16* (Suppl. 2), S581–S592. [CrossRef]
20. Abdo, J.; Nouby, M.; Mathivanan, D.; Srinivasan, K. Reducing disc brake squeal through FEM approach and experimental design technique. *Int. J. Veh. Noise Vib.* **2010**, *6*, 230–246. [CrossRef]
21. Kao, T.K.; Richmond, J.W.; Douarre, A. Brake disc hot spotting and thermal judder: An experimental and finite element study. *Int. J. Veh. Des.* **2000**, *23*, 276–296. [CrossRef]
22. Reimpel, J. *Braking Technology*; Vogel Verlag: Würzburg, Germany, 1998.
23. Belhocine, A.; Bouchetara, M. Structural and Thermal Analysis of Automotive Disc Brake Rotor. *Arch. Mech. Eng.* **2014**, *61*, 89–113. [CrossRef]
24. Belhocine, A.; Abu Bakar, A.R.; Abdullah, O.I. Structural and Contact Analysis of Disc Brake Assembly During Single Stop Braking Event. *Trans. Indian Inst. Met.* **2015**, *68*, 403–410. [CrossRef]
25. Ishak, M.R.; Abu Bakar, A.R.; Belhocine, A.; Mohd Tayeb, J.; Wan Omar, W.Z. Brake torque analysis of fully mechanical parking brake system: Theoretical and experimental approach. *Measurement* **2016**, *94*, 487–497. [CrossRef]
26. Belhocine, A.; Nouby, M.G. Effects of Young's Modulus on Disc Brake Squeal using Finite Element Analysis. *Int. J. Acoust. Vib.* **2016**, *21*, 292–300. [CrossRef]

27. Belhocine, A.; Wan Omar, W.Z. Three-dimensional finite element modeling and analysis of the mechanical behavior of dry contact slipping between the disc and the brake pads. *Int. J. Adv. Manuf. Technol.* **2017**, *88*, 1035–1051. [CrossRef]
28. Belhocine, A. FE prediction of thermal performance and stresses in an automotive disc brake system. *Int. J. Adv. Manuf. Technol.* **2017**, *89*, 3563–3578. [CrossRef]
29. Belhocine, A.; Wan Omar, W.Z. CFD analysis of the brake disc and the wheel house through air flow: Predictions of Surface heat transfer coefficients (STHC) during braking operation. *J. Mech. Sci. Technol.* **2018**, *32*, 481–490. [CrossRef]

 *materials*

Article

Effects of Substrate Preheating Temperatures on the Microstructure, Properties, and Residual Stress of 12CrNi2 Prepared by Laser Cladding Deposition Technique

Chenggang Ding *, Xu Cui *, Jianqiang Jiao and Ping Zhu

School of Materials Science and Engineering, Dalian Jiaotong University, Dalian 116028, China; jjq053190@163.com (J.J.); zhuping@djtu.edu.cn (P.Z.)
* Correspondence: dingcg@djtu.edu.cn (C.D.); cxtxwd@yahoo.com (X.C.)

Received: 23 October 2018; Accepted: 27 November 2018; Published: 28 November 2018

Abstract: The 12CrNi2 alloy steel powder studied in the present paper is mainly used to manufacture camshafts for nuclear power emergency diesel engines. Laser cladding deposition is of great significance for the manufacture of nuclear power emergency diesel camshafts, which has the advantages of reducing material cost and shortening the manufacturing cycle. However, due to the extremely uneven heating of the components during the deposition process, a complex residual stress field occurs, resulting in crack defects and residual deformation of the components. In the present paper, 12CrNi2 bulk specimens were prepared on the Q460E high-strength structural steel substrate at different preheating temperatures by laser cladding deposition technique, and a finite element residual stress analysis model was established to investigate the effects of different preheating temperatures on the microstructure, properties, and residual stress of the specimens. The results of the experiments and finite element simulations show that with the increase of preheating temperature, the content of martensite/bainite in the deposited layer decreases, and the ferrite content increases. The proper preheating temperature (150 °C) has good mechanical properties. The residual stress on the surface of each specimen decreases with the increase of the preheating temperature. The longitudinal stress is greater at the rear-end deposition part, and the lateral residual stress is greater on both sides along the scanning direction.

Keywords: laser cladding deposition; 12CrNi2 alloy steel powder; substrate preheating; microstructure and properties; residual stress

---

1. Introduction

Traditional technologies for the manufacturing of nuclear power emergency diesel engine camshafts are forging machining, post-forging heat treatment, etc. During machining of such parts, due to very dense surfaces, loose cores, insufficiently matched toughness, and forging cracks on the surface, manufacturing of nuclear power emergency diesel generator camshafts need to take new preparation methods into account. The 12CrNi2 metallic powder-based laser cladding deposition studied in the present paper will be mainly used for the production of camshafts for nuclear power emergency diesel engines, which provides a theoretical basis for ensuring safe operation and independent development. From a report by Murr et al. [1], it is known the recent progress has been made in the characterization and analysis of AM (additive manufacturing) prototypes fabricated by laser and electron beam melting technologies, referred to as direct metal laser sintering (DMLS), or selective laser melting (SLM) and electron beam melting (EBM), respectively [1]. Depending on the characteristics of high-energy density and non-contact processing in laser cladding deposition techniques, it can be used to effectively increase the material quality of refractory alloys, titanium alloys, nickel-based superalloys, and refractory metal materials; this has brought tremendous contributions to the aerospace, rail transit equipment manufacturing, and biomedical manufacturing industries [2]. The Hanover Laser Research Center in Germany selected the induction preheating method to study the direct forming of superalloy laser metals, which can effectively eliminate the defects of deposit cracking, and can be successfully applied to the repair and forming of superalloy blades. The Los Alamos Laboratory, Sandia Laboratory, and EADS have respectively optimized the process of laser metal deposition specimens such as ferroalloys and titanium alloys and produced fully dense and defect-free products with higher properties than forgings [3,4]. Ge et al. [5] have studied the effects of different laser power, scanning speed, and powder feeding rates on the forming quality and dimensional accuracy of DZ125L superalloy. From three aspects including the ratio of depth to width, laser energy density, and material dilution ratio, a characterization method of three-dimensional cladding layer was proposed, which provides a theoretical basis for the shape control of laser metal direct forming [5]. Laser cladding deposition technique is advantageous because of its high-machining precision, high-economic gain, simple modeling, short machining cycle, etc., so it holds great promise for research. Laser machining is featured by high brightness, strong directivity, and high-energy density. It can instantly form a high-temperature laser molten pool. However, due to the local heating of the substrate surface and the constant movement of the heat source, the overall heating of the component is extremely uneven. Solid–liquid phase changes are repeated in the laser molten pool. The thermal expansion during the formation of the molten pool produces a scompressive stress subject to the constraints of the surrounding cold end zone. During the cooling of the material after solidification, the metallic iron volumetrically changes through allotropic transformation to form structural stress. The superposition of the two force fields leads to a complicated residual stress field, which causes defects such as cracks and residual deformation [6]. For the residual stress problem of laser metal direct forming, substrate preheating is an effective stress control method for increasing the laser absorption rate of metal materials, reducing temperature gradients and cooling rate, and improving defects such as cracks. Masoud Alimardani [7], University of Waterloo, Canada, and others proposed a three-dimensional transient finite element analysis method. The stress field and temperature field of a 304 stainless-steel thin-walled wall after preheating and specimen restraint were studied. The multi-layer material addition of a 304 stainless-steel thin-walled wall was simulated, and the transient temperature distribution of molten pool and the real-time evolution of a stress field were obtained.

In the present paper, the substrate was preheated at different temperatures. The changes in microstructure, properties, and residual stress of the materials under different preheating conditions were explored, which effectively improved the defects due to the overall unevenness in heating during laser cladding deposition, and enhanced its performance to a certain extent while being applied in the process of manufacturing of nuclear power emergency diesel engine camshafts by laser cladding deposition.

2. Experimental Material and Methods

The experimental material was 12CrNi2 low-alloy steel powder with a particle size of 106 to 180 μm. The substrate material was a Q460E low-alloy, high-strength steel plate with a thickness of 10 mm, according to standard GB/T 1591-2018; the mechanical properties are shown in Table 1. The national standards for the chemical composition and mechanical properties of 12CrNi2 powder are shown in Table 2 [8].

Table 1. Mechanical composition of the Q460E substrate materials.

Grade	Yield Strength (MPa)	Tensile Strength (MPa)	Elongation after Fracture (%)
Q460E	460	540–720	17

Table 2. Chemical composition of the 12CrNi2 materials (quality components, %).

Grade	Fe	C	Ni	Cr	Mn	Ce
12CrNi2	Bal	0.132	1.68	0.763	0.484	0.879

In the experiment, an IPG YLS-6000 laser device (IPG Photonics Corporation, Beijing, China) with a KUKA-KR30 robot control system (Keller und Knappich Augsburg, Shanghai, China) was used to control the additive manufacturing process. The main process parameters were: laser power P = 1400 W; substrate preheating temperature 20 °C, 150 °C, and 300 °C; laser scanning speed V = 8 mm/s; laser spot diameter D = 3.0 mm; powder feeding amount = 2.5 r/min (8 g/min). The powder feeding gas and the shielding gas were both argon gas (gas flow rate of 15 L/min). The scanning path was S-type reciprocating scanning along the length direction. The laser multi-pass lap ratio was 40% [9,10]. The schematic diagram of the additive manufacturing process is shown in Figure 1.

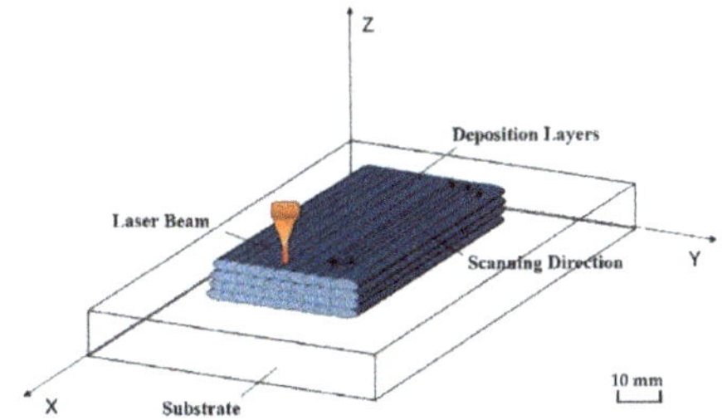

Figure 1. Schematic diagram of laser cladding deposition.

The microstructure of the deposited layer was observed using a Leica Dmi8 metallographic microscope. The microhardness of the additive specimen from the near-base layer to the surface layer was measured using a FM-700 microhardness tester (Future, Tokyo, Japan) with a load of 200 gf. A WDW-300 universal testing machine (Times Testing Instrument, Jinan, China) was used for a tensile test on the sample. The phase of the additive layer was analyzed by Empyrean X-ray powder diffraction (XRD, PANalytical, Almelo, the Netherlands). The residual stress on the surface of each test specimen was tested using a KJS-3 type indentation stress tester (Chinese Academy of Sciences, Shenyang, China).

3. Test Results and Analysis

3.1. Microscopic Analysis

Figure 2 shows the microstructure of the surface layer and the near-base layer of the deposited layer at different preheating temperatures. The surface microstructure of each test specimen consists of martensite, upper bainite, and acicular ferrite. As the preheating temperature of the substrate increases, the content of martensite and bainite decreases, and that of ferrite increases. The near-base microstructure consists of granular bainite and fine-grained ferrite. With the increase of preheating temperature, the residence duration of the high-temperature zone in the thermal cycle increases, which leads to the decomposition of some granular bainite into ferrite and carbide, the decrease in the content of whole granular bainite and the increase in the content of ferrite.

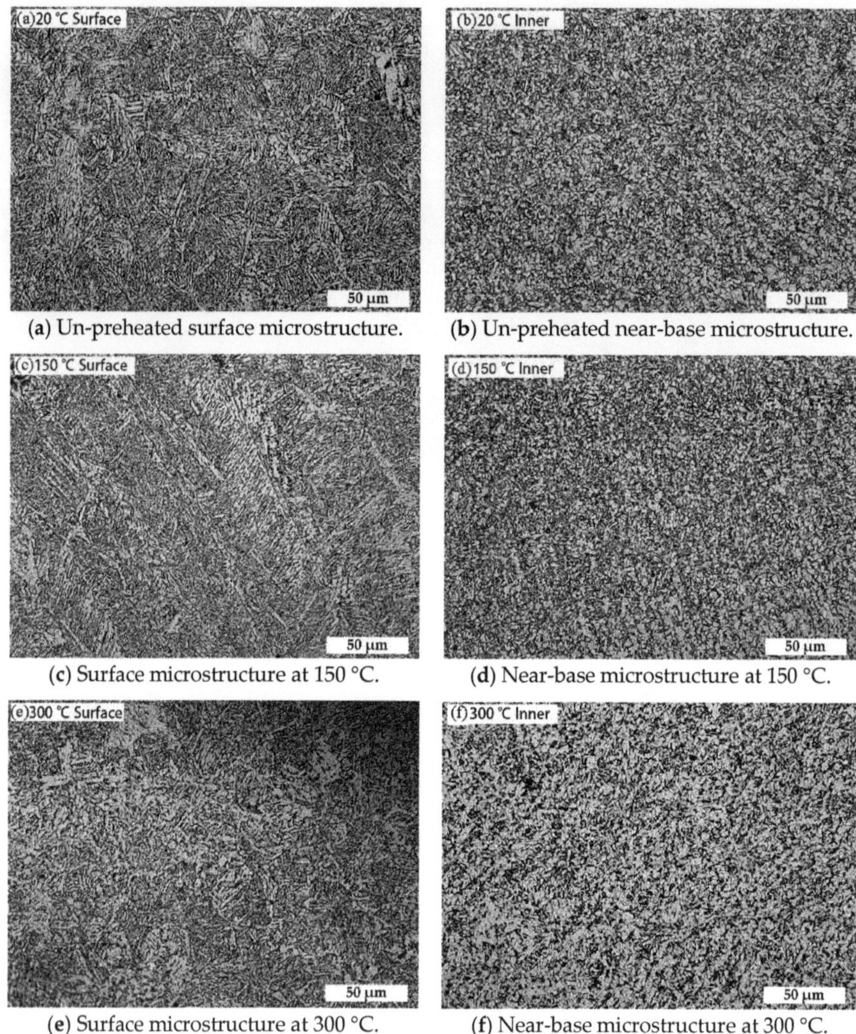

(**a**) Un-preheated surface microstructure.

(**b**) Un-preheated near-base microstructure.

(**c**) Surface microstructure at 150 °C.

(**d**) Near-base microstructure at 150 °C.

(**e**) Surface microstructure at 300 °C.

(**f**) Near-base microstructure at 300 °C.

Figure 2. Microstructure at different preheating temperatures.

Figure 3a–c shows the XRD diffraction peaks of different preheating temperature test specimens with the power of 1400 W and the scanning speed of 8 mm/s, compared with the standard PDF card, un-preheated, preheated at 150 °C, and preheated at 300 °C test specimens were consistent with the three cards of 35-1375, 34-0396, and 85-1410. It can be seen that there was no significant difference in the phase of each test specimen, which was composed of three phases of Fe, Cr-Fe, and Ni-Fe solid solution. According to the observation of the indices of crystallographic plane, the summation of the indices of crystallographic plane was even, which were all body-centered cubic lattices. With the increase of the preheating temperature, the diffraction peaks narrow and symmetrically taper, the crystallinity increases and the grain size increases.

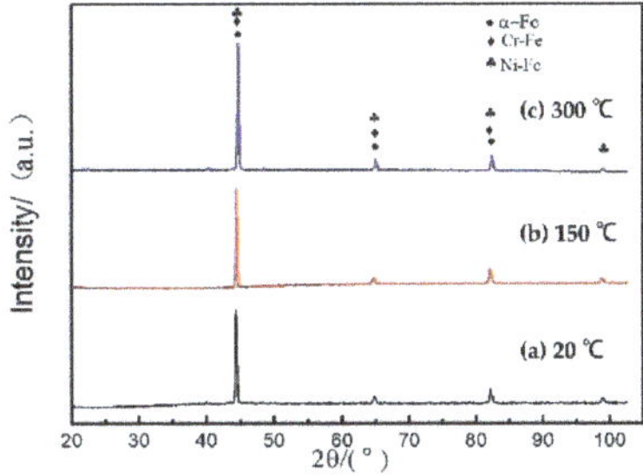

Figure 3. XRD image of different preheating temperatures.

3.2. Mechanical Performance Analysis

The hardness test position was from the surface to the inner layer, and the test interval was 0.2 mm. Then the hardness value was tested twice at the interval of 0.2 mm around each test point, and the average value of three points was taken as the hardness value of this position. Figure 4 shows the hardness distribution of the cross-section of each test specimen at different preheating temperatures. From the trend point of view, the hardness of each test specimen from the surface layer to the substrate was continuously decreasing. This was because the inner microstructure undergoes complex phenomena such as tempering, austenitizing, laser re-melting, and granular bainite decomposition under the repeated thermal cycling, resulting in a decrease in hardness. However, the top microstructure only undergoes a process of solidification and cooling solid-state phase transition, so the hardness was relatively high. The maximum hardness of the test specimen was 336 HV when it was not preheated. The hardness of the bottom layer of the test specimen was above 278 HV. The hardness of the test specimen preheated to 150 °C was slightly lower than that of the unheated test specimen. But when the preheating temperature rose to 300 °C, the temperature drop was slow due to the serious heat accumulation. The highest hardness of the surface layer was 313 HV, and the average hardness of the bottom layer under long-term high temperature was only 259 HV.

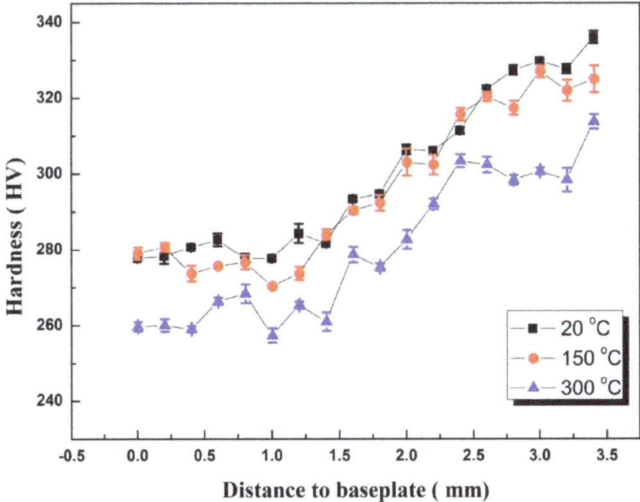

Figure 4. Hardness of different preheating temperature.

In the test of mechanical properties, the dimension of the additive manufacturing part of the specimen was 120 mm × 35 mm × 4 mm, and the dimension diagram of the standard tensile specimen is shown in Figure 5. Three samples were selected for each parameter test, at last, the average value was the final result.

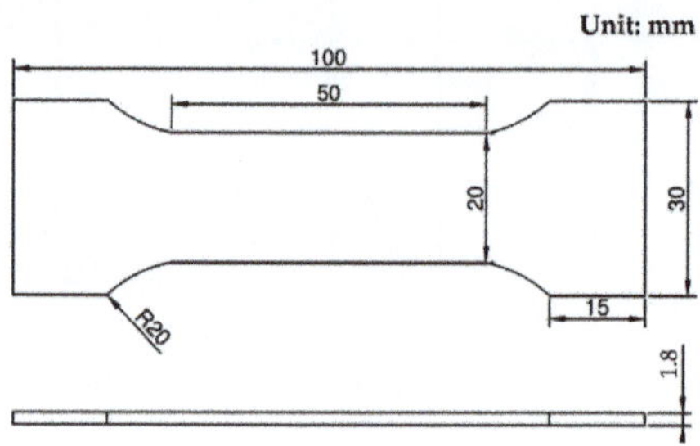

Figure 5. Dimension of tensile test specimen.

By comparison among the preheated test specimens of the substrate, the test specimen preheated at 150 °C and the test specimen preheated at 300 °C, the yield strength and tensile strength of the test specimen preheated at 150 °C were the highest, and those of the test specimen preheated at 300 °C were greatly reduced—10.2% lower than those of the test specimen preheated at 150 °C. However, the yield strength and tensile strength were still higher than the mechanical properties of 12CrNi2 in the national standard. The unheated specimen had the least ductility, and the elongation after fracture was 14.9%. The test specimen preheated at 150 °C had the greatest ductility, and the elongation after fracture was 19.8%. The yield-tensile strength ratios at the three substrate temperatures were between 0.84 and 0.86, which were required by the alloy structural steel. They met the requirements of alloy structural steel and had good resistance to deformation. The test specimen preheated at 150 °C had the highest tensile strength and elongation after fracture. The results of tensile properties are shown in Figure 6.

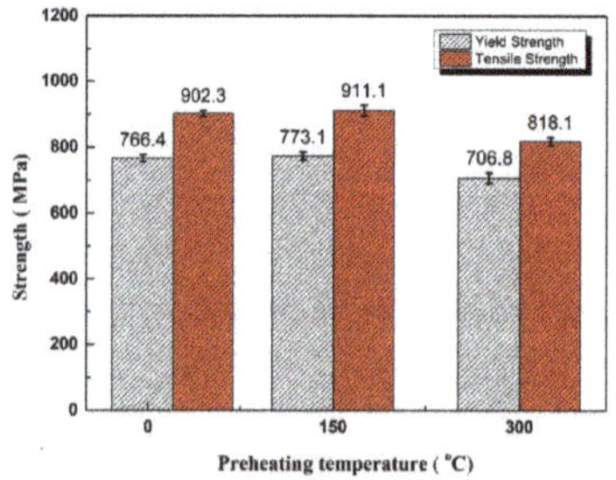

(**a**) Result of yield strength and tensile strength.

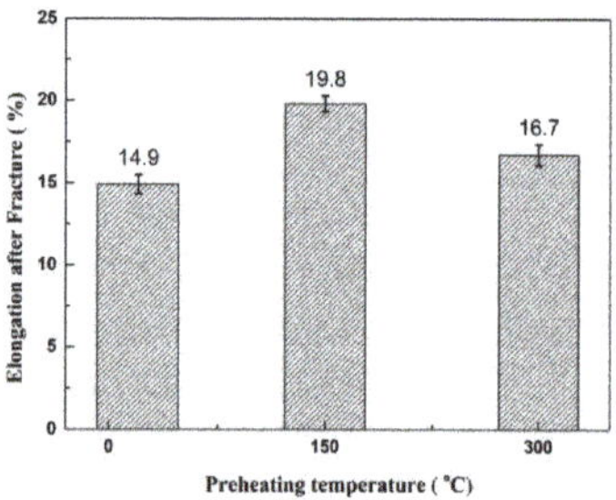

(**b**) Result of elongation after fracture.

Figure 6. Result of the tensile properties.

3.3. Residual Stress Analysis

The finite element analysis of temperature and residual stress field are based on "Abaqus" software (6.11-1, SIMULIA, Johnston, RI, USA). The model and mesh division are shown in Figure 7. The maximum mesh size of the substrate was 2 mm. The additive layer mesh size was 1 mm. The entire model had a total of 12,415 units and 15,168 nodes. The temperature field adopted the DC3D8 unit. The stress field adopted the C3D8R unit. Both units supported the dead-live unit technique, which was applicable to the finite element analysis for direct forming of laser metal powder.

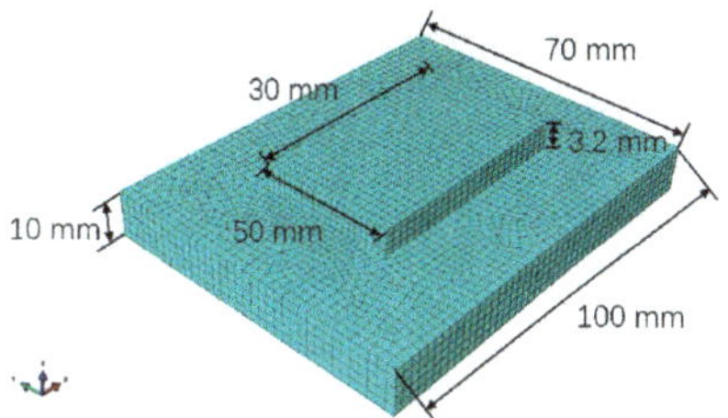

Figure 7. Finite element model and grid division.

In order to avoid the rigid displacement of the whole substrate and additive layer in the process of stress field calculation for direct forming of laser metal, it was necessary to impose constraints on the whole model. The finite element simulation part of the present paper applied full constraint on the bottom surface of XOY substrate. The thermophysical parameters of the 12CrNi2 material were tested by LFA (laser flash thermal conductivity apparatus) and a high temperature mechanical properties tester, (see Table 3).

Table 3. Material parameters of 12CrNi2 steel.

Temperature T (°C)	Thermal Conductivity λ ($W \cdot m^{-1} \cdot K^{-1}$)	Specific Heat Capacity C ($J \cdot kg^{-1} \cdot K^{-1}$)	Density ρ (kg/m^{-3})	Poisson's Ratio μ	Linear Expansion Coefficient ($10^{-6}/m \cdot °C^{-1}$)	Elastic Modulus E (GPa)	Maximum Yield (MPa)
20	44.5	475	7850	0.28	11.3	2.05	590
300	39	550	7850	0.28	11.9	1.85	490
500	30	690	7750	0.28	12.5	1.65	410
800	22	830	7700	0.28	13.4	1.42	20
1000	22	390	7500	0.28	14.8	1.13	17
1500	21	375	7350	0.28	14.9	1.13	17

Figure 8 is a cloud diagram of the residual stress distribution on the surface of each test specimen at different preheating temperatures. It can be seen that the position of the longitudinal tensile stress on the surface of each test specimen is the end of the deposition zone. The lateral tensile stress concentration area on the surface of each test specimen is on both sides in the scanning direction. The maximum longitudinal residual stress on the surface of the unheated test specimen was 398.7 MPa, wherein the longitudinal stress in the middle of the specimen parallel to the scanning direction it was between 210.8 and 336.1 MPa. The maximum lateral residual stress was 311.9 MPa, wherein the transverse stress in the middle of the specimen parallel to the scanning direction it was between 166.9 and 263.6 MPa. The maximum longitudinal residual stress on the surface of the test specimen preheated at 150 °C was 304.7 MPa, wherein the longitudinal stress in the middle of the test specimen parallel to the scanning direction was between 151.2 and 253.5 MPa. The maximum lateral residual stress was 282.2 MPa, wherein the transverse stress in the middle of the test specimen parallel to the scanning direction was between 164.8 and 243.1 MPa. The maximum longitudinal residual stress on the surface of the test specimen preheated at 300 °C was 275 MPa, wherein the longitudinal stress in the middle of the specimen parallel to the scanning direction was between 148.6 and 232.9 MPa. The maximum lateral residual stress was 251.4 MPa, wherein the longitudinal stress in the middle of the test specimen parallel to the scanning direction was between 144.8 and 215.9 MPa. After comparative analysis, the residual stress decreased with the increase of preheating temperature.

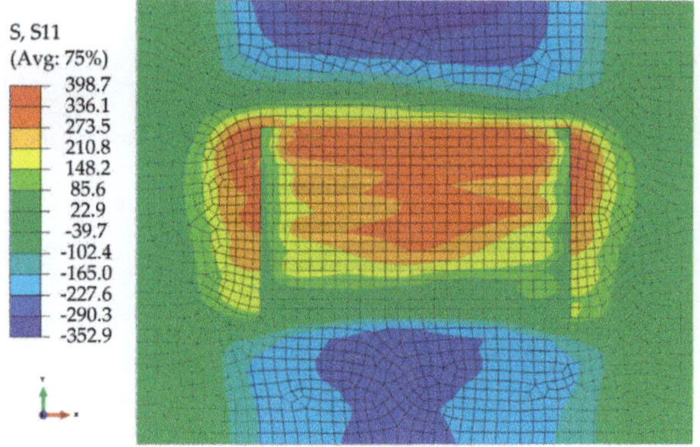

(**a**) Longitudinal stress of unpreheated test specimen.

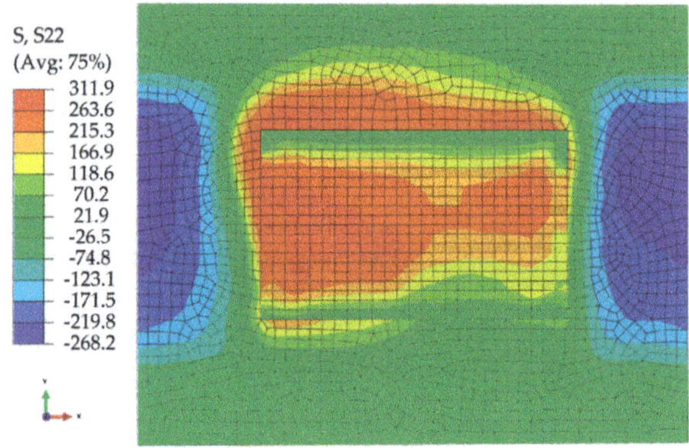

(**b**) Lateral stress of unpreheated test specimen.

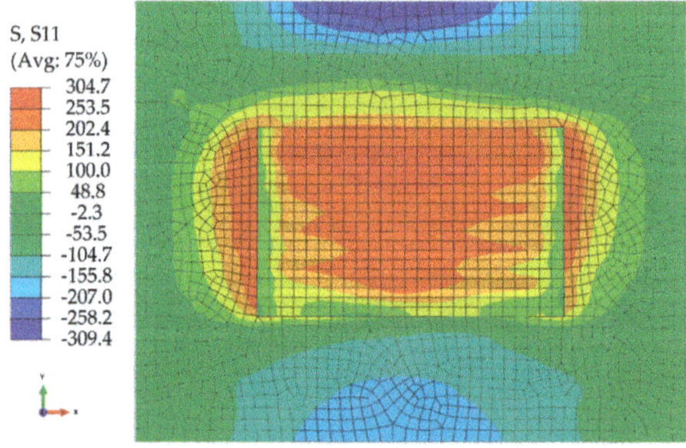

(**c**) Longitudinal stress of test specimen preheated at 150 °C.

Figure 8. *Cont.*

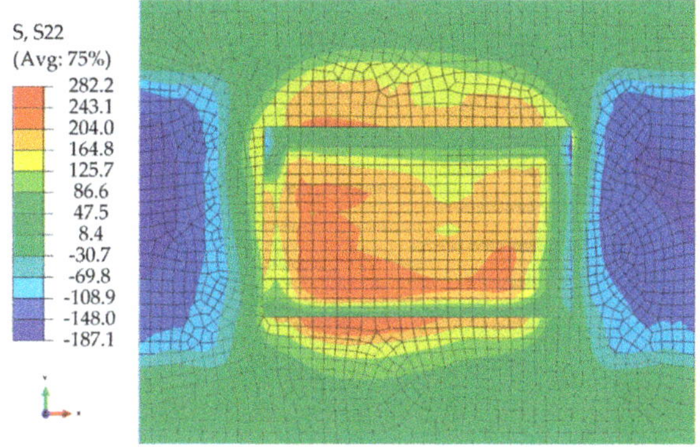

(**d**) Lateral stress of test specimen preheated at 150 °C.

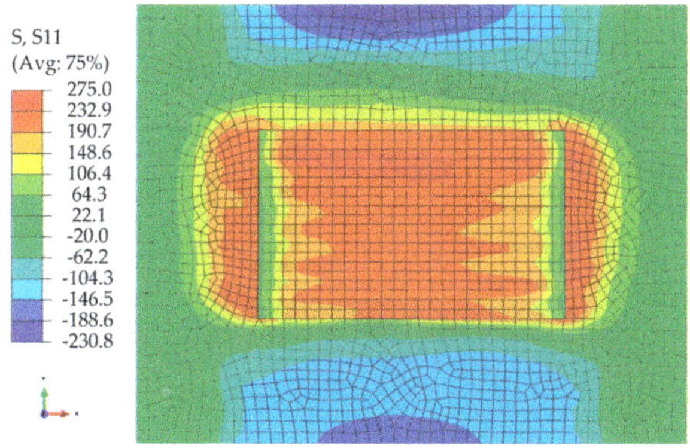

(**e**) Longitudinal stress of test specimen preheated at 300 °C.

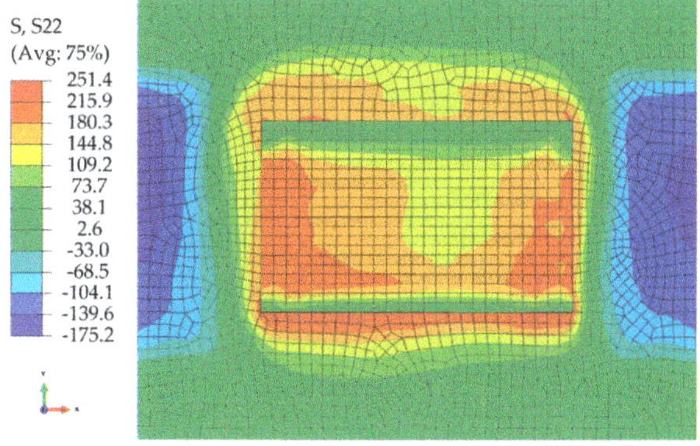

(**f**) Lateral stress of test specimen preheated at 300 °C.

Figure 8. Residual stress field of different preheating temperature.

The residual stress test points are shown in Figure 9. Figure 10 shows the results for testing of residual stress on the surface of each test specimens of un-preheated, preheated at 150 °C, and preheated at 300 °C. With the increase of preheating temperature, the longitudinal and lateral residual stresses were greatly decreased, and the longitudinal residual stress was slightly greater than the lateral residual stress. In full accordance with the numerical simulation law as shown in Figure 8, the numerical average deviation was 7.93%, and the simulation results agree well with the actual test values.

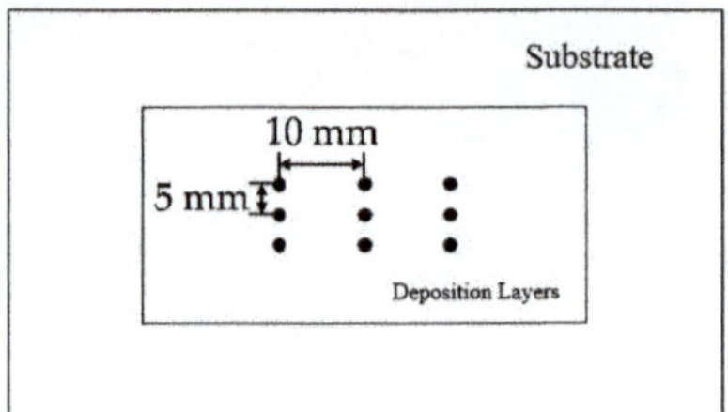

Figure 9. Diagram of residual stress test points.

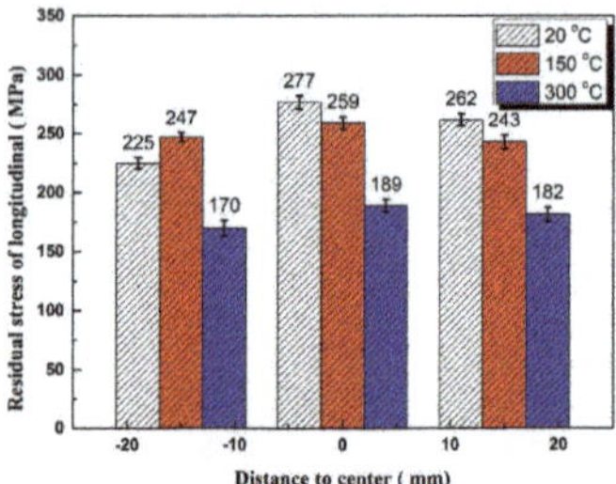

(**a**) Longitudinal stress of test specimen preheated at different temperatures.

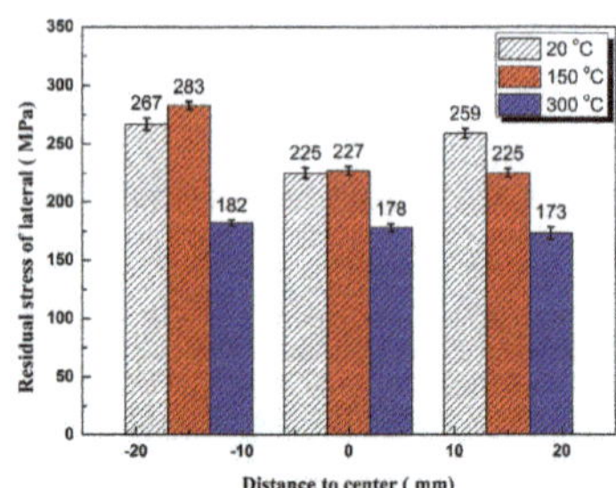

(**b**) Lateral stress of test specimen preheated at different temperatures.

Figure 10. Residual stress of different preheating temperature.

4. Conclusions

Firstly, as the preheating temperature of the substrate increased, the content of martensite and bainite in the surface layer of the deposited layer decreased, and the ferrite content increased. The granular bainite near the base layer decomposed into ferrite and carbide, and the content of bainite decreased, while the content of ferrite increased. Secondly, the hardness of the additive specimen gradually increased from the near-base layer to the surface layer, and the hardness was above 280 HV, and the maximum hardness reached 336 HV. The hardness of each test specimen decreased with the increase of the preheating temperature. When preheating temperature was up to 300 °C, due to the serious heat accumulation, the hardness decreased significantly, so too high preheating temperature could not be provided. Additionally, when the preheating temperature was 150 °C, tensile strength, yield strength, and elongation after fracture were higher than those under non-preheating. However, the properties were the lowest when preheating temperature was 300 °C. Proper preheating of the substrate can improve the mechanical properties. Lastly, the residual stress on the surface of each test specimen decreased with the increase of preheating temperature, which is based on the tensile stress. The longitudinal stress was particularly great at both ends in the scanning direction, and the lateral stress was obviously high at the back end of the deposition track.

Author Contributions: Investigation, C.D. and X.C.; Methodology, C.D. and X.C.; Resources, J.J. and P.Z.; Writing—original draft preparation, C.D. and X.C.; Writing—review and editing, C.D., X.C. All authors read and approved the final manuscript.

Funding: This research was funded by National Key R&D Program of China, grant number 2016YFB1100202, the Green Manufacturing System Integration Project of the Industry and Information Ministry of China (2017), and the Research and Development Plan for the Future Emerging Industries in Shenyang.

Conflicts of Interest: The authors declare no conflict of interest.

References

1. Murr, L.E.; Martinez, E.; Amato, K.N.; Gaytan, S.M.; Hernandez, J.; Ramirez, D.A.; Shindo, P.W.; Medina, F.; Wicker, R.B. Fabrication of metal and alloy components by additive manufacturing: examples of 3D materials science. *J. Mater. Res. Technol.* **2012**, *1*, 42–54. [CrossRef]

2. Qian, T.; Liu, D.; Tian, X.; Liu, C.; Wang, H. Microstructure of TA2/TA15 graded structural material by laser additive manufacturing process. *Trans. Nonferr. Met. Soc. China* **2014**, *24*, 2729–2736. [CrossRef]

3. Keicher, D.M.; Smugeresky, J.E.; Romero, J.A.; Griffith, M.L.; Harwell, L.D. Using the laser engineering net shapping process to produce complex components from a CAD solid model. *Int. Soc. Opt. Photonics* **1996**, *2993*, 91–97.

4. Noecker, F.F.; DuPont, J.N. Microstructural development and solidification cracking susceptibility of Cu deposits on steel. *J. Mater. Sci.* **2007**, *42*, 495–509. [CrossRef]

5. Ge, J.; Zhang, A.; Li, D.; Zhu, G.; Lu, Q.; He, B.; Lu, Z. Process research on DZ125L superalloy parts by laser metal direct forming. *Chin. J. Lasers* **2011**, *38*, 119–125.

6. Guan, Q. A survey of development in welding stress and distortion controlling in aerospace manufacturing in China. *Weld. World* **1999**, *43*, 64–74.

7. Alimardani, M.; Toyserkani, E.; Huissoon, J.P. A 3D dynamic numerical approach for temperaure and thermal stress distributions in multilayer laser solid freeform fabrication process. *Opt. Lasers Eng.* **2007**, *45*, 1115–1130. [CrossRef]

8. Dong, Z.; Kang, H.; Xie, Y.; Chi, C.; Peng, X. Effect of O content on microstructure and mechanical property of 12crni2 alloy steel prepared by laser additive manufacturing. *Appl. Laser* **2018**, *38*, 1–6.

9. Wu, X.; Mei, J. Near net shape manufacutring of components using direct laser fabireation technology. *J. Mater. Porcess. Technol.* **2003**, *135*, 266–270. [CrossRef]

10. Wei, K.; Gao, M.; Wang, Z.; Zeng, X. Effect of energy input on formability, microstructure and mechanical properties of selective laser melted AZ91D magnesium alloy. *Mater. Sci. Eng. A* **2014**, *611*, 212–222. [CrossRef]

Article

Influence of Vanadium on the Microstructure of IN718 Alloy by Laser Cladding

Kun Yang [1], Hualong Xie [1,*], Cong Sun [1], Xiaofei Zhao [1] and Fei Li [2]

[1] Department of Mechanical Engineering and Automation, Northeastern University, Shenyang 110819, China; 1770207@stu.neu.edu.cn (K.Y.); 1510096@stu.neu.edu.cn (C.S.); xiaofeizhao5@gmail.com (X.Z.)

[2] Department of Information Science and Engineering, Shenyang University of Technology, Shenyang 110870, China; lifei@sut.edu.cn

* Correspondence: hlxie@mail.neu.edu.cn

Received: 24 October 2019; Accepted: 19 November 2019; Published: 21 November 2019

Abstract: A deleterious Laves phase forms in the solidified structure of Inconel 718 (IN718) alloy during laser cladding. However, effective removal methods have not yet been identified. In this study, we first added the IN718 alloy cladding layers with a trace amount of vanadium (V, 0.066 wt.%). Then, we studied the solidification structure of cladding layers using a confocal laser scanning microscope and scanning electron microscopy. The microstructure and Laves phase morphology were investigated. The distribution of niobium (Nb) was observed by experiment as well. We found that V is evenly distributed in dendrites and interdendritic zones. A more refined dendrite structure, reduced second dendrite arm spacing and lower volume fraction of Laves phase were observed in the solidification structure. The results of linear energy-dispersive X-ray spectroscopy (EDS) indicate that the concentration of Nb decreases with an increasing of the distance from the Laves phase. The V-containing sample displayed a relatively slower decreasing tendency. The IN718 alloy sample was harder with the addition of V. In addition, the porosity of the sample decreased compared with the blank sample. The presented findings outline a new method to inhibit the Nb segregation in IN718 alloy during laser cladding, providing reference significance for improving the performance of IN718 alloy samples during actual processing.

Keywords: microstructure; element segregation; laves phase; vanadium; laser cladding

---

1. Introduction

Laser cladding (Figure 1) is a surface modification technology. Compared with traditional surface strengthening technology, laser cladding has high laser beam energy density, high process efficiency, and fast heating and cooling rates [1–3]. Nb is the most important element in IN718 alloy. The addition of Nb has a strong solid solution strengthening effect on Ni–Fe–Cr-based austenite and improves the elastic modulus of the alloy. Nb is the elemental basis for the main strengthening phase of IN718 alloy. Meanwhile, one of the most important microscopic characteristics of IN718 alloy is the distribution of Nb-rich Laves phase particles in the matrix during the laser cladding process [4,5]. Laves phase is a hard brittle phase that can provide conditions for nucleation and growth of the cracks under residual stress or other stress [6]. Therefore, the improvement in Nb segregation can benefit microstructure homogeneousness and enhance the performance of IN718 alloy cladding layer.

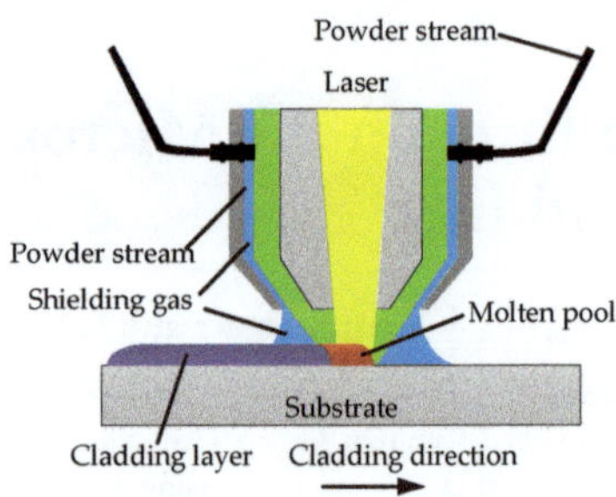

Figure 1. Laser cladding process schematic.

According to Han et al., IN718 alloy with Mo can reduce the solubility of Nb in the dendrite arm and Laves phases. The addition of Mo transforms the Laves phase morphology from eutectiform to granular and lessens the area of segregation zone around the Laves phase [7]. The effect of the addition of P and B on IN718 alloy as-cast microstructure was studied; the results indicated that the addition of these two alloying elements promoted the formation of a blocky Laves phase. A low melting B-bearing phase enriched in Nb, Mo, and Cr was observed [8]. Xin et al. investigated the effect of Co on precipitation behaviors of IN718 alloy, and the results showed Co was slightly segregated in the dendrite core and markedly increased the solubility of Mo in the dendrite core which resulted in reduced Mo in the residual liquid. Consequently, the Laves phase was retained while precipitation of Mo-depleted gray phase was promoted. The gray phase increased with increasing Co [9]. The effect of Zr on IN718 alloy was investigated as well. The addition of Zr not only inhibited the precipitation of Laves phase at the grain boundary, but also significantly promoted the precipitation of earlobe-like γ' and γ'' [10]. The addition of W, Ta, and Re other than Nb, was reported to reduce micro-segregation in the fusion zones of IN718 alloy [11]. As a result, adding alloying elements can significantly change the solidification behavior of IN718 alloy.

Vanadium has many desirable physical and chemical properties [12]. It was first used in steel to increase the grain coarsening temperature, which can improve the strength, toughness, and wear resistance of the steel. Afterwards, it was reported that V was found to occupy Al sites in the strengthening phase of Ni-based superalloy. The addition of V also can lead to a significant improvement in the material strength by forming stable nitrides and carbides [13].

The addition of V appears to positively influence IN718 alloy microstructures. Due to the lack of information of the influence of V on the solidification structure of IN718 alloy during laser cladding, this was our goal in this study. The previous work [7–10] on the influence of alloying elements on IN718 alloy focused on the solidification and precipitation behaviors. However, macroscopic and microscopic features were not compared between the sample with alloying elements and the blank sample. Furthermore, the addition of other elements can inhibit the formation of Laves phase. Unfortunately, previous studied neglected the quantitative analysis of the Laves phase concentration. Therefore, our focus was to study the influence of V on micro structure, Laves phase formation and performance of IN718 alloy. Figure 2 shows the workflow in this study.

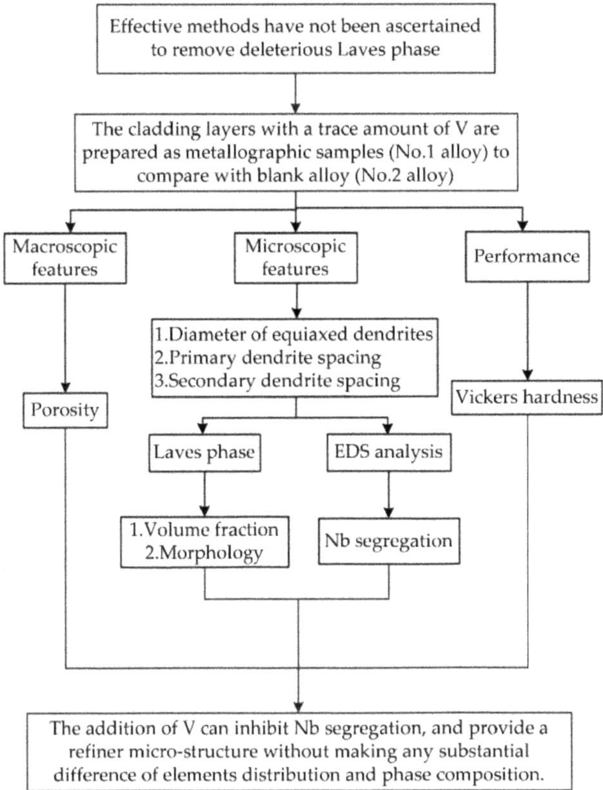

Figure 2. The workflow in this study.

2. Materials and Methods

2.1. Materials

The materials in this experiment included spherical IN718 alloy powder (Figure 3a) prepared by the plasma rotation electrode process (PREP), irregular V powder (Figure 3b) prepared by the atomization comminuting process (ACP), and IN718 alloy rolled substrate plate. Both powders had an average diameter of 100 μm. The size of substrate plate was 100 mm × 100 mm × 10 mm.

2.2. Laser Cladding Experiment Method

The experiment in this investigation was performed on the adding and subtracting material composite machining center, which is composed of the laser cladding head, powder feeder for laser processing, high purity nitrogen machine, etc., as shown in Figure 3a. The machine was equipped with a fiber laser, which is characterized by high precision, great power, and higher electro-optic conversion efficiency. The laser spot dimensions were 3 mm in length and 1 mm in width. The energy was distributed uniformly over the laser spot due to the property of the fiber laser source. Before the start of the experiment, we ensured that the V powder was evenly distributed in the IN718 alloy powder. To achieve this, the two powders were first stirred in a power agitator for 45 min. Afterwards, the other powders in the powder feeder were emptied to avoid impurities and contamination that would affect the accuracy of the experimental results. Then, 10 cladding layers were cladded on the substrate plate. The process parameters are shown in Table 1. The dimensions of the produced samples are shown in Figure 3d,e. Finally, the cladding layers were cut using wire along the scanning direction to prepare metallographic samples. The solidification structure samples were etched by a Kalling's etchant (40 mL HCl, 40 mL ethanol, 2 g CuCl$_2$). An OLYMPUS-OLS4100 Confocal Laser Scanning Microscope (CLSM, OLYMPUS, Tokyo, Japan) and a Zeiss ULTRA PLUS Scanning Electron Microscope (SEM, Zeiss, Oberkochen, Germany) with a X-Max 50 Energy Dispersive Spectrum (EDS, Oxford, UK) were used to characterize the microstructure and chemical composition. The average diameter of equiaxed dendrites (DED), secondary dendrite spacing (SDS), average size of columnar dendrites' primary dendrite spacing (PDS) and volume fraction of Laves phase (LPVF) measurements were counted and calculated by Image-Pro Plus6.0 software (Ropers Technologies, Sarasota, FL, USA) using secondary electron micrographs of the etched alloys captured by SEM. Three locations were selected along the height of the sample to measure the DED, SDS, and PDS (Figure 3f). We measured 4–6 dendrites at each location. A total of 15 dendrites were measured for each sample. The micro hardness test was performed using the micro hardness tester with a load (100 mN) and a microdiamond imprint. We selected 10 points along the length of the cross section to measure micro hardness (Figure 3f).

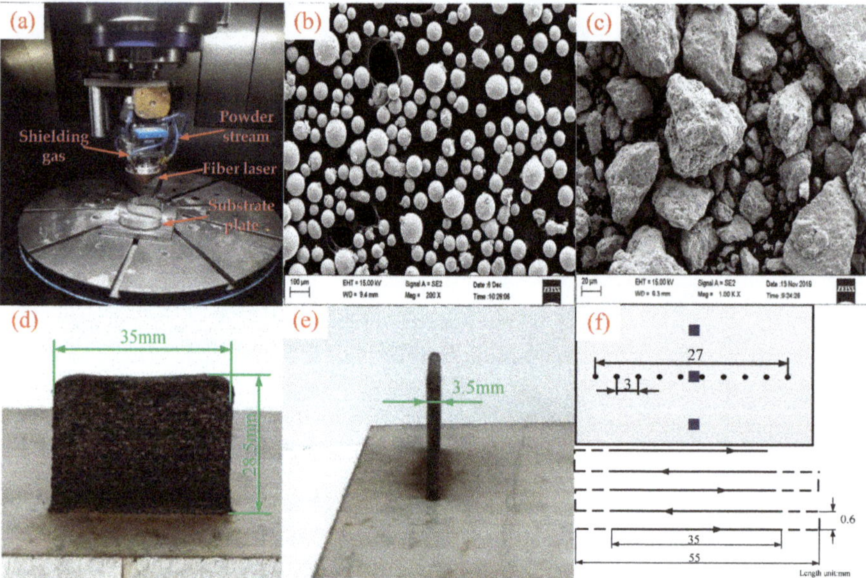

Figure 3. Experimental apparatus: (**a**) laser cladding system, (**b**) morphology of IN718 alloy powder (200×), (**c**) morphology of V powder (1000×), (**d**) main view of the sample, (**e**) left view of the sample, (**f**) Vickers hardness measurement points; secondary dendrite spacing (SDS), primary dendrite spacing (PDS), and diameter of equiaxed dendrites (DED), measurements locations and scan pattern.

Table 1. Processing parameters of laser cladding.

Parameters	Laser Power (W)	Scanning Speed (mm/s)	Powder Federate (g·min^{-1})	Shield Gas Flow (L·min^{-1})
-	1200	8	18	15

3. Results and Discussion

3.1. Solidification Structure Characteristics

In this study, we investigated the solidification structure of the V-containing cladding layer (No.1 alloy) from both macroscopic and microscopic aspects. In addition, the results were compared with the samples (No.2 alloy) without the addition of V. The chemical composition comparison between the two alloys is illustrated in Table 2.

Table 2. Chemical composition of two alloys (weight percentage wt.%).

Elements	Ni	Cr	Nb	Mo	Ti	Al	C	Fe	V
No.1 Alloy	53.1	18.43	5	3.18	1.06	0.54	0.014	18.61	0.066
No.2 Alloy	53.2	18.28	5	3.2	1.08	0.54	0.015	18.64	-

Figure 4 shows the macroscopic feature of the two samples via CLSM (50×). The difference in porosity between the two samples is obvious. The number and the size of pores in the No.1 alloy are much smaller than those of No.2 alloy (Figure 4). The existence of pores degrades the performance of the samples, especially the fatigue property [14]. Therefore, it is reasonable to expect that No.1 alloys would perform better compared to blank alloys.

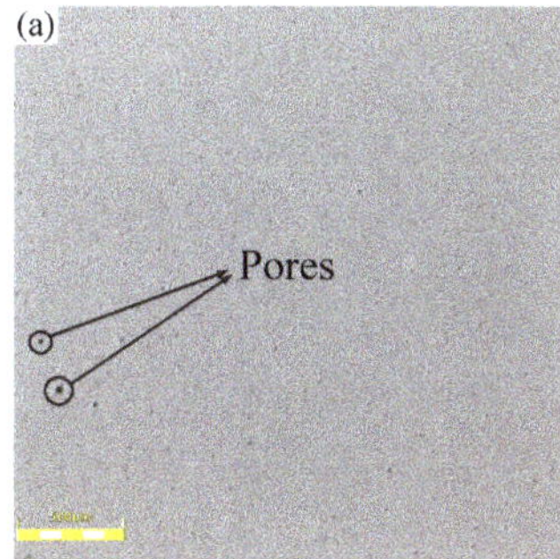

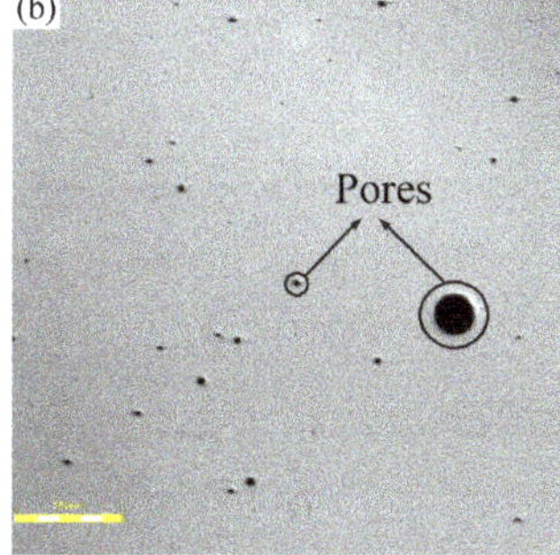

Figure 4. The macroscopic feature of (**a**) No.1 alloy and (**b**) No.2 alloy.

Figure 5 shows the micro structure of No.1 and No.2 alloys after solidification. The solidification structure of the two samples shows typical dendrite morphology at lower magnifications (200×). The black areas are dendrite, whereas the white areas are the interdendritic precipitation phase. As a result, the average diameter of equiaxed dendrites (DED) in V-containing cladding layers is smaller, as well as the average size of the columnar dendrites' primary dendrite spacing (PDS) when compared with the blank alloys. To further examine the influence of the addition of V on solidification structure, we measured the secondary dendrite spacing (SDS) of the two alloys. Because this parameter is a key index, it can be used to characterize the microstructure [11]. As shown in Figure 6, the SDS value of the No.1 alloy was 2.6 μm, which is smaller than the 3.8 μm of the No.2 alloy. Consequently, the addition of V can refine the dendrite structure and decrease the secondary dendrite spacing.

The addition of a trace amount of other powder into the original powder may change several parameters and mechanical properties of cladding layers. Generally, a more refined secondary dendrite spacing is desirable. According to Ahmadetal et al. [15], SDS depends on the composition and existence of additive elements, which is used to describe the scale of columnar dendritic structures [16].

Solidification structures with smaller secondary dendrite spacing limit the diffusion range of Nb Hence, decreasing the area ratio of the element segregation regions was more effective to achieve homogenization after heat treatment in No.1 alloy. In summary, the addition of V decreases the porosity in cladding layers and leads to a certain degree of dendrite refinement. Refined solidification and low porosity can enhance the performance of cladding layer.

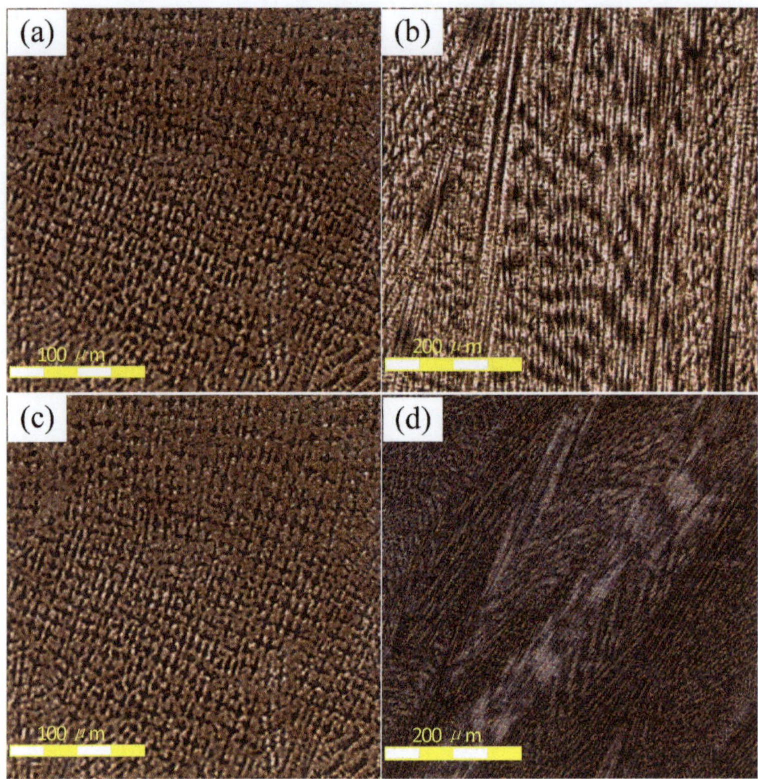

Figure 5. Microstructures of No.1 and No.2 alloy: (**a**) equiaxed dendrite in No.1 alloy; (**b**) columnar dendrite in No.1 alloy; (**c**) equiaxed dendrite in No.2 alloy; (**d**) columnar dendrite in No.2 alloy.

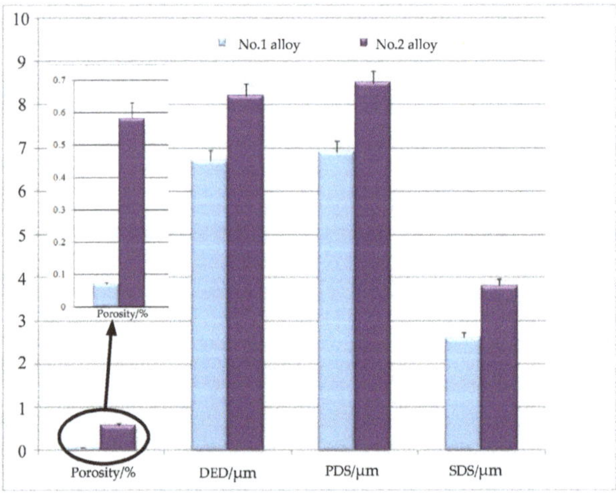

Figure 6. Comparison of macroscopic and microscopic morphology of two alloys.

3.2. Influence of V on Element Segregation

In this study, we performed a surface scan analysis on V-containing cladding layers. Figure 7 indicates the distribution of elements. The V-rich area in the analysis zone cannot be found. Conversely, the distribution of V was relatively uniform. In this investigation, V was solid-solved into the austenite matrix, which facilitates the formation of fine carbides at the grain boundaries [17]. These fine V-containing carbide particles create a pinning effect at the austenite grain boundaries, then the grain boundary migration, and the dendritic growth can be hindered [18]. This finding was confirmed by the reduction in the secondary dendrite arm spacing (Figure 6). According to a previous study [19], due to the redistribution property, Nb is prone to causing element segregation during the solidification of IN718 alloy. Detailed energy-dispersive X-ray spectroscopy (EDS) data of the Laves phase is shown in Figure 8. The insert shows an enlarged view of a lumpy Laves phase. The concentration of Nb is the second highest among all elements. Figure 8 also indicates that Nb is abundantly enriched in the interdendritic region. The segregation of Nb was found to be a key factor that controls the formation of Laves phase [20]. Therefore, we discussed whether V can improve Nb segregation. With this goal, we determined the chemical composition of the Laves phase in No.1 and No.2 alloy respectively. The results are listed in Table 3.

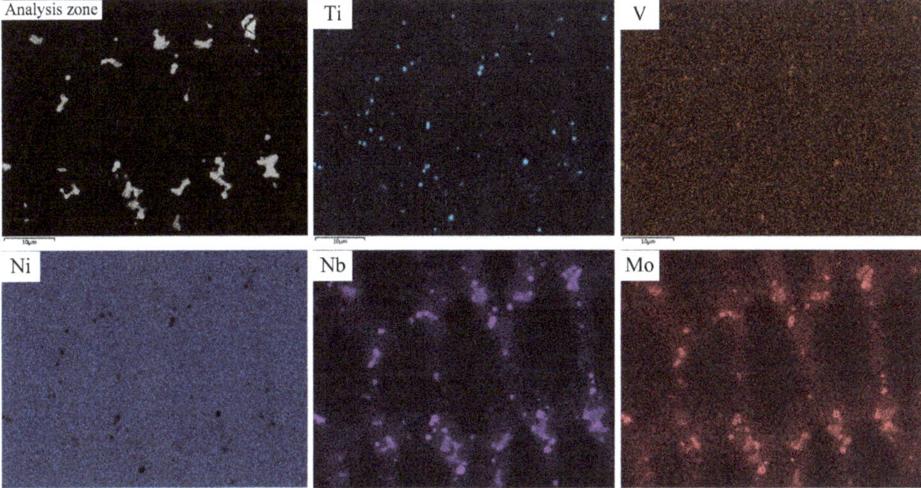

Figure 7. Elements distribution of No.2 alloy.

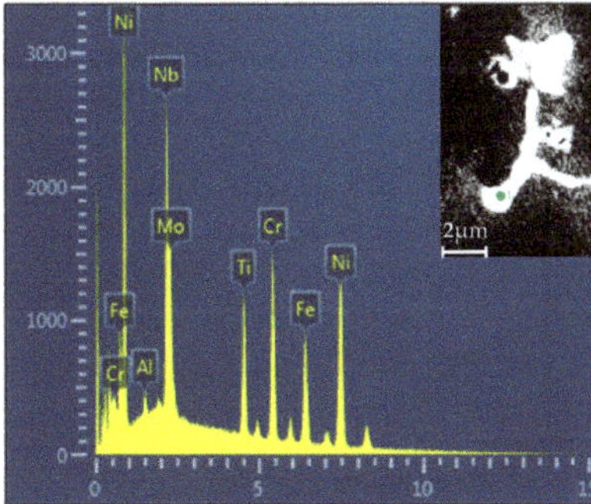

Figure 8. Energy-dispersive X-ray spectroscopy (EDS) data of Laves phase.

Table 3. Chemical composition of Laves phase (wt.%).

Alloy	Ti	Cr	Fe	Ni	Nb	Mo
No.1	1.64	14.14	12.95	45.04	21.24	5
No.2	2.24	13.18	16.98	34.97	26.15	6.28

Table 3 shows that the precipitated phase of two alloys have similar elements. However, the Nb concentration in the Laves phase decreases from 26.15 to 21.24 wt.% with the addition of V. This indicates that more Nb is solid-dissolved into the matrix to prepare for strengthening phase precipitation. We concluded that the addition of V improves Nb segregation in IN718 alloy during laser cladding. To further investigate the influence of V on element segregation, we conducted linear EDS analysis on No.1 and No.2 alloys (Figure 9), respectively. The result is shown in Figure 10. The path of the linear EDS passes through the Laves phase, the segregation zone, and the dendrite. Notably, the Nb concentration in the Laves phase of the No.2 alloy fluctuates considerably, which may be caused by the existence of pore.

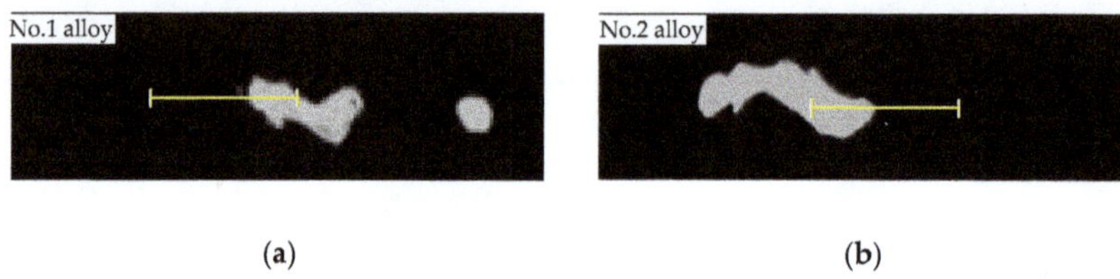

(a)   (b)

Figure 9. Liner EDS analysis of the Laves phase in (**a**) No.1 alloy and (**b**) No.2 alloy.

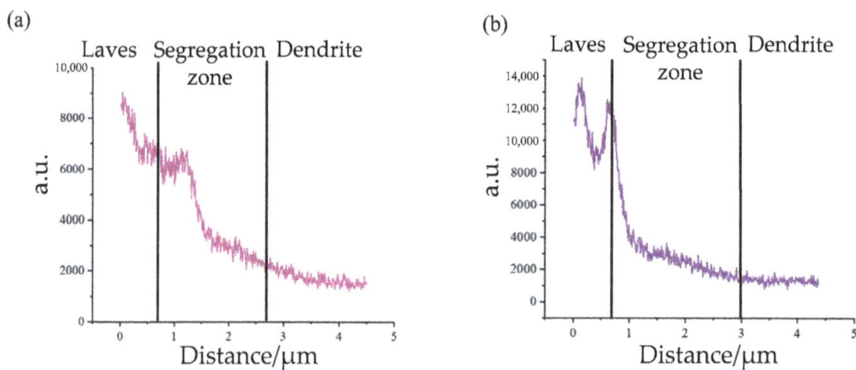

Figure 10. The distribution of Nb in (**a**) No.1 alloy and (**b**) No.2 alloy.

The results of linear EDS analysis indicated that Nb is heavily enriched in Laves phase. The Nb concentration decreases with increasing distance from the Laves phase. This tendency was observed in both No.1 and No.2 alloys. The difference was that the V-containing sample displayed a slower decreasing tendency. In addition, the average concentration of Nb in the analysis zone of No.1 alloy was less than that of the No.2 alloy, which is consistent with the results in Table 3. All the above results indicate the addition of V decreases the Nb concentration in the Laves phase, and evens out the distribution of Nb in cladding layers.

3.3. Influence of V on Laves Phase Formation

As is stated before, the existence of Laves phase can drastically degrade the performance of IN718 alloy. Under higher magnifications (2000×), a clear island-like segregation zone can be defined, in which the Laves phase particles are distributed (Figure 8). The addition of V can reduce the secondary dendrite spacing and decrease the concentration of Nb in the Laves phase. The microstructure becomes refined, reducing the segregation areas of elements. On this basis, we hypothesized that V can decrease the concentration of Laves phase and modify its morphology. To count the concentration of Nb-rich Laves phase and map its morphology, we conducted electron back scattered diffraction (EBSD) examination on both No.1 and No.2 alloys respectively. The Laves phase distribution of the two alloys in the equiaxial dendrite zones and columnar dendrite zones is shown in Figure 11. The white region represents Laves phase and the austenite appears as black. According to the theory of quantitative metallography [21], the area ratio of the white regions represents the volume fraction of Laves phase in the cladding layers.

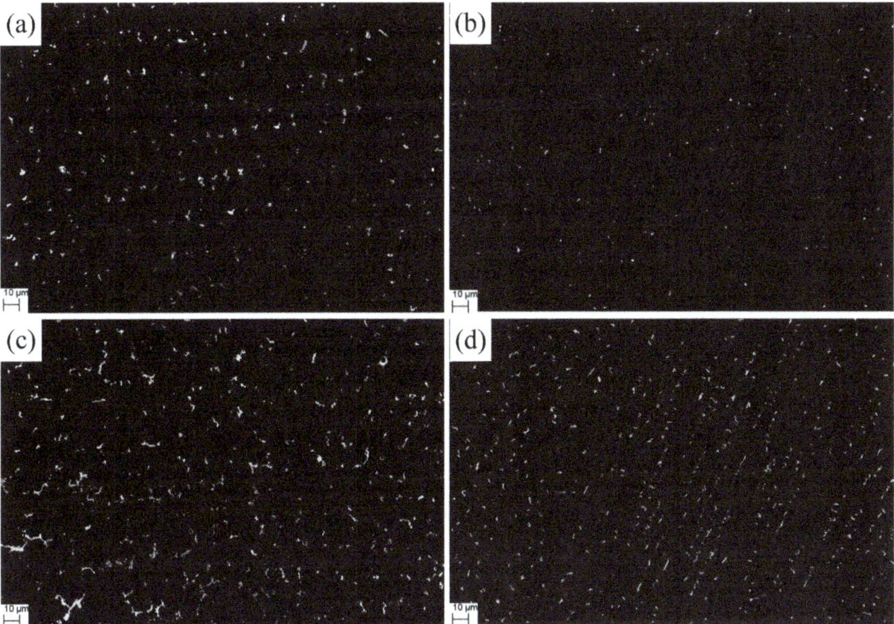

Figure 11. The morphology of Laves phase in the cladding layer: (**a**) in equiaxed interdendritic regions of No.1 alloy; (**b**) in columnar interdendriticregions of No.1 alloy; (**c**) in equiaxed interdendritic regions of No.2 alloy; (**d**) in columnar interdendritic regions of No.2 alloy.

Figure 11a,b show that the Laves phase morphology was modified dramatically with the addition of V. In the No.1 alloy, the Laves phase is particle-like. However, the Laves phase that formed in equiaxed interdendritic regions of No.2 alloy is reticular, as displayed in Figure 11c, and in columnar interdendritic region, the Laves phase is rod-like, as shown in Figure 11d. Normally, the particle-like Laves phase is the most desirable morphology feature [22], which could produce the preferred performance in IN718 alloy.

Figure 12 shows the average volume fraction of Laves phase (LPVF) in equiaxed interdendritic regions and in columnar interdendritic regions of No.1 and No.2 alloys respectively. We found that in the blank sample, the LPVF in equiaxed interdendritic regions was 24% higher than in columnar interdendritic regions, potentially due to the difference in the morphology of the two dendrites. With the addition of V, the difference in LPVF between different regions increased to 58%. Meanwhile, the statistical results indicate that the volume fraction of Laves phase in No.1 alloy decreased by 80%

compared with No.2 alloy, from 2.55% to 0.49%. If heat treatment is used to eliminate the Laves phase, the time required for homogenization of the No. 1 alloy would be much shorter than that of No. 2 alloy.

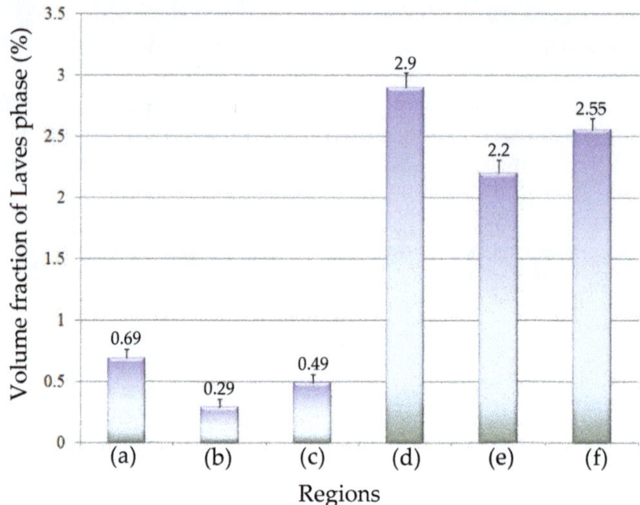

Figure 12. The average volume fraction of Laves phase (LPVF) (**a**) in equiaxed interdendritic regions of No.1 alloy; (**b**) in columnar interdendriticregions of No.1 alloy; (**c**) LPVF of No.1 alloy; (**d**) in equiaxed interdendritic regions of No.2 alloy; (**e**) in columnar interdendriticregions of No.2 alloy; (**f**) LPVF of No.2 alloy.

3.4. Influence of V on Hardness of Cladding Layer

Hardness is an important performance index used to measure the hardness degree of metal materials. Meanwhile, hardness has been widely used as a preliminary evaluation of the wear resistance of alloy [23]. Therefore the Vickers hardness (HV) values were obtained to tentatively estimate the V-addition influence on the wear resistance of IN718 alloy. Figure 13a depicts the typical indentation surface morphologies of the two alloys showing the smooth and regular rhombus shapes without any cracks or other defects, which indicates the fine metallurgical bonding and a superior relative density of both samples. Figure 13b shows that the average micro hardness of the No.1 alloy was 260.6 HV, whereas that of the No.2 alloy was 237.9 HV. We concluded that with the addition of V, the average micro hardness of the sample increased by 9.5%. According to previous studies [24,25], the Laves phase morphology and volume fraction are the main factors affecting the micro hardness of IN718 alloy. Our comparison of the average micro hardness between No.1 and No.2 alloy supports this finding. To study the dispersion of the micro hardness distribution in the samples, the coefficient of variation (CV) of the experimental data was calculated and compared. The CV of the micro hardness of the No.1 alloy was 0.017, which is slightly higher than that of theNo.2 alloy (0.011). The results illustrate the average micro hardness increases with the addition of a trace of V; however, the micro hardness distribution in the V-containing sample becomes uneven with respect to the blank sample. Based on the empirical correlation between hardness and wear resistance, it is reasonable to expect that the wear resistance of No.1 alloy increases with the addition of V.

The discussion above indicates that the addition of V can inhibit element segregation and change the morphology of the Laves phase, which is similar to adding other alloying elements. However, the decreased of porosity and increased micro hardness in the IN718 alloy sample fabricated by laser cladding was not mentioned in previous studies. Furthermore, a new phase precipitated in the interdendritic region of IN718 alloy with the addition of B or Co, such as B-bearing phase and Mo-depleted gray phase, and the influence on performance is unknown. The addition of V can provide more refined microstructure, without substantially affecting the element distributions and phase compositions.

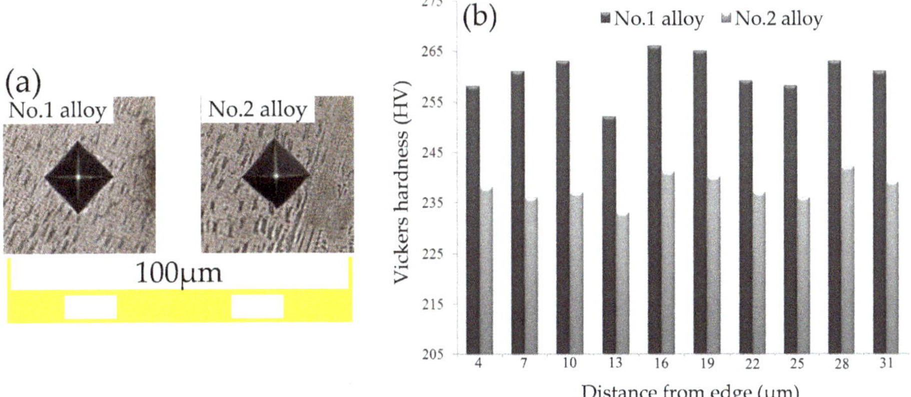

Figure 13. Vickers hardness testing. (**a**) Typical indentation surface morphologies of two alloys; (**b**) comparison of Vickers hardness between No.1 and No.2 alloy.

4. Conclusions

In this study, the influence of V on IN718 alloy solidification structure during laser cladding was investigated using experiments. Based on the obtained results our main conclusions were drawn:

(1) With the addition of V, the porosity of cladding layer decrease by 89%, and the average secondary dendrite arm spacing decreased from 3.8 to 2.6 μm, as did the average size of dendrite.

(2) The Nb concentration in the Laves phase of the V-containing sample decreased compared with blank sample. The results of linear EDS indicate that Nb concentration decreases with the increase in the distance from the Laves phase. However, the V-containing sample displayed a slower decreasing tendency.

(3) The morphology of the Laves phase in the V-containing cladding layer was modified dramatically, which changed from rod-like shape to a particle-like feature. The volume fraction of the Laves phase decreased by 80% with the addition of a trace amount of V.

(4) The hardness of the V-containing alloy was higher compared with the blank alloy. However, the distribution of micro hardness was uneven.

Consequently, the addition of V positively influence the microstructure and element segregation of IN718 alloy cladding layers. This investigation provides a new and effective method for inhibiting the formation of Laves phase and enhancing the performance of IN718 samples fabricated by laser cladding.

Author Contributions: Investigation, K.Y.; Formal Analysis, K.Y. and X.Z.; Writing-Original Draft Preparation, X.Z.; Writing-Review and Editing, C.S.; Supervision, F.L.; and Funding Acquisition, H.X.

Funding: The financial support from National Natural Science Foundation of China (ID61803272) and the Fundamental Research Funds for the Central Universities (IDN170313025) is highly appreciated.

Conflicts of Interest: The authors declare no conflict of interest.

References

1. Liu, J.; Yu, H.; Chen, C.; Weng, F.; Dai, J. Research and development status of laser cladding on magnesium alloys: A review. *Opt. Lasers Eng.* **2017**, *93*, 195–210. [CrossRef]
2. Sexton, L.; Lavin, S.; Byrne, G.; Kennedy, A. Laser cladding of aerospace materials. *J. Mater. Process.* **2002**, *122*, 63–68. [CrossRef]

3. Mokadem, S.; Bezençon, C.; Hauert, A.; Jacot, A.; Kurz, W. Laser repair of superalloy single crystals with varying substrate orientations. *Metall. Mater. Trans.* **2007**, *38*, 1500–1510. [CrossRef]
4. Ram, G.D.J.; Reddy, A.V.; Rao, K.P.; Reddy, G.M. Improvement in stress rupture properties of Inconel 718 gas tungsten arc welds using current pulsing. *J. Mater.* **2005**, *40*, 1497–1500. [CrossRef]
5. Reddy, G.M.; Murthy, C.S.; Rao, K.S.; Rao, K.P. Improvement of mechanical properties of Inconel 718 electron beam welds—Influence of welding techniques and postweld heat treatment. *Int. J. Adv. Manuf. Technol.* **2009**, *43*, 671–680. [CrossRef]
6. Zhu, L.; Xu, Z.F.; Liu, P.; Gu, Y.F. Effect of processing parameters on microstructure of laser solid forming Inconel 718 superalloy. *Opt. Laser Technol.* **2018**, *98*, 409–415. [CrossRef]
7. Han, D.W.; Sun, W.R.; Yu, L.X. Effects of molybdenum on segregation and diffusion of niobium during the solidification and homogenization of IN718 alloy. *Heat Treat.* **2018**, *33*, 6–12.
8. Miao, Z.; Shan, A.; Wu, Y. Effects of P and B addition on as-cast microstructure and homogenization parameter of Inconel 718 alloy. *Nonferr. Met. Soc. China* **2012**, *22*, 318–323. [CrossRef]
9. Xin, X.; Zhang, W.H.; Yu, L.X. Effects of Co on the solidification and precipitation behaviors of IN 718 alloy. *Mater. Sci. Forum* **2015**, *816*, 613–619. [CrossRef]
10. Li, Y.M.; Liu, H.J. Effect of Zr addition on precipitates in K4169 superalloy. *Res. Dev.* **2012**, *9*, 6–10.
11. Manikandan, S.G.K.; Sivakumar, D.; Rao, K.P. Laves phase in alloy 718 fusion zone—Microscopic and calorimetric studies. *Mater. Charact.* **2015**, *100*, 192–206. [CrossRef]
12. Muroga, T.; Nagasaka, T.; Abe, K.; Chernov, V.M.; Matsui, H.; Smith, D.L. Vanadium alloys—Overview and recent results. *J. Nucl. Mater.* **2002**, *307*, 547–554. [CrossRef]
13. Filipovic, M.; Kamberovic, Z.; Korac, M.; Jordovic, B. Effect of niobium and vanadium additions on the as-cast microstructure and properties of hypoeutectic Fe–Cr–C alloy. *ISIJ Int.* **2013**, *53*, 2160–2166. [CrossRef]
14. Ma, W.; Xie, Y.; Chen, C.; Fukanuma, H.; Wang, J.; Ren, Z.; Huang, R. Microstructural and mechanical properties of high-performance Inconel 718 alloy by cold spraying. *J. Alloy. Compd.* **2019**, *792*, 456–467. [CrossRef]
15. Ahmad, R.; Asmael, M.B.A.; Shahizan, N.R.; Gandouz, S. Reduction in secondary dendrite arm spacing in cast eutectic Al–Si piston alloys by cerium addition. *Int. J. Miner. Metall. Mater.* **2017**, *24*, 91–101. [CrossRef]
16. Miao, Z.; Shan, A.; Wang, W.; Lu, J.; Xu, W.; Song, H.W. Solidification process of conventional superalloy by confocal scanning laser microscope. *Trans. Nonferr. Met. Soc. China* **2011**, *21*, 236–242. [CrossRef]
17. Filipovic, M.; Kamberovic, Z.; Korac, M. Solidification of high chromium white cast iron alloyed with vanadium. *Mater. Trans.* **2011**, *52*, 386–390. [CrossRef]
18. Baker, T.N. Processes, microstructure and properties of vanadium microalloyed steels. *Mater. Sci. Technol.* **2009**, *25*, 1083–1107. [CrossRef]
19. Long, Y.T.; Nie, P.L.; Li, Z.G.; Huang, J.; Xiang, L.I.; Xu, X.M. Segregation of niobium in laser cladding Inconel 718 superalloy. *Trans. Nonferr. Met. Soc. China* **2016**, *26*, 431–436. [CrossRef]
20. Knorovsky, G.A.; Cieslak, M.J.; Headley, T.J.; Romig, A.D.; Hammetter, W.F. Inconel 718: A solidification diagram. *Metall. Mater. Trans.* **1989**, *20*, 2149–2158. [CrossRef]
21. Chen, M.R.; Lin, S.J.; Yeh, J.W.; Chuang, M.H.; Chen, S.K.; Huang, Y.S. Effect of vanadium addition on the microstructure, hardness, and wear resistance of Al 0.5 CoCrCuFeNi high-entropy alloy. *Metall. Mater. Trans.* **2006**, *37*, 1363–1369. [CrossRef]
22. Sozanska, M.; Maciejny, A.; Dagbert, C.; Galland, J.; Hyspecká, L. Use of quantitative metallography in the evaluation of hydrogen action during martensitic transformations. *Mater. Sci. Eng.* **1999**, *273*, 485–490. [CrossRef]
23. Sui, S.; Chen, J.; Fan, E.; Yang, H.; Lin, X.; Huang, W. The influence of Laves phases on the high-cycle fatigue behavior of laser additive manufactured Inconel 718. *Mater. Sci. Eng.* **2017**, *695*, 6–13. [CrossRef]
24. Zhang, Y.C.; Li, Z.G.; Nie, P.L.; Wu, Y.X. Effect of ultrarapid cooling on microstructure of laser cladding IN718 coating. *Surf. Eng.* **2013**, *29*, 414–418. [CrossRef]
25. Stevens, E.L.; Toman, J.; Chmielus, M. Variation of hardness, microstructure, and Laves phase distribution in direct laser deposited alloy 718 cuboids. *Mater. Des.* **2017**, *119*, 188–198. [CrossRef]

Article

W$_x$NbMoTa Refractory High-Entropy Alloys Fabricated by Laser Cladding Deposition

Qingyu Li [1,2], Hang Zhang [1,2,*], Dichen Li [1,2], Zihao Chen [1,2], Sheng Huang [1,2], Zhongliang Lu [1,2] and Haoqi Yan [1,2]

1   State Key Laboratory for Manufacturing Systems Engineering, Xi'an Jiaotong University, Xi'an 710049, China; liqingyu9206@126.com (Q.L.); dcli@mail.xjtu.edu.cn (D.L.); chenzihao18@stu.xjtu.edu.cn (Z.C.); ahhuangsheng@126.com (S.H.); zllu@mail.xjtu.edu.cn (Z.L.); yan21601122@163.com (H.Y.)

2   School of Mechanical Engineering, Xi'an Jiaotong University, Xi'an 710049, China

*   Correspondence: zhanghangmu@hotmail.com

Received: 20 January 2019; Accepted: 8 February 2019; Published: 11 February 2019

Abstract: W$_x$NbMoTa refractory high-entropy alloys with four different tungsten concentrations (x = 0, 0.16, 0.33, 0.53) were fabricated by laser cladding deposition. The crystal structures of W$_x$NbMoTa alloys are all a single-phase solid solution of the body-centered cubic (BCC) structure. The size of the grains and dendrites are 20 μm and 4 μm on average, due to the rapid solidification characteristics of the laser cladding deposition. These are much smaller sizes than refractory high-entropy alloys fabricated by vacuum arc melting. In terms of integrated mechanical properties, the increase of the tungsten concentration of W$_x$NbMoTa has led to four results of the Vickers microhardness, i.e., H_v = 459.2 ± 9.7, 476.0 ± 12.9, 485.3 ± 8.7, and 497.6 ± 5.6. As a result, NbMoTa alloy shows a yield strength (σ_b) and compressive strain (ε_p) of 530 Mpa and 8.5% at 1000 °C, leading to better results than traditional refractory alloys such as T-111, C103, and Nb-1Zr, which are commonly used in the aerospace industry.

Keywords: W$_x$NbMoTa; refractory high-entropy alloy; laser cladding deposition; rapid solidification

---

1. Introduction

With the rapid development of materials and manufacturing technologies in the aerospace industry, the established macroscopic thermal protection theory and the existing traditional types of the refractory alloy are difficult to meet the harsh requirements of the advanced aerospace industry. Traditional alloy design is based on a metallic element with a mass fraction of more than 50% and the comprehensive properties of the alloys are improved by adding trace elements. This system of alloy design has reached the bottleneck after thousands of years of development.

In 2004, Taiwan scholar professor Yeh JW and his team [1] proposed the concepts of multi-principal element alloys (MPEAs) and high-entropy alloys (HEAs). Generally, HEAs can be defined as a simple and disordered structure of solid solution mixed by multi-elements of 5%–35% atom ratio, showing specific characteristics, such as the high-entropy effect in thermodynamics and hysteresis diffusion effect in dynamics [2,3]. These characteristics then contribute to the advantages of HEAs in such aspects as high-temperature resistance [4], high strength and ductility [5,6], corrosion and radiation resistance [7,8], providing more possible applications in the aerospace industry.

In the research field of refractory HEAs, researchers have conducted a series of studies on the different refractory HEAs in strength and ductility at small scales [9,10], the relationship between grain size and mechanical properties [11], thermodynamic properties [12] and the interplay among lattice distortions, vibrations, and phase stability [13]. In 2011, Senkov, O.N. [14] produced the refractory

HEAs with equiatomic concentrations, WNbMoTa and WNbMoTaV, via vacuum arc melting (VAM) for the first time, obtaining a single-phase body-centered cubic (BCC) structure. Later on, he [4] found that WNbMoTa and WNbMoTaV HEAs have better high-temperature yield strength than traditional superalloys, namely Inconel 718 and Haynes 230. In 2017, after adding the equal mole titanium element, Han [15] improved the ductility of WNbMoTa and WNbMoTaV through VAM at room temperature. As a result, existing research is mainly aimed at revealing the nature or the characteristics of HEAs, instead of being driven by engineering applications.

At present, refractory HEAs are mainly produced by the methods of VAM or powder metallurgy (PM). Traditional manufacturing methods have difficulties in forming the HEAs with a large size, a complex structure, and a variable composition. The technology of laser cladding deposition (LCD) is the process that fuses the metal powders from points to layers and finally fabricates the parts according to the three-dimensional model data. As an emerging technology, it has many unparalleled advantages in forming refractory HEAs. One the one hand, refractory alloys can be melted rapidly due to the high energy density of the laser, thus not limiting the forming size and structure. On the other hand, the functionally gradient structure of the HEAs can be manufactured in any direction of the parts by controlling the feeding rate of the different hopper, which can realize the macrostructure and micro metallurgy synchronous manufacturing [16]. In 2016, Dobbelstein [17] prepared WNbMoTa HEA by means of remelting in the process of LCD, analyzing its composition uniformity and macroscopic segregation. Then, in 2018, Zhang [18] formed WNbMoTa HEA through selective laser melting (SLM) and carried out the simulation of the finite difference-finite element (FD-FE) coupling in the SLM forming process.

Special requirements of "high strength, low density" are put forward due to the severe environment in aerospace. The high density of the tungsten element limits the application of the refractory HEA with equiatomic concentrations, WNbMoTa, in the aerospace industry. Presented in current work with the tungsten mole of 0%, 5%, 10% and 15%, and the remaining elements of the equal mole, refractory alloy materials were formed by LCD, respectively. The purpose of the present work is to recognize the relationships between mechanical behavior and the content of tungsten in the WNbMoTa HEAs and to provide a feasible scheme for the possibility of large-size and complex shaped gradient structures in the aerospace application.

2. Experimental Procedures

The four W_xNbMoTa HEAs are referred to as $x = 0$, 0.16, 0.33 and 0.53, respectively. The deposition of the four different tungsten mole alloys were fabricated by using the LCD-1000-a type of the coaxial LCD system, which was independently designed and developed by the State Key Laboratory for Manufacturing Systems Engineering at Xi'an Jiaotong University. The system was equipped with two powder feeders, a 25 kW intermediate frequency induction heating auxiliary device and a standard, industrial pulse JK802/1002 Nd:YAG laser (JK Lasers, Rugby, UK) with 1000 W maximum power, which had a spot diameter of 500 μm at a central emission wavelength of 1064 nm. The diagram of the LCD forming process is shown in Figure 1.

All experimental materials, including particles with sizes ranging from 45 μm to 125 μm elemental tungsten (99.5 wt.%), niobium (99.78 wt.%), molybdenum (99.84 wt.%), tantalum (99.54 wt.%) metal powders and Φ 20 mm × 80 mm of pure molybdenum substrates, were provided by Beijing AMC Powder Metallurgy Technology Co., Ltd. (Beijing, China). The scanning electron microscope photos of the four powders are shown in Figure 2 and the basic physical properties of the four elements are shown in Table 1.

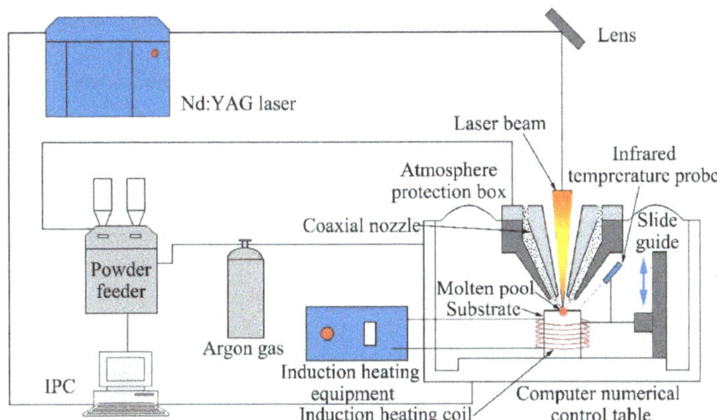

Figure 1. Schematic of the forming process of laser cladding deposition (LCD).

Figure 2. Scanning electron microscope (SEM) images of four metal powders. (**a**) Tungsten powders, (**b**) niobium powders, (**c**) molybdenum powders, (**d**) tantalum powders.

Table 1. The basic physical parameters of the four materials used in the alloy.

Metallic Element.	W	Nb	Mo	Ta
Relative Atomic Mass, u	183.84	92.9	95.94	180.9
r, Å	1.37	1.43	1.36	1.43
ρ, g/cm^3	19.35	8.57	10.2	16.65
Hv	350	135	156	89
T_m, K	3695	2750	2896	3290

The elemental metal powders of the four different tungsten mole alloys were accurately measured by the electronic, analytical balance before the experiment. Powders were mixed for 4 h via SYH three-dimensional motion mixer series to ensure that they were mixed evenly. Then, the mixed powders were put into a vacuum oven at 120 °C to dry for 8 h in order to remove moisture and enhance the liquidity of the powder. The atmosphere chamber was filled with Argon gas to ensure that the oxygen content of the environment was lower than 80 ppm for the purpose of protecting the powders from being oxidized during the forming process. The LCD processing parameters in the experiments after optimizing were 565 W laser power, 8 mm/s scanning speed, 0.08 mm thickness of the deposited layer, and 600 °C induction heating temperature. The substrates and the formed area were heated during the forming process to reduce the internal stress.

The crystal structures of the four different tungsten mole alloys were identified on the cross-section surfaces, using Bruker D8 Advanced A25 X-ray diffraction (XRD, Bruker AXS GmbH, Karlsruhe, Germany) equipment. Taking advantage of a METTLER TOLEDO XS105 (Mettler-Toledo GmbH, Zurich, Switzerland) analytical balance through the drainage method, the densities of the alloys were measured. Microstructures were analyzed with the use of a TESCAN MIRA3 LMH scanning electron microscope (SEM, TESCAN, a.s., Brno-Kohoutovice, Czech Republic) equipped with a backscatter electron (BSE) and an energy dispersive spectroscopy (EDS) detector. The sizes and orientations of the grains were determined by the electron back-scatter diffraction (EBSD) using an SU3500 SEM (Techcomp (China) Ltd, Beijing, China). Vickers microhardness was measured on polished longitudinal-section surfaces using an HXD-2000TMSC/LCD tester (Shanghai Taiming Optical Instrument CO., Ltd, Shanghai, China) and the test parameter was applied for 30 s under a 500 g load. Before the compression mechanical performance testing, the samples were cut into pieces that were 4 mm in diameter and 6 mm in height, giving an aspect ratio of 1.5. The room temperature compression test was conducted at 25 °C by a Sans CMT4304 multi-function static experiment machine (MTS Systems (China) CO., Ltd, Shenzhen, China) at a strain rate of $0.001~\mathrm{s}^{-1}$, and the high temperature compression test was conducted at 1000 °C by Gleeble 3500 equipment. In high temperature compression, the chamber was evacuated to 10^{-3} torr. The sample was heated to 1000 °C in 3.5 min, soaked at this temperature for 15 min, and then compressed at a strain rate of $0.001~\mathrm{s}^{-1}$.

3. Results and Discussions

The W_xNbMoTa HEAs were fabricated by LCD, shown in Figure 3. NbMoTa, $W_{0.16}$NbMoTa, $W_{0.33}$NbMoTa and $W_{0.53}$NbMoTa ranged from left to right. Four groups of samples deposited by LCD had a rectangular geometry with the size of 15 mm × 9 mm × 6 mm.

Figure 3. Macroscopic photograph of the W_xNbMoTa alloys.

3.1. Chemical and Phase Composition

The composition of the alloys, determined by EDS detector, is reported in Table 2. It can be found that the composition of each alloy is close to the ratio of design, and the biggest deviation is the mole of niobium in alloy 3, up to 2.6%. It shows that the composition uniformity is ensured and there is no obvious macroscopic burning loss to the elements during the process of LCD.

Table 2. Chemical composition (in at.%) of four refractory alloys produced by LCD.

Alloy ID/element	W	Nb	Mo	Ta
NbMoTa	0%	31.26%	32.90%	35.84%
$W_{0.16}$NbMoTa	5.80%	29.76%	30.86%	33.58%
$W_{0.33}$NbMoTa	12.42%	27.40%	30.01%	30.17%
$W_{0.53}$NbMoTa	14.95%	26.80%	27.62%	30.63%

The phase compositions of W_xNbMoTa HEAs were checked by XRD patterns as shown in Figure 4. All peaks on these X-ray patterns have been indexed and are congruent with a single BCC phase. Meanwhile, there is no alternative phase found under the analysis of the XRD result. This result indicates that these elements could still form the same BCC structures in spite of the varied lattice parameters of the four elements. Four different tungsten mole alloys fabricated by LCD all have a single-phase BCC crystal structure. Due to the texture effects caused by the grains within the X-ray excited volume, the same peaks in the four alloys have shown the different relative intensities.

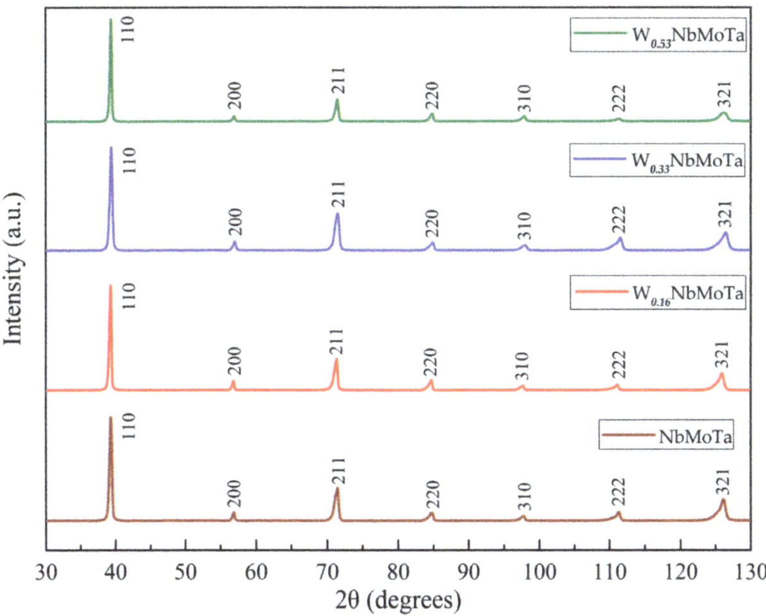

Figure 4. X-ray diffraction (XRD) patterns of the W_xNbMoTa alloys.

Yeh [1] proposed that HEAs should contain at least five elements, and that the entropy of mixing (ΔS_{mix}) is the main factor that promotes the formation of a multicomponent solid solution. With further research, WNbMoTa [14], NbTiVZr [19] quaternary equiatomic alloys and the ZrNbHf [20] ternary equiatomic alloy are proven to be able to form typical HEAs. Therefore, the formation of the solid solution in HEAs systems can be predicted in terms of the atomic size difference (δ), enthalpy of mixing (ΔH_{mix}) and other factors in addition to the entropy of mixing (ΔS_{mix}).

Atomic size difference (δ) is determined by:

$$\delta = 100\sqrt{\sum_{i=1}^{n} c_i\left(1 - \frac{r_i}{\sum_{j=1}^{n} c_j r_j}\right)^2} \tag{1}$$

where c_i and c_j are the atomic fractions of the i-th and j-th components, r_i and r_j are the atom radii of the i-th and j-th components.

Entropy and enthalpy of mixing could be obtained by:

$$\Delta S_{mix} = -R\sum_{1}^{n} c_i \ln c_i \tag{2}$$

$$\Delta H_{mix} = \sum_{i=1, i\neq j}^{n} \Omega_{ij} c_i c_j, \Omega_{ij} = 4\Delta H_{AB}^{mix} \tag{3}$$

where ΔS_{mix} is mixing entropy change, ΔH_{mix} is mixing enthalpy mixing, R is constant gases, ΔH_{AB}^{mix} is the enthalpy change of the binary liquid alloy composed of the components A and B in the regular solution, which is calculated based on Miedema macroscopic model [21,22].

Zhang [3] proposed a new parameter to determine the ability of the solid solution formation of the multi-component alloys. It consists of ΔS_{mix}, ΔH_{mix} and can be defined as follows:

$$\Omega = \frac{T_m \Delta S_{mix}}{|\Delta H_{mix}|} \tag{4}$$

$$T_m = \sum_{i=1}^{n} c_i (T_m)_i \tag{5}$$

where T_m is the average melting point of the alloy, $(T_m)_i$ is the melting point of *i-th* component. Combined with δ and Ω parameters, statistical analysis has been made for a large number of known solid solution HEAs. Then, the criteria for forming staple HEAs solid solution are [23,24]:

$$\delta\% \leq 6.6\%, \Omega \geq 1.1 \tag{6}$$

The relevant parameters of four different tungsten mole HEAs are shown in Table 3. The calculated parameters meet the criteria of solid solution formation and are consistent with the experimental XRD patterns.

Table 3. The parameters, δ, ΔS_{mix}, ΔH_{mix}, T_m and Ω for W$_x$NbMoTa high-entropy alloys (HEAs).

Parameter/Alloy ID	NbMoTa	W$_{0.16}$NbMoTa	W$_{0.33}$NbMoTa	W$_{0.53}$NbMoTa
δ, %	2.334	2.359	2.365	2.364
ΔS_{mix}, J·mol^{-1}·K^{-1}	9.14	10.33	10.93	11.28
ΔH_{mix}, J·mol^{-1}	−4.67	−5.22	−5.69	−6.07
T_m, K	2979	3015	3050	3086
Ω,	5.83	5.96	5.86	5.74

3.2. Density and Microstructure

The density of the alloys measured by the drainage method is reported in Table 4. According to the theoretical density of the alloys, it can be calculated by the formula:

$$\rho_{other} = \frac{\sum c_i A_i}{\sum \frac{c_i A_i}{\rho_i}} \tag{7}$$

where c_i, A_i and ρ_i are the atomic fraction, atomic weight and density of element i, respectively, and calculated ρ_{theor} value for the four different tungsten mole high-entropy alloys is also reported in Table 4.

Table 4. Theoretical and experimental densities of the W_xNbMoTa HEAs.

Alloy ID/Density	Theoretical Density; g/cm^3	Experimental Density; g/cm^3
NbMoTa	11.913	10.486
$W_{0.16}$NbMoTa	12.205	10.572
$W_{0.33}$NbMoTa	12.595	10.634
$W_{0.53}$NbMoTa	12.940	11.044

It can be seen that the experimental densities are lower than the theoretical densities in that many spherical gas porosities appear inside the alloys during the process of LCD, which is shown in Figure 5. In addition, the quality of the powders is different, owing to the different preparation methods of metal powders. Figure 1 shows that the mechanically pulverized tantalum powders were spheroidized through the technology of radio frequency plasma spherification, showing a good property in sphericity and granule density, while the tungsten and niobium powders were not spheroidized after mechanical grinding, leading to a higher granule density but a poor sphericity. Molybdenum powders were prepared through granulation reunion. The sphericity of molybdenum powders is much better than tungsten and niobium powders, but all of them are hollow, having nodular surfaces and containing significant amounts of pores.

Figure 5. SEM image of the gas porosities with spherical shape in HEAs.

The studies have shown that the granule density of powders is the main factor affecting the gas porosities. Conclusions have been verified in Ti-6Al-4V, Inconel 718 and other powders' forming process [25–27]. During the forming process, the protective Argon gas from the powder feed system is adsorbed in the pores of molybdenum powders, drawn into the forming part, forming a certain amount of spherical and randomly distributed gas porosities. The appearance of porosities has no correlation with microstructure features. In order to verify the above conclusions, the molybdenum powders with high granule density and good sphericity were used to replace that used in the previous experiment, shown in Figure 6. Then, the HEA was formed with the same process parameters to control a single variable. The SEM image of the formed sample is shown in Figure 7 and the porosities disappeared. This evidence strongly suggests that the granule density, instead of sphericity of powders, has a great influence on the defects of porosities. Particle pores may directly contribute to deposit porosities.

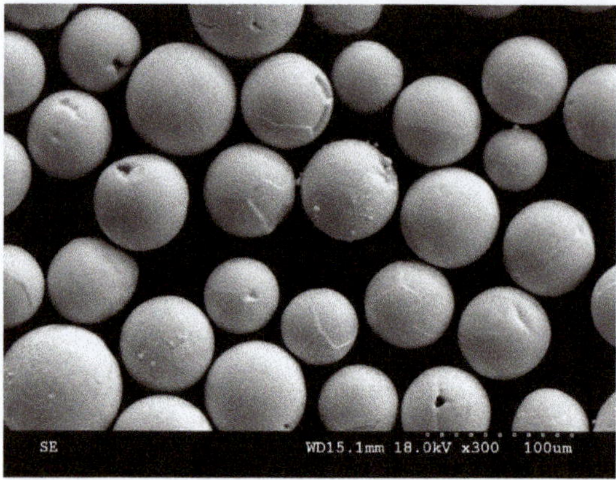

Figure 6. SEM image of the molybdenum powders with high granule density.

Figure 7. SEM image of the HEAs used powders with high granule density.

BSE images of the W_xNbMoTa HEAs are shown in Figure 8. The microstructures of the W_xNbMoTa are similar. The interior of the grain contains spheroidal dendrites (as the sub-structure of grain). Grains of about 20 µm in diameter and dendrites of about 4 µm in diameter can be seen. EBSD images of the NbMoTa HEAs formed in the process of LCD and VAM are shown in Figure 9. The grain size of NbMoTa fabricated by LCD is much smaller than that of NbMoTa fabricated by VAM and the growth way of grains is equiaxed crystals. Due to the characteristics of rapid solidification, the laser with high energy density fuses the metallic powders with a high melting point in a moment (10^{-3}–10^{-2} s). The phase transition of powder materials from the liquid phase to the solid phase is so fast that it is easy to form fine grain, low constitutional segregation microstructure.

This is compared to the HEAs produced by vacuum arc melting. According to the Hall-Petch relation, the yield strength or microhardness of materials has a negative linear correlation with grain size, when the grain size of testing materials is not ultrafine (≥ 1 µm) [28–32]. In engineering manufacture, large yield strength or hardness of the material may be attained by decreasing the grain size, while the characteristics of rapid solidification in LCD facilitate the formation of fine grain. Compared with traditional VAM manufacturing, the yield strength and microhardness of alloys by LCD are usually much better.

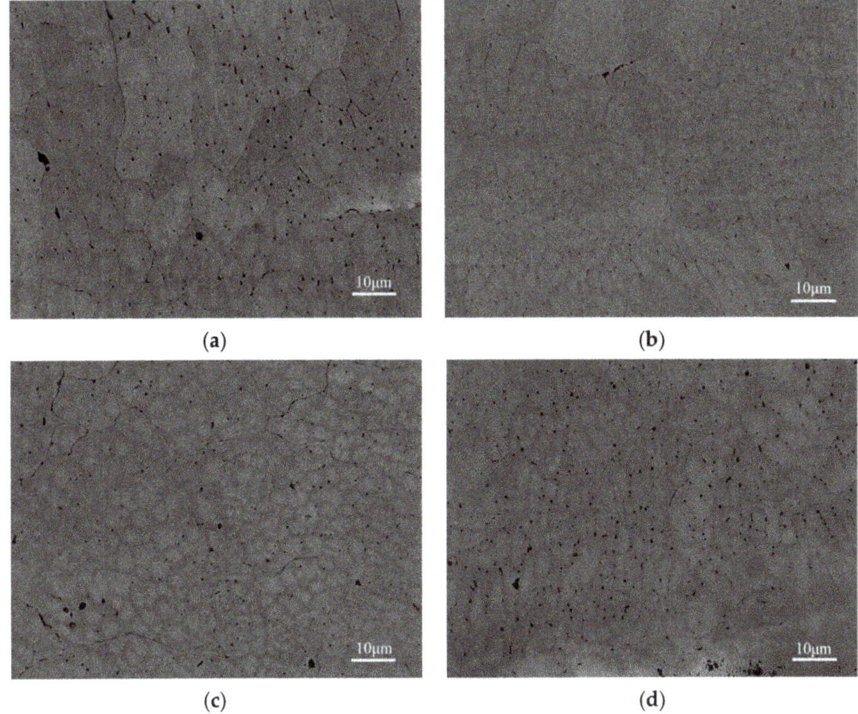

Figure 8. Backscatter electron (BSE) of polished longitudinal-sections of the W_xNbMoTa HEAs. (**a**) NbMoTa, (**b**) $W_{0.16}$NbMoTa, (**c**) $W_{0.33}$NbMoTa, (**d**) $W_{0.53}$NbMoTa.

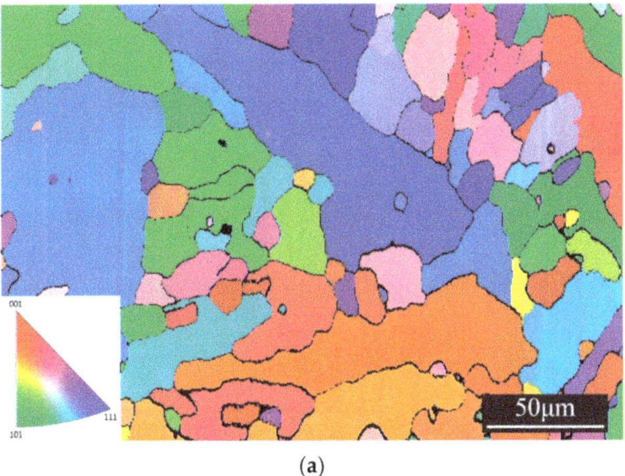

(**a**)

Figure 9. *Cont.*

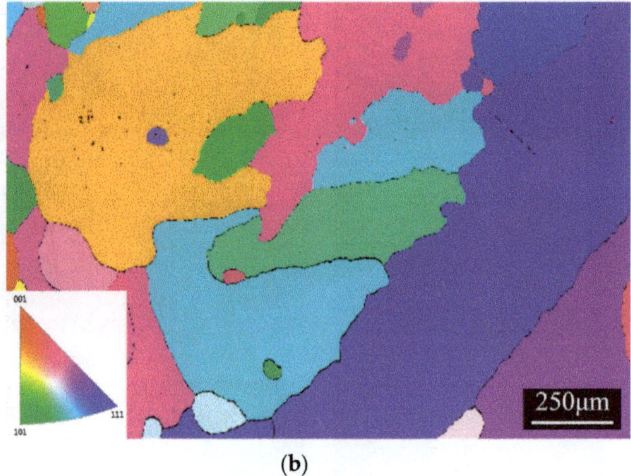

(b)

Figure 9. EBSD of polished longitudinal-sections of the NbMoTa HEAs. (**a**) Laser cladding deposition (LCD), (**b**) vacuum arc melting (VAM).

3.3. Mechanical Properties

The Vickers microhardness of the different tungsten mole HEAs were measured from the substrate to the metallurgical bonding surface, and then the cladding layer. With the increase of the cladding height, no obviously regular change was found in microhardness. The results are demonstrated below in Figure 10. The average values of microhardness are shown in Table 5. With the increase of the content of tungsten, the hardness shows an increasing tendency.

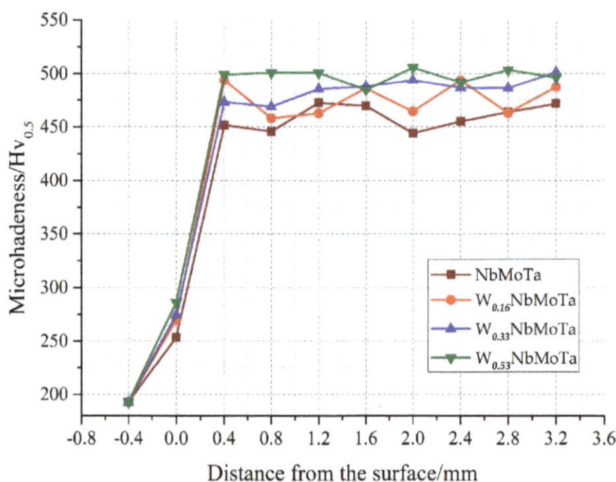

Figure 10. Microhardness of the W$_x$NbMoTa HEAs fabricated in LCD.

Table 5. Experimental H_v and theoretical H_v values of the W$_x$NbMoTa HEAs.

Hv/Alloy ID	NbMoTa	W$_{0.16}$NbMoTa	W$_{0.33}$NbMoTa	W$_{0.53}$NbMoTa
Experimental H_v	459.2 ± 9.7	476.0 ± 12.9	485.3 ± 8.7	497.6 ± 5.6

The bulk modulus of the W$_x$NbMoTa (x = 0, 0.16, 0.33, 0.53) were measured to be 174 GPa, 182 GPa, 186 GPa and 192 GPa. The compressive stress-strain curve of the alloys at room temperature

(T = 25 °C) is illustrated in Figure 11a. NbMoTa showed a yield strength (σ_b), compressive strength (σ_m) and compressive strain (ε_p) of 874 MPa, 1140 MPa and 5.8%, respectively. The compressive strain was improved compared with the WNbMoTa HEAs by VAM (1.5%) [4], indicating that NbMoTa has certain machinability at room temperature. $W_{0.16}$NbMoTa, $W_{0.33}$NbMoTa and $W_{0.53}$NbMoTa were characterized as brittle materials, and their maximum compressive strength (σ_b) showed 840 MPa, 895 MPa, and 890 MPa.

The compressive stress-strain curve for the NbMoTa at 1000 °C is illustrated in Figure 11b, showing a yield strength (σ_b), compressive strength (σ_m) and compressive strain (ε_p) of 530 MPa, 684 MPa and 8.5%, respectively. It reflects a better effect than such traditional refractory alloys, as T-111, Nb-1Zr and C103 are commonly used in aerospace industry [33], as shown in Table 6. The result indicates that NbMoTa fabricated by LCD shows potential for application in the aerospace industry.

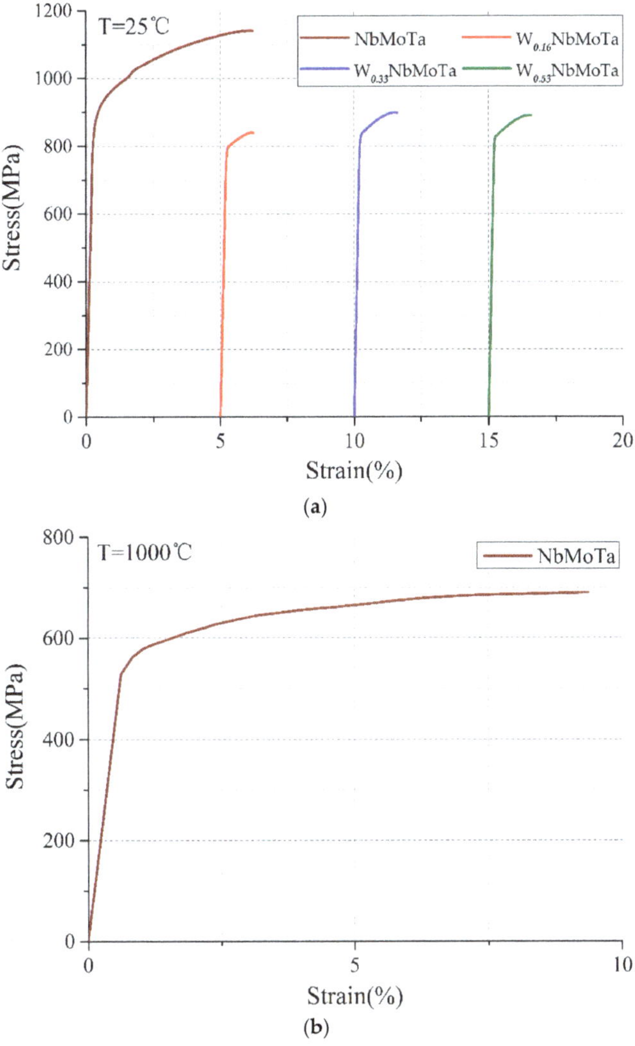

Figure 11. Compressive stress-strain curves of different temperature for the W_xNbMoTa HEAs. (**a**) Room temperature, (**b**) high temperatures.

Table 6. The comparison of high temperature (1000 °C) performance of aerospace materials.

Alloy ID	Yield Strength at 1000 °C/MPa
Nb-1Zr	113
C103	144
ODS-MA754	212
Mo-14Re	371
T111	505
NbMoTa	530

4. Conclusions

The purpose of the work is to study the dependence of the yield strength of the refractory HEAs on their composition fabricated by LCD for the potential applications in aerospace industry. The following conclusions are drawn:

(1) The crystal structures of each W_xNbMoTa (x = 0, 0.16, 0.33, 0.53) alloys are all a single-phase solid solution of the BCC structure analyzed by XRD.

(2) Due to the characteristic of rapid solidification, the size of the grains and dendrites on the microcosmic of W_xNbMoTa refractory HEAs was 20 μm and 4 μm on average, smaller than that of the HEAs fabricated by VAM.

(3) The increase of the tungsten concentration of W_xNbMoTa led to four results of the Vickers microhardness, i.e., H_v = 459.2 ± 9.7, 476.0 ± 12.9, 485.3 ± 8.7, 497.6 ± 5.6, respectively.

(4) The NbMoTa alloy has a compressive strain (ε_p) of 5.8% at room temperature and its yield strength (σ_b), compressive strength (σ_m) and compressive strain (ε_p) of 530 MPa, 684 Mpa and 8.5% respectively at 1000 °C. The effects show better performance than many traditional refractory metals such as T-111, Nb-1Zr, and C103, which are commonly used in aerospace.

(5) The content of tungsten has no effect on the formation of a single-phase solid solution and the microstructure of the HEAs. In terms of mechanical behavior, the microhardness shows an increasing tendency with the increase of the content of tungsten. As a result, the yield strength and plasticity of the W-free alloy is improved compared with alloys containing tungsten at room temperature.

In this study, with excellent yield strength at high temperature, the NbMoTa alloy shows potential for application in aerospace industry. The present work provides a theoretical basis for LCD manufacturing of aerospace parts of refractory HEAs with a large size, complex structure, and variable composition.

Author Contributions: Conceptualization, Q.L., H.Z., D.L. and S.H.; Methodology, Q.L.; Software, Q.L.; Validation, H.Z. and D.L.; Formal Analysis, Q.L.; Investigation, Q.L.; Resources, Q.L.; Data Curation, Q.L.; Writing-Original Draft Preparation, Q.L.; Writing-Review & Editing, Q.L., H.Y. and Z.C.; Visualization, Q.L.; Supervision, H.Z. and D.L.; Project Administration, H.Z.; Funding Acquisition, H.Z., D.L. and Z.L.

Funding: The research was funded by the National Natural Science Foundation of China (No.51505366), China Postdoctoral Science Foundation (No.2015M570827) and the Guangdong Province Science and Technology Project (No.2017B090911006).

Acknowledgments: The authors would like to acknowledge their appreciation of the technical support provided by Jiuhong Wang, Zhaoxiang Chen, Hong Wang and Jinzhi Zhang in sample experiments and microstructure analysis. The high-temperature compression test was conducted at School of Materials Science and Engineering, Shanghai University.

Conflicts of Interest: The authors declare no conflict of interest. The funders had no role in the design of the study; in the collection, analyses, or interpretation of data; in the writing of the manuscript, and in the decision to publish the results.

References

1. Yeh, J.W.; Chen, S.K.; Lin, S.J.; Gan, J.Y.; Chin, T.S.; Shun, T.T.; Tsau, C.H.; Chang, S.Y. Nanostructured high-entropy alloys with multiple principal elements: Novel alloy design concepts and outcomes. *Adv. Eng. Mater.* **2004**, *6*, 299–303. [CrossRef]
2. Miracle, D.B.; Senkov, O.N. A critical review of high entropy alloys and related concepts. *Acta Mater.* **2017**, *122*, 448–511. [CrossRef]
3. Zhang, Y.; Zuo, T.T.; Tang, Z.; Gao, M.C.; Dahmen, K.A.; Liaw, P.K.; Lu, Z.P. Microstructures and properties of high-entropy alloys. *Prog. Mater. Sci.* **2014**, *61*, 1–93. [CrossRef]
4. Senkov, O.N.; Wilks, G.B.; Scott, J.M.; Miracle, D.B. Mechanical properties of $Nb_{25}Mo_{25}Ta_{25}W_{25}$ and $V_{20}Nb_{20}Mo_{20}Ta_{20}W_{20}$ refractory high entropy alloys. *Intermetallics* **2011**, *19*, 698–706. [CrossRef]
5. Moravcik, I.; Gouvea, L.; Hornik, V.; Kovacova, Z.; Kitzmantel, M.; Neubauer, E.; Dlouhy, I. Synergic strengthening by oxide and coherent precipitate dispersions in high-entropy alloy prepared by powder metallurgy. *Scr. Mater.* **2018**, *157*, 24–29. [CrossRef]
6. Li, Z.; Pradeep, K.G.; Deng, Y.; Raabe, D.; Tasan, C.C. Metastable high-entropy dual-phase alloys overcome the strength-ductility trade-off. *Nature* **2016**, *534*. [CrossRef] [PubMed]
7. Tang, Z.; Huang, L.; He, W.; Liaw, P.K. Alloying and processing effects on the aqueous corrosion behavior of high-entropy alloys. *Entropy* **2014**, *16*, 895–911. [CrossRef]
8. Chou, Y.L.; Yeh, J.W.; Shih, H.C. The effect of molybdenum on the corrosion behaviour of the high-entropy alloys $Co_{1.5}CrFeNi_{1.5}Ti_{0.5}Mo_x$ in aqueous environments. *Corros. Sci.* **2010**, *52*, 2571–2581. [CrossRef]
9. Zou, Y.; Maiti, S.; Steurer, W.; Spolenak, R. Size-dependent plasticity in an $Nb_{25}Mo_{25}Ta_{25}W_{25}$ refractory high-entropy alloy. *Acta Mater.* **2014**, *65*, 85–97. [CrossRef]
10. Zou, Y.; Ma, H.; Spolenak, R. Ultrastrong ductile and stable high-entropy alloys at small scales. *Nat. Commun.* **2015**, *6*, 7748. [CrossRef] [PubMed]
11. Juan, C.C.; Tsai, M.H.; Tsai, C.W.; Hsu, W.L.; Lin, C.M.; Chen, S.K.; Lin, S.J.; Yeh, J.W. Simultaneously increasing the strength and ductility of a refractory high-entropy alloy via grain refining. *Mater. Lett.* **2016**, *184*, 200–203. [CrossRef]
12. Song, H.; Tian, F.; Wang, D. Thermodynamic properties of refractory high entropy alloys. *J. Alloys Compd.* **2016**, *682*, 773–777. [CrossRef]
13. Körmann, F.; Sluiter, M.H.F. Interplay between lattice distortions, vibrations and phase stability in NbMoTaW high entropy alloys. *Entropy* **2016**, *18*, 403. [CrossRef]
14. Senkov, O.N.; Wilks, G.B.; Miracle, D.B.; Chuang, C.P.; Liaw, P.K. Refractory high-entropy alloys. *Intermetallics* **2010**, *18*, 1758–1765. [CrossRef]
15. Han, Z.D.; Chen, N.; Zhao, S.F.; Fan, L.W.; Yang, G.N.; Shao, Y.; Yao, K.F. Effect of Ti additions on mechanical properties of NbMoTaW and VNbMoTaW refractory high entropy alloys. *Intermetallics* **2017**, *84*, 153–157. [CrossRef]
16. Li, D.C.; He, J.K.; Tian, X.Y.; Liu, Y.X.; Zhang, A.F.; Lian, Q.; Jin, Z.M.; Lu, B.H. Additive manufacturing: integrated fabrication of macro/microstructures. *J. Mech. Eng.* **2013**, *49*, 129–135. [CrossRef]
17. Dobbelstein, H.; Thiele, M.; Gurevich, E.L.; George, E.P.; Ostendorf, A. Direct metal deposition of refractory high entropy alloy MoNbTaW. *Phys. Procedia* **2016**, *83*, 624–633. [CrossRef]
18. Zhang, H.; Xu, W.; Xu, Y.; Lu, Z.; Li, D. The thermal-mechanical behavior of WTaMoNb high-entropy alloy via selective laser melting (SLM): experiment and simulation. *Int. J. Adv. Manuf. Technol.* **2018**, *96*, 1–14. [CrossRef]
19. Senkov, O.N.; Senkova, S.V.; Woodward, C.; Miracle, D.B. Low-density, refractory multi-principal element alloys of the Cr-Nb-Ti-V-Zr system: Microstructure and phase analysis. *Acta Mater.* **2013**, *61*, 1545–1557. [CrossRef]
20. Guo, W. Molecular Dynamics Simulation of Irradiation Damage in Multicomponent Alloys. Ph.D. Thesis, University of Tennessee, Knoxville, TN, USA, 2015.
21. Takeuchi, A.; Inoue, A. Classification of Bulk Metallic Glasses by Atomic Size Difference, Heat of Mixing and Period of Constituent Elements and Its Application to Characterization of the Main Alloying Element. *Mater. Trans.* **2005**, *46*, 2817–2829. [CrossRef]
22. Zhang, R.F.; Rajan, K. Statistically based assessment of formation enthalpy for intermetallic compounds. *Chem. Phys. Lett.* **2014**, *612*, 177–181. [CrossRef]

23. Zhang, Y.; Yang, X.; Liaw, P.K. Alloy design and properties optimization of high-entropy alloys. *JOM* **2012**, *64*, 830–838. [CrossRef]

24. Yang, X.; Zhang, Y. Prediction of high-entropy stabilized solid-solution in multi-component alloys. *Mater. Chem. Phys.* **2012**, *132*, 233–288. [CrossRef]

25. Qi, H.; Azer, M.; Ritter, A. Studies of standard heat treatment effects on microstructure and mechanical properties of laser net shape manufactured Inconel 718. *Metall. Mater. Trans. A Phys. Metall. Mater. Sci.* **2009**, *40*, 2410–2422. [CrossRef]

26. Susan, D.F.; Puskar, J.D.; Brooks, J.A.; Robino, C.V. Quantitative characterization of porosity in stainless steel LENS powders and deposits. *Mater. Charact.* **2006**, *57*, 36–43. [CrossRef]

27. Kobryn, P.A.; Moore, E.H.; Semiatin, S.L. Effect of laser power and traverse speed on microstructure, porosity, and build height in laser-deposited Ti-6Al-4V. *Scr. Mater.* **2000**, *43*, 299–305. [CrossRef]

28. Chokshi, A.H.; Rosen, A.; Karch, J.; Gleiter, H. On the validity of the hall-petch relationship in nanocrystalline materials. *Scr. Metall.* **1989**, *23*, 1679–1683. [CrossRef]

29. Furukawa, M.; Horita, Z.; Nemoto, M.; Valiev, R.Z.; Langdon, T.G. Microhardness measurements and the hall-petch relationship in an Al-Mg alloy with submicrometer grain size. *Acta Mater.* **1996**, *44*, 4619–4629. [CrossRef]

30. Hansen, N. Hall-petch relation and boundary strengthening. *Scr. Mater.* **2004**, *51*, 801–806. [CrossRef]

31. Yoshida, S.; Bhattacharjee, T.; Bai, Y.; Tsuji, N. Friction stress and Hall-Petch relationship in CoCrNi equi-atomic medium entropy alloy processed by severe plastic deformation and subsequent annealing. *Scr. Mater.* **2017**, *134*, 33–36. [CrossRef]

32. Gwalani, B.; Soni, V.; Lee, M.; Mantri, S.A.; Ren, Y.; Banerjee, R. Optimizing the coupled effects of Hall-Petch and precipitation strengthening in a $Al_{0.3}CoCrFeNi$ high entropy alloy. *Mater. Des.* **2017**, *121*, 254–260. [CrossRef]

33. El-Genk, M.S.; Tournier, J.M. A review of refractory metal alloys and mechanically alloyed-oxide dispersion strengthened steels for space nuclear power systems. *J. Nucl. Mater.* **2005**, *340*, 93–112. [CrossRef]

 *materials*

Article

Mapping the Tray of Electron Beam Melting of Ti-6Al-4V: Properties and Microstructure

E. Tiferet [1,2,*], M. Ganor [1], D. Zolotaryov [3], A. Garkun [3], A. Hadjadj [1], M. Chonin [2], Y. Ganor [2], D. Noiman [1], I. Halevy [4,5], O. Tevet [1] and O. Yeheskel [6,*]

[1]   Materials Department, Nuclear Research Center Negev, P.O. Box 9001, Be'er Sheva 8419001, Israel; ganorm55@gmail.com (M.G.); sivan.moore82@gmail.com (A.H.); noimand@hotufi.net (D.N.); tevet.ofer@gmail.com (O.T.)

[2]   AM Center, Rotem Ind. Mishor Yamin 86800, Israel; michaelc@rotemi.co.il (M.C.); Yarong@rotemi.co.il (Y.G.)

[3]   Israel Institute of Metals, Technion Haifa 32000, Israel; denisz@trdf.technion.ac.il (D.Z.); agar@trdf.technion.ac.il (A.G.)

[4]   Nuclear Engineering Unit, Ben Gurion University, Be'er Sheva 8410501, Israel; halevy.itzhak.dr@gmail.com

[5]   Physics Department, Nuclear Research Center Negev, P.O. Box 9001, Be'er Sheva 8419001, Israel

[6]   Materials Engineering Department, Ben Gurion University, Be'er Sheva 8410501, Israel

*   Correspondence: tiferete@gmail.com (E.T.); oriyehe@gmail.com (O.Y.)

Received: 1 April 2019; Accepted: 29 April 2019; Published: 7 May 2019

Abstract: Using an electron beam melting (EBM) printing machine (Arcam A2X, Sweden), a matrix of 225 samples (15 rows and 15 columns) of Ti-6Al-4V was produced. The density of the specimens across the tray in the as-built condition was approximately 99.9% of the theoretical density of the alloy, ρ_T. Tensile strength, tensile elongation, and fatigue life were studied for the as-built samples. Location dependency of the mechanical properties along the build area was observed. Hot isostatic pressing (HIP) slightly increased the density to 99.99% of ρ_T but drastically improved the fatigue endurance and tensile elongation, probably due to the reduction in the size and the distribution of flaws. The microstructure of the as-built samples contained various defects (e.g., lack of fusion, porosity) that were not observed in the HIP-ed samples. HIP also reduced some of the location related variation in the mechanical properties values, observed in the as-printed condition.

Keywords: Powder bed; fatigue; Hot Isostatic Pressure; Electron Beam Melting

---

1. Introduction

Additive manufacturing (AM) is a method of transforming digital design files into functional engineering products. Offering flexibility in design and environmental advantages, the technology also enables engineers to produce low volumes of unique objects in an economical way. The methods and materials may vary, but all AM processes typically work by building their components up one layer at a time. One of the most commonly used methods is powder bed fusion AM (PBF-AM), which involves focusing an energetic (laser or electron) beam in order to melt specific locations of a powder layer spread on a base plate. A second powder layer is then spread on top of the first one, and the fusion process is repeated [1]. The energy emitted is absorbed by powder particles via both bulk-coupling and powder-coupling mechanisms [2]. The resulting transient temperature field, characterized by high temperatures and rapid solidification rates, is formed concomitantly during the interaction between the beam and the powder bed and has a significant effect on defect formation, final microstructure, and mechanical properties of the components [3]. In addition, the transient thermal behavior is controlled by processing parameters such as material properties, beam characteristics, and scan speed and strategy. In laser powder bed fusion (L-PBF), due to the fact that there is relatively little preheating, the high cooling rates during melting and solidification cause internal residual stresses, which in turn induce plastic deformation within thin metal layers [4]. Sacrificial support structures do not always compensate for this local plastic deformation, leading to the formation of cracks and product or support structure failure. In fact, the complex metallurgical nature of L-PBF involves multiple modes of heat, mass, and momentum transfer that often lead to uncertainty concerning the final part mechanical properties [5,6]. Furthermore, post-processing heat treatments are inevitable in the case of L-PBF, whereas in electron beam melting (EBM), it is not mandatory to stress relief after AM builds. A common post process is hot isostatic pressing (HIP), which is used mainly to eliminate pores in castings [7,8]. Since pores and other defects affect the fatigue life of materials [9], HIP can improve the fatigue endurance limit of samples produced by both L-PBF and EBM [10–12]. Several materials, including steel, titanium, and aluminum alloys, manufactured by AM have been studied for defect formation as well as for mechanical performance [13,14]. The most well-studied titanium alloy applicable in the aerospace and biomedical fields is Ti-6wt%Al-4wt%V (Ti-6Al-4V); these studies have shown that the quasi-static tensile strength of AM manufactured Ti–6Al–4V is comparable to that of conventionally-manufactured alloys [3,9,15–17]. In addition, it has been shown that with conventional manufacturing, elongation of Ti-6Al-4V alloys gradually decreases as oxygen concentration increases [16,17]. In contrast, for AM using direct energy deposition (DED), an increase in oxygen concentration promotes the formation of Ti-α at grain boundaries, Ti-β inside grains, and finer Ti-α laths. Consistent experimental results have shown an increase in tensile properties with increased oxygen, with no significant effect on elongation [18]. When utilizing EBM, it can be assumed that objects can be placed at any location in the powder bed tray.

HIP as a Post-Process for Additively Manufactured End Products

Many engineering parts once produced from powders using conventional methods such as cold compaction followed by sintering are now produced using AM [19]. However, these products often contain material imperfections in the form of pores and/or lack of melting [3]. In certain applications, mostly those where cyclic loading is incorporated, these imperfections are detrimental to the material's mechanical properties and thus must be eliminated. The HIP process consists of the concomitant application of high temperature and pressure on parts that are placed in a pressure vessel. The temperatures employed in metals are in the range of 0.5 Tm to 0.9 Tm, where Tm is the melting temperate [K] of the material, while the typical pressure utilized in most HIP processes of metallic objects is 100 MPa [7]. Applying high temperature and pressure on a material results in the initiation of creep mechanisms in addition to enhancing the bonding of adjacent surfaces by diffusion while also closing existing porosity [7,20]. AM offers intricate geometries that are designed for a specific demand. Residual flaws such as lack of fusion, cracks, or spherical porosity affect the fatigue life. In order to reduce such flaws population, post processing HIP is required to enhance the properties of products built via AM, especially those subjected to fatigue conditions [10]. Yet, the fact that the HIP process takes place at elevated temperatures may actually have adverse effects on the properties that it was originally intended to enhance. For example, the unique microstructure that is typical of materials built via AM may be completely altered as a consequence of the high temperatures employed. Recently, a Laser-PBF-AM study conducted on Inconel showed improved mechanical properties with an HIP post-process [21], as well as the partial elimination of inherently imposed pores. Since different locations in the build tray impose different thermal histories [22], it may be that defect variation is correlated with their position on the build plate.

The objective of the current study is to present a map dependency of the physical and the mechanical properties and the microstructure of AM-EBM Ti-6Al-4V within the powder bed volume were it is located. Additional aspects of HIP influence on microstructure and mechanical properties are also presented and discussed.

2. Experimental

2.1. Research Methodology

The methodology used in this study was based on relationships between the raw material AM process parameters and the microstructure properties with and without post treatment, as seen in Figure 1. This was done by utilizing powder after multiple uses (see Section 2.2), single printing parameters (see Section 2.3), and a single post-treatment cycle (see Section 2.4), measuring microstructure mainly via SEM (see Section 2.5) and neutron diffraction, reported elsewhere [23]. Physical properties, density, elastic properties, tensile strength, and fatigue were correlated to the location on the tray.

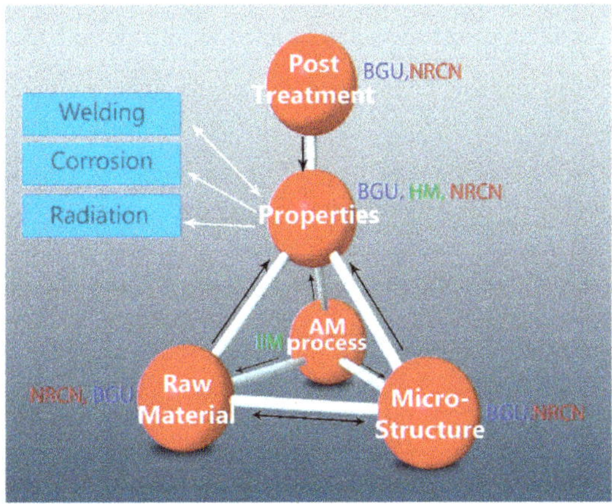

Figure 1. Material-processing properties and microstructure relationships examined.

2.2. Powder Characteristics

The Ti-6Al4V powder was supplied by Arcam (batch 1410, in compliance with the American Society for Testing and Materials, ASTM F2924 standard), and used in the previous 38 cycles. Particle size distributions of the powder in cycle 39 (C-39) for D_{10}, D_{50}, and D_{90} values were 66, 84, and 120 µm, respectively. The sphericity ratio was 0.9 (where 1 represents a perfect sphere), and powder density was determined using pycnometry to be 4427 ± 15 kg/m^3. As expected [24], reuse of powder did not affect particle size or shape distribution.

2.3. Printing Theme

The Ti-6Al-4V alloy samples were produced by an Arcam A2X machine (Arcam AB, Mölndal, Sweden) using recycled grade 5 powder. The acceleration voltage remained constant at 60 kV. Vacuum conditions were 10^{-5} mbar of the initial vacuum with a needle valve providing a constant 10^{-3} mbar helium environment.

Powder was spread on a platform of a pre-heated plate. Each powder layer was pre-heated in order to attain a temperature between 550 and 700 °C (~0.5 T_M). The main reason for conducting the pre-heating stage was to improve the energy deposition efficiency. When an electron beam interacts with a cold powder, part of the energy is not deposited (and transferred to heat) as intended. Instead, some powder gets charged by the electrons [25], and a kinematic phenomenon of recoiled powder particles emerging from the powder bed occurs. The pre-heat ensures that the efficiency of the energy deposition is optimal.

The tray consisted of a matrix of 15 rows (labeled A to O) by 15 columns (labeled numerically). Some of these rows contained cuboid samples (10 mm by 12 mm by 70 mm height) intended for non-destructive evaluation of the build process quality. Most of the samples were rods with a diameter of ~10 mm that were used for testing mechanical properties (Figure 2). The height of all samples was ~70 mm, each one designated by a row and a column notation, e.g., A3 refers to the third sample (column) in row A. The intersection of lines A, H, and O with columns 2, 4, 6, 10, 12, and 14 marks specific samples that were tested.

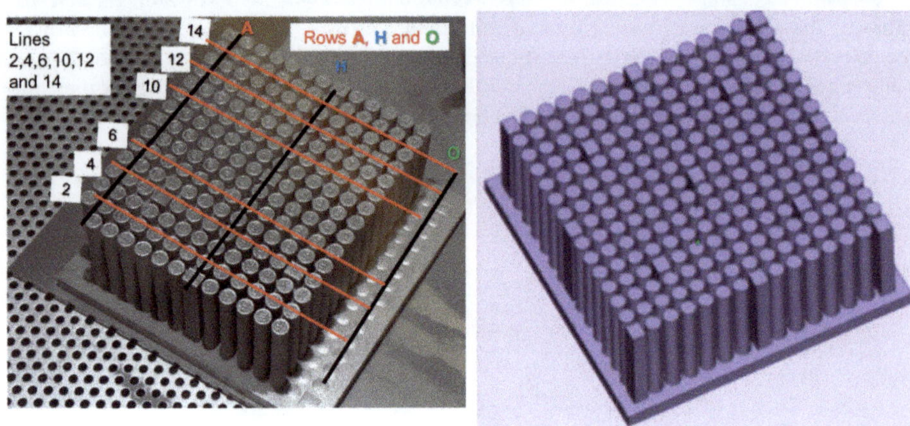

Figure 2. Samples and coordinates designating their location (left) and the computer-assisted design (CAD) file used for the electron beam melting (EBM) build (right).

The electron beam path and the parameters (e.g., beam speed and current) were guided by an algorithm (proprietary) formulated by Arcam. The purpose of this algorithm was to maintain a constant heat deposition (J/mm^2) in each area in a single layer; additionally, all layers should have had a similar heat deposition. In the current work, the geometries were simple, therefore equal heat deposition was achievable. Nevertheless, since the melted area in each layer was relatively large (i.e., more than 50%), it was necessary to manage the melt order in an attempt to keep the heat distribution as even as possible. Various STL (stereo lithography) builds were tried, yet most of them resulted in the smoke [25] that evolves when an electron beam interacts with cold powder. For the successful build, 15 STLs were used for the various rows, and 1 STL was used for the rectangular rods.

2.4. HIP Post-Processing

Post-processing of AM specimens was carried out using a laboratory-sized HIP apparatus under high purity (99.99% pure) Ar gas at a pressure of 120 MPa at 920 °C for two hours in accordance with an appropriate standard [26]. The printed rods were wrapped with proprietary protective foil to reduce the likelihood of reaction with minute impurities of the gas.

2.5. Characterization Methods

Powder PSSD (particle size and shape distribution) was evaluated using a Qicpic instrument (Sympatec, Germany), while microstructure and fracture surface were studied utilizing high resolution SEM. Chemical composition of the metallic elements was measured by EDS (energy dispersive spectroscopy); evolved gas was measured by Leco (™). The density of the built specimens was measured using the Archimedes method described previously [27,28], and gas pycnometery was utilized to measure powder density. Samples for tensile and fatigue testing were machined to a surface finish of Ra = 0.4 − 0.6 < 1 μm. Tensile testing was carried out according to ASTM F2924–14 [26,29] at 24 °C using an Instron 3369 testing machine (50 kN load cell) and an Instron 2620-602. Elongation was measured based on a 20 mm length (about four times the diameter of the specimen). The tensile test was carried out at a strain rate of 0.005 min^{-1} until 2% total elongation was achieved. Then, the crosshead speed was changed to 1 mm per minute. The fatigue test was performed according to the ASTM E466 Standard with a focus on the high cycle section. The force controlled constant amplitude uniaxial loading was carried out using an Instron 8801 testing machine (Dynacell, Dynamic Load Cell +/− 100 kN) under load control with a sinusoidal waveform. Specimens were fixed using Instron fatigue-rated mechanical wedge grips. The Young's moduli of the as-built and the HIP samples were calculated based on OLS (ordinary least squares) regression as per Section 0.2–0.5 of the proof stress. The elastic moduli of limited samples were also measured according to the pulse echo method described previously [25,30]. For fatigue, cyclic loading was applied in air at 23 ± 2 °C with a load ratio of R = 0.1 and a frequency of 30 Hz through the end of the test. Most specimens were tested to failure, while for some samples following HIP, the tests were stopped (i.e., run out) at 3×10^6 cycles, and a few were stopped after 1×10^7 cycles had elapsed.

3. Results

3.1. Bulk Density

The average density, $\bar{x}$, and the standard deviation, S, of more than 140 as-built samples were 4427.6 ± 1.2 kg/m^3, while the average density and the standard deviation of more than 50 samples that underwent HIP were 4432.1 ± 1.45 kg/m^3. The theoretical density (ρ_T) of the material was calculated by adding three standard deviations, $3\sigma = 3 \times 1.45/\sqrt{50}$, to the average of the HIP-ed samples, resulting in $\rho_T = 4432.7$ kg/m^3. The relative density, ρ^*, was $\rho^* = \rho/\rho_T$, where ρ was the density. Figure 3 shows a map of the relative density of AB (as-built) samples for which the average was 0.9991 with no dependency in location and with a variation of less than 0.2%. After HIP treatment, ρ^* improved to a value of 0.9998 of the full density with a variation of less than 0.035%.

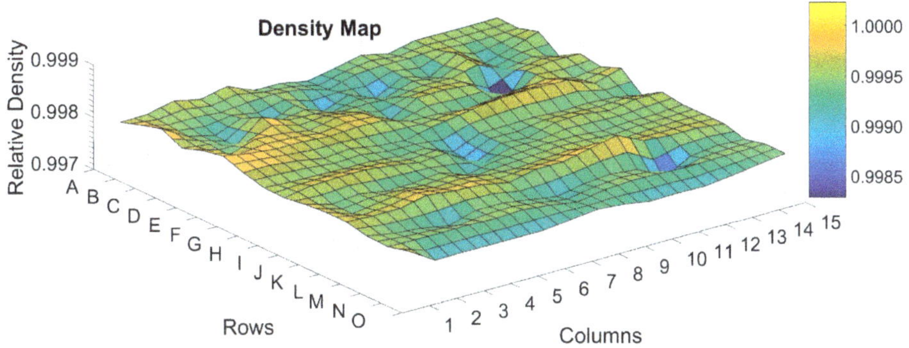

Figure 3. Mapping the variation of relative density across the built tray.

3.2. Chemical Composition

Results were obtained from specimens A3, H3, and O3; averages and standard deviation are shown in Table 1. Note that the oxygen content exceeded ASTM F2924 standard requirements.

3.3. Microstructure

The microstructure of the as-built Ti-6Al-4V samples is shown in Figure 4. As expected, an α-β mixture with a very fine lamellar morphology was observed, in which the β phase is the white phase between lamellae of α, the gray phase. In samples taken from the HIP-ed samples, the microstructure was still lamellar with coarser lamellae (Figure 4B). It was recently shown using neutron diffraction that the β content in samples from a similar EBM process was <1 wt%. [31] This β phase content was analyzed using the SEM micrographs (Figure 4A,B), and it was determined that for the as-built sample and for the HIP-ed samples, the contents were about 2.5 ± 1.4% and about 5.8 ± 1.5% wt%, respectively. Such an increase is consistent with the increase in beta phase content upon heating up to 1000 °C [26,29].

Table 1. Averages and standard deviations, $S^\dagger$, of elemental composition of the sample across the tray compared to the requirements of a proper ASTM standard.

Element	Ti	Al	V	Fe	O	C	H
ASTM F2924	Balance	5.5–6.75	3.5–4.5	<0.3	<0.2	<0.08	<0.0015
Average ±S † [%]	89.0 ± 0.22	6.5 ± 0.13	3.86 ± 0.04	0.27 ± 0.05	0.34 ± 0.01	0.031 ± 0.001	0.0027 ± 0.0008

$\dagger\ S = \sqrt{\frac{(x^2 - \bar{x}^2)}{n-1}}$ where $\bar{x}$ is the average, n is the number of measurements, and x is the result.

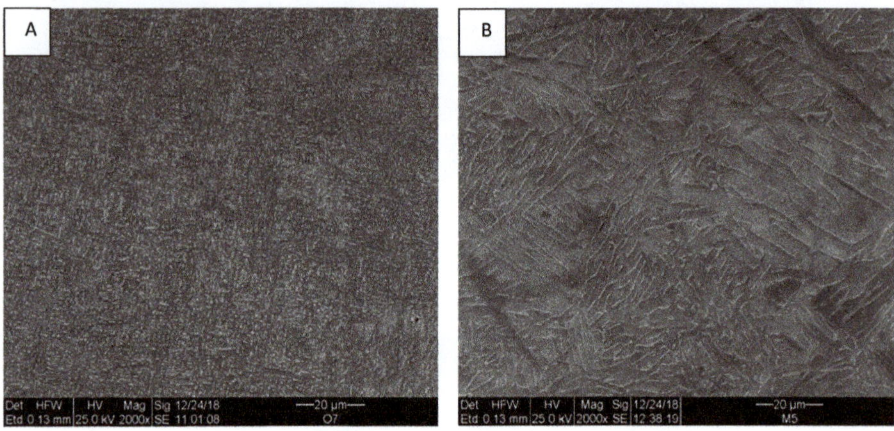

Figure 4. Microstructure of as-built (**A**) and hot isostatic pressing (HIP)-ed (**B**) samples.

3.4. Tensile Properties and Fractography

Results of the tensile tests for the as-built samples presented in Table 2 show that the proof stress, or the plastic strain stress of 0.2% ($R_{p0.2}$), and the ultimate tensile stress (R_m) were higher by ~200 MPa than both ASTM F3001 and ASTM F2924 standard requirements. Low ductility in some of the samples that did not reach these required standards was observed. Variation in properties depending upon location was pronounced for the reduction in area and elongation, both with relatively high standard deviations, while the standard deviations of R_m and $R_{p0.2}$ were much lower (Table 2). Upon correlating the properties with the printing order, further insights were revealed. Figure 5 shows the relationship between the engineering ultimate strength and the fracture strain; it should be noted that the middle line possesses a greater elongation compared to that in samples located closer to the circumference. Minor differences were observed when applying the ultrasound pulse echo method to measure the elastic moduli in small slices (~5 mm) taken from the bottom of selected rods (Table 3).

Following HIP treatment, the specimens' reduction of area and elongation improved significantly, while dependency of the location across the tray became insignificant (Figure 6). Hence, homogeneity of mechanical properties improved. In a single sample yielding the lowest value for reduction area (A4), a surface defect caused during the machining of the specimen was observed. Excluding this one sample, variation in the reduction of area was about 43%±3%, whereas for the AB samples, variation was ±11% (Table 2). Thus, it appears that HIP significantly improved average elongation and reduction of area, while at the same time slightly decreased the R_m and $R_{p0.2}$ (about 5%), suggesting in this composition a Hall-Petch-like behavior. Additionally, Figure 7 presents typical tensile results of AB and HIP samples that failed prematurely. This early fracture in AB samples was initiated by three

AM-related phenomena: lack of fusion, overheating, and porosity (Figure 8). Sample A2 showed large porosity and one lack of fusion defect (Figure 8A), sample A14 showed several lack of fusion defects and small size porosity (Figure 8B), while sample B8's overheated area was evident (Figure 8C). No such defects appeared in the HIP sample fracture surface (Figure 8D). Finally, the difference in strain-to-fracture in samples A2 and A14 is discussed in Section 4.3.

Table 2. Averages and standard deviations, S, of engineering tensile test results: Young's modulus, E, proof strength, $R_{p0.2}$, ultimate tensile strength, R_m, elongation, e, and reduction of area, A, for the as-built and the HIP samples. The results for HIP samples, excluding sample A4, are given in the last raw.

Property	Young's Modulus, E, GPa	Proof Stress, $R_{p0.2}$, MPa	Tensile Strength R_m, MPa	Elongation, e, %	Reduction of Area, A, %	Number of Samples
ASTM F2914 Requirements		825 min	895 min	10 min	15 min	
Average and standard deviation, As-built	118.8 ± 3.8	1036 ± 17	1122 ± 22	9.8 ± 3.8	11.5 ± 5.6	12
Average and standard deviation, HIP	119.5 ± 5.3	971 ± 20	1086 ± 19	18.9 ± 3.1	38.3 ± 11	6
Average and standard deviation, HIP	120.0 ± 5.8	976 ± 19	1090 ± 18	20.1 ± 0.7	42.8 ± 2.2	5

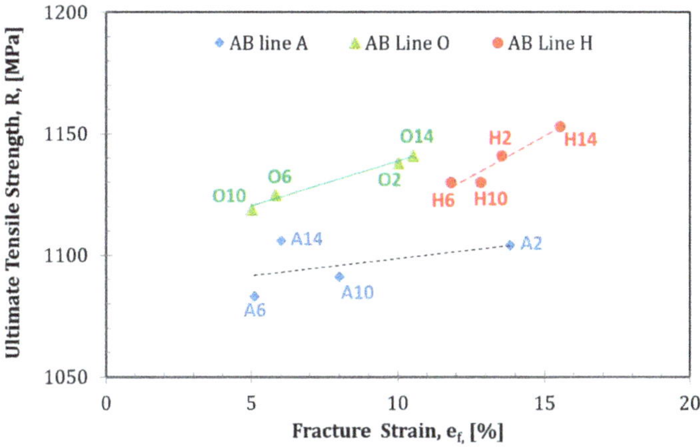

Figure 5. Change in engineering ultimate tensile strength with engineering fracture strain in as-built (AB) samples as a function of printing location in rows A, H, and O.

Table 3. Averages and standard deviations, S, of the dynamic elastic moduli of as-built samples A3, H3, and O3. The columns indicate density, longitudinal velocity, V_L, shear velocity, V_S, Young's modulus, E, shear modulus, G, and Poisson's ratio, ν.

Sample Designation	Density [kg/m³]	V_L [m/s]	V_S [m/s]	E [GPa]	G [GPa]	ν
A3	4422 ± 3	6201 ± 1	3213 ± 2	120.2 ± 0.3	45.7 ± 0.1	0.316 ± 0.005
H3	4422 ± 3	6233 ± 4	3222 ± 2	121.0 ± 0.3	45.9 ± 0.1	0.318 ± 0.006
O3	4222 ± 3	6212 ± 14	3211 ± 2	120.1 ± 0.4	45.6 ± 0.1	0.318 ± 0.007

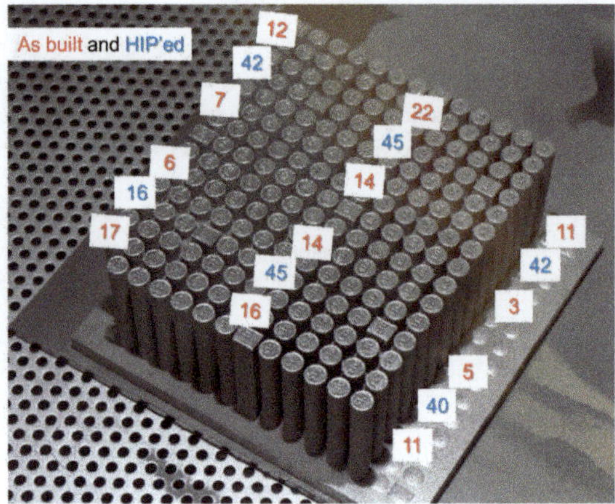

Figure 6. The reduction of area (%) as a function of location of the samples on the tray. The numbers represent the reduction of area in the as-built state (red) and after HIP (blue).

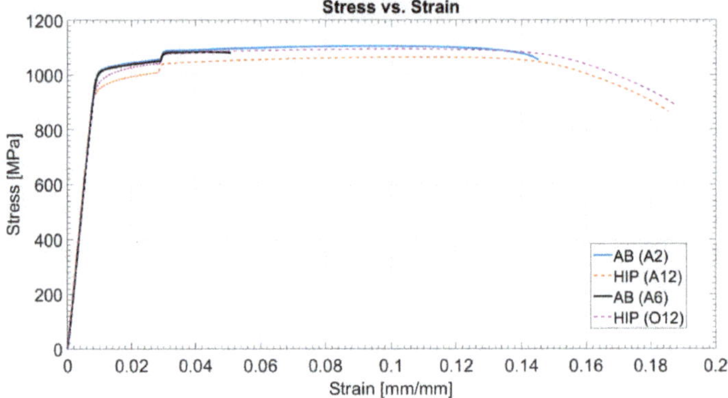

Figure 7. Typical engineering stress–strain curves for AB and HIP samples.

3.5. Fatigue Tests and Fractography

The results of the fatigue tests for both as-built and HIP-ed samples (Figure 9) indicate that the fatigue limit ($N_f = 10^7$ cycles) for the HIP-ed samples was about 550 MPa, and the estimated fatigue limit for the AB state was below 300 MPa. Similar to the tensile results, a pronounced improvement in fatigue properties after HIP was observed. Figure 10 shows the comparison in number of cycles between HIP and as-built samples when tested at a stress of 623 MPa. As demonstrated recently [10], voids were most often the source of fatigue crack initiation in EBM as-built samples; the improvement in fatigue strength could be attributed to the reduction in the number of initiation sites through the internal pore and the void closure during HIP. Figure 11 presents fracture surfaces of typical samples. It should be noted that remnants of printing defects were not observed on the HIP-ed samples. However, in the as-built samples, a crack was initiated by a fusion defect (overheating) in the bulk of the rod that, incidentally, was close in proximity to the machined fatigue sample.

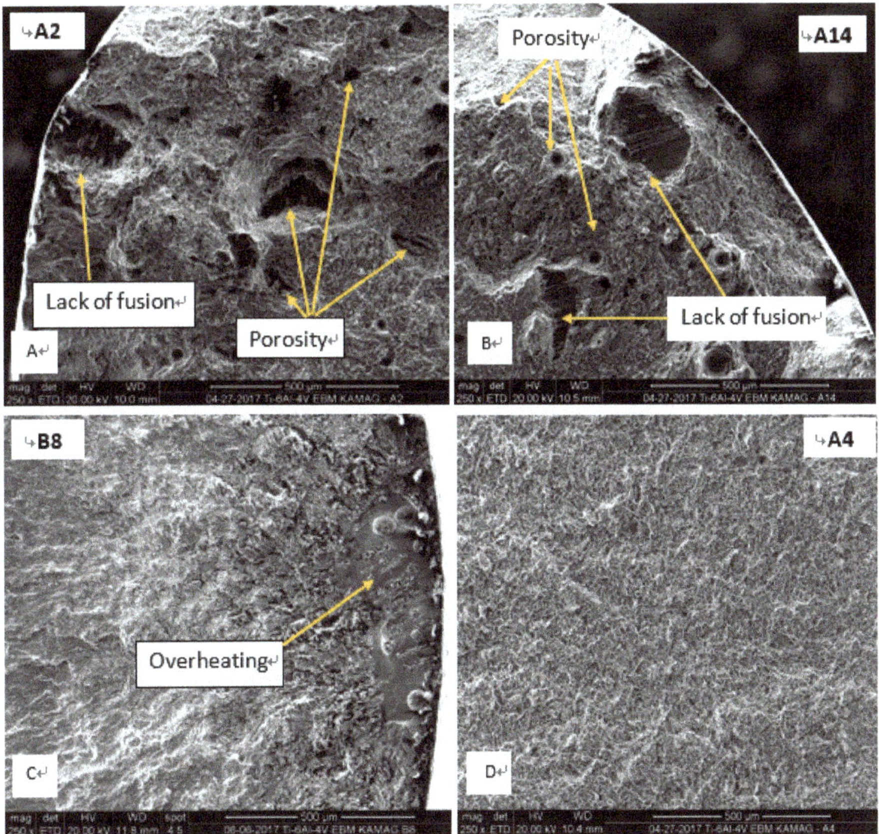

Figure 8. Fractography of AB samples depicting large porosity (**A**), large lack of fusion (**B**), large overheating (**C**), and of HIP-ed sample (**D**), A4, showing a much sounder microstructure than the AB.

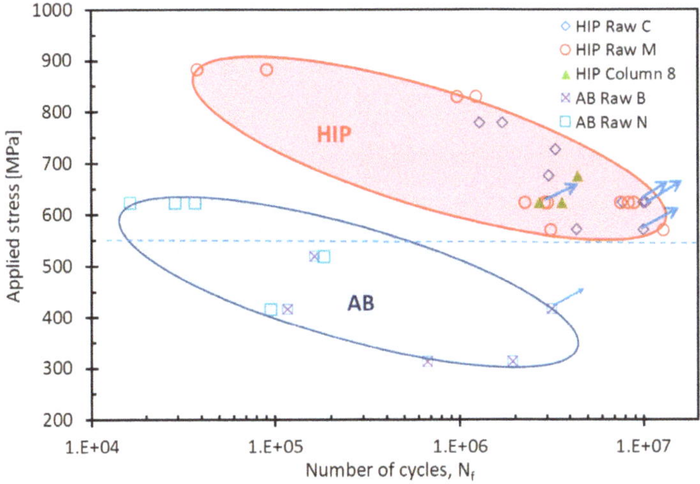

Figure 9. Fatigue tests on AB samples and HIP. Arrows indicate deliberate run outs.

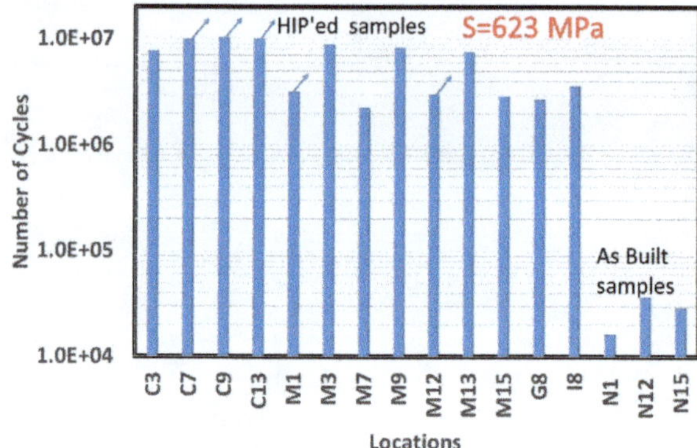

Figure 10. Number of cycles to failure at stress of 623 MPa in HIP-ed samples from row C and row M and a few samples from column 8; AB samples are from row N. Arrows indicate "run outs" at either 3×10^6 or 1×10^7 cycle.

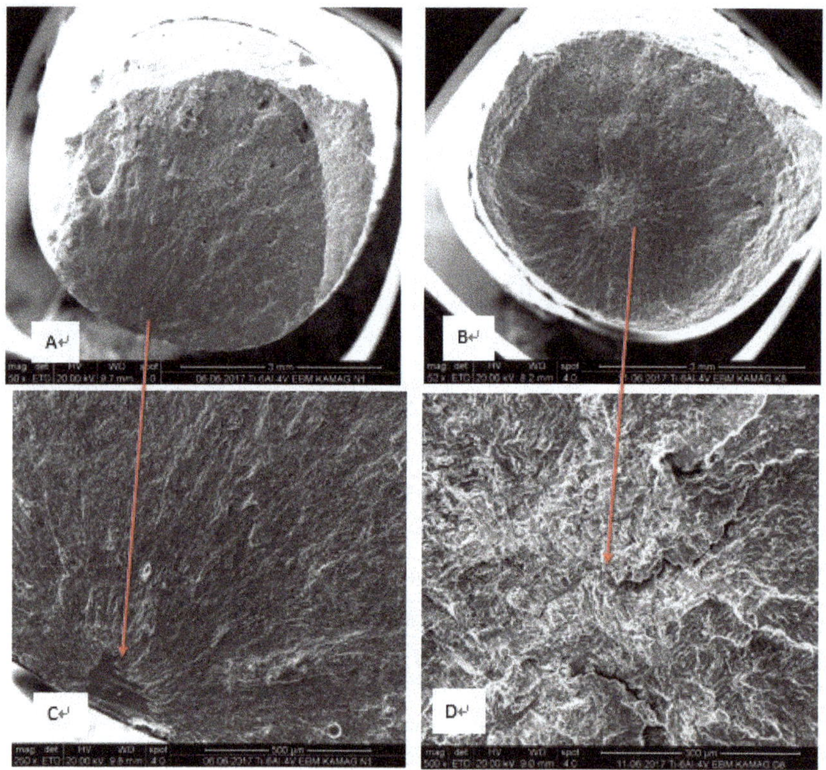

Figure 11. Macrographs of the fracture surface of the fatigue samples as-built (**A**) and HIP-ed (**B**). Fractography reveals a lack of fusion defect in the as-built sample as a source of early fracture (**C**), while the fracture of the HIP sample was initiated in the bulk (**D**).

4. Discussion

4.1. Thermal Management of Built Tray

The geometry of the specimens in the current study was simple, thus the average of the energy deposition between layers or between the areas in each layer was inherently imposed by the CAD (computer assisted design) model. Even though the heat depositions in different locations for a given layer were aimed at being similar, the thermal history of specimens created in the center of the tray was different from the samples originating at the edges of the tray. This variation in temperature was the main driving force for the variation in the mechanical properties. Nevertheless, since the melted area in each layer was relatively large (i.e., more than 50%), it was necessary to manage the melt order in an attempt to keep the heat distribution in every layer as even as possible. Upon splitting the build geometry to a single STL for each sample, the melt sequence became much more even (The reported tray was "printable" only when splitting the built file into eight individual STL (this work was done with the Arcam 4.2 version, and today, with 5.2 version, this is done automatically)). This even distribution was the key factor that prevented relatively cold areas on the built tray. Heat management prevented the electron beam from interacting with cold powder, and smoke generation was limited. Therefore, managing the heat during the build is important not only in order to maintain a constant melt pool size to achieve homogeneous microstructure and mechanical properties [9], but also to sustain production when generating large volume builds. With all the caution taken to achieve homogeneous printing, Figure 8C shows an example of an uncontrolled, overheated area.

4.2. Location Dependency of As-Built Samples

As described in Section 4.1, an attempt to maintain similar heat deposition along the built was done, yet AM-PB-EBM had a noticeable location dependency that dictated mechanical property variation across the build area (Figure 6). Front and rear built borders (rows O and A) were subjected to relatively fast cooling rates, which might explain why the inner part of the built tray possessed better physical properties. Differing thermal histories, a direct result of location, is the subject of ongoing research.

4.3. HIP Influence

As expected, the microstructure of the AB EBM Ti-6Al-4V samples consisted of an α-β mixture with a very fine lamellar morphology in which the β phase was located between the lamellae of α. When executing an HIP heat treatment, a sufficiently high temperature was necessary to eliminate pores and lack of fusion defects with minor changes in the unique AM microstructure. At 920 °C, the microstructure of the HIP samples was still lamellar, although the lamellae size coarsened, which led to a slight decrease in yield strength as well as in ultimate tensile strength [11]. Proof, or the yield stress (YS), and the ultimate tensile strength (UTS) of the as-built samples were measured to be more than 200 MPa, greater than required by both ASTM F3001 and ASTM F2924 standards (Table 2). Ductility of the as-built samples was low, ~10% on average, and in five of the 12 samples, it was lower than standards dictate. The YS and the UTS of the HIP-ed samples were 7% lower compared to the AB samples. When excluding the result of the one sample that failed due to a machining defect, the averages of YS and UTS were 976 MPa and 1090 MPa, respectively, or about 20% higher than standards require. In addition, remarkable ductility improvement of HIP samples was observed, from an average of ~10% in the AB samples to ~20% in the HIP-ed samples. The main reason for this increase was the reduction in size and population of defects of the as-built samples. Defects such as porosity, lack of fusion, and overheating were successfully healed by the HIP process. Moreover, fatigue limit for HIP-ed samples was much higher with respect to that in the as-built samples, above 550 MPa for the HIP-ed samples versus around or even below 300 MPa for the as-built ones (Figure 11). The fatigue endurance limit ($N_f = 10^7$ cycles) for the as-built samples was found to be~250 MPa [10]; other studies have previously shown that the stress for this limit is much higher after HIP [10,32]. Whereas the fatigue crack initiation of the HIP samples started inside the bulk, for as-built samples, it started from the printing defect located below the sample surface. A similar observation held true for tensile tests. A few examples are related herein, e.g., in the fracture of sample A2, large porosity and one large lack of fusion defect were detected (Figure 9A). This sample failed with a strain of ~13%, while sample A14, which contained two large lack of fusion defects and small porosity (Figure 8B), failed at 6%. The reason for this inconsistency may be related to the critical flaw size of the two combined lack of fusion defects in A14 as compared to the combined lack of fusion and large porosity in A2. Further discussion of this hypothesis is irrelevant here due to a scarcity of data. However, supporting evidence on the importance of lack of fusion as a source of fracture of PBF-AM also existed for sample N4 (Figure 11C), which prematurely failed in fatigue after ~10^5 cycles at a stress of 415 MPa (Figure 9). It is suggested here that HIP heals not only porosity—a rounded defect—but also planar defects such as lack of fusion or overheating. Thus, HIP reduces the population of internal defects below a certain critical level. This reduction leads to dramatic improvement in elongation and fatigue cycles at any applied level of stress. Furthermore, HIP demonstrates a dramatic reduction in variation of mechanical properties depending upon location on the tray. These values were achieved in the presence of relatively high (0.34%) oxygen content. In commercial Ti64 [15], elongation was reduced linearly from 9% in 0.125% oxygen content to ~7% for 0.35% oxygen content. The high elongation and fatigue results of PBF-EBM in the current study were likely an outcome of the fine microstructure, which compensated for the deleterious effects of the elevated presence of oxygen in pure Ti and commercial Ti64 [15].

5. Summary

In this study, it was demonstrated that a very high areal density (>50%) was successfully printed using PBF electron beam melting with an average sample density of ~0.999 of the theoretical value. Mapping the mechanical properties across the printed tray revealed that, while density variation across samples was very small (<0.5%) and variation of tensile strength was small (~2%), the tensile elongation of the as-built samples showed large scatter (~40%) and some dependency on the location across the tray. The fatigue properties of the as-built samples were found to be low, with an estimated fatigue stress at 10^7 cycles below 300 MPa. Fractography of the samples following tensile and fatigue testing showed that planar defects such as lack of fusion and overheating were the major defects causing premature failure. However, while performing HIP at 920 °C only slightly increased density to 0.9999 of the theoretical density, it drastically improved elongation by ~200% with only a slight decrease (~5%) in proof stress and ultimate stress. In addition, the variation in properties relative to the location on the tray diminished, with the strength and the elongation of HIP-ed samples far surpassing the requirements dictated by ASTM F2924. Finally, and most importantly, was the improvement in fatigue life, whereby the fatigue stress for failure after 10^7 cycles was above 550 MPa. Our main claim here, therefore, is that HIP not only reduces the size and the distribution of flaws, but also significantly improves elongation and fatigue limit. Mechanical property variation decreased significantly as well, and no dependency in location could be noted after HIP treatment.

Author Contributions: E.T. and O.Y. made equal contributions of scientific supervision and publication of this work. D.Z. conducted the fractography, M.G. and O.T. contributed to designing of experiments and to the discussion, A.G. conducted the mechanical tests, A.H. conducted the HIP, M.C. designed the AM tray, Y.G. and D.N. conducted the density measurements, and I.H. conducted the EDS experiments.

Funding: This work was supported by Research Grant 2022851 from the Israel Ministry of Science, Technology and Space.

Acknowledgments: The authors wish to thank Max Shrager for his assistance in printing this tray and for fruitful discussions.

Conflicts of Interest: The authors certify that they have no commercial or associative interest that represents a conflict of interest in connection with the manuscript.

References

1. Ian, G.; David, R.; Brent, S. *Additive Manufacturing Technologies*; Springer: New York, NY, USA, 2014; ISBN 978-1-4419-1120-9.
2. Rosenthal, I.; Tiferet, E.; Ganor, M.; Stern, A. Post-processing of am-slm alsi10mg specimens: Mechanical properties and fracture behaviour. *Ann. Dunarea de Jos Univ. Galati Fascicle XII Weld. Equip. Technol.* **2015**, *26*, 33–38.
3. DebRoy, T.; Wei, H.L.; Zuback, J.S.; Mukherjee, T.; Elmer, J.W.; Milewski, J.O.; Beese, A.M.; Wilson-Heid, A.; De, A.; Zhang, W. Additive manufacturing of metallic components—Process, structure and properties. *Prog. Mater. Sci.* **2018**, *92*, 112–224. [CrossRef]
4. Murr, L.E.; Martinez, E.; Amato, K.N.; Gaytan, S.M.; Hernandez, J.; Ramirez, D.A.; Shindo, P.W.; Medina, F.; Wicker, R.B. Metal fabrication by additive manufacturing using laser and electron beam melting technologies. *J. Mater. Sci. Technol.* **2012**, *28*, 1–14. [CrossRef]
5. Siddique, S.; Imran, M.; Rauer, M.; Kaloudis, M.; Wycisk, E.; Emmelmann, C.; Walther, F. Computed tomography for characterization of fatigue performance of selective laser melted parts. *Mater. Des.* **2015**, *83*, 661–669. [CrossRef]
6. Li, Y.; Gu, D. Parametric analysis of thermal behavior during selective laser melting additive manufacturing of aluminum alloy powder. *Mater. Des.* **2014**, *63*, 856–867. [CrossRef]
7. Atkinson, H.V.; Davies, S. Fundamental aspects of hot isostatic pressing: An overview. *Metall. Mater. Trans. A* **2000**, *31*, 2981–3000. [CrossRef]
8. Steffens, H.-D.; Lebkuchner-Neugebauer, J.; Wielage, B. Hot Isostatic Pressing. *Mat.-Wiss. u. Werkstofftech* **1990**, *21*, 28–31. [CrossRef]
9. Harbe, N.; Gnäupel-Herold, T.; Quinn, T. Fatigue properties of a titanium alloy (Ti–6Al–4V) fabricated via electron beam melting (EBM): Effects of internal defects and residual stress. *Int. J. Fatigue* **2017**, *94*, 202–210.
10. Qian, M.; Xu, W.; Brandt, M.; Tang, H. Additive manufacturing and postprocessing of Ti-6Al-4V for superior mechanical properties. *MRS Bull.* **2016**, *41*, 775–784. [CrossRef]
11. LU, S.L.; TANG, H.P.; NING, Y.P.; LIU, N.; STJOHN, D.H.; QIAN, M. Microstructure and Mechanical Properties of Long Ti6Al-4V Rods Additively Manufactured by Selective Electron Beam Melting Out of a Deep Powder Bed and the Effect of Subsequent Hot Isostatic Pressing. *Metall. Mat. Trans. A* **2015**, *46*, 3824. [CrossRef]
12. Günther, J.; Krewerth, D.; Lippmann, T.; Leuders, S.; Tröster, T.; Weidner, A.; Biermann, H.; Niendorf, T. Fatigue life of additively manufactured Ti–6Al–4V in the very high cycle fatigue regime. *Int. J. Fatigue* **2017**, *94*, 236–245. [CrossRef]
13. Olakanmi, E.O.; Cochrane, R.F.; Dalgarno, K.W. Densification mechanism and microstructural evolution in selective laser sintering of Al-12Si powders. *J. Mater. Process. Technol.* **2011**, *211*, 113–121. [CrossRef]
14. Jia, Q.B.; Gu, D.D. Selective laser melting additive manufacturing of Inconel 718 superalloy parts: Densification, microstructure and properties. *J. Alloys Compd.* **2014**, *585*, 713–721. [CrossRef]
15. Greitemeier, D.; Palm, F.; Syassen, F.; Melz, T. Fatigue performance of additive manufactured TiAl6V4 using electron and laser beam melting. *Int. J. Fatigue* **2017**, *94*, 211–217. [CrossRef]
16. Khan, A.S.; Kazmi, R.; Farrokh, B.; Zupan, M. Effect of oxygen content and microstructure on the thermo-mechanical response of three Ti–6Al–4V alloys: Experiments and modeling over a wide range of strain-rates and temperatures. *Int. J. Plast.* **2007**, *23*, 1105–1125. [CrossRef]
17. Oh, J.M.; Lee, B.G.; Cho, S.W.; Lee, S.W.; Choi, G.S.; Lim, J.W. Oxygen effects on the mechanical properties and lattice strain of Ti and Ti-6Al-4V. *Met. Mater. Int.* **2011**, *17*, 733–736. [CrossRef]
18. Nicolas, J.; Rousseau, A.; Brochu, B.; Blais, C. Effect of oxygen content in new and reused powder on microstructural and mechanical properties of Ti6Al4V parts produced by directed energy deposition. *Addit. Manuf.* **2018**, *23*, 197–205.
19. Barre, C. Hot Isostatic Pressing. *Adv. Mater. Process.* **1999**, *155*, 47–48.
20. Arzt, E.; Ashby, M.F.; Easterling, K.E. Practical applications of hotisostatic pressing diagrams: Four case studies. *Met. Trans. A* **1983**, *14*, 211–221. [CrossRef]
21. Tillmanna, W.; Schaak, C.; Nellesen, J.; Schaper, M.; Aydinöz, M.E.; Hoyer, K.-P. Hot isostatic pressing of IN718 components manufactured by selective laser melting. *Addit. Manuf.* **2017**, *13*, 93–102. [CrossRef]

22. Pesach, A.; Tiferet, E.; Vogel, S.; Rivin, O.; Yeheskel, O.; Caspi, E.N. Microstructure Study of Ti-6Al-4V Additively Manufactured using Neutron Diffraction. *Addit. Manuf.* **2018**. submitted. [CrossRef]

23. Tiferet, E.; Rivin, O.; Ganor, M.; Ettedgui, H.; Ozeri, O.; Caspi, E.N.; Yeheskel, O. Structural investigation of selective laser melting and electron beam melting of T6Al-4V using neutron diffraction. *Addit. Manuf.* **2016**, *10*, 43–46. [CrossRef]

24. Tang, H.P.; Qian, M.; Liu, N.; Zhang, X.Z.; Yang, G.Y.; Wang, J. Effect of Powder Reuse Times on Additive Manufacturing of Ti-6Al-4V by Selective Electron Beam Melting. *J. Met.* **2015**, *67*, 555–563. [CrossRef]

25. Cordero, Z.C.; Meyer, H.M.; Nandwana, P.; Dehoff, R.R. Powder bed charging during electron-beam additive manufacturing. *Acta Mater.* **2017**, *124*, 437–445. [CrossRef]

26. *ASTM F2924-14, Standard Specification for Additive Manufacturing Titanium-6 Aluminum-4 Vanadium with Powder Bed Fusion*; ASTM International: West Conshohocken, PA, USA, 2014.

27. Yeheskel, O.; Shokhat, M.; Salhov, S.; Tevet, O. Effect of initial particle and agglomerate size elastic moduli of porous yttria (Y2O3). *J. Am. Ceram. Soc.* **2009**, *93*, 2028–2034. [CrossRef]

28. Sol, T.; Hayun, S.; Noiman, D.; Tiferet, E.; Yeheskel, O.; Tevet, O. Nondestructive ultrasonic evaluation of additively manufactured AlSi10Mg samples. *Addit. Manuf.* **2018**, *22*, 700–707. [CrossRef]

29. *ASTM E8/E8M-15a, Standard Test Methods for Tension Testing of Metallic Materials*; ASTM International: West Conshohocken, PA, USA, 2015.

30. Yeheskel, O.; Tevet, O. Elastic moduli of transparent yttria. *J. Am. Ceram. Soc.* **1999**, *82*, 136–144. [CrossRef]

31. Vogel, S.C.; Takajo, S.; Kumar, M.A.; Caspi, E.N.; Pesach, A.; Tiferet, E.; Yeheskel, O. Ambient and High Temperature Bulk Characterization of Additively Manufactured Ti-6Al-4V Using Neutron Diffraction. *JOM* **2018**, *70*, 1714–1722. [CrossRef]

32. Draper, S.L.; Lerch, B.A.; Telesman, J.; Martin, R.E.; Locci, I.E.; Ring, A.G.A.J. Materials Characterization of Electron Beam Melted Ti-6Al-4V. U.S. Patent NASA/TM—2016-219136, 1 September 2016.

Article

Superior Wear Resistance in EBM-Processed TC4 Alloy Compared with SLM and Forged Samples

Weiwen Zhang [1,2], **Peiting Qin** [1,2], **Zhi Wang** [1,2,*], **Chao Yang** [1,2], **Lauri Kollo** [3], **Dariusz Grzesiak** [4] **and Konda Gokuldoss Prashanth** [3,5]

[1]   Guangdong Key Laboratory for Processing and Forming of Advanced Metallic Materials, South China University of Technology, Guangzhou 510640, China; mewzhang@scut.edu.cn (W.Z.); 201720101729@mail.scut.edu.cn (P.Q.); cyang@scut.edu.cn (C.Y.)

[2]   National Engineering Research Center of Near-net-shape Forming for Metallic Materials, South China University of Technology, Guangzhou 510640, China

[3]   Department of Mechanical and Industrial Engineering, Tallinn University of Technology, Ehitajete tee 5, 19086 Tallinn, Estonia; lauri.kollo@taltech.ee (L.K.); prashanth.konda@taltech.ee (K.G.P.)

[4]   Department of Mechanical Engineering and Mechatronics, West Pomeranian University of Technology, Aleja Piastów 17, 70-310 Szczecin, Poland; dariusz.grzesiak@zut.edu.pl

[5]   Erich Schmid Institute of Materials Science, Austrian Academy of Sciences, Jahnstraße 12, A-8700 Leoben, Austria

*   Correspondence: Wangzhi@scut.edu.cn

Received: 12 January 2019; Accepted: 4 March 2019; Published: 7 March 2019

Abstract: The wear properties of Ti-6Al-4V alloy have drawn great attention in both aerospace and biomedical fields. The present study examines the wear properties of Ti-6Al-4V alloy as prepared by selective laser melting (SLM), electron beam melting (EBM) and conventional forging processes. The SLM and EBM samples show better wear resistance than the forged sample, which correlates to their higher hardness values and weak delamination tendencies. The EBM sample shows a lower wear rate than the SLM sample because of the formation of multiple horizontal cracks in the SLM sample, which results in heavier delamination. The results suggest that additive manufacturing processes offer significantly wear-resistant Ti-6Al-4V specimens in comparison to their counterparts produced by forging.

Keywords: Ti-6Al-4V; wear; additive manufacturing; properties

---

1. Introduction

Additive manufacturing (AM), commonly known as 3D printing, is a process of joining materials to make objects from 3D computer aided design (CAD) data, usually layer upon layer, as opposed to subtractive manufacturing methodologies [1–3]. Powder bed fusion processes, like laser-based powder bed fusion (selective laser melting (SLM)), electron beam-based powder bed fusion processes or electron beam melting (EBM), have gained increasing attention for the fabrication of metallic components [4,5]. EBM uses a high-energy electron beam to selectively melt a conductive metal powder bed directed by a CAD model under a high vacuum. EBM is capable of producing fully dense, near-net-shape complex parts with exceptional mechanical properties [6,7]. SLM, which emerged in the late 1980s and 1990s, uses a laser beam during the fabrication process for the selective melting of metallic powders [8–10]. Both SLM and EBM offer the flexibility to produce parts of any shape (theoretically) without restrictions [11–13]. SLM can process a wider spectrum of alloys, whereas EBM can process only limited alloys (e.g., Ti-6Al-4V, pure-Ti, CoCrMo and Ni-based alloys). Some other differences exist between the SLM and EBM processes. The difference in the heat source used means that the focus spot size is different—typically ~80 μm in diameter for SLM and ~100 μm in diameter for EBM [9]. Furthermore, typical particle sizes used are 10–60 μm for SLM and 60–105 μm for EBM processes, respectively [9]. The differences in spot size of the heat source and particle size can lead to differences in the size and shape of the melt pool, which subsequently affect the solidification behavior. Even though SLM and EBM share many common features, the minor differences between these two processes could lead to significant variations in the microstructure and, in turn, their mechanical properties [4,6,9]. Ti-6Al-4V is the most prevalent Ti-based alloy and one of the most important engineering materials. Due to its high strength-to-weight ratio, good biocompatibility and outstanding corrosion resistance, Ti-6Al-4V has been widely used in aerospace, biomedical implants, marine and offshore, etc. [13–16]. With the development of AM, Ti-6Al-4V alloy has been fabricated using both EBM and SLM techniques and has been extensively studied. Compared to the Ti-6Al-4V alloy produced by traditional forging, the SLM/EBM-fabricated materials show a unique microstructure and properties [17–21]. The wear property of Ti-based alloys is one of the most important properties to be considered for particular applications [22–26]. The wear properties are not, however, extensively studied and compared. It is important to know the variation in wear properties of the Ti-6Al-4V alloy produced by both SLM and EBM processes, which means differences in their microstructure. Hence, this work aims to investigate the wear properties of Ti-6Al-4V alloy produced by forging, SLM and EBM. The differences in the wear properties are studied and a possible mechanism for the correlating different microstructures is explained.

2. Experimental Details

Conventional forged samples of Ti-6Al-4V were used to compare the properties with the SLM and EBM samples. The condition of the forged sample is designated as forging a rod with a diameter of 520 mm, hot rolled to 200 mm × 200 mm long squares, subsequently forged to 60 mm × 60 mm long squares, and finally rolled into 14 straight strips. The rolled samples were then annealed at a temperature of 700~800 °C for 1~3 h, and cooled in the air so as the obtain the best possible properties out of the Ti-6Al-4V alloy. The SLM samples were fabricated using a Realizer 50 device with the standard parameter set (laser power—25 W, spot size—~100 μm, laser scan speed—200 mm/s and layer thickness—50 μm). The samples were processed in an argon atmosphere. The powder used for the SLM samples was supplied by Realizer 50 (Realizer, Borchen, North Rhine-Westphalia, Germany) with an average diameter of 30 μm. An Arcam A2X device (ARCAM, Mölndal, Västra Götaland, Sweden) was used to fabricate the sample using the EBM process. Conventional EBM process parameters included: vacuum—10^{-4}–10^{-5} (mbar), accelerating voltage—60 kV, layer thickness—50 μm, scan speed—0.50 m/s and the process chamber was maintained at 973 K (the powder bed was pre-heated to 973 K before melting each layer). A core shell scan strategy (with 0.5 mm contour) was used with the melting sequence varying between 0° and 90° between layers [10] for both SLM and EBM samples. The powder used for the EBM samples was supplied by Arcam with an average diameter of 65 μm.

Friction and wear tests were performed on a friction wear testing machine (MM-2000) produced by Jinan Sida Test Technology Company Limited in Jinan, China. The test configuration is shown in Figure 1 below. The wear tests were performed using a ring made of GCr15 alloy with a diameter of 47.15 mm and a thickness of 10 mm, under an applied load of 50 N for a distance of 592.5 m in an air environment with relative humidity ranging from 50% to 60% and at ambient temperature (about 25 °C). Four friction and wear tests were performed for each material type. The GCr15 alloy refers to a high carbon chromium-bearing steel, which has a chemical composition (unit wt. %) as follows: C 0.95–1.05, Mn 0.20–0.40, Si 0.15–0.35, S ≤ 0.020, P ≤ 0.027, Cr 1.30–1.65, Mo ≤ 0.10, Ni ≤ 0.30, Cu ≤ 0.25, Ni + Cu ≤ 0.50, the balance is Fe. This alloy has a tensile strength of 861.3 MPa, yield strength of 518.4 MPa, elongation of 27.95%, bending strength of 1821.6 MPa and hardness of ~630 Hv. The wear test samples were cut to 8 mm × 8 mm × 8 mm by wire cutting, and then they were polished down to 1.5 μm using SiC abrasive paper (wet) and ultrasonically cleaned with ethanol. The sample surface was flat. The samples were weighed to an accuracy of 0.0001 g. The tribological test direction was vertical to the building and forging direction. A new wheel was used for each test. Before and after the test, the samples were cleaned ultrasonically in an alcohol bath for 3 min and dried by air blower. A computerized system was used to record the test parameters such as frequency, load, duration, friction coefficient and speed. The frequency is the number of rotations in one minute of the wheel rotating. The wear rate (WR) was evaluated by the following equation [27,28]:

$$WR = V_S/L_S \tag{1}$$

where L_s is the sliding distance and V_s is the sliding volume loss. The volume loss was calculated from the wear loss determined by measuring the weight of samples before and after tests. The sliding distance is given by $L_S = \pi d_s v_s t_s$, where d_s is the diameter of the ring, v_s is speed (400 rpm), and t_s is the time (10 min) [27,28]. The wear scar was measured according to the depth and width of the longitudinal midsplit face. The microstructure and wear tracks were studied by optical microscopy (OM) using DMI5000 device as well as scanning electron microscopy (SEM) using NOVA NANSEM 430 equipped with an energy dispersive spectroscopy (EDS) analysis. Phase analysis was done by X-ray diffraction (XRD) using a Bruker D8 ADVANCE X-ray diffractometer fitted with Cu-kα radiation (step size—0.01), which was performed on the surfaces vertical to the forging or building directions. The Vickers microhardness was performed using a Vickers microhardness tester from China with a load of 100 N and a dwelling time of 15 s.

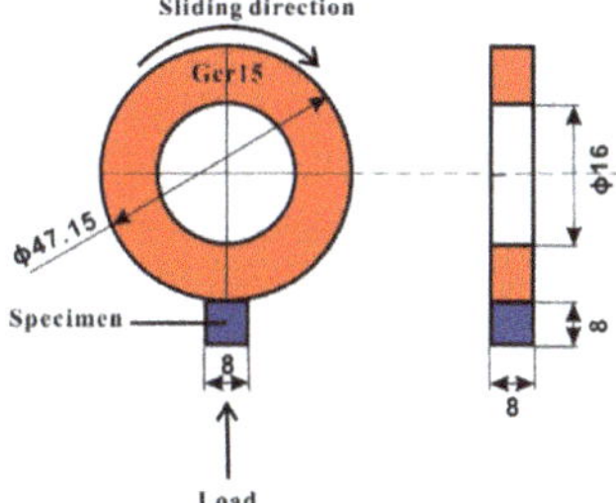

Figure 1. A schematic showing the wear test configuration including the position of the specimen, sliding direction and dimensions of the wheel.

3. Results

3.1. Microstructural Observation

The structural characterization of the Ti-6Al-4V forged, SLM and EBM samples is shown in Figure 2. The XRD pattern shows the presence of a hexagonally close-packed (hcp) phase in all the samples produced by forging, SLM and EBM processes. Here, diffraction peaks corresponding to hcp Ti (JCPDS Card No. 89-3725) were used for comparison, which has lattice parameters a = 0.294 nm and c = 0.467 nm, and shows the standard diffraction peaks for hcp Ti (α phase), located at 2θ = 38.446° and 2θ = 40.177°. The XRD diffraction patterns of the SLM sample indicate the presence of a hexagonal phase with diffraction peaks at 2θ = 38.581° and 2θ = 40.478°, which has lattice parameters a = 0.292 nm and c = 0.467 nm. These values correspond more to the lattice parameters for the α′ phase, i.e., a = 0.2931 nm and c = 0.4681 nm reported in the Materials Properties Handbook by Boyer and Collings (1994) [29]. On the other hand, the EBM sample shows a hexagonal phase with lattice parameters a = 0.299 nm and c = 0.476 nm, indicating the presence of a supersaturation α phase. The matensitic α′ phase in the SLM sample and α phase in the EBM sample are confirmed by the SEM micrograph, where the EBM sample shows a singular shape phase and the SLM sample shows a finer acicular shape phase, which are identified as the α phase and α′ martensitic phase, respectively. These findings are in accordance with the reported work [5,21,25,30,31], revealing that the α′ martensitic phase is typical for SLM-processed Ti-6Al-4V samples. The SLM process is said to offer a very high cooling rate in the order of 10^5–10^6 °C/sec [32–34]. Such high cooling rates are not, however, observed during conventional forging/EBM processes, hence the difference in the microstructure.

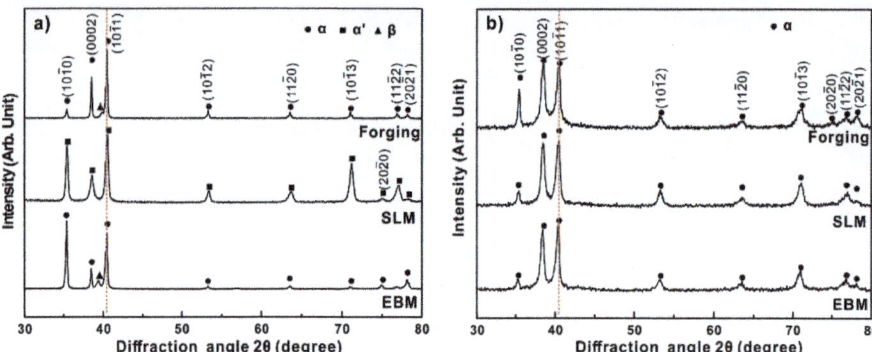

Figure 2. X-ray diffraction (XRD) analysis of Ti-6Al-4V samples produced by forging, selective laser melting (SLM) and electron beam melting (EBM) processes: (**a**) before sliding wear tests, (**b**) the tribo-layer surface after sliding wear tests.

Furthermore, the XRD patterns seen in Figure 2 reveal that the SLM sample shows high intensities at (10-10) and (10-11) planes, indicating the presence of texture of (10-10) and (10-11) planes in the SLM samples. The EBM sample shows high intensities at the (10-10) plane, indicating the presence of texture of the (10-10) plane. In contrast, the forging sample shows a relatively random distribution of planes. XRD patterns of both forged and EBM samples display peaks of α and β phases. However, the peaks of the β phase are not observed in the SLM samples, suggesting the absence of a β phase or, alternatively, the possible presence of β phase in low concentrations (<5%), which makes it difficult to deduct by XRD [28].

It is worth noting that some XRD peaks of the SLM and EBM samples are slightly shifted from the originally expected 2θ positions, and broadened compared to the forged sample. Among the three XRD patterns, the SLM pattern shows the widest peaks, which may be due to the presence of nano-crystalline phases obtained from high cooling rates during solidification or due to the presence of a high degree of internal stresses [10].

The OM and SEM images of Ti-6Al-4V alloy manufactured by forging, SLM and EBM processes are shown in Figure 3. Both the SLM and EBM samples have long columnar prior β grains growing along the building direction, but it shows more equiaxed grain in the forged sample, as seen in Figure 3a,b,e,f,i,j. Due to the different solidification conditions, the SLM sample showed acicular α' martensite filling the columnar prior β grains, while the EBM sample showed singular α phase filling the columnar prior β grains, which is in accordance with the published work [35–37]. Black lines are clearly visible inside the columnar prior β grains (as seen in Figure 3e,i), which were identified as the interface between the acicular α' martensitic and singular α phases for the SLM and EBM samples, respectively. The average width of the columnar prior β grains was ~106 ± 11 μm and ~162 ± 25 μm for the SLM and EBM samples, respectively, indicating that the SLM sample had smaller columnar grains than the EBM sample. The width of the acicular α' martensite in the SLM sample was found to be 1.1 ± 0.4 μm and the singular α bulge in the EBM sample was found to be 1.4 ± 0.3 μm. Meanwhile, the forged sample exhibited equiaxed α grains with irregular β phase distributed in the α grains or at the boundaries.

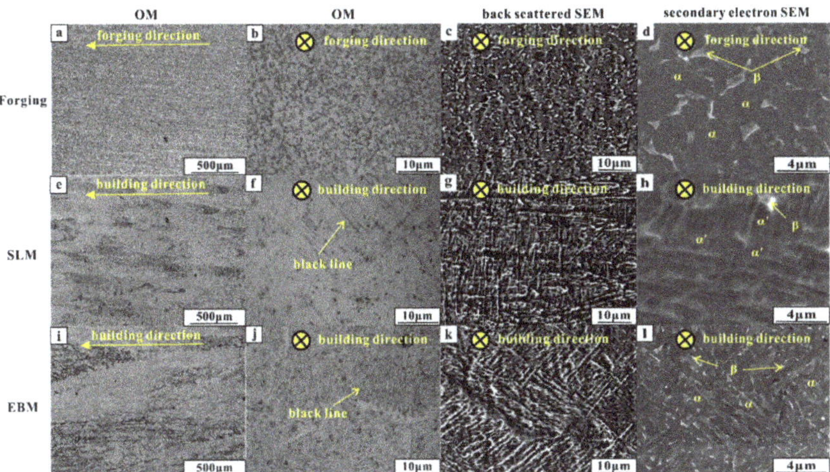

Figure 3. OM, back scattered SEM and secondary electron SEM images of samples produced by (**a–d**) forging; (**e–h**) SLM; (**i–l**) EBM.

3.2. Friction and Wear Properties

The XRD analysis of the tribo-layer is shown in Figure 2b. It shows that the diffraction peaks correspond more to the hcp α phase rather than the martensite α' phase, indicating that the martensite phase underwent transformation to the α phase during the test. Meanwhile, it reveals that the texture of (10-10) and (10-11) in the SLM sample and the texture of the (10-10) plane in the EBM sample disappeared or significantly weakened after testing, while the peak of (0002) plane for all the samples became much stronger, showing that the texture of the (0002) plane occurred during tribological testing. In addition, all the peaks became wider, indicating added residual strain in these samples.

The coefficient of friction (COF) as a function of sliding distance for Ti-6Al-4V samples is shown in Figure 4. The SLM sample showed the lowest COF, which is 0.4 ± 0.2, the EBM sample had a COF of 0.5 ± 0.1 and the forged sample had the highest COF of 0.6 ± 0.3. The higher the COF, the worse the wear of the property will be. Similarly, in the present case, the forged sample exhibited the highest COF and, hence, the worst wear property amongst the forged, SLM and EBM samples. The COF was determined in relation to the alloying element and composition of the alloy. This caused a difference in the microstructure, hardness and strength in the alloy, thereby affecting its COF [30,38,39].

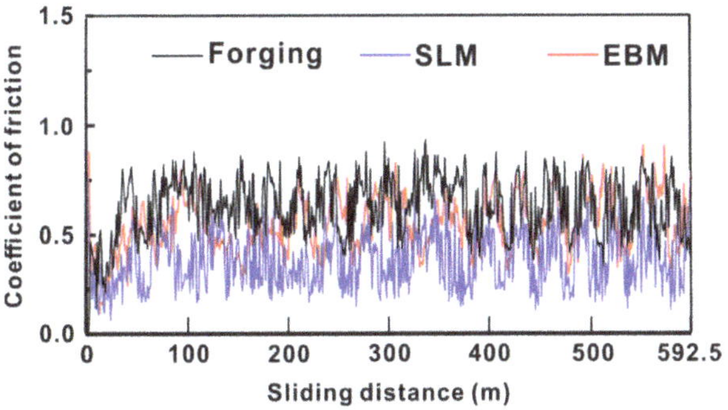

Figure 4. Coefficient of friction values for Ti-6Al-4V samples produced by forging, SLM and EBM.

The hardness, weight loss and wear rate for the forging, SLM- and EBM-fabricated Ti-6Al-4V samples is shown in Figure 5. Microstructural differences lead to different hardness levels (as seen in

Figure 5a) in these samples, where the highest hardness of 399 $\pm$ 14 HV was observed for the SLM sample as compared to the EBM (383 $\pm$ 13 HV) and forged (368 $\pm$ 12 HV) samples. SEM micrographs of the samples after Vickers indentation are shown in Figure 6, which shows that the smallest indentation size obtained in the SLM sample indicated the highest hardness compared to the other samples by forging and EBM processes. Finer microstructure (fine needle martensite) of samples produced by additive manufacturing had higher hardness than the forged sample. As expected from the hardness plot, the forged sample showed the highest wear rate ($23.9 \pm 4.6 \times 10^{-5}$ mm^3 N^{-1} m^{-1}). The EBM sample, however, showed the lowest wear rate ($16.6 \pm 4.2 \times 10^{-5}$ mm^3 N^{-1} m^{-1}) as compared to the SLM and the forged sample. The SLM sample showed a wear rate of $19.0 \pm 3.7 \times 10^{-5}$ mm^3 N^{-1} m^{-1}, proving that the EBM sample showed the highest wear resistance among the three considered samples.

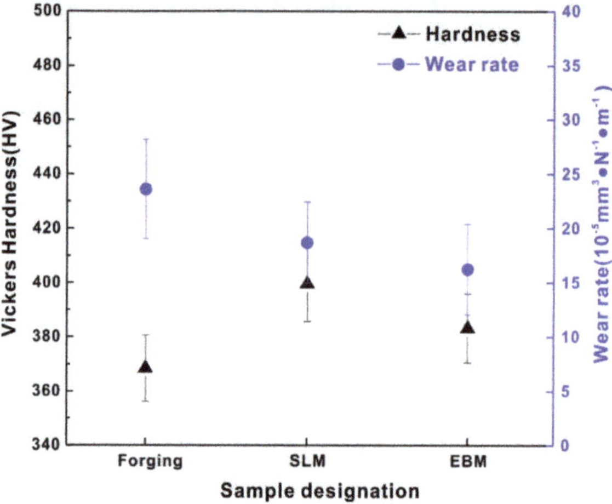

Figure 5. Hardness and wear rate of Ti-6Al-4V samples fabricated by forging, SLM and EBM.

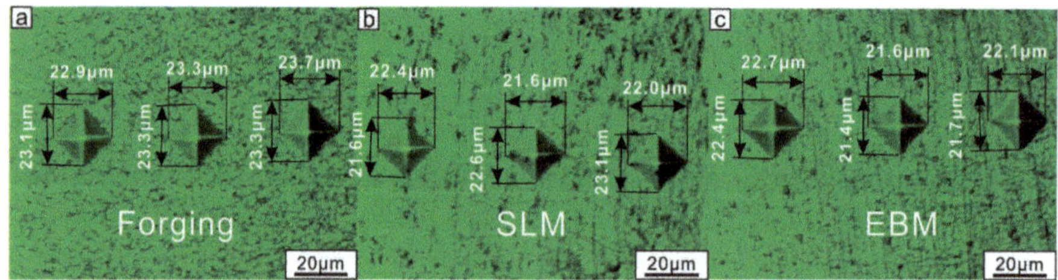

Figure 6. SEM micrographs of the samples after Vickers indentation: (**a**) forged, (**b**) SLM and (**c**) EBM.

SEM images of the worn surfaces can be seen in Figure 7. All of the Ti-6Al-4V samples fabricated by forging, SLM and EBM processes showed the presence of typical wear scars and ploughing grooves along the sliding direction. The wear scars observed on the forged samples (depth of 0.24 $\pm$ 0.01 mm) were deeper than the SLM (depth of 0.21 $\pm$ 0.03 mm) and EBM (depth of 0.17 $\pm$ 0.02 mm) samples. The EBM sample showed the shallowest wear scar. Hard particles, such as metallic debris and oxide particles, formed and were reduced to a fine debris under continuous sliding, which can lead to ploughing grooves as observed in the tribo-layer (Figure 7a,d,g). The wear test with high strain raised the temperature of the surface in contact with the counter disc. Subsequently, the debris was compacted on the worn surfaces to generate the tribo-layer, as seen in Figures 8 and 9. The EDX results are shown in Table 1.

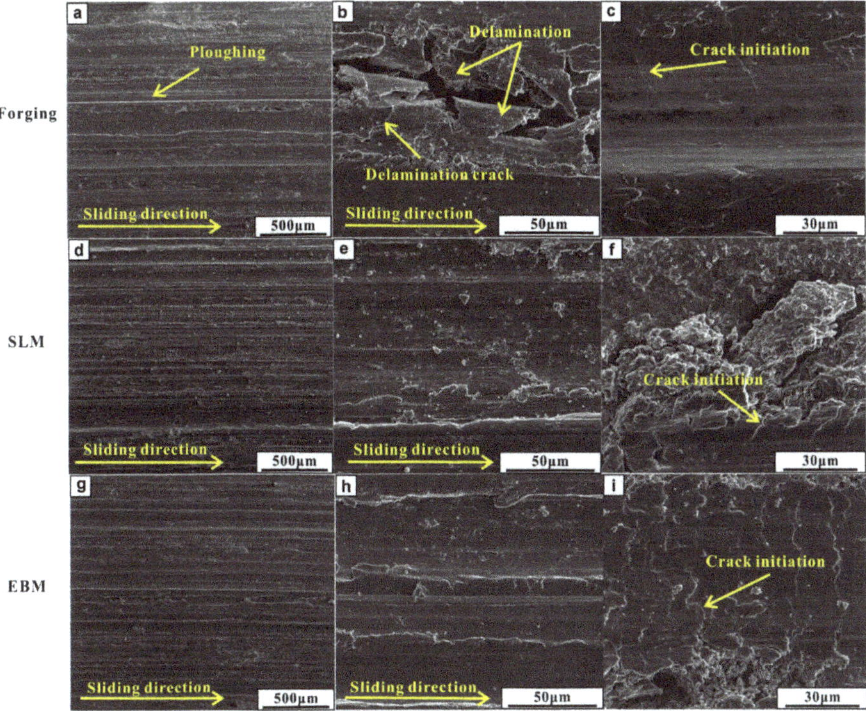

Figure 7. SEM images of wear tracks after the sliding wear tests for the Ti-6Al-4V samples: (**a–c**) forged, (**d–f**) SLM and (**g–i**) EBM.

Table 1. Element analysis (wt. %) of the tribo-layer and matrix of the samples produced by forging, SLM and EBM processes.

Process	Position	C (%)	O (%)	Al (%)	Ti (%)	V (%)	Fe (%)
Forging	tribo-layer	6.10 ± 3.85	5.73 ± 4.90	2.24 ± 0.85	77.39 ± 7.00	3.45 ± 0.28	4.86 ± 1.92
	matrix	1.30 ± 0.40	−2.26 ± 1.17	5.96 ± 0.19	91.14 ± 1.24	3.72 ± 0.27	0.19 ± 0.10
SLM	tribo-layer	5.91 ± 1.49	10.43 ± 0.99	3.40 ± 0.69	72.37 ± 2.62	2.97 ± 0.40	4.93 ± 0.11
	matrix	0.90 ± 0.31	−1.63 ± 0.40	5.25 ± 0.36	91.90 ± 0.75	3.49 ± 0.13	0.06 ± 0.09
EBM	tribo-layer	2.47 ± 2.14	12.37 ± 5.10	3.29 ± 0.34	75.08 ± 6.80	3.02 ± 0.37	3.77 ± 0.46
	matrix	1.27 ± 0.35	−1.68 ± 1.27	5.60 ± 0.24	91.51 ± 0.96	3.18 ± 0.43	0.07 ± 0.01

High oxygen content was observed in the tribo-layer, suggesting that the oxidation occurred during the sliding test. Meanwhile, a high Fe content presented in the tribo-layer, which was mostly from the steel wheel during testing. The tribo-layer showed high carbon content as seen in Table 1. In addition, there were higher carbon content presences beneath the tribo-layer as seen in Figure 9. The high carbon content observed in and beneath the tribo-layer may have formed during the wear testing and/or may have come from the original sample surface that was exposed to the laboratory environment and handling. A much deeper and systematic study is, however, needed to address this issue. A continuous deep horizontal crack was observed at the bottom of the tribo-layer in the forged sample (as seen in Figure 8), resulting in high wear rates due to the delamination initiated by these sub-surface cracks. More horizontal cracks were also observed in the SLM sample but these were discontinuous in nature. On the other hand, the EBM sample showed few horizontal cracks but many small vertical cracks.

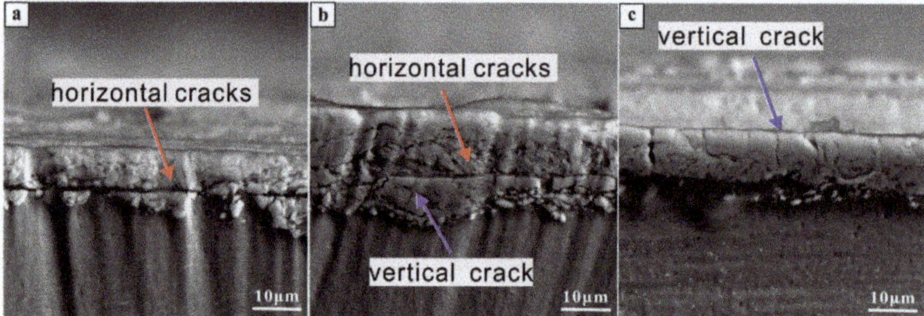

Figure 8. Cross-section comparison of Ti-6Al-4V alloy produced by (**a**) forging, (**b**) SLM and (**c**) EBM.

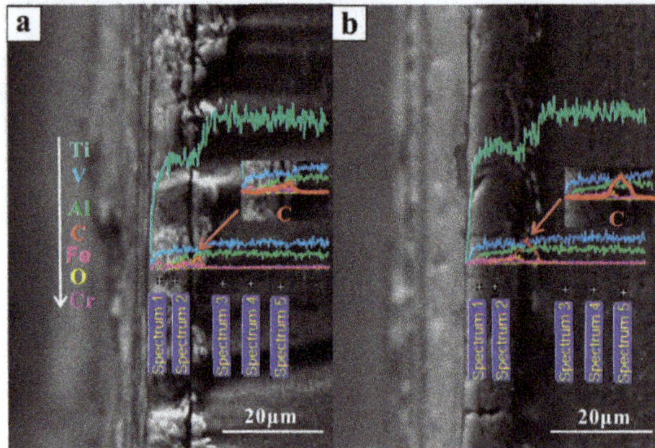

Figure 9. Energy dispersive spectroscopy (EDS) lines and points analysis of tribo-layer of the cross-sections of the (**a**) forged, (**b**) EBM samples.

4. Discussion

Based on the above results, it can be observed that the SLM and EBM Ti-6Al-4V samples showed predominantly similar wear mechanisms. An abrasive wear due to the loading from the GCr15 ring and tribo-layer was due to plastic deformation and oxidation. The SLM and EBM samples showed better wear resistance than the forged sample. This result can be correlated to the higher hardness and weaker delamination tendency in the sub-surface of the SLM and EBM samples during the wear tests. Moreover, the SLM and EBM samples exhibited higher hardness than the forged sample, which can be ascribed to the presence of fine microstructure and martensitic phase (as seen in Figure 3). Severe delamination of the tribo-layer was developed in the forged sample due to the formation of the primary crack during the cycling loading, leading to significant material removal (as seen in Figure 7b), which is in accordance with other reported works [40–43]. Unlike the forged sample, the SLM and EBM samples showed more uniformly distributed cracks, which can avoid the formation of primary cracks.

Although the EBM samples had lower hardness in comparison to the SLM samples, they showed a lower wear rate (as seen in Figure 5). This may be mainly ascribed to the formation of several horizontal cracks in the SLM sample resulting in the relatively severe delamination of the tribo-layer—in comparison to the EBM sample, which had more vertical cracks and only a few horizontal cracks. After the delamination of the tribo-layer, a new tribo-layer was formed at the nascent surface of the Ti-6Al-4V SLM sample during continuous loading, leading to relatively severe mass loss in the SLM sample in comparison to the EBM sample.

A schematic illustration of the different wear behavior in Ti-6Al-4V produced by the different processes is shown in Figure 10. The tribo-layer and the cracks were highlighted in the figures, which are the most important facts affecting the wear resistance of the forged, SLM and EBM samples. Most of the cracks propagated horizontally in the SLM samples, which could lead to delamination during the continuous wear loading. Meanwhile, most of the cracks propagated vertically in the EBM samples, giving them better resistance than the SLM samples. The main reason for the crack propagation behavior resulted from the different microstructure, in which a more brittle martensitic phase existed in the SLM sample.

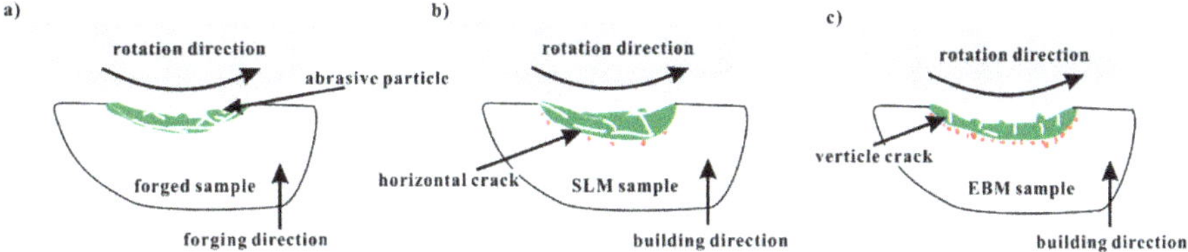

Figure 10. Schematic illustration of the wear mechanisms in Ti-6Al-4V samples produced by (**a**) forging, (**b**) SLM and (**c**) EBM processes.

5. Conclusions

In this work, Ti-6Al-4V samples were produced by forging, SLM and EBM processes, and their resultant microstructure, hardness and wear behaviors were compared. The microstructural characterization revealed that the forged sample had equiaxed α grains with an irregular β phase distributed in α grains or at their boundaries. The SLM sample showed the presence of an acicular α' martensite phase, while the EBM sample showed a singular α bulge phase distributed in the columnar prior β grains. The diameter of the columnar prior β grains was smaller in the SLM sample than in the EBM sample. The SLM sample showed the highest hardness owing to the fine microstructure and α' martensitic phase. The SLM and EBM samples showed better wear resistance than the forged sample, which correlate to their higher hardness and weaker delamination in the AM samples. Furthermore, the EBM sample showed a lower wear rate than the SLM sample owing to the fact that a number of horizontal cracks were formed in the SLM sample, resulting in the heavier delamination of the tribo-layer.

Author Contributions: Z.W. and K.G.P. conceived and designed the experiments; P.Q., L.K. and D.G. performed the experiments; W.Z., P.Q., Z.W., C.Y. and K.G.P. analyzed the data; P.Q. and Z.W. wrote the manuscript; Z.W., W.Z. and K.G.P. revised the manuscript.

Funding: This research was funded by National Natural Science Foundation of China (Grant No. 51701075), the Science and Technology Program of Guangzhou (Grant No. 201804010365), the Fundamental Research Funds for the Central Universities (Grant No. 2017ms009) and the Team Project of Natural Science Foundation of Guangdong Province (2015A030312003). The authors are grateful to the Estonian Research foundation for their support (Grant Nr.: MOBERC15).

Conflicts of Interest: The authors declare no conflict of interest.

References

1. Harun, W.S.W.; Kamariah, M.S.I.N.; Muhamad, N.; Ghani, S.A.C.; Ahmad, F.; Mohamed, Z. A review of powder additive manufacturing processes for metallic biomaterials. *Powder Technol.* **2018**, *327*, 128–151. [CrossRef]
2. Prashanth, K.G.; Scudino, S.; Klauss, H.J.; Surreddi, K.B.; Löber, L.; Wang, Z.; Chaubey, A.K.; Kühn, U.; Eckert, J. Microstructure and mechanical properties of Al–12Si produced by selective laser melting: Effect of heat treatment. *Mater. Sci. Eng. A* **2014**, *590*, 153–160. [CrossRef]
3. Prashanth, K.G.; Damodaram, R.; Maity, T.; Wang, P.; Eckert, J. Friction welding of selective laser melted Ti6Al4V parts. *Mater. Sci. Eng. A* **2017**, *704*, 66–71. [CrossRef]
4. Gokuldoss, P.K.; Kolla, S.; Eckert, J. Additive manufacturing processes: Selective laser melting, electron beam melting and binder jetting-selection guidelines. *Materials* **2017**, *10*, 672. [CrossRef] [PubMed]

5. Van Hooreweder, B.; Moens, D.; Boonen, R.; Kruth, J.P.; Sas, P. Analysis of fracture toughness and crack propagation of Ti6Al4V produced by selective laser melting. *Adv. Eng. Mater.* **2012**, *14*, 92–97. [CrossRef]

6. Frazier, W.E. Metal Additive Manufacturing: A Review. *J. Mater. Eng. Perform.* **2014**, *23*, 1917–1928. [CrossRef]

7. Sing, S.L.; An, J.; Yeong, W.Y.; Wiria, F.E. Laser and electron-beam powder-bed additive manufacturing of metallic implants: A review on processes, materials and designs. *J. Orthop. Res.* **2016**, *34*, 369–385. [CrossRef] [PubMed]

8. Scudino, S.; Unterdörfer, C.; Prashanth, K.G.; Attar, H.; Ellendt, N.; Uhlenwinkel, V.; Eckert, J. Additive manufacturing of Cu–10Sn bronze. *Mater. Lett.* **2015**, *156*, 202–204. [CrossRef]

9. Herzog, D.; Seyda, V.; Wycisk, E.; Emmelmann, C. Additive manufacturing of metals. *Acta Mater.* **2016**, *117*, 371–392. [CrossRef]

10. Prashanth, K.G.; Scudino, S.; Eckert, J. Defining the tensile properties of Al-12Si parts produced by selective laser melting. *Acta Mater.* **2017**, *126*, 25–35. [CrossRef]

11. Ataee, A.; Li, Y.; Brandt, M.; Wen, C. Ultrahigh-strength titanium gyroid scaffolds manufactured by selective laser melting (SLM) for bone implant applications. *Acta Mater.* **2018**, *158*, 354–368. [CrossRef]

12. Prashanth, K.; Löber, L.; Klauss, H.-J.; Kühn, U.; Eckert, J. Characterization of 316L Steel Cellular Dodecahedron Structures Produced by Selective Laser Melting. *Technologies* **2016**, *4*, 34. [CrossRef]

13. Attar, H.; Löber, L.; Funk, A.; Calin, M.; Zhang, L.C.; Prashanth, K.G.; Scudino, S.; Zhang, Y.S.; Eckert, J. Mechanical behavior of porous commercially pure Ti and Ti–TiB composite materials manufactured by selective laser melting. *Mater. Sci. Eng. A* **2015**, *625*, 350–356. [CrossRef]

14. Banerjee, D.; Williams, J.C. Perspectives on titanium science and technology. *Acta Mater.* **2013**, *61*, 844–879. [CrossRef]

15. Williams, J.C. Titanium alloys: Production, behavior and application. In *High Performance Materials in Aerospace*; Flower, H.M., Ed.; Springer: Dordrecht, The Netherlands, 1995; pp. 85–134. ISBN 978-94-011-0685-6.

16. Bruschi, S.; Bertolini, R.; Bordin, A.; Medea, F.; Ghiotti, A. Influence of the machining parameters and cooling strategies on the wear behavior of wrought and additive manufactured Ti6Al4V for biomedical applications. *Tribol. Int.* **2016**, *102*, 133–142. [CrossRef]

17. Koike, M.; Greer, P.; Owen, K.; Lilly, G.; Murr, L.E.; Gaytan, S.M.; Martinez, E.; Okabe, T. Evaluation of titanium alloys fabricated using rapid prototyping technologies-electron beam melting and laser beam melting. *Materials* **2011**, *4*, 1776–1792. [CrossRef] [PubMed]

18. Attar, H.; Bönisch, M.; Calin, M.; Zhang, L.C.; Scudino, S.; Eckert, J. Selective laser melting of in situ titanium-titanium boride composites: Processing, microstructure and mechanical properties. *Acta Mater.* **2014**, *76*, 13–22. [CrossRef]

19. Weißmann, V.; Drescher, P.; Bader, R.; Seitz, H.; Hansmann, H.; Laufer, N. Comparison of Single Ti6Al4V Struts Made Using Selective Laser Melting and Electron Beam Melting Subject to Part Orientation. *Metals* **2017**, *7*, 91. [CrossRef]

20. Zhao, X.; Li, S.; Zhang, M.; Liu, Y.; Sercombe, T.B.; Wang, S.; Hao, Y.; Yang, R.; Murr, L.E. Comparison of the microstructures and mechanical properties of Ti-6Al-4V fabricated by selective laser melting and electron beam melting. *Mater. Des.* **2016**, *95*, 21–31. [CrossRef]

21. Vastola, G.; Zhang, G.; Pei, Q.X.; Zhang, Y.W. Active Control of Microstructure in Powder-Bed Fusion Additive Manufacturing of Ti6Al4V. *Adv. Eng. Mater.* **2017**, *19*, 1–6. [CrossRef]

22. Attar, H.; Prashanth, K.G.; Chaubey, A.K.; Calin, M.; Zhang, L.C.; Scudino, S.; Eckert, J. Comparison of wear properties of commercially pure titanium prepared by selective laser melting and casting processes. *Mater. Lett.* **2015**, *142*, 38–41. [CrossRef]

23. Saravanan, I.; Perumal, A.E. Wear behavior of γ-irradiated Ti6Al4V alloy sliding on TiN deposited steel surface. *Tribol. Int.* **2016**, *93*, 451–463. [CrossRef]

24. Wang, Y.; Li, J.; Dang, C.; Wang, Y.; Zhu, Y. Influence of carbon contents on the structure and tribocorrosion properties of TiSiCN coatings on Ti6Al4V. *Tribol. Int.* **2017**, *109*, 285–296. [CrossRef]

25. Balla, V.K.; Soderlind, J.; Bose, S.; Bandyopadhyay, A. Microstructure, mechanical and wear properties of laser surface melted Ti6Al4V alloy. *J. Mech. Behav. Biomed. Mater.* **2014**, *32*, 335–344. [CrossRef] [PubMed]

26. Bartolomeu, F.; Buciumeanu, M.; Pinto, E.; Alves, N.; Silva, F.S.; Carvalho, O.; Miranda, G. Wear behavior of Ti6Al4V biomedical alloys processed by selective laser melting, hot pressing and conventional casting. *Trans. Nonferr. Met. Soc. China* **2017**, *27*, 829–838. [CrossRef]

27. Prashanth, K.G.; Debalina, B.; Wang, Z.; Gostin, P.F.; Gebert, A.; Calin, M.; Kühn, U.; Kamaraj, M.; Scudino, S.; Eckert, J. Tribological and corrosion properties of Al-12Si produced by selective laser melting. *J. Mater. Res.* **2014**, *29*, 2044–2054. [CrossRef]

28. Wang, Z.; Tan, J.; Sun, B.A.; Scudino, S.; Prashanth, K.G.; Zhang, W.W.; Li, Y.Y.; Eckert, J. Fabrication and mechanical properties of Al-based metal matrix composites reinforced with Mg65Cu20Zn5Y10 metallic glass particles. *Mater. Sci. Eng. A* **2014**, *600*, 53–58. [CrossRef]

29. Welsch, G.; Boyer, R.; Collings, E.W. *Materilas Properties Handbook: Titanium Alloy*; ASM International: Metals Park, OH, USA, 1994.

30. Tang, B.; Wu, P.Q.; Fan, A.L.; Qin, L.; Hu, H.J.; Celis, J.P. Improvement of corrosion-wear resistance of Ti-6Al-4V alloy by plasma Mo-N surface modification. *Adv. Eng. Mater.* **2005**, *7*, 232–238. [CrossRef]

31. Gu, D.; Hagedorn, Y.; Meiners, W.; Meng, G. Densification behavior, microstructure evolution, and wear performance of selective laser melting processed commercially pure titanium. *Acta Mater.* **2012**, *60*, 3849–3860. [CrossRef]

32. Jung, H.Y.; Choi, S.J.; Prashanth, K.G.; Stoica, M.; Scudino, S.; Yi, S.; Kühn, U.; Kim, D.H.; Kim, K.B.; Eckert, J. Fabrication of Fe-based bulk metallic glass by selective laser melting: A parameter study. *Mater. Des.* **2015**, *86*, 703–708. [CrossRef]

33. Prashanth, K.G.; Eckert, J. Formation of metastable cellular microstructures in selective laser melted alloys. *J. Alloy. Compd.* **2017**, *707*, 27–34. [CrossRef]

34. Hooper, P.A. Melt pool temperature and cooling rates in laser powder bed fusion. *Addit. Manuf.* **2018**, *22*, 548–559. [CrossRef]

35. Xu, W.; Brandt, M.; Sun, S.; Elambasseril, J.; Liu, Q.; Latham, K.; Xia, K.; Qian, M. Additive manufacturing of strong and ductile Ti-6Al-4V by selective laser melting via in situ martensite decomposition. *Acta Mater.* **2015**, *85*, 74–84. [CrossRef]

36. Xu, W.; Lui, E.W.; Pateras, A.; Qian, M.; Brandt, M. In situ tailoring microstructure in additively manufactured Ti-6Al-4V for superior mechanical performance. *Acta Mater.* **2017**, *125*, 390–400. [CrossRef]

37. Zhang, L.-C.; Liu, Y.; Li, S.; Hao, Y. Additive Manufacturing of Titanium Alloys by Electron Beam Melting: A Review. *Adv. Eng. Mater.* **2017**, *20*, 1700842. [CrossRef]

38. Chapala, P.; Sunil Kumar, P.; Joardar, J.; Bhandari, V.; Ghosh Acharyya, S. Effect of alloying elements on the microstructure, coefficient of friction, in-vitro corrosion and antibacterial nature of selected Ti-Nb alloys. *Appl. Surf. Sci.* **2019**, *469*, 617–623. [CrossRef]

39. Rodgers, G.W.; Herve, R.; MacRae, G.A.; Chanchi Golondrino, J.; Chase, J.G. Dynamic Friction Coefficient and Performance of Asymmetric Friction Connections. *Structures* **2018**, *14*, 416–423. [CrossRef]

40. Li, Y.; Kang, G.; Wang, C.; Dou, P.; Wang, J. Vertical short-crack behavior and its application in rolling contact fatigue. *Int. J. Fatigue* **2006**, *28*, 804–811. [CrossRef]

41. Under, S.; Tractive, M.; Contact, R. Vertical Short Crack Initiation in Medium Carbon Bainitic Steel Under Mild Tractive Rolling Contact. *J. Iron Steel Res. Int.* **2008**, *15*, 37–41.

42. Dou, P.; Suo, S.; Yang, Z.; Li, Y.; Chen, D. Ratcheting short crack behavior in medium carbon bainitic back-up roll steel under mild tractive rolling contact. *Wear* **2010**, *268*, 302–308. [CrossRef]

43. Rapoport, L.; Salganik, R.L.; Gotlib, V.A. A model of brittle-dominated wear due to delayed fracture (vertical crack growth). *Wear* **1995**, *188*, 166–174. [CrossRef]

 *materials* 

Article

Reducing Porosity and Refining Grains for Arc Additive Manufacturing Aluminum Alloy by Adjusting Arc Pulse Frequency and Current

Donghai Wang, Jiping Lu, Shuiyuan Tang *, Lu Yu, Hongli Fan, Lei Ji and Changmeng Liu

School of Mechanical Engineering, Beijing Institute of Technology, Beijing 100081, China;
wdh4043@163.com (D.W.); jipinglu@bit.edu.cn (J.L.); yulu2219@163.com (L.Y.); fanhongli2000@sina.com (H.F.);
jl12199@163.com (L.J.); liuchangmeng@bit.edu.cn (C.L.)
* Correspondence: shuiyuantang@bit.edu.cn; Tel.: +86-10-6891-1652

Received: 9 July 2018; Accepted: 27 July 2018; Published: 3 August 2018

Abstract: Coarse grains and gas pores are two main problems that limit the application of additive manufacturing aluminum alloys. To reduce porosity and refine grains, this paper presents a quantitative investigation into the effect of pulse frequency and arc current on the porosity and grains of arc additive manufacturing Al–5Si alloy. The experiment results show that pulse frequency and arc current have a significant impact on the macrostructure, microstructure, porosity, and tensile properties of the samples. Fine grains and a uniform microstructure can be obtained with low pulse frequency and low arc current as a result of the rapid cooling of the molten pool. With the increase of pulse frequency, density shows a trend that firstly escalates and attains the maximum value at 50 Hz, but later declines as a result of the relation between pores formation and gas escape. Moreover, better tensile properties can be obtained at low pulse frequency and low arc current because of the finer grains.

Keywords: arc additive manufacturing; Al–5Si alloy; pulse frequency; arc current; microstructure; porosity

1. Introduction

Over the last few decades, researchers have already made great strides in metal additive manufacturing technologies [1–3], which can be classified by the employed heat source, including laser additive manufacturing, electron beam additive manufacturing, and arc additive manufacturing. Williams et al. [4] pointed out that, compared with other additive manufacturing methods, arc additive manufacturing has distinct advantages: High manufacturing efficiency, low cost (machine and materials), and good structural integrity. The arc additive manufacturing process often employs wire as feedstock, compared with other powder-feed additive manufacturing process, it can produce large scale metallic parts with a much higher deposition rate [5]. Consequently, the research of arc additive manufacturing has become a hotspot. Arc additive manufacturing has been widely used to fabricate titanium alloy, stainless steel, and nickel alloy [6,7]. For aluminum alloys, which are widely used in the aerospace and automobile industry owing to their excellent performance, such as high strength-to-weight ratio, high formability, and high durability [8–10], their applications of additive manufacturing are still limited. Some recent studies have been devoted to the process parameters on the quality and properties of arc additive manufacturing aluminum alloy [11,12]. However, the main reason that limited the aluminum's application of additive manufacturing is that gas pores and coarse grains easily form during arc additive manufacturing, which largely reduce the mechanical properties of aluminum alloys [13–16]. Some researchers employed different cold metal transfer processes and pure argon flow rates in order to reduce the pores and obtain refine microstructure in arc additive manufacturing of Al–Cu alloy [17,18].

Among the alumina alloys, Al–Si alloys have been widely used in automotive and aerospace industries because of their excellent characteristics such as excellent cast ability and mechanical properties [19]. However, studies focused on the reducing porosity and improving properties of Al–Si alloys have rarely been reported. Thus, the investigation on reducing the gas pores and refining grains is meaningful, which allows a broader application of Al–Si alloys.

The formation of gas pores and coarse grains are intensely associated with the melting pool, which is influenced by the arc pulse frequency and arc current [20]. Hence, in this study, the effect of pulse frequency and arc current on the gas pores and grains has been systematically investigated. The mechanism of reducing porosity and refining grains was revealed by adjusting the pulse frequency and arc current. Meanwhile, the geometry, microstructures, and tensile properties under different pulse frequencies and arc currents will be investigated.

2. Experimental Details

Figure 1 schematically expressed the experimental setup for arc additive manufacturing. The system consists of a gas tungsten arc welding (GTAW) equipment (Miller Dynasty 350, Miller Electric Manufacturing Co., Appleton, WI, USA), a wire feeder (Jetline 9700 W, Miller Electric Manufacturing Co., Appleton, WI, USA), a computer, a three-dimensional workbench, and a working chamber.

Al–5Si alloy wire with a diameter of 1.2 mm was employed in this work (ER4043, chemical composition is shown in Table 1). The rolled pure aluminum substrate was mechanically cleaned and fixed on the workbench before the deposition process. Based on careful analysis of previous experiments about the GTAW of Al–5Si alloy, optimized fabrication parameters were determined (Table 2), then ten thin walls with 15 layers were fabricated, which were built layer by layer.

Table 1. Composition of the ER4043 (Al–5Si) filler wire.

Element	Si	Mn	Mg	Cu	Fe	Zn	Al
wt %	5	<0.05	<0.05	<0.05	<0.4	<0.1	Balance

Table 2. Deposition parameters for arc additive manufacturing in this study.

Deposition Parameters	Values I	Values II
Pulse frequency	2, 5, 10, 50, 200, 500 Hz	2 Hz
Peak current	100 A	100, 125, 150, 175 A
Peak time ratio	30%	30%
Base-to-peak current ratio	30%	30%
Wire feed speed	100 cm/min	100 cm/min
Deposition speed	100 mm/min	100 mm/min
Shield gas flow rate	20 L/min	20 L/min

The following analyses are performed to investigate the specimens manufactured under different pulse frequencies and currents in terms of macrostructure, microstructure, porosity, and tensile properties. The sketch of layer thickness and width is shown in Figure 2a. Generally speaking, there are two different definitions of the width of deposition alloy. TWW refers to the total wall width and EWW refers to effective wall width as shown in Figure 2a. In this study, TWW is selected as the width of deposited wall, layer thickness is referred to the average of 15 layers. For microstructural observations, test samples were taken from the middle of deposited walls, inlayed, and ground with silicon carbide papers, then the samples were polished by electro-polishing. After that, samples were etched using a modified Keller's reagent (except the specimens used for the porosity characterization), and observed using an optical microscope (OM, Leica DM4000M, Leica Microsystems Inc., Buffalo Grove, IL, USA). Then, the dendrite arm spacing, and the numbers and cross-section areas of gas pores were measured by 10 OM images with a magnification of 100 from five cross-sections for each deposited wall, and the statistical data were obtained by Image Pro Plus software (6.0, Media Cybernetics, Warrendale, PA, USA). Here, density is represented by the ratio of total area of gas pores to total area of 10 OM images. For tensile properties, three tested tensile specimens (parallel to the longitudinal direction of the deposited walls) were selected from each deposited wall and then tested by tensile test machine (Instron5966, Instron, Norwood, CO, USA) at room temperature (the cross-head speed was 0.01 mm/s, and a dynamic strain gauge extensometer was applied to record the strain). The average tensile data was adopted and the fractured samples were observed by scanning electron microscope (SEM, JSM-6610LV, JEOL, Tokyo, Japan). The dimensions of tensile specimens are shown in Figure 2c.

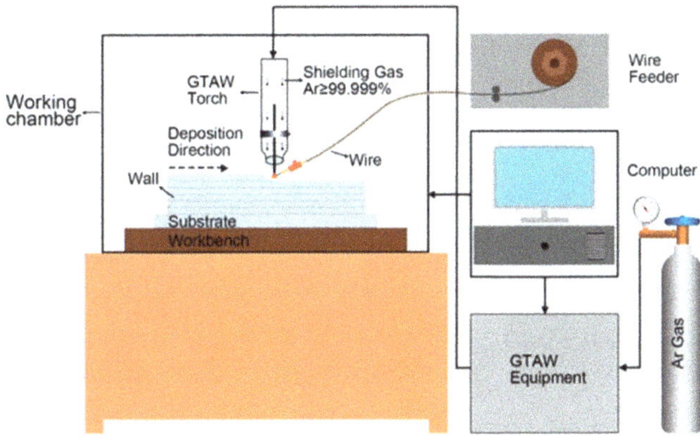

Figure 1. Experimental set-up for arc additive manufacturing.

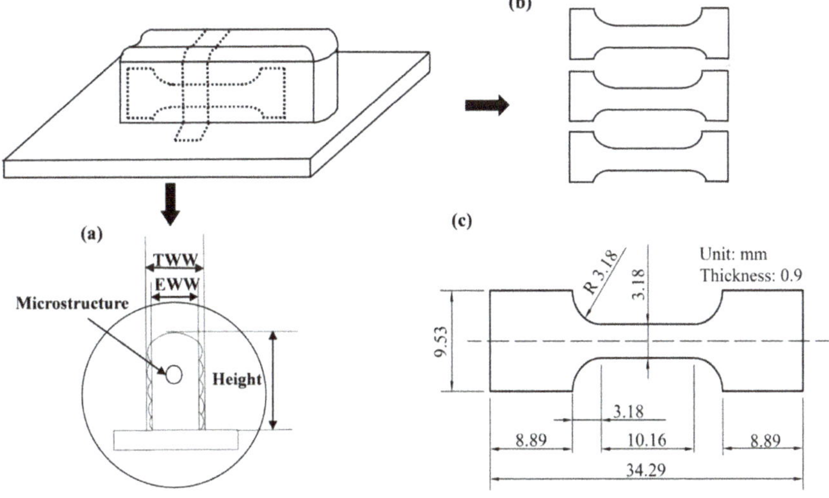

Figure 2. Test samples: (**a**) Sketch of layer thickness and width; (**b**) tensile specimens; (c) dimensional sketch of tensile specimens. TWW—total wall width; EWW—effective wall width.

3. Results and Discussion

3.1. Effects of Pulse Frequency

3.1.1. Macrostructure

From Figure 3, it is found that the geometry and surface morphology of the deposited walls change significantly with the alteration of pulse frequency. Much more coarse periodic ripples are observed at the surface under low pulse frequency, as shown in Figure 3a,b. These coarse periodic ripples are indicative of the typical patterns of interfaces caused by solidified droplet. Besides, fine striations are observed on the surface of the deposited walls at high pulse frequency, as shown in Figure 3c–f. Only very few surface ripples with a low amplitude are found on the surface of the deposited part under the high pulse frequency. The effect of pulse frequency on the surface of deposited walls can be understood as follows: During the manufacturing process, Al–5Si alloy melted at peak current period and droplet will drip on the substrate or deposited wall as a result of the surface tension during the background current. Hence, it takes more time to become bigger at low pulse frequency, so the droplets are usually larger and the number of the droplets is fewer than that at high pulse frequency. This results in coarse periodic ripples on the surface.

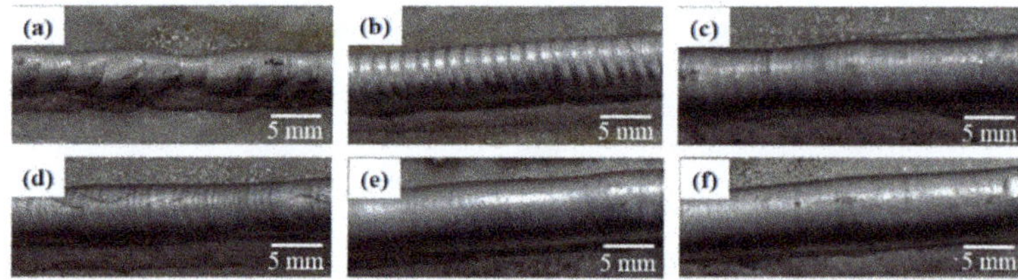

Figure 3. Surface morphology of the walls deposited by (**a**) 2 Hz; (**b**) 5 Hz; (**c**) 10 Hz; (**d**) 50 Hz; (**e**) 200 Hz; and (**f**) 500 Hz.

Table 3 lists the effect of the pulse frequency on the layer thickness and width of the deposited walls. In general, with the pulse frequency increasing from 2 Hz to 500 Hz, the layer thickness decreases and the width increases. As known, the variation of geometry mainly depends on the size of the molten pool that is strongly affected by heat input and heat loss [21]. In this study, under different pulse frequencies, the total value of heat input is kept constant while the conditions of heat loss are different. During the peak current period, high heat input results in the formation of the melting pool [22]. During the background current period, the melting pool cools down rapidly with lots of heat loss. For low pulse frequency, the background current time during a single pulse is long, which is beneficial for heat loss. However, for high pulse frequency, the peak current time happens more frequently, which will lead to the concentration of heat input and inhibit the heat loss, so the molten pool will obtain more continuous heat input and higher temperature. Based on the above mentioned factors, the high temperature under high pulse frequency can be conducive to broadening the molten pool. Hence, the width of deposited wall increases and layer thickness decreases as the pulse frequency increases.

Table 3. Effect of pulse frequency on the geometry of deposited layers.

Pulse Frequency (Hz)	Layer Thickness (mm)	Width (mm)
2	0.92 ± 0.02	5.81 ± 0.01
5	0.96 ± 0.03	5.84 ± 0.03
10	0.88 ± 0.02	6.51 ± 0.02
50	0.86 ± 0.03	6.37 ± 0.05
200	0.83 ± 0.04	6.46 ± 0.02
500	0.84 ± 0.05	6.73 ± 0.03

3.1.2. Microstructure

The microstructures of arc additive manufacturing Al–5Si alloy under different pulse frequency are shown in Figure 4. It can be clearly seen that all the grains exhibit dendrites morphology and the grains become coarser with the increase of pulse frequency. The quantitative statistics results of dendrite arm spacing are shown in Figure 5. It shows that the dendrite arm spacing increases with the increase of pulse frequency from 2 Hz to 500 Hz in both directions (horizontal and vertical). Specifically, with the increase of pulse frequency, the horizontal dendrite arm spacing changes from about 30 μm to 60 μm, and the vertical dendrite arm spacing changes from about 80 μm to 170 μm. It is well-known that the grain size is strongly affected by cooling rate, which also determines the tensile properties of the deposited alloy [23]. The heat input is more concentrated at high pulse frequency, which will reduce cooling rate efficiently. So, much more coarse grains can be found at the high pulse frequency. However, low pulse frequency generates the rapid cooling rate in the molten pool. Hence, far more fine grains can be observed at the low pulse frequency.

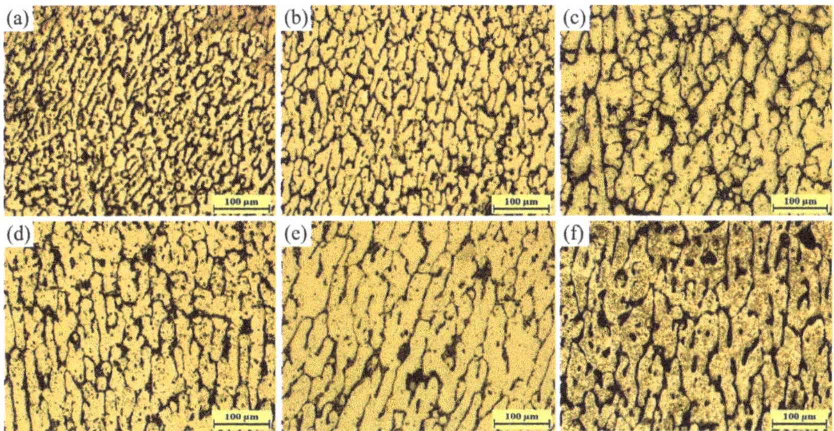

Figure 4. Microstructures of the Al–5Si walls deposited by (**a**) 2 Hz; (**b**) 5 Hz; (**c**) 10 Hz; (**d**) 50 Hz; (**e**) 200 Hz; and (**f**) 500 Hz.

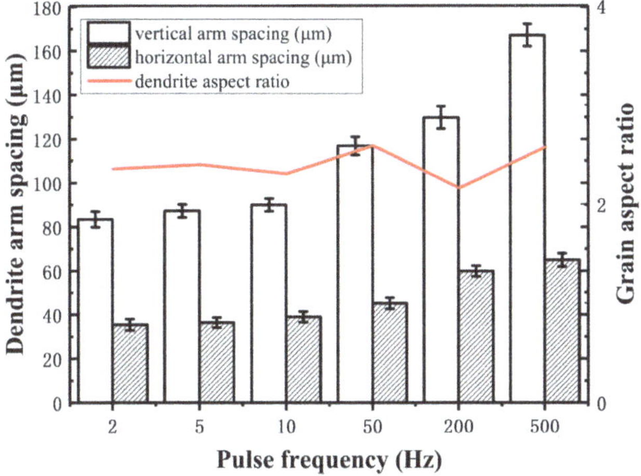

Figure 5. Effect of pulse frequency on the dendrite arm spacing in Al–5Si alloy.

Besides, it is found that the morphology of dendrites tends to be columnar, and the horizontal arm spacing is much smaller than the vertical arm spacing, as shown in Figure 5. This is because of the high temperature gradient caused by the heat loss through the substrate or previously deposited layers during arc additive manufacturing [24]. The high temperature gradient will promote the formation of columnar grains from the bottom of the molten pool [17]. For the additive manufacturing Al–5Si alloy in this study, the grains exhibit short columnar morphology because the formation of columnar grains is not only associated with the temperature gradient, but also with the alloy composition. Moreover, although the dendrites become coarser with the increase of the pulse frequency, the aspect ratio of dendrites seems to be stable at about 2.2. As known, the dendrites morphology mainly depends on the ratio of temperature gradient to solidification velocity [25]. Hence, the results indicate that the change of pulse frequency has not had significant effect on the ratio of temperature gradient to solidification velocity.

3.1.3. Porosity

Figure 6a shows the gas pores in the additive manufacturing Al–5Si alloy samples. The cross section of gas pores are nearly circular. Then, the quantitative analysis of the relationship between the gas pores and pulse frequency is carried out as follows.

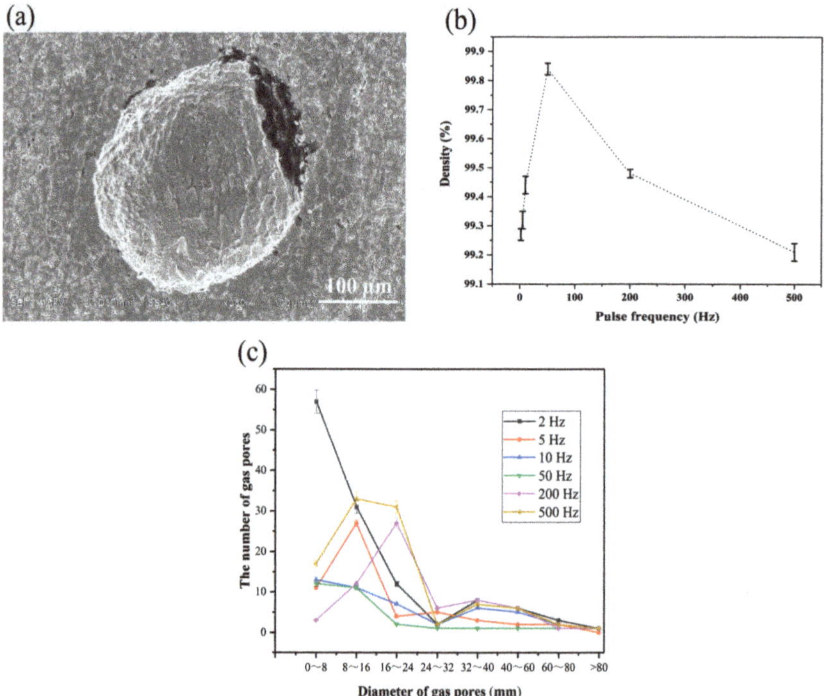

Figure 6. Effect of pulse frequency on porosity: (**a**) Scanning electron microscope (SEM) image shows the gas pores; (**b**) density of the Al–5Si alloy sample; and (**c**) distribution of gas pores with different size.

As shown in Figure 6b, the turning point is around 50 Hz. At the low frequency condition (under 50 Hz), the density of samples increases with the increase of the pulse frequency in general. However, at high pulse frequency (above 50 Hz), the density decreases with the increase of pulse frequency. At this pulse frequency, the density is at a maximum (99.85%). Moreover, some relationships between the area of gas pores and the pulse frequency can be found, as shown in Figure 6c. At low frequency, much smaller pores are found in the samples. However, at high pulse frequency, there are more large pores than the small pores. It shows that the pulse frequency can affect the proportion of the size of pores on arc additive manufacturing process.

Apparently, pulse frequency has an appreciable influence on porosity. According to the previous work as reported in the literature [26,27], there are four stages in the formation of gas pores in the manufacturing process, namely, nucleation, growth, detaching, and escaping. Gas pore growth needs to follow the condition that was represented in Equation (1). The detachability mainly depends on infiltration angle θ, which is shown in Equation (2). It determines whether the gas pore detaches wholly or not. For the escaping stage, it is recognized as a complex effect of density, curvature radius, and liquid viscosity. The expression of escape speed is shown in Equation (3).

$$P_h > 1 + 2\sigma_{2,g}/r \tag{1}$$

where P_h is the pressure inside of gas pore, $\sigma_{2,g}$ is the surface tension between liquid and gas pore, and r is the curvature radius.

$$\cos\theta = (\sigma_{1,g} - \sigma_{1,2})/\sigma_{2,g} \tag{2}$$

where θ is the infiltration angle, $\sigma_{1,g}$ is the surface tension between wall and gas pore, $\sigma_{2,g}$ is the surface tension between liquid and gas, and $\sigma_{1,2}$ is the surface tension between wall and liquid.

$$v = 2\,(\rho_1 - \rho_2)\,g\,r^2/9\eta \tag{3}$$

where v is the escape speed, ρ_1 is the density of liquid, ρ_2 is the density of gas, g is the acceleration of gravity, r is the curvature radius of gas pore, and η is the liquid viscosity.

This phenomenon can be analyzed as follows:

Firstly, increased pulse frequency can cause the increase of the arc force and arc pressure, so the liquid pressure will increase. Then, the curvature radius of gas pores will be larger, as shown in Figure 7a, surface tension between gas and liquid is constant during the process of gas growth. According to Equation (1), higher pulse frequency will reduce the value of the right side of the inequation. Obviously, higher pulse frequency helps gas pores grow, as reported by Cong B. et al. [15].

Secondly, increased pulse frequency with a larger heat input, which will increase the temperature of molten pool, and the cooling rate will lower simultaneously [28]. We can conclude that high pulse with high temperature will reduce the $\sigma_{2,g}$ (surface tension between liquid and gas pore) and $\sigma_{1,2}$ (surface tension between wall and liquid). Because of the larger curvature, gas pores will gain a larger touching area. As known, larger touching area will increase the $\sigma_{1,g}$. Above all, as shown in Equation (2), with the increase of pulse frequency, $\sigma_{1,g}$ will increase while $\sigma_{2,g}$ and $\sigma_{1,2}$ will decrease. Therefore, high pulse frequency with small infiltration angle will help pores detach, as shown in Figure 7b. According to previous work, liquid viscosity is also affected by temperature. With the temperature increasing, liquid viscosity will decrease. The variation of pulse frequency has no obvious effect on the density of liquid and gas and curvature radius of gas pores increase with the increase of the pulse frequency. It can be seen that with the increase of pulse frequency, the cooling rate will reduce and the escape speed will increase. Gas pores can escape from the molten pool on the condition that escape speed is greater than cooling rate. Nonetheless, the higher the temperature of the molten pool is, the more intense the chemical reacts in molten pool. Besides, higher temperature of the molten pool will increase its capacity of H^+ absorption, which leads to more nucleation sites of gas pores in the molten pool, as shown in Figure 7c.

Based on the above mentioned reasons, there exists an optimum pulse frequency of porosity minimization. With the increase of pulse frequency (under 50 Hz), it is easier for gas pores primarily to detach and escape from the molten pool. Therefore, density will increase with the increase of pulse frequency (under 50 Hz). For high pulse frequency (above 50 Hz), with the increase of pulse frequency, the gas pores' capacity to detach will be larger, nonetheless, the number of gas pores' nucleation sites will increase, more gas pores will form, although the escaping speed is bigger, there is still leaving gas pores. Thus, density decreases with the pulse frequency increasing. At the pulse frequency of 50 Hz, nucleation and escaping of gas pores seem to be more balanced, leading to high-density additive manufacturing aluminum alloys.

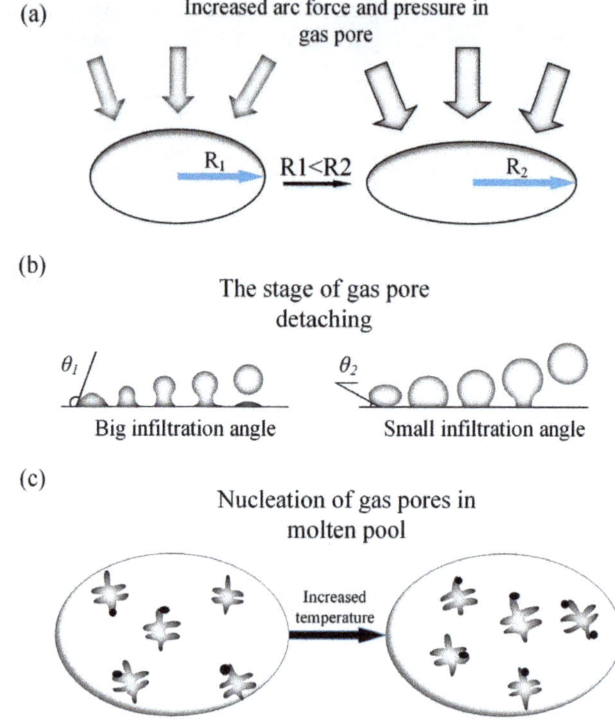

Figure 7. The formation of gas pores with increased pulse frequency: (**a**) Curvature radius; (**b**) infiltration angel; and (**c**) the number of nucleation sites.

3.1.4. Tensile Properties

From Figure 8, with the pulse frequency decreasing, a good strength was achieved. Concretely speaking, the ultimate tensile strength (UTS) of the samples decreases from 125.1 MPa to 108.6 MPa with an increase in the pulse frequency from 2 Hz to 500 Hz. Meanwhile, the yield strength (YS) of the samples also decreases with an increase in the pulse frequency from 2 Hz to 500 Hz. It is claimed that the finer microstructure will lead to a higher tensile strength [22,29]. When the pulse frequency increases, the grains become coarser, thus the strength decreases. Furthermore, the fracture surface micrographs of the tensile samples at different pulse frequencies are given in Figure 9. It can be seen that far more dimples can be found at low pulse frequency than at high pulse frequency, and the dimples appear deeper at low pulse frequency, which supports the results that the samples fabricated at low frequency have higher strength.

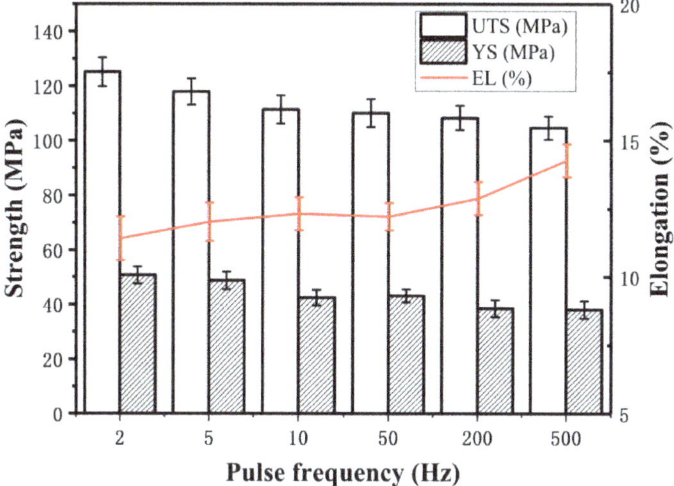

Figure 8. Effect of pulse frequency on tensile properties of arc additive manufacturing Al–5Si alloy. UTS—ultimate tensile strength, YS—yield strength; EL—elongation.

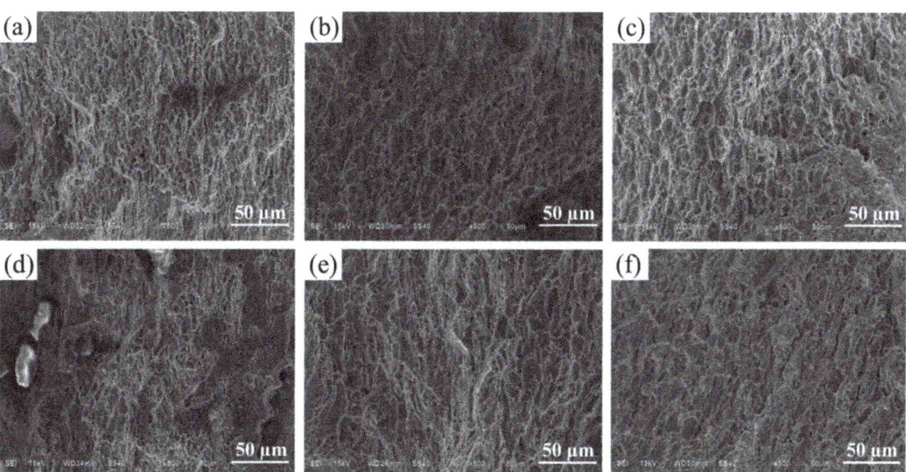

Figure 9. SEM fractographs of tensile samples fabricated with different pulse frequencies: (**a**) 2 Hz; (**b**) 5 Hz; (**c**) 10 Hz; (**d**) 50 Hz; (**e**) 200 Hz; and (**f**) 500 Hz.

3.2. Effect of Arc Current

3.2.1. Macrostructure

Similar to pulse frequency, arc current can also influence the geometry of Al–5Si alloy manufactured by arc additive manufacturing, as shown in Table 4. As arc current increases from 100 A to 175 A, layer thickness decreases from 1.04 mm to 0.79 mm, while the width of samples increases from 5.72 mm to 9.56 mm. This is because with the increase of arc current, molten pool will absorb more heat, which broadens the molten pool and decreases the layer thickness. Besides, the surface morphologies of samples vary only a little under different arc current.

Table 4. Effect of arc current on the geometry of deposited layers.

Arc Current (A)	Layer Thickness (mm)	Width (mm)
100	1.04 ± 0.07	5.72 ± 0.02
125	0.83 ± 0.02	5.70 ± 0.01
150	0.80 ± 0.03	8.14 ± 0.07
175	0.79 ± 0.02	9.56 ± 0.03

3.2.2. Microstructure

Compared with pulse frequency, arc current can affect grain size and secondary dendrite arm spacing more strongly, as shown in Figure 10. It is obvious that grain size increases with the increase of arc current, as shown in Figure 11. Specifically, both the horizontal arm spacing and vertical arm spacing of grain increase. However, vertical arm spacing appears to have a more clear growth than horizontal arm spacing, which can been seen in Figure 11. Because of the variation of arm spacings in two different directions, grains change from near equiaxed morphology into columnar morphology, as shown in Figure 10.

Obviously, arc current has significant effect on the microstructures of Al–5Si alloys because it controls the heat input to material [30]. Small arc current leads to low heat input, high cooling rate and high nucleation rate, which results in fine grains with near equiaxed morphology. For large arc current, the cooling rate and temperature gradient both decrease, thus the grains exhibit coarse and long columnar morphology.

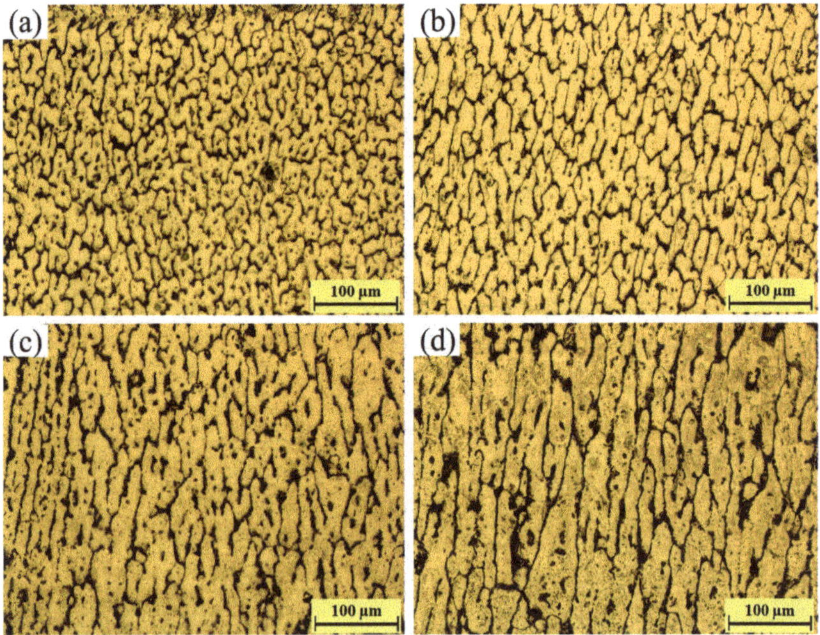

Figure 10. Microstructures of the walls deposited by (**a**) 100 A; (**b**) 125 A; (**c**) 150 A; and (**d**) 175 A.

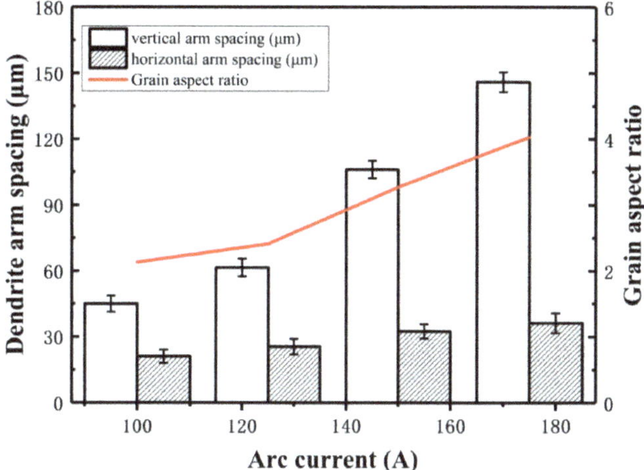

Figure 11. Effect of arc current on the microstructures of deposited layers.

3.2.3. Tensile Properties

Figure 12 shows that with an increase of the arc current from 100 A to 175 A, the ultimate tensile strength (UTS) of the samples clearly decreases from 146.5 MPa to 110.5 MPa. The yield strength (YS) at failure also decreases in general, but the elongation (EL) increased. Obviously, this is caused by the microstructure changes. According to the Hall–Petch equation [31], the mechanical property of the alloy is directly affected by the average grain size. In other words, the fine microstructures under small arc current leads to high strength but poor tensile ductility [32]. Figure 13 shows the fracture surface micrograph of the tensile samples with different arc currents. It can be seen that far more dimples can be found at low arc current than that at large arc current. Compared with the dimples at large arc current (see Figure 13c,d), the dimples appear deeper at low arc current (Figure 13a,b), which is consistent with the tensile properties.

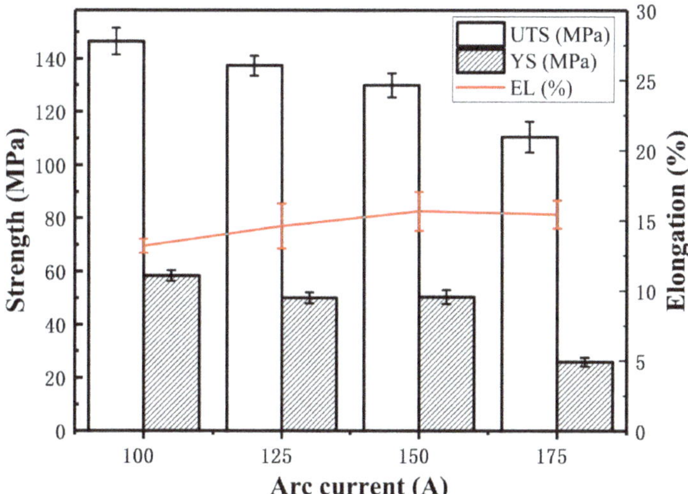

Figure 12. Effect of arc current on tensile properties of deposited of Al–5Si alloy.

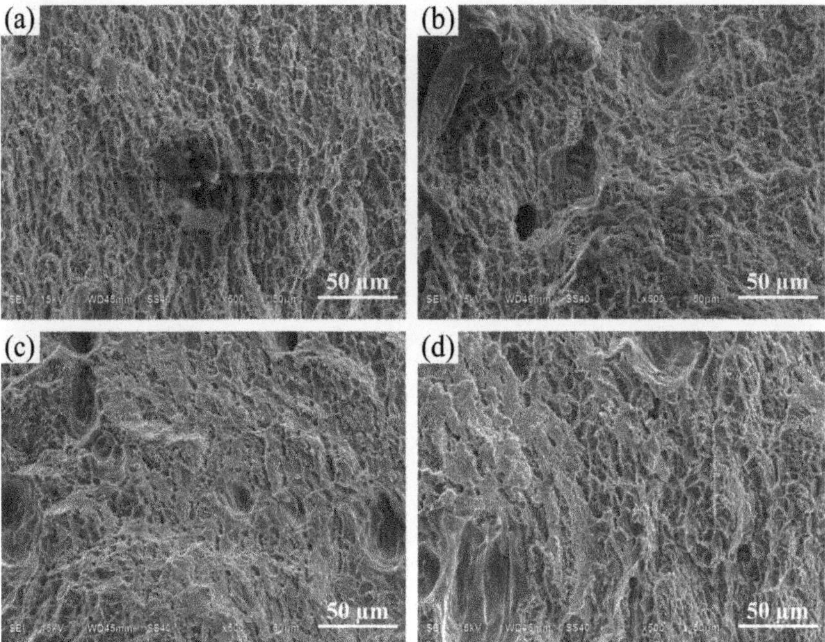

Figure 13. SEM fractographs of tensile samples manufactured with different pulse frequencies: (**a**) 100 A; (**b**) 125 A; (**c**) 150 A; and (**d**) 175 A.

4. Conclusions

In this study, Al–5Si alloy samples were fabricated by arc additive manufacturing with different arc currents and pulse frequencies. The macrostructure, microstructure, porosity, and tensile properties were studied. From this investigation, a few key points can be concluded, as below:

(1) Pulse frequency strongly affects the porosity of arc additive manufacturing Al–5Si alloy. With the increase of pulse frequency, density shows a trend that escalates firstly and attains its maximum in 50 Hz, but declines later because of the relation between pores formation and gas escape.

(2) With the increase of pulse frequency, grains become coarser because of the more centralized heat input, but the grains still exhibit short columnar morphology. Because of the variation of microstructure, tensile strength decreases from 125.1 MPa to 108.6 MPa with the increase of pulse frequency.

(3) Arc current also has a significant impact on microstructure and tensile properties. With the increase of arc current, grains become coarser and change from short columnar morphology to long columnar morphology as a result of more heat input, and hence leads to the decrease of tensile strength from 146.5 MPa to 110.5 MPa.

In summary, coarse grains and gas pores are two main problems that limit the application of arc additive manufacturing Al–Si alloys. Combined with the above results, fine grains and uniform microstructure can be obtained. This paper illustrate that pulse frequency and arc current have a significant impact on the macrostructure, microstructure, porosity, and tensile properties of the samples, which is meaningful for the further application of arc additive manufacturing aluminum alloy.

Author Contributions: D.W. performed all experiments and wrote this manuscript. C.L. designed the research and gave some constructive suggestions. J.L., H.F. and L.Y. helped analyze the experimental data. S.T. and L.J. participated in the discussion on the results and guided the writing of the article.

Funding: This research was funded by (National Natural Science Foundation of China) grant number (51505033) and (Beijing Natural Science Foundation) grant number (3162027).

Acknowledgments: The work was financially supported by the National Natural Science Foundation of China (51505033), Beijing Natural Science Foundation (3162027).

Conflicts of Interest: The authors declare no conflict of interest.

References

1. Frazier, W.E. Metal additive manufacturing: A review. *J. Mater. Eng. Perform.* **2014**, *23*, 1917–1928. [CrossRef]
2. Singh, S.; Ramakrishna, S.; Singh, R. Material issues in additive manufacturing: A review. *J. Manuf. Process.* **2017**, *25*, 185–200. [CrossRef]
3. Herzog, D.; Seyda, V.; Wycisk, E.; Emmelmann, C. Additive manufacturing of metals. *Acta Mater.* **2016**, *117*, 371–392. [CrossRef]
4. Williams, S.W.; Martina, F.; Addison, A.C.; Ding, J.; Pardal, G.; Colegrove, P. Wire + arc additive manufacturing. *Mater. Sci. Technol.* **2016**, *32*, 641–647. [CrossRef]
5. Zhang, C.; Li, Y.; Gao, M.; Zeng, X. Wire arc additive manufacturing of al-6mg alloy using variable polarity cold metal transfer arc as power source. *Mater. Sci. Eng. A* **2018**, *711*, 415–423. [CrossRef]
6. Ding, D.; Pan, Z.; Cuiuri, D.; Li, H. Wire-feed additive manufacturing of metal components: Technologies, developments and future interests. *Int. J. Adv. Manuf. Technol.* **2015**, *81*, 465–481. [CrossRef]
7. Wu, Q.; Ma, Z.; Chen, G.; Liu, C.; Ma, D.; Ma, S. Obtaining fine microstructure and unsupported overhangs by low heat input pulse arc additive manufacturing. *J. Manuf. Process.* **2017**, *27*, 198–206. [CrossRef]
8. Nakai, M.; Eto, T. New aspect of development of high strength aluminum alloys for aerospace applications. *Mater. Sci. Eng. A* **2000**, *285*, 62–68. [CrossRef]
9. King, J.F. 3—Alumina. In *The Aluminium Industry*; Woodhead Publishing: Cambridge, UK, 2001; pp. 3–41.
10. Balasubramanian, V.; Ravisankar, V.; Reddy, G.M. Effect of pulsed current welding on mechanical properties of high strength aluminum alloy. *Int. J. Adv. Manuf. Technol.* **2008**, *36*, 254–262. [CrossRef]
11. Ortega, A.G.; Galvan, L.C.; Deschaux-Beaume, F.; Mezrag, B.; Rouquette, S. Effect of process parameters on the quality of aluminium alloy Al–5Si deposits in wire and arc additive manufacturing using a cold metal transfer process. *Sci. Technol. Weld. Join.* **2018**, *23*, 316–332. [CrossRef]
12. Geng, H.; Li, J.; Xiong, J.; Lin, X.; Zhang, F. Geometric limitation and tensile properties of wire and arc additive manufacturing 5A06 aluminum alloy parts. *J. Mater. Eng. Perform.* **2017**, *26*, 621–629. [CrossRef]
13. Silva, C.L.M.D.; Scotti, A. The influence of double pulse on porosity formation in aluminum GMAW. *J. Mater. Process. Technol.* **2006**, *171*, 366–372. [CrossRef]
14. Wang, H.; Jiang, W.; Ouyang, J.; Kovacevic, R. Rapid prototyping of 4043 Al-alloy parts by VP-GTAW. *J. Mater. Process. Technol.* **2004**, *148*, 93–102. [CrossRef]
15. Cong, B.; Ding, J.; Williams, S. Effect of arc mode in cold metal transfer process on porosity of additively manufactured Al-6.3%Cu alloy. *Int. J. Adv. Manuf. Technol.* **2015**, *76*, 1593–1606. [CrossRef]
16. Gu, J.; Cong, B.; Ding, J.; Williams, S.W.; Zhai, Y. Wire + arc additive manufacturing of aluminium. In Proceedings of the 25th Annual International Solid Freeform Fabrication Symposium, Austin, TX, USA, 4–6 August 2014.
17. Cong, B.; Qi, Z.; Qi, B.; Sun, H.; Zhao, G.; Ding, J. A comparative study of additively manufactured thin wall and block structure with Al-6.3%Cu alloy using cold metal transfer process. *Appl. Sci.* **2017**, *7*, 275. [CrossRef]
18. Cong, B.; Ding, J. Influence of CMT process on porosity of wire arc additive manufactured Al-Cu alloy. *Rare Met. Mater. Eng.* **2014**, *43*, 3149–3153.
19. Zamani, M. Al-Si Cast Alloys—Microstructure and Mechanical Properties at Ambient and Elevated Temperature. Ph.D. Thesis, Jonkoping University, Jonkoping, Sweden, 2015.
20. Traidia, A.; Roger, F. Numerical and experimental study of arc and weld pool behaviour for pulsed current gta welding. *Int. J. Heat Mass Transf.* **2011**, *54*, 2163–2179. [CrossRef]
21. Do-Quang, M.; Amberg, G. *Modelling of Time-Dependent 3d Weld Pool Due to a Moving Arc*; Bock, H.G., Phu, H.X., Kostina, E., Rannacher, R., Eds.; Springer: New York, NY, USA, 2005; pp. 127–138.
22. Guo, J.; Zhou, Y.; Liu, C.; Wu, Q.; Chen, X.; Lu, J. Wire arc additive manufacturing of AZ31 magnesium alloy: Grain refinement by adjusting pulse frequency. *Materials* **2016**, *9*, 823. [CrossRef] [PubMed]

23. Pourkia, N.; Emamy, M.; Farhangi, H.; Ebrahimi, S.H.S. The effect of Ti and Zr elements and cooling rate on the microstructure and tensile properties of a new developed super high-strength aluminum alloy. *Mater. Sci. Eng. A* **2010**, *527*, 5318–5325. [CrossRef]

24. Ding, J.; Colegrove, P.; Mehnen, J.; Williams, S.; Wang, F.; Almeida, P.S. A computationally efficient finite element model of wire and arc additive manufacture. *Int. J. Adv. Manuf. Technol.* **2014**, *70*, 227–236. [CrossRef]

25. Bontha, S.; Klingbeil, N.W.; Kobryn, P.A.; Fraser, H.L. Effects of process variables and size-scale on solidification microstructure in beam-based fabrication of bulky 3d structures. *Mater. Sci. Eng. A* **2009**, *513–514*, 311–318. [CrossRef]

26. Yang, M.X.; Yang, Z.; Cong, B.Q.; Qi, B.J. How ultra high frequency of pulsed gas tungsten arc welding affects weld porosity of Ti-6Al-4V alloy. *Int. J. Adv. Manuf. Technol.* **2015**, *76*, 955–960. [CrossRef]

27. Uehara, K.; Katayama, K.; Date, H.; Fukada, S. Hydrogen gas driven permeation through tungsten deposition layer formed by hydrogen plasma sputtering. *Fusion Eng. Des.* **2015**, *98–99*, 1341–1344. [CrossRef]

28. Merchant Samir, Y. Investigation on effect of heat input on cooling rate and mechanical property (hardness) of mild steel weld joint by mmaw process. *Int. J. Mod. Eng. Res.* **2015**, *5*, 34–41.

29. Zuo, Y.B.; Cui, J.Z.; Zhao, Z.H.; Zhang, H.T.; Qin, K. Effect of low frequency electromagnetic field on casting crack during DC casting superhigh strength aluminum alloy ingots. *Mater. Sci. Eng. A* **2005**, *406*, 286–292.

30. Morita, Y.; Iwao, T.; Yumoto, M. Heat transfer of pulsed arc as a function of current frequency. In Proceedings of the 2009 IEEE International Conference on Plasma Science–Abstracts, San Diego, CA, USA, 1–5 June 2009; p. 1.

31. Zhu, Y.Z.; Wang, S.Z.; Li, B.L.; Yin, Z.M.; Wan, Q.; Liu, P. Grain growth and microstructure evolution based mechanical property predicted by a modified hall–petch equation in hot worked Ni76Cr19AlTiCo alloy. *Mater. Des.* **2014**, *55*, 456–462. [CrossRef]

32. Cheng, S.; Zhao, Y.H.; Zhu, Y.T.; Ma, E. Optimizing the strength and ductility of fine structured 2024 Al alloy by nano-precipitation. *Acta Mater.* **2007**, *55*, 5822–5832. [CrossRef]

Article

Research on Mechanisms and Controlling Methods of Macro Defects in TC4 Alloy Fabricated by Wire Additive Manufacturing

Lei Ji [1], Jiping Lu [1], Shuiyuan Tang [1,*], Qianru Wu [1], Jiachen Wang [2], Shuyuan Ma [1], Hongli Fan [1] and Changmeng Liu [1]

[1] School of Mechanical Engineering, Beijing Institute of Technology, Beijing 100081, China; jl12199@163.com (L.J.); jipinglu@bit.edu.cn (J.L.); qrwu@foxmail.com (Q.W.); bitmc@bit.edu.cn (S.M.); fanhongli2000@sina.com (H.F.); liuchangmeng@bit.edu.cn (C.L.)

[2] Institute of Advanced Structure Technology, Beijing Institute of Technology, Beijing 100081, China; jcwang666@163.com

* Correspondence: shuiyuantang@bit.edu.cn; Tel.: +86-10-6891-1652

Received: 24 May 2018; Accepted: 26 June 2018; Published: 28 June 2018

Abstract: Wire feeding additive manufacturing (WFAM) has broad application prospects because of its advantages of low cost and high efficiency. However, with the mode of lateral wire feeding, including wire and laser additive manufacturing, gas tungsten arc additive manufacturing etc., it is easy to generate macro defects on the surface of the components because of the anisotropy of melted wire, which limits the promotion and application of WFAM. In this work, gas tungsten arc additive manufacturing with lateral wire feeding is proposed to investigate the mechanisms of macro defects. The results illustrate that the defect forms mainly include side spatters, collapse, poor flatness, and unmelted wire. It was found that the heat input, layer thickness, tool path, and wire curvature can have an impact on the macro defects. Side spatters are the most serious defects, mainly because the droplets cannot be transferred to the center of the molten pool in the lateral wire feeding mode. This research indicates that the macro defects can be controlled by optimizing the process parameters. Finally, block parts without macro defects were fabricated, which is meaningful for the further application of WFAM.

Keywords: wire feeding additive manufacturing; wire lateral feeding; macro defects; side spatters

---

1. Introduction

Wire feeding additive manufacturing is a very promising technology [1]. Compared with powder additive manufacturing, all the materials are sent to the welding pool in a lateral wire feeding mode. Consequently, the utilization of materials is extremely high, which can reduce wastage of materials and improve fabricating efficiency [2–4]. Moreover, materials in the form of wire are cheaper than powders. Therefore, WFAM is particularly suitable for manufacturing large scale structures or other complicated structures in aerospace [5].

According to the types of heat source, WFAM can be divided into wire and laser additive manufacturing (WLAM) [6,7], electron beam freeform fabrication (EBF) [8,9], gas tungsten arc welding (GTAW)-based wire arc additive manufacturing (WAAM) [10–12] and pulse arc welding (PAW)-based wire arc additive manufacturing [5,13].

However, in these WFAM methods, owing to the anisotropy of the lateral wire feeding method, the wire is not melted uniformly [14], as the wire is sent into the welding pool from the side of the arc zone (as Figure 1a [3]). In the lateral wire feeding mode, the uneven melting of wire could lead to macro defects of the components, such as spatters and unmelted wire (Figure 1b), which are common and inevitable phenomena in WFAM. Hagqvist found that the process is hard to control during laser metal deposition, creating parts with spatters and unmelted phenomena that he called "the ugly" [15]. Norsk Titanium is the pioneering supplier of aerospace-grade, additive manufactured, structural titanium components [16]. Spatters and rough surfaces also exist in its components, as indicated in Figure 1c [17]. Much follow-up processing is needed even if the component is a thin-walled part. Yongzhe Li [18] devised a layers-overlapping strategy to improve the accuracy of components fabricated by wire arc additive manufacturing. Temperature field control [19] and passive-vision [20] were used to improve the accuracy of the sample. Meanwhile, if the wire is heated under improper parameters, the wire cannot be heated uniformly, and non-uniform melting will lead to changes in the size of the welding pool, which may also lead to other defects. Much research has been conducted on the microstructure and mechanical properties of components fabricated using WFAM [7,8,10,21], but little research on the macro defects of WFAM has been reported. However, these kinds of macro defects always restrict the development of WFAM. Reducing or eliminating those defects through adjusting process parameters would significantly promote the development of WFAM.

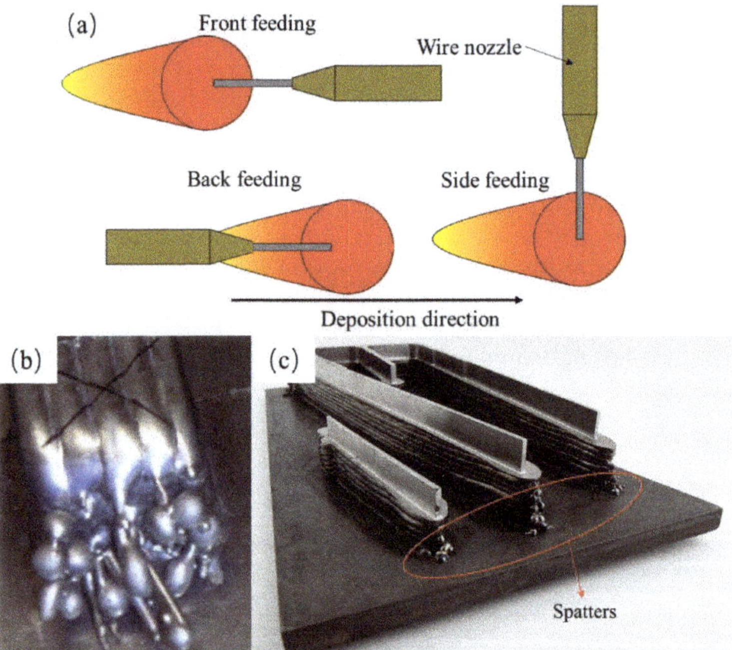

Figure 1. Schematic of the (**a**) lateral wire feeding in different feeding direction [3], "the ugly" samples (**b**) referred by P. Hagqvist, and component [15] (**c**) fabricated by Norsk Titanium [16].

This work aimed to find the mechanisms and controlling methods of the macro defects fabricated by wire arc additive manufacturing, based on previous research on high-efficiency fabricating processes of GTAW-based wire arc additive manufacturing for TC4 titanium alloy. Four sets of experiments were set up to investigate the effects of heat input, layer thickness, tool path, and wire curvature. The effect of these parameters on the defects was analyzed systematically. The formation mechanisms of side spatters, collapse, unmelted wire and poor flatness were investigated, and controlling methods based on the process are presented.

2. Experimental Details

The equipment used in this research was developed independently by Beijing Institute of Technology (Figure 2a), a schematic is shown in Figure 2b [11]. The developed wire arc additive manufacturing system mainly consists of a wire feeder, a computer, a working chamber, and a GTAW machine. Normally, the parts were fabricated under protective conditions with an argon atmosphere of 99.99%.

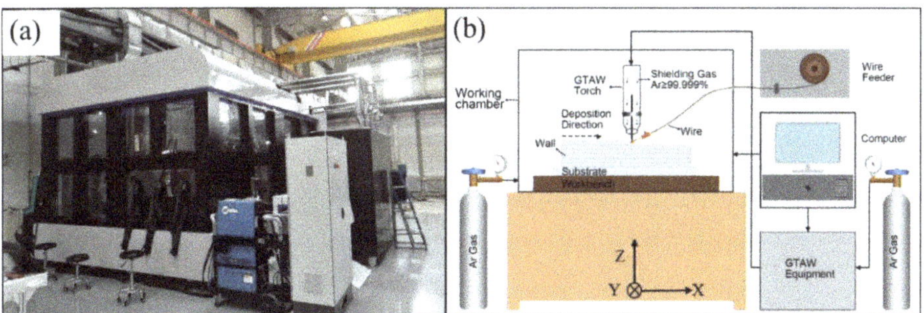

Figure 2. (**a**) Overview of wire and arc additive manufacturing system and (**b**) schematic of the equipment. GTAW = gas tungsten arc welding.

Based on previous research, in this research we obtained a relatively mature weld bead process and a thin-walled part molding process. The parts in this research were fabricated based on these processes. A large amount of research on the wire arc additive manufacturing process with 1.6-mm wire has been conducted. Figure 3a [4] shows a typical part with thin walls fabricated previously. Although numerous parameters may result in macro defects, according to actual fabricating experience, four parameters had a great influence on the macro defects and were chosen to investigate the mechanisms and controlling methods. As given in Table 1, the variables of heat input, layer thickness, tool path, and wire curvature were used to design four sets of experiments. The wire curvature is defined as the bending degree of the wire in its unrestrained state (Figure 3c). In mathematics, the curvature of the circle can be regarded as the reciprocal of the radius approximatively. So the radii of wires with different curvature were measured to calculate the curvature (Figure 3d). Block parts with a size of about $80 \times 40 \times 32$ mm were fabricated by wire arc additive manufacturing in this work, and the macro defects existing in them were studied. To describe the research expediently, the relationship between the part, substrate, and the welding torch is shown in Figure 3b. The welding torch with the wire attached at a fixed angle ($45°$) moves under a computer numerical control (CNC) system. The wire feeding method is lateral paraxial feeding. In the following description, surfaces 1, 2, and 3 are named for the convenience of description. The parameters used in this research are listed in Table 1. Finally, ten samples (1–10) of four experiment groups are shown in Figure 4. Especially, the surfaces of samples 1, 4, and 9 look dark gray because of poor argon protection. This poor argon protection was unintentional and was the result of the chamber being opened too early, causing the parts to oxidize during processing. However, each layer was fabricated in an argon atmosphere without oxidation, so only the surface of the parts was oxidized. Because this research is aimed at the existence of defects on the macro level, this factor can be ignored in the research.

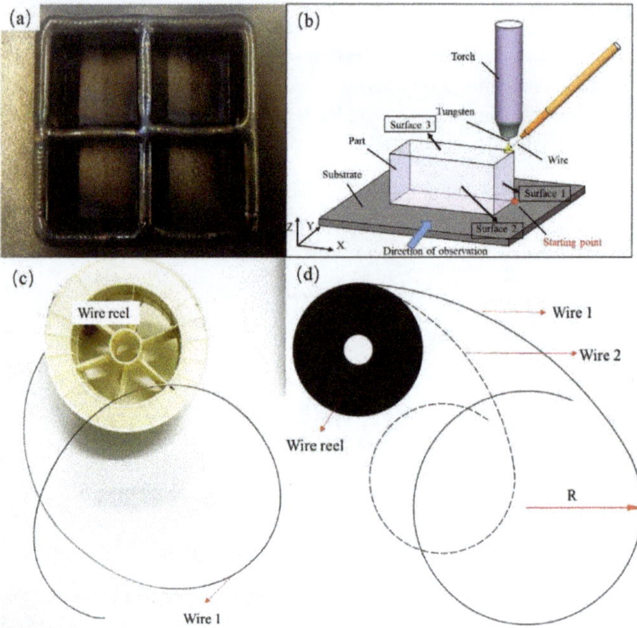

Figure 3. (**a**) Sample fabricated previously [4]; (**b**) schematic of the relationship between the part, base plate, and the welding torch; (**c**) wire reel and wire with curvature and (**d**) schematic of the wire with different curvature.

Table 1. Deposition parameters used in this research.

Sample	1	2	3	4	5	6	7	8	9	10
Experimental group	Group 1	Group 1	Group 1 / Group 2	Group 1 / Group 2	Group 2 / Group 3	Group 3	Group 3	Group 3 / Group 4	Group 4	Group 4
Heat input (J/mm)	262	173	217	217	217	217	217	217	217	217
Layer thickness (mm)	1.05	1.05	1.05	1.1	1.05/1.1	1.05/1.1	1.05/1.1	1.05/1.1	1.05/1.1	1.05/1.1
Path	Z	Z	Z	Z	Z	Vertical	Contour filling	Z	Z	Z
Wire curvature (mm^{-1})	0	0	0	0	0	0	0	1/150	1/300	0

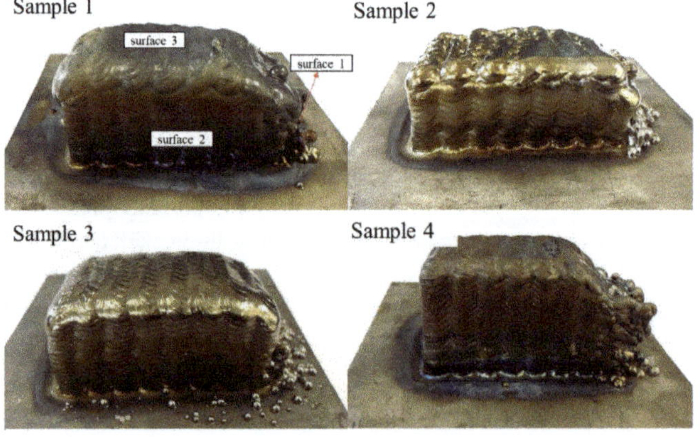

Figure 4. *Cont.*

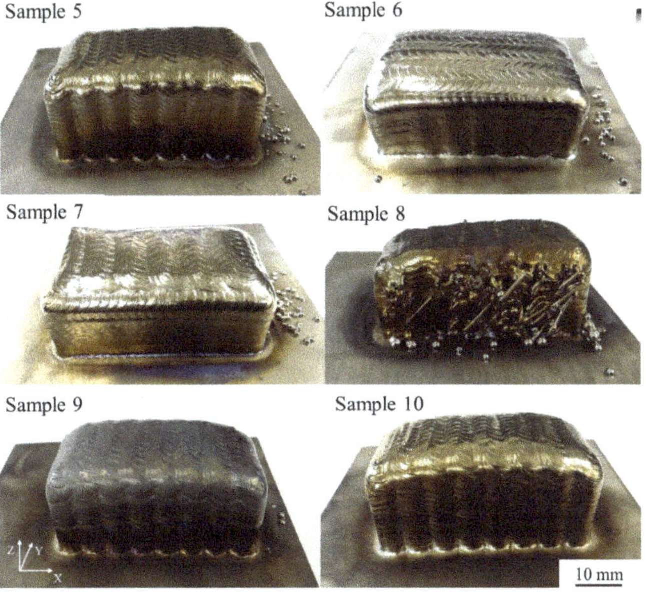

Figure 4. Macro images of the ten samples fabricated by using wire and arc additive manufacturing.

3. Results and Discussion

3.1. Effect of Heat Input on Macro Defects

3.1.1. The Form of Defects

Figure 5 shows the defects existing in three samples with different heat inputs. It can be seen that the defects are mainly located on surface 1. For sample 1, spatter of a large size occurs on the first few layers of surface 1, and there is a serious metal flowing phenomenon that causes the collapse of surface 1. For sample 2, spatter occurs a bit later, and its size appears to be small compared to that of sample 1. The degree of collapse has also been significantly reduced. However, the wire is not melted sufficiently on surface 3, as the enlarged view of Figure 5b shows. In sample 3, the spatter occurs in the last few layers and the number of spatters decreases significantly, thereby reducing the degree of collapse of surface 1.

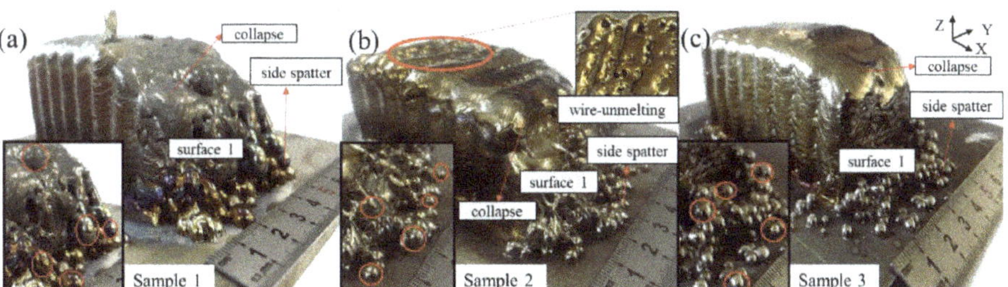

Figure 5. Macro images of three samples with different heat inputs: (**a**) 261 J/mm; (**b**) 173 J/mm; and (**c**) 217 J/mm.

3.1.2. Macro Defect Mechanisms with Different Heat Inputs

In Figure 5, there are three types of defects, i.e., side collapse, side spatters, and unmelted wire, respectively. To analyze these phenomena, the following equations [10,22,23] are used to calculate the value of the heat input by regarding the difference in heat input as the difference in current.

$$Hi\frac{J}{mm} = \eta I_{av}\frac{V_{av}}{TS} \tag{1}$$

$$I_{av}(A) = \frac{I_p t_p + I_b t_b}{t_p + t_b} \tag{2}$$

where Hi is the heat input per unit length, η is the arc efficiency (assumed to be 0.83) for tungsten inert gas (TIG) welding [10], V_{av} is the average of the instantaneous arc voltage, I_{av} is the average current for pulse current TIG welding, V_{av} is the average of the instantaneous arc voltage, TS is the travel speed, t_p is the peak time, t_b is the base current duration, I_p is the peak current, and I_b is the base current.

As a consequence, the heat input to samples 1, 2, and 3 are 261, 173, and 217 J/mm, respectively. Combined with these values, the defect mechanisms are discussed as follows:

1. Side collapse is mainly caused by an overlapping process and excessive heat input. When fabricating the multi-layer, multi-bead blocks, the paths used are coincident in every layer. Therefore, the blocks' edge is lower than the blocks' main body. As Figure 6a shows, Li [24] demonstrated that material shortage areas are generated at the edges of the blocks when two layers overlap. With the increase of the number of layers, material shortage areas can accumulate, and finally the blocks begin to collapse. Moreover, excessive heat input can cause serious collapse because of the metal flow along the wall. As Figure 6c,d shows, with decreasing heat input, the degree of collapse can be significantly reduced and the actual size of the part becomes closer to the designed size.

2. Side spatters are mainly caused by the lateral wire feeding mode and excessive heat input. The wire is always inserted from one side, which can easily generate side spatters. When the process parameters are not matched with each other, the metal droplets can easily be splashed under the arc force in a fastigiated direction. As Figure 6c shows, when the heat input is 261 J/mm, the wire melts away from the center of the arc. Because the arc temperature is fairly high, the wire can easily be melted and the droplets generated at the front of the wire are larger and larger. Once at the pulsed current stage, the droplet falls under the arc force in a fastigiated direction, causing the droplet to be blown to the outside. The part material is lessened as well, especially at the edge of the block, so the droplets cannot fall into the center of the path with the lower surface. Therefore, the droplets are splashed as side spatters. When the heat input is decreased to 216 and 173 J/mm, the wire is more likely to be sent to the center of the arc and the size of the droplet is diminished, as Figure 6d shows. Once at the pulsed current stage, the droplet can be blown to the path more accurately. As shown in Figure 5, five droplets with a clear appearance are selected for each sample and the average size is calculated. The spatter sizes of samples 1, 2, and 3 in Figure 5 are 6.39, 4.21, and 4.25 mm. That is, even if the heat input is reduced, spatter still exists because of the special side feeding mode, but the size of the spatters and their number can be reduced efficiently.

3. The wire cannot be fully melted when the heat input is too small. As Figure 6b shows, a large amount of wire material is not melted at the base current, with the wire traveling against the formed surface. Once at the peak current, the wire is directly fused to the formed surface, causing the wire not to be fully melted and resulting in poor bonding between the contiguous weld beads. When the heat input is insufficient, the parts are not well formed and the wire feeding system becomes unstable during manufacturing and the stability of part forming is reduced, owing to the front of the wire touching the surface.

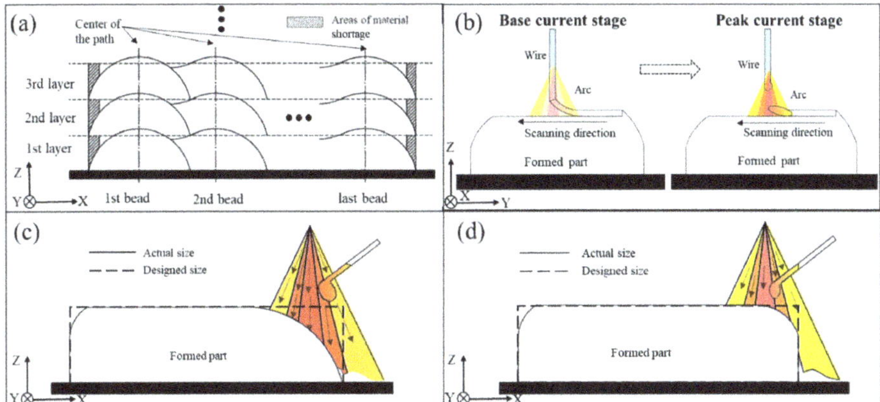

Figure 6. Schematics of (**a**) multi-layer overlapping; (**b**) the wire melted under a low heat input; (**c**) deposition process with excessive heat input and (**d**) deposition process with a low heat input.

3.2. Effect of Layer Thickness on Macro Defects

3.2.1. The Form of Defects

Figure 7 shows the defects existing in three samples with different layer thicknesses. It can be clearly seen that the defects are mainly located on surface 1 in the form of side spatters. The side spatters of sample 3 occur early and the number of spatters is great. Sample 4 has fewer and smaller spatters than sample 3. Surface 1 in sample 5 has a small number of spatters with smaller size, and the outline of the sample is clear as well.

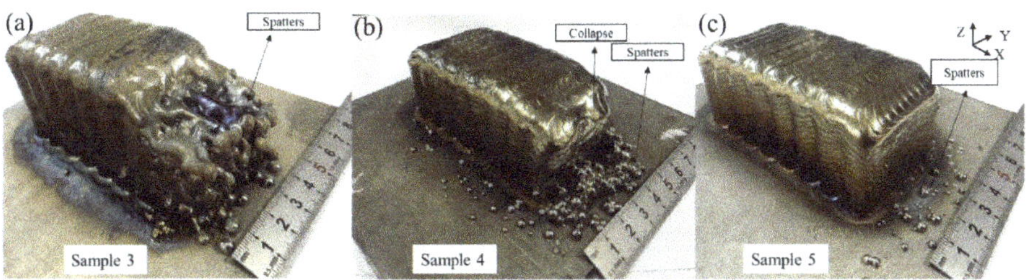

Figure 7. Macro images of samples 3, 4, and 5 with different layer thicknesses: (**a**) 1.1 mm; (**b**) 1.05 mm; and (**c**) 1.05/1.1 mm.

3.2.2. Macro Defect Mechanisms with Different Layer Thicknesses

The type of transfer mode depends on the arc length (AL), which is the distance from the front of the tungsten to the fabricated surface (Figure 8). The definition of the layer thickness is the height lifted by the machine of every layer in the fabricating process. Therefore, in wire arc additive manufacturing, the setting of the layer thickness has an accumulative effect on the AL. The value of the layer thickness is regarded with the height of the weld bead. There is no doubt that the height of the weld bead is determined by the heat input and wire feed speed, but these other process parameters are the same in each sample. Therefore, the following equation [25] is used to calculate the height of the weld bead, *h*, in these process parameters:

$$h = \frac{3\pi v_w d_w^2}{8 w v_t} \tag{3}$$

where w is the width of the weld bead, v_w is the wire feed speed, d_w is the wire diameter, v_t is the welding speed. The layer thickness is then calculated as 1.12 mm. Given that the condition of heat dissipation is different from that for an ideal weld bead, the preset layer thicknesses are 1.05, 1.05/1.1, and 1.1 mm, which lead to three different modes (Figure 8).

1. Figure 8a shows the no-droplet mode with small AL. The wire is fed into the bottom of the molten pool during the fabricating process, which means that the wire is in contact with the bottom of the molten pool during the feeding as in continuous molding. Opderbecke and Guiheux mentioned the existence of a liquid bridge, which contrasts with the no-contact mode. In this mode, the metal in the molten pool is mainly controlled by surface tension and gravity and remains stable, so each weld bead has a relatively flat appearance [26]. Compared with other modes, this mode can lower the height of the weld bead, reflected as collapse when the block part is fabricated.

2. As the AL increases, as shown in the Figure 8b, the tangent-droplet mode dominates. Here the wire does not touch the bottom of the molten pool during the feeding process. The front of the wire is melted, then a droplet is formed that adheres to the front of the wire owing to its own surface tension. The droplet becomes larger and larger until it is in contact with the molten pool. Owing to its own gravity and the arc force under the peak current, the droplet falls off the front of the wire and falls onto the formed surface to complete the metal transition from the wire smoothly.

3. Figure 8c shows the no-contact mode with large AL. The wire is further away from the bottom of the molten pool. Similarly, a droplet forms at the base current and sticks to the front of the wire. Necking phenomenon occurs as a result of gravity, and the droplet is stretched but cannot be in contact with the molten pool. Such droplets are blown to the formed surface by the arc force until the pulsed current works. In this mode, magnetic bias blowing occurs, and this easily leads to the splash of the droplets. At the same time, droplets in this mode are relatively close to the tungsten, and may pollute the tungsten and reduce the service life and fabricating stability.

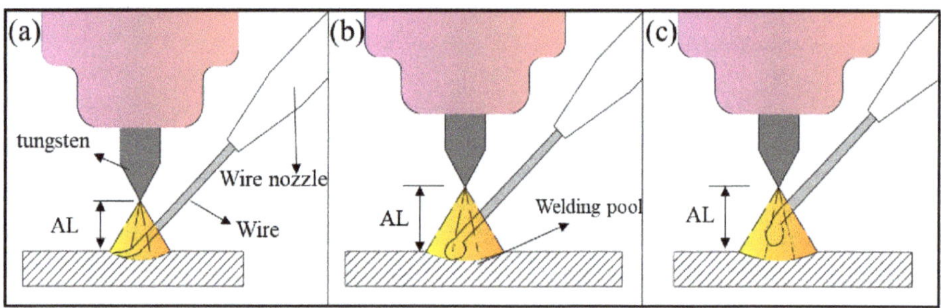

Figure 8. Schematic of (**a**) no-droplet mode; (**b**) tangent-droplet mode and (**c**) no-contact mode.

As sample 3 in Figure 7 shows, a layer thickness of 1.1 mm is selected, although the metal transfer mode of the first few layers is the tangent-droplet mode (Figure 8b). As the number of layers increases, the cumulative error becomes larger, leading to the AL becoming larger in the later layers. The transfer mode changes to the no-contact mode (Figure 8c), resulting in the occurrence of spatters. When the layer thickness is reduced to 1.05 mm, as sample 4 in Figure 7 shows, the initial layers are still via the tangent-droplet mode (Figure 8b), and the layers are still well formed at a height of 1.05 mm. Similarly, as the number of layers increases, the cumulative error makes the AL become too small, and the metal transfer mode is changed to the no-droplet mode (Figure 8a). Although this can reduce the number of droplets, the wire is always at the bottom of the molten pool. As a result, unfused wire may exist on the surface or inside of the part. Moreover, the wire always touches the fabricated surface in the process, which can produce a certain reaction force on the wire feeding system.

After taking the above effects into account, a variable layer thickness is adopted. When the first layers are fabricated at the initial stage, the substrate temperature is low, so a 1.05 mm layer thickness is set to ensure the tangent-droplet (Figure 8b) mode, which can ensure that the droplet falls on the predefined path and lays a good foundation of appearance for the later layers. To ensure that the AL is maintained at a certain distance in tangency-droplet mode (Figure 8b) in the remaining layers, a layer thickness of 1.1 mm is used for the rest of the layers. As sample 5 in Figure 7 shows, after changing the layer thickness, it can be clearly seen that the spatters of surface 2 are pretty rare and the collapse phenomenon is also improved a lot. Obviously, the adjustment of the layer thickness plays a significant role in controlling the existence of side spatters.

3.3. Effect of Tool Path on Macro Defects

3.3.1. The Form of Defects

Figure 9 shows the defects existing in three samples with different tool paths. It can be clearly seen that the defects are mainly in the form of poor flatness, where samples 5, 6, and 7 of Figure 9 are formed in a Z-shaped path, a vertical path and a contour-filling path respectively. For surface 1, the three samples all look smooth with little side spatters. For surface 2, sample 5 appears to be regularly corrugated and rough. Sample 6 appears slightly rippled, but the accuracy is not high. Sample 7 has no apparent corrugations in the longitudinal direction. There are only slight traces of overlapping layers. For surface 3, all the samples look smooth, but the main body of the samples is higher than the edge of the samples.

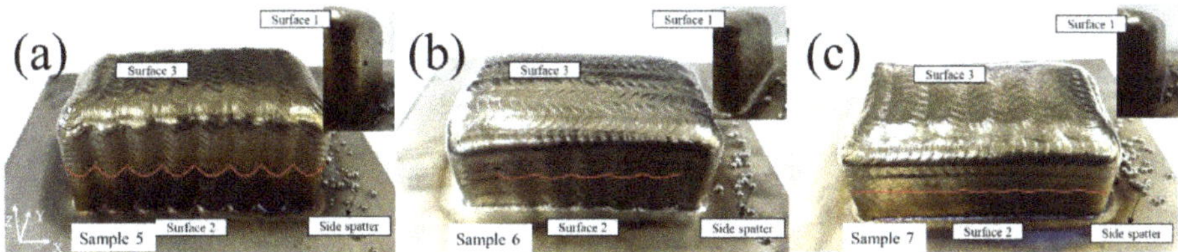

Figure 9. Macro images of samples 5, 6, and 7 with different tool paths: (**a**) Z-shaped path; (**b**) vertical path; and (**c**) contour-filling path.

3.3.2. Macro Defect Mechanisms with Different Tool Paths

The defects in this set of experiments mainly occurred in the form of poor flatness and side spatters on surface 2. The side spatters are the result of the lateral wire feeding which was discussed in Section 3.1. Different layer-to-layer overlapping methods lead to great differences in the appearance of the flatness.

From the path schematic (Figure 10a–c), there are mainly three types of layer-to-layer overlapping (Figure 10d–f). The trajectory of each layer in Figure 10a is repetitive. The paths of odd layers are perpendicular to those of the even layers in Figure 10b. The paths in Figure 10c are frame with zigzag filled, that is, the frame is fabricated for every layer at first, then the zigzag paths are used to fill the frame. Three overlapping types were analyzed combined with the value of the flatness as follows:

1. The first overlapping type is shown in Figure 10d. The surface is composed of the metal that is located on the turning of each weld bead. The movement of the machine tool is continuous, and the weld bead has a certain width. Therefore, the path in the 180° corner of each layer is always in the form of a circle, and because the paths of each layer are repetitive, surface 2 becomes obviously corrugated during the fabricating process. Figure 11 reflects the flatness values of surface 1, surface 2, and surface 3 of the three paths. Surface 2 of sample 5 in Figure 9 is used with this overlapping method with a flatness value of 1.689 mm. The second overlapping type is shown

in Figure 10e. The metal of odd layers is at the corners of each weld bead, and the metal of even layers is covered by a straight weld bead directly. In this way, the overlapping gaps produced by the odd layers are supplemented by the metal of even layers, and the even-numbered layers of molten metal complement the voids from the odd-numbered layers, reducing the amount of ripple phenomenon of surface 2.

2. The overlap patterns of surface 1 and surface 2 of sample 6 in Figure 9 are the same, but the values are 0.329 and 0.628 mm, a difference of nearly a factor of 2. This difference occurs mainly because the weld bead of surface 1 is much shorter than that of surface 2. The heat input is relatively concentrated, causing the even layer to fill the odd layers with more metal, thus reducing the flatness value. Furthermore, the accumulation of heat input tends to cause collapse on surface 1, which corresponds to the previous result. The third overlapping type is shown in Figure 10f. Each layer is a complete straight weld bead without turning of the path, so the surface can be smoother. There are only small scale patterns in the horizontal direction, which is the effect of overlapping of welding pool, an inherent phenomenon in wire arc additive manufacturing.

3. Surface 1 of sample 5 and surfaces 1 and 2 of sample 7 of Figure 9 are of the same overlapping type, according to the above description. Their flatness values are 0.434, 0.405, and 0.354 mm, which are almost the same and confirms the above-mentioned overlap mechanism.

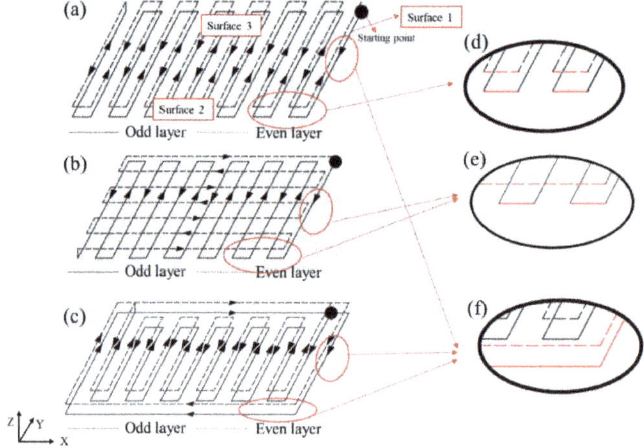

Figure 10. Schematic of (**a**) Z-shaped path; (**b**) vertical path; and (**c**) contour filling path; (**d**–**f**) mean three kinds of layer-to-layer overlapping methods.

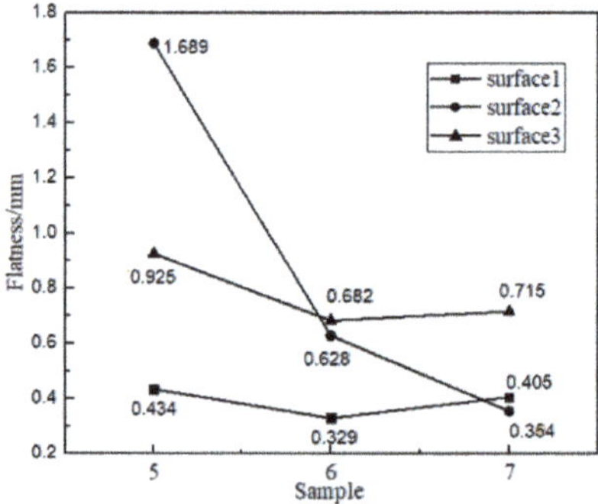

Figure 11. Flatness values of surfaces 1, 2, and 3 from samples 5, 6, and 7.

In conclusion, under the three paths, the flatnesses of the surfaces 3 and 1 are almost the same, but the flatnesses of surface 2 of samples 5, 6, and 7 in Figure 9 are successively decreased from 1.689 mm to 0.628 mm to 0.354 mm. Therefore, different paths still have obvious regulation effects on the flatness defect of the surface. Choosing a reasonable path for parts helps to improve surface accuracy, reduce the amount of subsequent processing and improve the efficiency of wire arc additive manufacturing.

3.4. Effect of Wire Curvature on Macro Defects

3.4.1. The Form of Defects

Figure 12 shows the defects existing in three samples with different wire curvatures. It can be seen clearly that the defects are mainly in the form of spatters and unmelted state. Samples 8 and 9 are fabricated using wires with curvatures of 1/150 and 1/300, and sample 10 is fabricated using a wire with a curvature of nearly zero after straightening, which can be regarded as a straight line. Surface 2 of sample 8 in Figure 12 shows that a large amount of wire poked out from the side of the sample in an unmelted state, and there are also droplets hanging onto surface 2 or spilling out. Surface 2 of sample 9 shows only a small amount of spatter. Surface 2 of sample 10 is free from spatters, and only droplets are present at the intersection of surface 2 and surface 1.

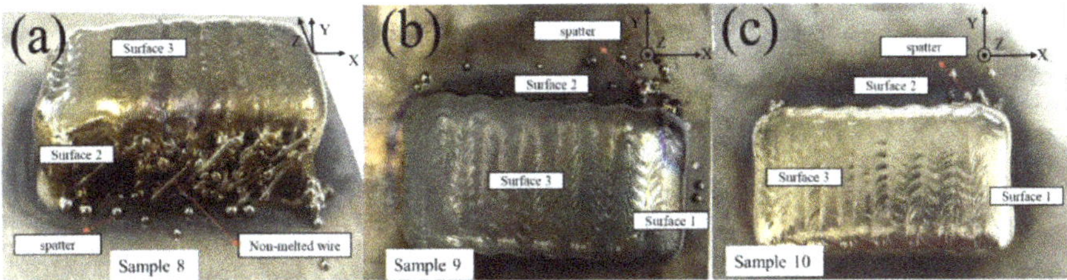

Figure 12. Macro images of samples 8, 9, and 10 with different wire curvatures: (**a**) 1/150; (**b**) 1/300; and (**c**) 0.

3.4.2. Macro Defect Mechanisms with Different Wire Curvatures

The defects of this experimental set are mainly in the form of unmelted wire and side spatters on surface 2. The wire used in the fabricating process was provided by the wire manufacturer, and different heat treatment processes resulted in different curvatures when the wire was wrapped around the wire reel. As shown in Figure 13, the larger the curvature of the wire, the more bent is the extended wire. This problem is not noticeable when fabricating a part with a few layers. When fabricating a block whose height is much higher than the substrate, wire with different curvatures can bring different defects once the path is scanned to the edge.

As Figure 13 shows, when the wire is straight (with a curvature of 0), the front of the wire is located on the center of the arc where the heat input is the highest and the droplets can fall accurately on the trajectory of the movement. Almost no spatter appears. Spatter from sample 10 occurs because the position of the spatter is the starting point of each layer where the welding torch is lifted suddenly, making the AL longer at that moment and causing the spatters.

When the curvature of the wire is 1/300, droplets do not form at the center of the arc under the tungsten. The droplets are easily blown out of the sample under the arc force in a fastigiated direction. The spatters of sample 9 are caused by this.

When the wire is excessively bent with a curvature of 1/150, the front of the wire stays away from the concentrated area of the heat input, causing the heat input to be too low so that the wire cannot be melted sufficiently. As a result, the spatters appear at surface 2, and the wire cannot be melted and poked out from surface 2. Therefore, the poor melting of the wire makes the wire adhere to the formed surface easily. It also has a certain influence on the wire feeding system and it can impede the stability of the equipment.

The three samples of Figure 12 show three kinds of wires whose curvatures are 1/300, 1/150, and 0. It can be clearly seen that the unmelted wire of surface 2 is gradually reduced, and the the number of spatters likewise, which confirms that the wire produced by different processes has a significant effect on the spatters. The wire used in sample 10 is based on the selection of a lower curvature wire, and a straightener is applied, so the wire from the wire nozzle is nearly straight. Therefore, choosing a suitable wire curvature also provides an effective method of controlling the generation of macro defects such as spatters, which is beneficial to reducing defects in the block parts.

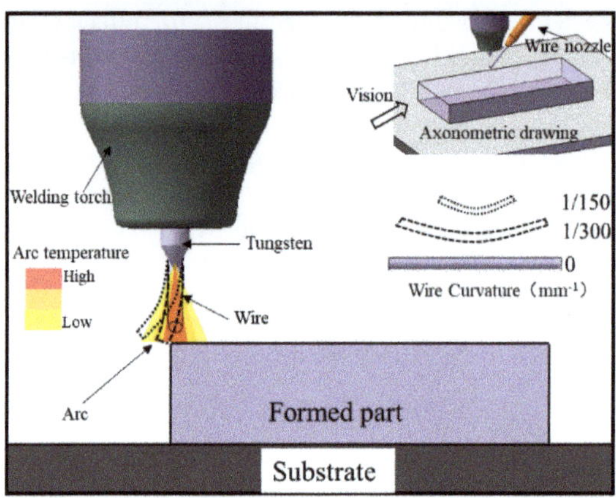

Figure 13. Schematic of the wire at the edge of the part for different curvatures.

4. Conclusions

In this paper, four variables were selected to find the defect forms of the sample fabricated by wire arc additive manufacturing. The effects of heat input, layer thickness, tool path, and wire curvature on macro defects were investigated to establish the mechanisms of macro defects. The controlling methods also could be obtained by adjusting these parameters. According to the results and discussions, the main conclusions of this paper are as follows:

1. Improper heat input brings effect of side spatters, collapse, and unmelted wire. When the heat input is decreased from 261 to 173 J/mm, the size and number of spatters are reduced obviously and less degree of collapse is observed as well. However, unmelted wire occurs as the heat input is decreased. Even if decreasing the heat input can reduce defects, the main reason for the side spatters is the lateral wire feeding mode.

2. Improper layer thickness brings effects of side spatters and collapse. Different layer thicknesses affect the arc length, which can determine the metal transfer mode: no-droplet mode, tangent-droplet mode, or no-contact mode. Based on lateral wire feeding, the no-contact mode easily results in side spatters and the no-droplet mode easily results in collapse. Choosing a changeable layer thickness can effectively reduce those defects.

3. Different tool paths mainly lead to different surface flatness levels. Three overlapping types at the edge of the sample are proposed. It is meaningful to choose the appropriate tool path to fabricate the components in WFAM, which can efficiently reduce post-processing time and improve manufacturing efficiency.

4. Different wire curvatures bring side spatters and unmelted wire. In this paper, three kinds of wire with curvatures of 0, 1/300, and 1/150 were investigated. It was found that the

more bent the wire, the more frequent are the spatters and the appearance of unmelted wire. Therefore, the development of wire standards or the use of a straightener in the manufacturing process can effectively improve the straightness of the wire and the reliability of manufacturing, which is of significance to the development of WFAM.

In summary, the most serious defects in WFAM are side spatters and collapse. Combined with the above results, it can be concluded that the main reason for these defects is that the wire is not melted in the arc accurately with improper variables. This paper illustrates the mechanism and proposes controlling methods so that these defects can be effectively reduced by choosing the appropriate heat input, layer thickness, and wire curvature, which is meaningful for the further application of WFAM.

Author Contributions: Conceptualization, L.J. and C.L.; Methodology, S.M. and H.F.; Formal Analysis, L.J., Q.W., and J.W.; Writing-Review & Editing, S.T and J.L.

Funding: This research was funded by [the National Natural Science Foundation of China] grant number [51505033] and [the Beijing Natural Science Foundation] grant number [3162027].

Conflicts of Interest: The authors declare no conflict of interest.

References

1. Fousova, M.; Vojtech, D.; Doubrava, K.; Daniel, M.; Lin, C.F. Influence of inherent surface and internal defects on mechanical properties of additively manufactured Ti6Al4V alloy: Comparison between selective laser melting and electron beam melting. *Materials* **2018**, *11*, 537. [CrossRef] [PubMed]

2. Brandl, E.; Schoberth, A.; Leyens, C. Morphology, microstructure, and hardness of titanium (Ti-6Al-4V) blocks deposited by wire-feed additive layer manufacturing (ALM). *Mater. Sci. Eng. A* **2012**, *532*, 295–307. [CrossRef]

3. Ding, D.; Pan, Z.; Cuiuri, D.; Li, H. Wire-feed additive manufacturing of metal components: Technologies, developments and future interests. *Int. J. Adv. Manuf. Technol.* **2015**, *81*, 465–481. [CrossRef]

4. Wu, Q.; Lu, J.; Liu, C.; Shi, X.; Ma, Q.; Tang, S.; Fan, H.; Ma, S. Obtaining uniform deposition with variable wire feeding direction during wire-feed additive manufacturing. *Mater. Manuf. Process.* **2017**, *32*, 1881–1886. [CrossRef]

5. Williams, S.W.; Martina, F.; Addison, A.C.; Ding, J.; Pardal, G.; Colegrove, P. Wire + arc additive manufacturing. *Mater. Sci. Technol.* **2016**, *32*, 641–647. [CrossRef]

6. Moures, F.; Cicală, E.; Sallamand, P.; Grevey, D.; Vannes, B.; Ignat, S. Optimisation of refractory coatings realised with cored wire addition using a high-power diode laser. *Surf. Coat. Technol.* **2005**, *200*, 2283–2292. [CrossRef]

7. Mok, S.H.; Bi, G.; Folkes, J.; Pashby, I. Deposition of Ti–6Al–4V using a high power diode laser and wire, part I: Investigation on the process characteristics. *Surf. Coat. Technol.* **2008**, *202*, 3933–3939. [CrossRef]

8. Thomsen, P.; Malmstrom, J.; Emanuelsson, L.; Rene, M.; Snis, A. Electron beam-melted, free-form-fabricated titanium alloy implants: Material surface characterization and early bone response in rabbits. *J. Biomed. Mater. Res. B Appl. Biomater.* **2009**, *90*, 35–44. [CrossRef] [PubMed]

9. Hafley, R.; Taminger, K.; Bird, R. Electron beam freeform fabrication in the space environment. In Proceedings of the 45th AIAA Aerospace Sciences Meeting and Exhibit, Reno, NV, USA, 8–11 January 2007.

10. Yilmaz, O.; Ugla, A.A. Microstructure characterization of SS308LSi components manufactured by GTAW-based additive manufacturing: Shaped metal deposition using pulsed current arc. *Int. J. Adv. Manuf. Technol.* **2016**, *89*, 13–25. [CrossRef]

11. Guo, J.; Zhou, Y.; Liu, C.; Wu, Q.; Chen, X.; Lu, J. Wire arc additive manufacturing of AZ31 magnesium alloy: Grain refinement by adjusting pulse frequency. *Materials* **2016**, *9*, 823. [CrossRef] [PubMed]

12. Frazier, W.E. Metal additive manufacturing: A review. *J. Mater. Eng. Perform.* **2014**, *23*, 1917–1928. [CrossRef]

13. Bai, X.; Colegrove, P.; Ding, J.; Zhou, X.; Diao, C.; Bridgeman, P.; Roman Hönnige, J.; Zhang, H.; Williams, S. Numerical analysis of heat transfer and fluid flow in multilayer deposition of PAW-based wire and arc additive manufacturing. *Int. J. Heat Mass Transf.* **2018**, *124*, 504–516. [CrossRef]

14. Brandl, E.; Palm, F.; Michailov, V.; Viehweger, B.; Leyens, C. Mechanical properties of additive manufactured titanium (Ti–6Al–4V) blocks deposited by a solid-state laser and wire. *Mater. Des.* **2011**, *32*, 4665–4675. [CrossRef]

15. Hagqvist, P. Laser Metal Deposition with Wire-Enabling New Manufacturing Solutions through Process Control. Available online: http://www.lortek.es/files/merlin/03-P_Hagqvist-UW&GKN-Aero-Case-study.pdf (accessed on 27 June 2018).
16. Boeing 787 Dreamliner Will Be First Commercial Airplane to Fly with Certified Additive-Manufactured Titanium Parts in Structural Applications. Available online: http://www.norsktitanium.com/media/press/norsk-titanium-to-deliver-the-worlds-first-faa-approved-3d-printed-structural-titanium-components-to-boeing (accessed on 27 June 2018).
17. Civil Aviation Demand Flying High. Available online: https://www.metalbulletin.com/Article/3552854/3650683/ (accessed on 27 June 2018).
18. Li, Y.; Han, Q.; Zhang, G.; Horváth, I. A layers-overlapping strategy for robotic wire and arc additive manufacturing of multi-layer multi-bead components with homogeneous layers. *Int. J. Adv. Manuf. Technol.* **2018**, *96*, 3331–3344. [CrossRef]
19. Kwak, Y.-M.; Doumanidis, C.C. Geometry regulation of material deposition in near-net shape manufacturing by thermally scanned welding. *J. Manuf. Process.* **2002**, *4*, 28–41. [CrossRef]
20. Xiong, J.; Zhang, G.; Qiu, Z.; Li, Y. Vision-sensing and bead width control of a single-bead multi-layer part: Material and energy savings in gmaw-based rapid manufacturing. *J. Clean. Prod.* **2013**, *41*, 82–88. [CrossRef]
21. Wu, Q.; Ma, Z.; Chen, G.; Liu, C.; Ma, D.; Ma, S. Obtaining fine microstructure and unsupported overhangs by low heat input pulse arc additive manufacturing. *J. Manuf. Process.* **2017**, *27*, 198–206. [CrossRef]
22. Giridharan, P.K.; Murugan, N. Optimization of pulsed GTA welding process parameters for the welding of AISI 304L stainless steel sheets. *Int. J. Adv. Manuf. Technol.* **2008**, *40*, 478–489. [CrossRef]
23. Kishore Babu, N.; Ganesh Sundara Raman, S.; Mythili, R.; Saroja, S. Correlation of microstructure with mechanical properties of TIG weldments of Ti–6Al–4V made with and without current pulsing. *Mater. Charact.* **2007**, *58*, 581–587. [CrossRef]
24. Ding, D.; Pan, Z.; Cuiuri, D.; Li, H. A multi-bead overlapping model for robotic wire and arc additive manufacturing (WAAM). *Robot. Comput.-Integr. Manuf.* **2015**, *31*, 101–110. [CrossRef]
25. Suryakumar, S.; Karunakaran, K.P.; Bernard, A.; Chandrasekhar, U.; Raghavender, N.; Sharma, D. Weld bead modeling and process optimization in hybrid layered manufacturing. *Comput.-Aided Des.* **2011**, *43*, 331–344. [CrossRef]
26. Opderbecke, T.; Guiheux, S. Toptig: Robotic tig welding with integrated wire feeder. *Weld. Int.* **2009**, *23*, 523–529. [CrossRef]

Article

Effect of Mg Content on Microstructure and Properties of Al–Mg Alloy Produced by the Wire Arc Additive Manufacturing Method

Lingling Ren [1], Huimin Gu [1,*], Wei Wang [2], Shuai Wang [1], Chengde Li [1], Zhenbiao Wang [3], Yuchun Zhai [1,3] and Peihua Ma [1]

[1] College of Metallurgy, Northeastern University, Shenyang 110000, China; renlingling27@126.com (L.R.); wangshuai106123@126.com (S.W.); lichengde20031698@126.com (C.L.); zhaiyc@smm.neu.edu.cn (Y.Z.); mapeihuadbdx@126.com (P.M.)

[2] Inner Mongolia Metal Research Institute, Baotou 014000, China; wwneu@hotmail.com

[3] North East Industrial Materials & Metallurgy Co., Ltd, Fushun 113000, China; zhenbiaowang@hotmail.com

* Correspondence: guhm@smm.neu.edu.cn

Received: 13 November 2019; Accepted: 5 December 2019; Published: 11 December 2019

Abstract: In this study, an Al–Mg alloy was fabricated by wire arc additive manufacture (WAAM), and the effect of Mg content on the microstructure and properties of Al–Mg alloy deposits was investigated. The effects on the deposition surface oxidation, geometry, burn out rate of Mg, pores, microstructure, mechanical properties and fracture mechanisms were investigated. The results show that, when the Mg content increased, the surface oxidation degree increased; a "wave"-shaped deposition layer occurred when the Mg content reached 8%. When the Mg content was more than 6%, the burning loss rate of the Mg element increased significantly. With the increase of Mg content, the number of pores first decreased and then increased, and the size first decreased and then increased. When the Mg content reached 7% or above, obvious crystallization hot cracks appeared in the deposit bodies. When the Mg content increased, the precipitated phase (FeMn)Al$_6$ and β(Mg$_2$Al$_3$) increased, and the grain size increased. When the Mg content was 6%, the comprehensive mechanical properties were best. The horizontal tensile strength, yield strength and elongation were 310 MPa, 225 MPa and 17%, respectively. The vertical tensile strength, yield strength and elongation were 300 MPa, 215 MPa and 15%, respectively. The fracture morphology was a ductile fracture.

Keywords: arc additive manufacture; Al–Mg alloy; Mg content; microstructure; mechanical properties

1. Introduction

Wire arc additive manufacture (WAAM) is an additive manufacturing technology which is based on the discrete additive forming principle to form 3D physical parts suitable for the rapid manufacturing of medium and large-scale parts with medium complexity [1–3]. The WAAM method cannot realize net forming at present, which requires subsequent machining. At present, most WAAM aluminum alloys are Al–Cu alloys and Al–Si–Mg alloy, which need a solid solution and aging heat treatment to be strengthened [4–6]. In actual production, parts—especially large parts—will undergo severe deformation after quenching treatment, which makes the product accuracy difficult to control and subsequent machining difficult. Therefore, the pursuit of an Al–Mg alloy with excellent mechanical properties without heat treatment and strengthening has attracted the attention of WAAM manufacturing technology researchers [7,8].

At present, the research into WAAM Al–Mg alloys is not in-depth and is limited to traditional brands. Jiang [9] studied the rapid forming process of 5356 aluminum alloy based on CMT (Cold Metal Transfer) and proposed the anisotropy of mechanical properties, but did not explain the reason behind this. Horgar et al. [10] prepared AA5183 aluminum alloy by using the short pulse arc additive manufacturing method, but the mechanical properties of the deposits were not high, and the tensile strength and yield strength were 293 MPa and 145 MPa, respectively. Geng et al. [11] fabricated 5A06 aluminum alloy with a GTAW (gas tungsten arc welding) additive and obtained deposits with poor properties. The tensile strength, yield strength and elongation were 273 MPa, 124 MPa and 34%, respectively. As Mg is the main strengthening element of the Al–Mg alloy, it has high activity and is easy to burn and oxidize; thus, Mg content exerts an important influence on the performance of WAAM Al–Mg alloy deposits.

In this paper, Al–Mg alloy deposits with different Mg contents were prepared by the WAAM method, and the surface oxidation degree, geometric morphology, burning loss of Mg elements, pores, microstructure, mechanical properties and tensile fracture mechanism of Al–Mg alloy deposits were studied, laying a foundation for the further study of WAAM Al–Mg alloys.

2. Experimental Method

The Al–Mg alloy welding wire used in this paper was provided by North East Industrial Materials & Metallurgy Co., Ltd. (Fushun, China) and had a diameter of 1.2 mm. In this experiment, four kinds of welding wires with different Mg contents were prepared, and the target mass percentages of Mg content were 5%, 6%, 7% and 8%, with corresponding numbers of 1#, 2#, 3# and 4#, respectively. The Mg content mentioned in this paper was the target mass percentage of wire, and the measured chemical compositions of wires are shown in Table 1. The chemical composition of the deposit was measured from the upper, middle and lower points of the deposit and is shown in Table 2. The CMT + Advance forming process in Fronius Advanced CMT [12] power supply is adopted, and the equipment is shown in Figure 1. Because the Mg element is active, the Al–Mg alloy is greatly affected by process parameters (such as interpass temperature, etc.), and the experimental results are obtained under specific process parameters. The deposition process parameters are shown in Table 3, and the size of the deposition body is 200 mm × 150 mm.

Table 1. Chemical composition of welding wire with different Mg contents.

	Si	Fe	Cu	Mn	Mg	Zn	Zr
1#	0.046	0.107	0.0034	0.84	5.03	0.0083	0.098
2#	0.049	0.102	0.002	0.85	6.10	0.0073	0.092
3#	0.041	0.120	0.0019	0.85	6.88	0.0074	0.087
4#	0.042	0.137	0.0063	0.84	7.91	0.0098	0.087

Table 2. Chemical composition of deposits with different Mg contents.

	Position	Si	Fe	Cu	Mn	Mg	Zn	Zr
	Upper	0.047	0.114	0.0040	0.83	4.67	0.0093	0.087
1#	Middle	0.055	0.113	0.0044	0.79	4.58	0.0096	0.090
	Lower	0.048	0.109	0.0045	0.78	4.61	0.0087	0.087
	Average	**0.050**	**0.112**	**0.0043**	**0.80**	**4.62**	**0.0092**	**0.088**
	Upper	0.052	0.105	0.0030	0.81	5.68	0.0083	0.086
2#	Middle	0.057	0.109	0.0029	0.85	5.65	0.0087	0.088
	Lower	0.050	0.113	0.0025	0.80	5.59	0.0082	0.078
	Average	**0.053**	**0.109**	**0.0028**	**0.82**	**5.64**	**0.0084**	**0.084**
	Upper	0.042	0.131	0.0028	0.74	6.18	0.0074	0.079
3#	Middle	0.043	0.124	0.0023	0.82	6.23	0.0079	0.080
	Lower	0.047	0.126	0.0027	0.81	6.19	0.0078	0.084
	Average	**0.044**	**0.127**	**0.0026**	**0.79**	**6.20**	**0.0077**	**0.081**
	Upper	0.050	0.141	0.0062	0.78	7.00	0.0094	0.077
4#	Middle	0.045	0.138	0.0063	0.85	6.97	0.0099	0.082
	Lower	0.043	0.145	0.0070	0.80	7.06	0.0086	0.078
	Average	**0.046**	**0.141**	**0.0065**	**0.81**	**7.01**	**0.0093**	**0.079**

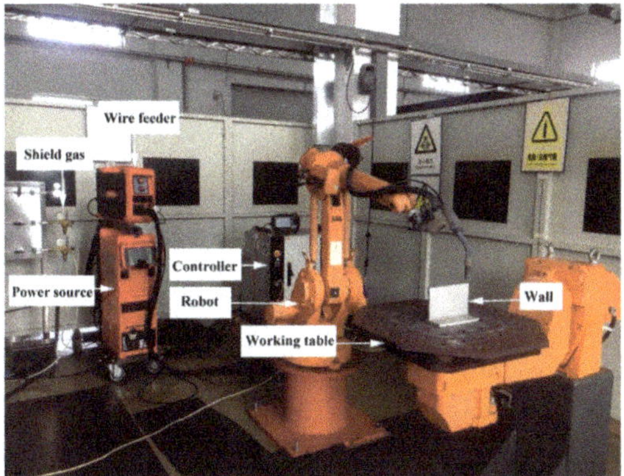

Figure 1. Cold Metal Transfer (CMT)-wire arc additive manufacture (WAAM) system.

Table 3. Deposition process parameters.

Process Parameters	
Current	90 A
Arc voltage	10 V
Travel speed	8 mm/s
Wire feed speed	5.5 mm/min
Interlayer wait time	90 s
99.999% argon flow rate	25 L/min

The sampling location of the deposition body and the specification of tensile samples are shown in Figure 2. Two tensile test samples perpendicular to the deposition direction (horizontal samples) and two tensile test samples parallel to the deposition direction (vertical sample) were extracted from each deposit body. Tensile samples were processed by using a milling machine; the size and roughness are shown in Figure 2b. Tensile tests were performed at room temperature using a wdw-300 micro-controlled electronic universal testing machine. An ICAP7400 plasma spectrometer (Thermo Scientific, Waltham, MA, USA) was used for component detection. In this paper, the chemical composition of the deposit is the average of the measured values at the top, middle and bottom of the deposit. The metallographic specimens were ground and polished to a mirror finish and then etched in mixed acid reagent containing 0.5 vol% HF, 1.5 vol% HCl and 2.5 vol% HNO_3, with the balance consisting of H_2O. The etching time is 20 s. A metallographic microscope (OM), scanning electron microscope (SEM) and energy spectrum (EDS) were used to observe the microstructure and analyze its composition. The width of the deposition body was measured by the Vernier caliper. Three parallel samples were taken and the average values of the measurement results were obtained.

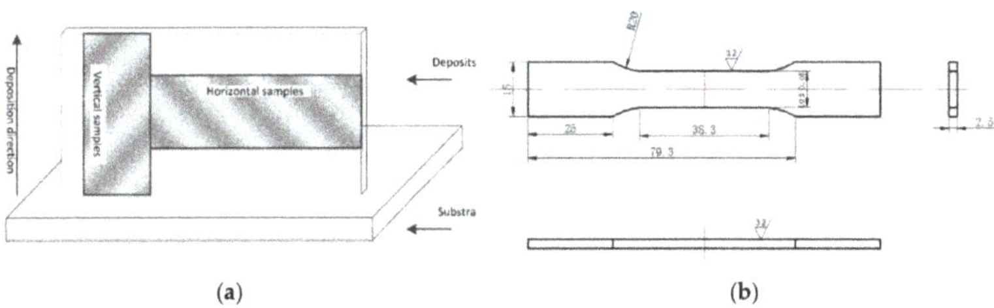

(**a**)
(**b**)

Figure 2. Schematic diagram of extracted tensile sample ((**a**): Sampling position of tensile samples; (**b**): Tensile sample specification (the units for coupon dimension are mm)).

3. Results and Discussion

3.1. Surface Oxidation and Deposition Geometry

Figure 3 shows the surface morphologies of deposits with different Mg contents. It can be seen that, with the increase of Mg content, the surface colour of the deposits gradually deepens, and the lines of deposits on the surface of the deposition body are clearly visible. When the Mg content is less than 7%, the surface texture is smooth, and when the Mg content is 8%, the deposition layer shows a "wave" shape. This indicates that the increase of Mg content will increase the surface oxidation degree. The "wave"-shaped layer is caused by the increasing viscosity of the molten pool, reducing its fluidity when the Mg content is too high [13].

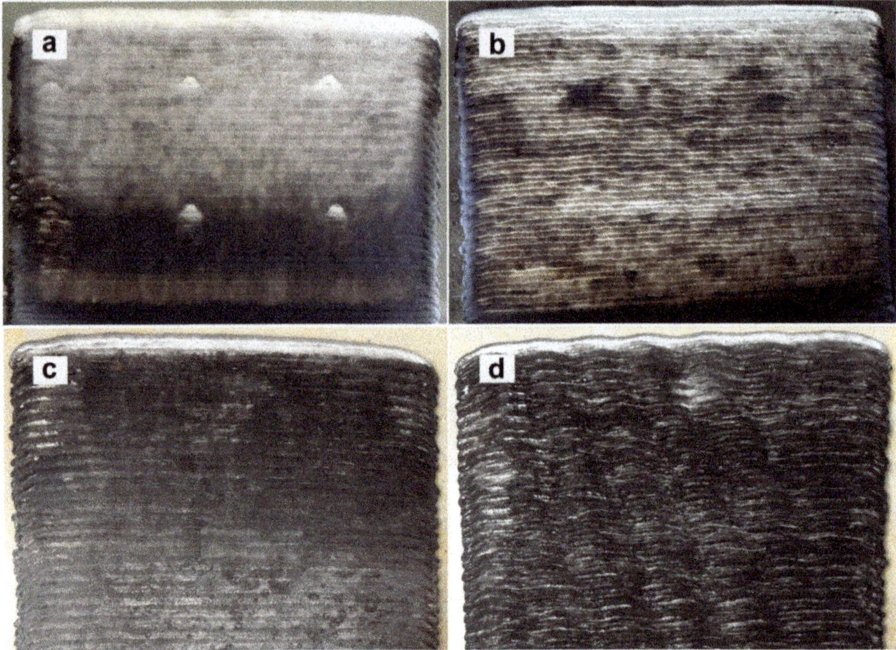

Figure 3. The surface oxidation appearance of WAAM Al–Mg deposits with different Mg contents ((**a**): 5% Mg; (**b**): 6% Mg; (**c**): 7% Mg; (**d**): 8% Mg).

Figure 4 shows the width of deposition bodies with different Mg contents. Due to the increase of Mg content, the viscosity of the molten pool increases, the fluidity and spreading property become worse, and the width of the deposition body slightly decreases; the width of the deposition body decreases by 6.7% when the Mg content is 8% compared with 5%.

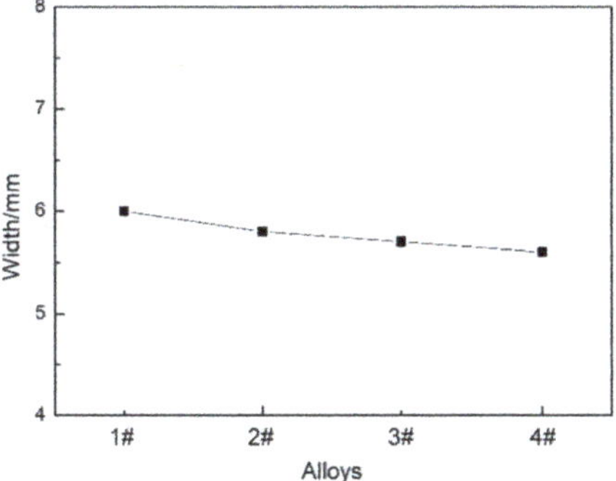

Figure 4. Width of deposits with different Mg contents.

3.2. Burn Loss Rate of Mg Elements

Figure 5 is the comparison diagram of the Mg burning loss in deposition bodies with different Mg contents. The burn loss rate is calculated by Equation (1):

$$A = \frac{X_1 - X_2}{X_1} \tag{1}$$

where A is the burn loss rate of the element, X_1 is the measured element content in the wire, and X_2 is the measured element content in the deposit.

It can be seen that the deposition bodies all show different degrees of Mg burning loss. When the Mg content increased from 5% to 6%, the burning rate of Mg increased slightly. When the content of Mg was more than 6%, the burning rate of Mg increased significantly. This is because Mg is more active, and its activity is proportional to its concentration.

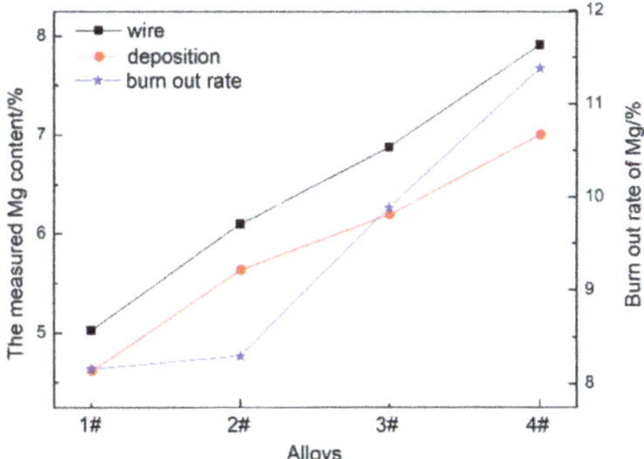

Figure 5. The burning loss rate of Mg in deposits with different Mg contents.

3.3. Microstructure

Figure 6a–d shows the pores in the deposition bodies with different Mg contents. It can be seen that the pores in the deposition body are all round. With the increase of Mg content, the number of pores first decreases and then increases, and the size first decreases and then increases. When the Mg content is 6%, the number and size of pores are the least. There are three stages in the formation of pores: nucleation of bubbles, growth of bubbles and rise of bubbles. The nucleation probability of bubbles depends on Equation (2) [14]:

$$j = Ce^{-\frac{4\pi r\sigma}{3KT}} \tag{2}$$

where j is the number of bubble nuclei formed per unit time, r is the critical radius of the bubble core, and σ is the surface tension between the bubbles and the metal liquid. K is the Boltzmann constant ($K = 1.38 \times 10 - 16$ erg/K). T is kelvin (K), and C is constant. When the Mg content increases, the viscosity of the molten pool increases, and the surface tension between the bubbles and metal liquid increases. It can be seen from Equation (1) that the nucleation probability of the bubbles decreases. In addition, the growth of bubbles needs to satisfy the relational Equation (3) [14]:

$$P_h > P_o \tag{3}$$

where P_h is the internal pressure of the bubble and P_o is the external pressure of the bubble. It can be seen from Equation (2) that, when the Mg content increases, the viscosity of the molten pool increases and the external pressure P_o of bubbles increases, which hinders the growth of bubbles.

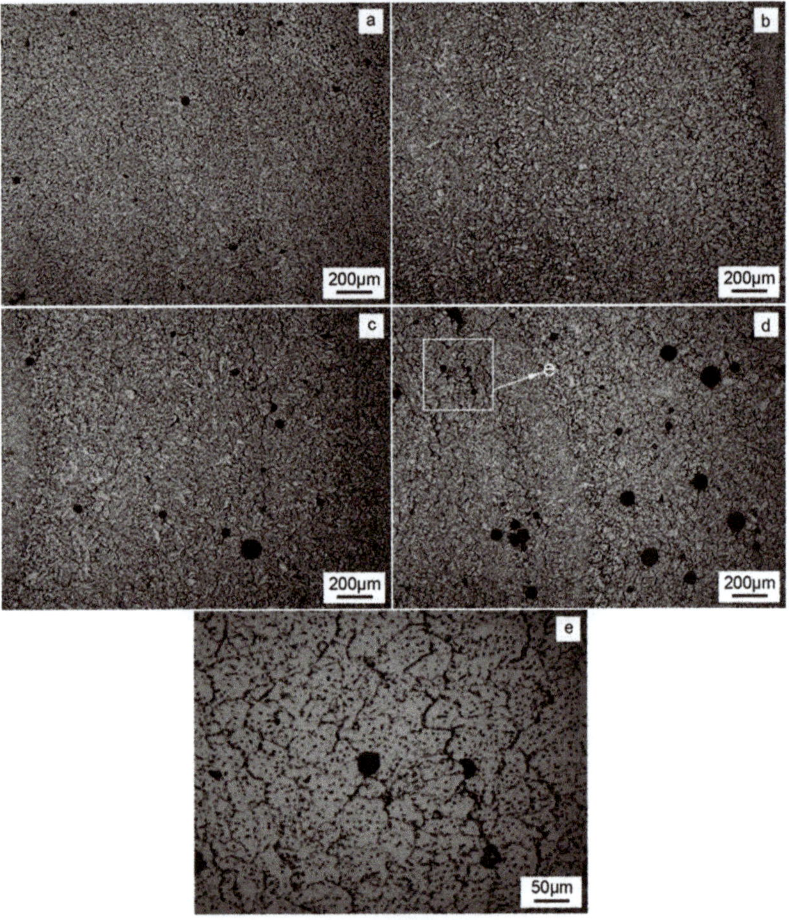

Figure 6. Optically observed porosity and cracking for the deposits with different Mg contents ((**a**): 5% Mg; (**b**): 6% Mg; (**c**): 7% Mg; (**d**): 8% Mg; (**e**): enlarged view of the crack in Figure 6d).

From the above analysis, it can be seen that increasing the Mg content will reduce the nucleation probability of bubbles and hinder the growth of bubbles, which is why the number and size of bubbles decrease when the Mg content increases from 5% to 6%. However, when the Mg content continues to increase, the number and size of pores increase, which is related to the origin of pores. Hydrogen is the main reason for the existence of pores in the WAAM aluminum alloy. Water in the arc column atmosphere and water absorbed by the wire and substrate are important sources of hydrogen. Mg has high activity and is easy to oxidize. When the content of Mg is increased, the following two sources of hydrogen will be increased: first, the oxide film on the surface of the wire thickens and increases the water absorption; second, during the processing of WAAM, the deposition body surface forms Al_2O_3 due to the oxidation and burning loss of Mg, which easily absorbs water in the air. This can be easily seen in Figures 3 and 5. When the Mg content is more than 6%, the surface oxidation is serious, and the burning rate of Mg increases significantly, greatly increasing the water absorption of the body. Therefore, when the Mg content is higher than 6%, the number and size of pores increase.

Cracks are clearly visible in Figure 6c,d, while no obvious cracks are found in Figure 6a,b. Figure 6e is a magnification of cracks in Figure 6d. It can be seen that cracks occur and develop along grain boundaries, and the main extension direction is perpendicular to the deposition direction, which belongs to the crystal thermal crack, which has a serious impact on the vertical mechanical properties. When the Mg content of the alloy is increased, the viscosity of the molten pool increases and the fluidity is poor. During the rapid solidification process of the molten pool, it cannot feed in a timely manner, and the thermal crystallization cracking occurs due to the influence of tensile stress.

Figure 7 shows the microstructure of layers and interlayers (remelting parts) of depositions with different Mg contents. As can be seen, the layer and interlayer tissues can be clearly distinguished. When the Mg content is 5% and 6%, the microstructures of the layers and interlayers are finely equiaxed crystals, and the interlayer grains are smaller. With the increase of Mg content, the grains grew gradually. When the Mg content was 8%, columnar crystals with la arger size appeared in the interlayer. With the increase of Mg content, the number of precipitated phases increased along with the precipitation phase aggregation phenomenon.

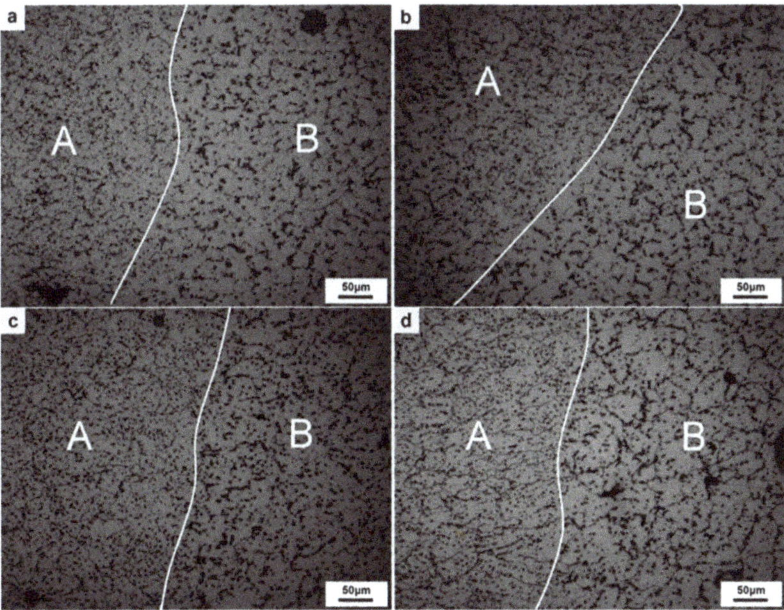

Figure 7. The metallographic structure of deposits with different Mg contents ((**a**): 5% Mg; (**b**): 6% Mg; (**c**): 7% Mg; (**d**): 8% Mg; A: interlayer; B: layer).

Figure 8 shows the two main precipitated phases in the deposition body. According to spectrogram A and spectrogram B, the structure and image of the aluminum alloy [15] and the binary phase diagrams of Al–Mg and Al–Mn show that the two precipitated phases are the $(FeMn)Al_6$ phase and $\beta(Mg_2Al_3)$ phase, respectively. The $(FeMn)Al_6$ phase is insoluble, hard and brittle, and is a thick sheet segregation polymer. The $\beta(Mg_2Al_3)$ phase is face-centered cubic and brittle at room temperature, meaning that the more of this phase the alloy has, the less plastic it is. Al–Mg alloy is a solid solution strengthening alloy [15]. Parts of the Mg elements are dissolved in the $\alpha(Al)$, and the rest are precipitated in the form of $\beta(Mg_2Al_3)$ phase. Figure 9 shows the contents of Mg and Mn within the grains. As can be seen, with the increase of Mg content, the $\beta(Mg_2Al_3)$ phase is precipitated more. In addition, with the increase of solid solution of Mg element in the $\alpha(Al)$, the Mn solid solution decreases, thus increasing the precipitation of the $(FeMn)Al_6$ phase. The precipitation of these two phases results in the coarsening of deposition and the growth of grain.

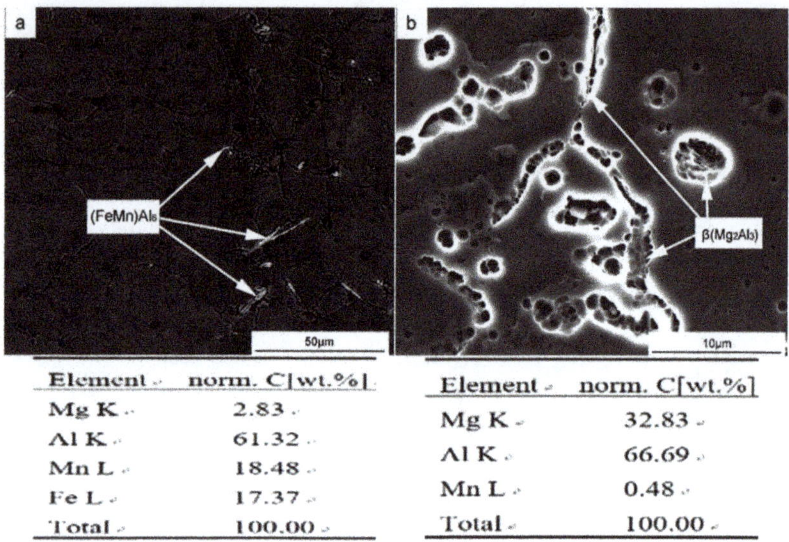

Element	norm. C[wt.%]
Mg K	2.83
Al K	61.32
Mn L	18.48
Fe L	17.37
Total	100.00

Element	norm. C[wt.%]
Mg K	32.83
Al K	66.69
Mn L	0.48
Total	100.00

Figure 8. The morphology and composition of the precipitated phase in WAAM Al–Mg alloy deposits ((**a**): $(FeMn)Al_6$ phase; (**b**): $\beta(Mg_2Al_3)$ phase).

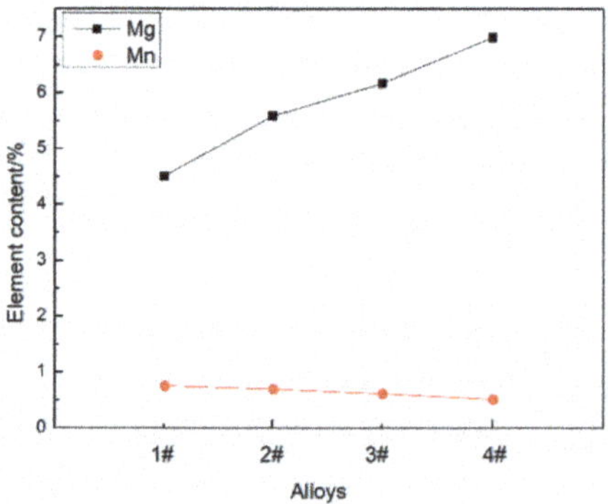

Figure 9. The contents of Mg and Mn within the grains.

3.4. Mechanical Properties

Figure 10 shows the horizontal and vertical mechanical properties of depositions with different Mg contents. As shown in Figure 10a, when the Mg content in the deposition body is less than 7%, the horizontal tensile strength and yield strength increase with the increase of Mg content. When the Mg content was more than 7%, both the tensile strength and yield strength decreased. The elongation decreases with the increase of Mg content. The strengthening mechanism of the Mg element is solid solution strengthening; therefore, with the increase of Mg content, the content of the solid solution Mg in α(Al) increases, and the tensile strength and yield strength increase. However, when the Mg content was more than 7%, the tensile strength and yield strength decreased due to the coarsening of the tissues, the increase of pores and the severe thermal cracks in the crystals. The decrease of elongation was mainly caused by the increase of the precipitated phase and the thermal cracking of the crystal.

As shown in Figure 10b, the vertical tensile strength, yield strength and elongation all peak when the Mg content is 6%. The mechanical properties of a Mg content greater than 6% were significantly reduced—especially the elongation. The increase of Mg content can improve the vertical mechanical properties, but when the Mg content is more than 6%, the occurrence of thermal cracks causes a sharp decline in the vertical mechanical properties.

Comparing the horizontal and vertical mechanical properties of Figure 10a,b, when the Mg content is 5–6%, the horizontal and vertical properties are more uniform. The difference was smallest when the Mg content was 6%. When the Mg content was more than 6%, the difference in mechanical properties between the horizontal and vertical directions increased significantly. There are two reasons for this difference in mechanical properties: one is the heterogeneity of the layer and the interlayer tissues. The grains of the layer were larger than those in the interlayer, and with the increase of Mg content, $\beta(Mg_2Al_3)$ and $(FeMn)Al_6$ phase segregated and aggregated in the layer. The nonuniformity of the layered structure is detrimental to the vertical mechanical properties. The second reason for this difference is the generation of thermal cracking. It can be seen from Figure 6c,d that the extension direction of the crack is perpendicular to the deposition direction—that is, parallel to the layer—which can seriously reduce the vertical mechanical properties.

A comprehensive data analysis shows that when the Mg content is 5–6%, the mechanical properties are better. When the Mg content is 6%, the comprehensive mechanical performance is the best, the horizontal tensile strength, yield strength and elongation are 310 MPa, 225 MPa and 17% respectively, and the vertical tensile strength, yield strength and elongation are 300 MPa, 215 MPa and 15%, respectively. The mechanical property data are consistent with the trend of the microstructure.

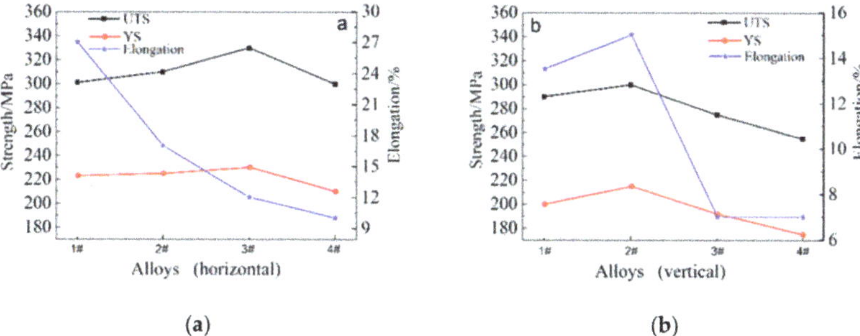

Figure 10. The mechanical properties of deposits with different Mg contents. (**a**): horizontal mechanical properties; (**b**): vertical mechanical properties.

3.5. Fracture Behaviour

Figure 11 shows the fracture morphology of the tensile specimen. Letters a, b, c and d show the fracture morphology of the horizontal tensile specimen, while e, f, g and h are fracture surfaces of the vertical tensile specimen. It can be seen that all fracture mechanisms are ductile fractures with obvious dimples. When the Mg content was 5% and 6%, the dimples were fine and uniform, and the horizontal and vertical fracture morphology was basically the same. When the Mg content increased to more than 7%, the number of dimples decreased, and more "grape grain"-like loose tissues appeared due to delayed supplementation, especially vertical fractures. This kind of shrunk loose tissue appeared in a large area, corresponding to the crack in Figure 6c,d, which further explained the cause of the crack.

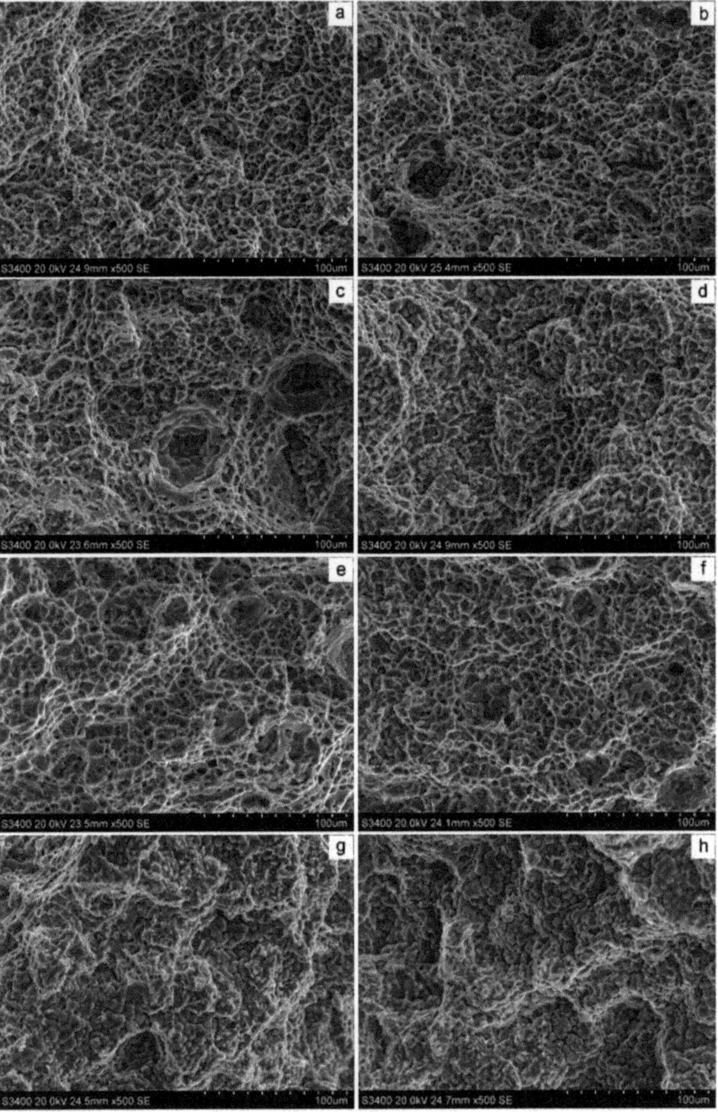

Figure 11. The fracture morphology of tensile simples with different Mg contents ((**a**): 5% Mg; (**b**): 6% Mg; (**c**): 7% Mg; (**d**): 8% Mg; (**e**): 5% Mg; (**f**): 6% Mg; (**g**): 7% Mg; (**h**): 8% Mg. (**a–d**): horizontal fracture; (**e–h**): vertical fracture).

4. Conclusions and Future Prospects

(1)   The content of Mg affects the surface oxidation degree and geometric size of WAAM Al–Mg alloy deposits. The surface oxidation degree increased with the increase of Mg content. When the Mg content reached 8%, a "wave"-shaped deposition layer appeared.

(2)   The effect of Mg content on the mechanical properties of the WAAM Al–Mg alloy is significant. The mechanical properties were excellent when the Mg content was controlled at 5–6%. When the Mg content is 6%, the comprehensive mechanical properties were optimized, with the horizontal tensile strength, yield strength and elongation being 310 MPa, 225 MPa and 17% respectively, and the vertical tensile strength, yield strength and elongation being 300 MPa, 215 MPa and 15%, respectively.

(3)   The effect of Mg contents on the properties of WAAM Al–Mg alloy deposits is mainly attributed to three points: first, with the increase of Mg content, the number of pores first decreases, then increases, and the size first decreases, then increases. When the Mg content is 6%, the number of pores is the least and the size is the smallest. Secondly, when the Mg content reaches 7% or above, a serious shrinkage will appear due to the poor fluidity of the molten pool, which will lead to crystallization heat cracking. Third, with the increase of Mg content, the precipitated phase $(FeMn)Al_6$ and $\beta(Mg_2Al_3)$ increased, and the grain size increased, and larger columnar crystals appeared in the layer when the Mg content was 8%.

In this paper, the influence of Mg content on WAAM Al–Mg alloy deposits was systematically described, and the optimal Mg content range was obtained, which is of guiding significance for the development of WAAM Al–Mg alloys. The WAAM technology's technological characteristics determine that the composition of an alloy fabricated by WAAM is special. In future studies, other alloy elements of Al–Mg alloys will be further investigated to finally obtain the composition range of Al–Mg alloys suitable for the WAAM process, in order to promote the engineering application of WAAM Al–Mg alloys.

Author Contributions: Conceptualization, Y.Z. and P.M.; methodology, W.W.; software, C.L. and Z.W.; validation, L.R., S.W. and H.G.; formal analysis, L.R.; investigation, S.W.; resources, W.W.; data curation, H.G.; writing—original draft preparation, L.R.; writing—review and editing, L.R.; visualization, L.R. and C.L.; supervision, W.W.; project administration, W.W.

Funding: This research received no external funding

Acknowledgments: The authors would like to express appreciation for the support provided by the National Key Research and Development Plan "additive manufacturing and laser manufacturing" (project no. 2018yfb1106300-5).

Conflicts of Interest: The authors declare no conflict of interest.

References

1.   Derekar, K.S. A review of wire arc additive manufacturing an advances in wire arc additive manufacturing of aluminium. *Mater. Sci. Technol.* **2018**, *34*, 859–916. [CrossRef]

2.   Gu, J.L.; Ding, J.L.; Cong, B.Q.; Bai, J.; Gu, H.M.; Williams, S.W.; Zhai, Y.C. The influence of wire properties on the quality and performance of wire+arc additive manufactured aluminium part. *Adv. Mater. Res.* **2014**, *1081*, 210–214. [CrossRef]

3.   Williams, S.W.; Martina, F.; Addison, A.C.; Jialuo, D.; Goncalo, P.; Paul, A.C. Wire+ arc additive manufacturing. *Mater. Sci. Technol.* **2016**, *32*, 641–647. [CrossRef]

4.   Shuai, W.; Gu, H.; Wei, W.; Li, C.; Ren, L.; Wang, Z.; Zhai, Y.; Ma, P. Microstructure and Mechanical Properties of Aluminum Alloy (ZL205A) Wall Produced by Wire Arc Additive Manufacturing Method. Available online: http://www.rmme.ac.cn/rmme/ch/reader/create_pdf.aspx?file_no=20181225&flag=1&journal_id=rmme&year_id=2019 (accessed on 1 October 2019).

5.   Li, C.; Gu, H.; Wang, W.; Zhai, Y.; Ming, Z.; Wang, S.; Ren, L.; Wang, Z. Investigation on Microstructure and Properties of Aluminum Alloy (ZL114A) by Wire Arc Additive Manufacturing (WAAM). Available online: http://www.rmme.ac.cn/rmme/ch/reader/create_pdf.aspx?file_no=20181148&flag=1&journal_id=rmme&year_id=2019 (accessed on 1 October 2019).

6.	Cong, B.; Qi, Z.; Qi, B.; Sun, H.; Zhao, G.; Ding, J. A Comparative Study of Additively Manufactured Thin Wall and Block Structure with Al–6.3%Cu Alloy Using Cold Metal Transfer Process. *Appl. Sci.* **2017**, *7*, 275. [CrossRef]

7.	Jone, R.H.; Baer, D.R.; Danielson, M.J.; Vetrano, J.S. Role of Mg in the stress corrosion cracking of an Al-Mg alloy. *Metall. Mater. Trans. A* **2001**, *32A*, 1699–1711. [CrossRef]

8.	Gu, J.; Wang, X.; Bai, J.; Ding, J.; Williams, S.; Zhai, Y.; Liu, K. Deformation microstructures and strengthening mechanisms for the wire+ arc additively manufactured Al-Mg4.5Mn alloy with inter-layer rolling. *Mater. Sci. Eng. A* **2017**, *712*, 292–301. [CrossRef]

9.	Jiang, Y. *Study on Rapid Forming Technology and Process of Aluminum Alloy Based on Cold Metal Transition Technology*; Harbin Institute of Technology: Harbin, China, 2013.

10.	Horgar, A.; Fostervoll, H.; Nyhus, B.; Ren, X.; Eriksson, M.; Akselsen, O.M. Additive manufacturing using WAAM with AA5183 wire. *Mater. Process. Technol.* **2018**, *259*, 68–74. [CrossRef]

11.	Geng, H.; Li, J.; Xiong, J.; Lin, X.; Zhang, F. Geometric limitation and tensile properties of wire and arc additive manufacturing 5A06 aluminum alloy parts. *Mater. Eng. Perform.* **2017**, *26*, 621–629. [CrossRef]

12.	Liu, Y.; Sun, Q.; Jiang, Y.; Wu, L.; Sang, H.; Feng, J. Rapid prototyping process based on cold metal transfer arc welding technology. *Trans. China Weld. Inst.* **2014**, *35*, 1–5.

13.	Li, C.; Liu, H.; He, W.; Hao, Z.; Zhao, T.; Zheng, L. Viscosity and Effect Factors of Aluminium Alloy Melt. *Light Alloy Fabr. Technol.* **2005**, *33*, 22–25. [CrossRef]

14.	Zhang, W. *Metal Fusion Welding Principle and Technology*; Machinery Industry Press: Beijing, China, 1983; pp. 160–169. (In Chinese)

15.	Li, X. *Microstructure and Metallographic Atlas of Aluminum Alloy Materials*; Metallurgical Industry Press: Beijing, China, 2010; pp. 257–261.

Article

Fatigue Performance of ABS Specimens Obtained by Fused Filament Fabrication

Miquel Domingo-Espin [1], J. Antonio Travieso-Rodriguez [2,*], Ramn Jerez-Mesa [3] and Jordi Lluma-Fuentes [4]

1   Fundació Eurecat, Av. Universitat Autònoma, 23, 08290 Cerdanyola del Vallès, Spain;
    miquel.domingo@eurecat.org
2   Mechanical Engineering Department, Escola d'Enginyeria de Barcelona Est,
    Universitat Politècnica de Catalunya, Avinguda d'Eduard Maristany, 10-14, 08019 Barcelona, Spain
3   Engineering Department, Faculty of Sciences and Technology, Universitat de Vic-Universitat Central de
    Catalunya, C. Laura, 13, 08500 Vic, Spain; ramon.jerez@uvic.cat
4   Materials Science and Metallurgical Engineering Department, Escola d'Enginyeria de Barcelona Est,
    Universitat Politècnica de Catalunya, Avinguda d'Eduard Maristany, 10-14, 08019 Barcelona, Spain;
    jordi.lluma@upc.edu
*   Correspondence: antonio.travieso@upc.edu

Received: 7 November 2018; Accepted: 30 November 2018; Published: 11 December 2018

Abstract: In this paper, the fatigue response of fused filament fabrication (FFF) Acrylonitrile butadiene styrene (ABS) parts is studied. Different building parameters (layer height, nozzle diameter, infill density, and printing speed) were chosen to study their influence on the lifespan of cylindrical specimens according to a design of experiments (DOE) using the Taguchi methodology. The same DOE was applied on two different specimen sets using two different infill patterns—rectilinear and honeycomb. The results show that the infill density is the most important parameter for both of the studied patterns. The specimens manufactured with the honeycomb pattern show longer lifespans. The best parameter set associated to that infill was chosen for a second experimental phase, in which the specimens were tested under different maximum bending stresses so as to construct the Wöhler curve associated with this 3D printing configuration. The results of this study are useful to design and manufacture ABS end-use parts that are expected to work under oscillating periodic loads.

Keywords: parts design; additive manufacturing; fused filament fabrication; fatigue; Taguchi; ABS

---

1. Introduction

Additive manufacturing (AM) technologies were, for years, considered only to manufacture prototypes, not end-use or functional objects. However, since the growth of the industry in the past years due to the improvement in technologies, the increasing quantity of materials, and the ease of access to the technologies, interest in manufactured functional parts has increased [1].

In order to manufacture a 3D object with AM, a virtual design is needed. Normally, the virtual design is done using computer aided design (CAD) software. After modeling the CAD file, the geometry is exported to an STL file, which describes the surface geometry of a three-dimensional object without any representation of color, texture, or other common model attributes. The STL file must be prepared before it is 3D printed, as it must be sliced. Slicing is dividing the 3D model into the horizontal layers that the printer will stack to form the part.

The first step before slicing is to orientate the part, which means how to place the part referred to the printer axis (X, Y, and Z). The orientation affects the surface roughness and/or dimensional accuracy [2–10], printing time [4–6], and part strength [7,10–26].

Slicing allows you to set several other printing parameters whose values affect the performance and characteristics of the part. Their values are critical in FFF technologies, as they affect the surface finish [27–35], cost [28,30–32,34,36], and mechanical performance [12,35,37–41].

The mechanical properties of FFF manufactured parts are difficult to predict, mainly because the parts present anisotropic mechanical behavior [16,23,25,42–44], and the printing parameters affect their mechanical response, the most studied being the layer height [19,35,39,41,45–50], infill orientation [13,18,21,23,35,44,48,50–54], infill pattern [13,24,41,46,53,55–57], infill density [13,35,38,41, 44–46,49,57], wall thickness [22,23,45], and nozzle diameter [41,44].

Not many fatigue studies on AM manufactured parts have been reported. Most of them focus on metallic parts, as their applications require knowing the number of cycles to failure [15,26,58–60]. The combination of platform heating and peak-hardening on the selective laser melting (SLM) parts of AlSi10Mg increased the fatigue resistance and neutralized the differences in the fatigue life for different building orientations [15]. Also, the fatigue life of Ti-6Al-4V alloys fabricated by electron beam melting (EBM) and laser beam melting (LBM) was investigated. The results indicated that the LBM Ti-6Al-4V parts exhibited a longer fatigue life than the EBM parts. The difference in the fatigue life behavior was attributed to the presence of the rough surface features that acted as fatigue crack initiation sites in the EBM material [58]. The same material was tested using SLM technology. The fatigue life was significantly lower compared to similar specimens manufactured with the same wrought material. This reduction in the fatigue performance was attributed to a variety of issues, such as the microstructure, porosity, surface finish, and residual stress. Also, a high degree of anisotropy in the fatigue performance was found and was associated with the specimen build orientation [26]. Different SLM stainless steel parts were tested under fatigue regimes. Depending on the material and the post-treatment, the resulting lifetimes were different [60]. Fatigue tests were also performed on the parts manufactured with Stratsys® Polyjet technology using a printed elastomer material. The findings showed the relationship between elongation and expected fatigue life, and that the better surface finish that this technology delivers, contributed to improving the fatigue life of the components [61].

The fatigue life of polylactic acid (PLA) was also investigated, as it is becoming a commonly used thermoplastic in open-source FFF machines for various engineering applications. The samples manufactured in three different orientations were tested. The results showed that the 45° build orientation parts showed a higher fatigue life than the parts built along the X and Y axis [22]. A DOE using different building parameters was used to determine their optimal values on the fatigue performance of the PLA FFF manufactured specimens. It was found that the infill density was the most important parameter, followed by the nozzle diameter and the layer height. Two different infill patterns were compared, with the honeycomb pattern being the best one. The fracture examination evidenced the necessity of post processing the outer layers to maximize the lifespan of the PLA parts [41]. The infill orientation of the FFF ABS parts was investigated by Zieman et al. [52]. The spciments built with the ±45° strategy had the longest fatigue life, followed by the 0, 45, and 90° orientations. The difference between the average cycles to failure was statistically significant for all of the infill orientations at each stress level. The failure modes are similar to those observed in the static tension tests.

During the last years, researchers have tackled the time dependence of the mechanical properties of parts manufactured through AM, specifically their fatigue behavior. Lee and Huang [62] studied the fatigue behavior for different part build orientations of two different materials, ABS and ABS plus. They analyzed the total strain energy absorbed by the specimens, but only one piece at each stress level was tested. Ziemian et al. [63] also published their results regarding the fatigue behavior of fused deposition modelling (FDM) ABS pieces. A fatigue damage analysis and an empirical model of an effective elastic modulus were presented. Senatov et al. [64] published a low cycle fatigue test for the PLA porous scaffolds for bone implants manufactured by FDM, functioning under cyclic loading. The Ultem FDM specimens for several build orientations were investigated by Fischer and Schöppner [65]. Puigoriol-Forcada et al. [66] recently published a study about the flexural fatigue properties of polycarbonate FDM parts.

To carry out fatigue studies, tests of different types can be selected, where parts of the different configurations are also used. Some examples are those used in the papers previously referred to as bending fatigue tests [66] and tensile fatigue tests [21,63]. In this article, a rotating flexural fatigue test was carried out. The detected lack of references about the influence of other parameters on the fatigue life, and a comprehensive study about the fatigue behavior of the FFF ABS parts has motivated the realization of this study. The innovative approach of this paper lays on the fact that ABS is an almost unexplored material for FFF in terms of fatigue, and the study is performed including a high number of factors in the experimental procedure. The results of the study shall deliver a recommended parameter set in order to maximize the service life of the ABS FFF parts. Furthermore, the influence of the maximum stress characterizing that load shall be studied by constructing the Wöhler curves for the defined optimal parameter set.

2. Materials and Methods

The experimental procedure is divided into four parts. First, the experimental factors will be chosen to perform a design of experiments (DOE), so a statistical analysis of the results can be performed. Then, the specimens will be designed and manufactured according to the related experimental matrix. Afterwards, they will be tested and the results will be statistically analyzed. Finally, 24 specimens will be manufactured using the optimal parameters found previously, to represent the Wöhler curve, also known as the S–N fatigue diagram.

3. Experimental Factors and Design of Experiments

The fixed manufacturing parameters' values were selected following different criteria, as shown below:

- Printing temperature. It is the target temperature at which the extruder must operate, in order to have a proper extrusion and to guarantee cohesion in the workpiece. It has been selected according to the manufacturer's datasheet recommendations.
- Platform temperature. The printing bed must keep this temperature during the whole extrusion process, to improve the quality of the printed pieces by preventing the warping caused by thermal stresses. It was selected according to the manufacturer's datasheet recommendation.
- Infill angle. It defines the direction of the trajectory that the nozzle follows to fill the internal section limited by the perimeter of the piece. We considered an infill angle of 45°, because it proved to be the best orientation in previous studies [41,67].
- Solid shell. It defines the number of contours present in every layer of the workpieces. The higher the number of layers in the solid shell, the higher the stiffness of the obtained part. The number of solid layer shells were selected so that the specimens had the smallest number of contours, so that it would affect at its minimum the results of the experimental campaign (the influence of the number of contours was not a target parameter in this study).

As for the parameters included in the DOE, they were selected by taking into account both the manufacturer datasheet and the previous investigations concerning the mechanical properties in terms of the fatigue life of other AM parts [21,41,52,67]. They are described below:

- Layer height. It determines the thickness of the layers. Thinner layer heights increase the part quality, leading to a smoother surface but a higher building time. Thicker layers have the opposite effect.
- Nozzle diameter. It determines the diameter of the extruded plastic. This parameter affects the mechanical performance, surface roughness, and cost of the manufactured parts.
- Infill density. It defines the amount of plastic used on the interior part of the print. A higher infill density means more plastic inside the part, leading to a stronger object. This parameter also affects the building time.

- Printing speed. It determines at which speed the print head and the platform move while printing. This setting also determines how fast the filament must be extruded in order to obtain the desired extruded filament width. A higher print speed will lead to a shorter print time.

These variable parameters have been selected by considering both the manufacturer datasheet and the previous investigations concerning the mechanical properties in terms of the fatigue life of other AM parts [21,41,52,67]. The selected fabrication parameters, as well as each of their levels, are shown in Table 1.

Table 1. Fabrication factors considering levels for experimentation.

Fixed Manufacturing Factors			Variable Manufacturing Factors					
Factor	Value	Unit	Factor	Symbol	Level			Unit
					1	2	3	
Printing temperature	230	°C	Layer height	A	0.1	0.2	0.3	mm
Platform temperature	100	°C	Nozzle diameter	B	0.3	0.4	0.5	mm
Infill angle	45	°	Infill density	C	25	50	75	%
Number of perimeters	2	-	Printing speed	D	25	30	35	mm/s
Solid layers shell	3	-	Fill Pattern	E	Rectilinear	-	Honeycomb	

A full factorial DOE involving four factors at three levels would consist of 81 experiments (3^4). The Taguchi method reduces the number of experimental tests and still allows for a statistical analysis of the process parameters and their interactions. Taguchi proposes an experimental plan, in terms of an orthogonal array, giving a certain combination of parameters for each experiment [34,41,45,53].

In this study, the influence of the four factors and the interaction between three of them are studied ($A \times B$, $B \times C$, and $A \times C$). This combination leads to 16 degrees of freedom, therefore the most appropriate orthogonal array is L_{27}. The assignment of factors and interactions into the orthogonal matrix was performed using the linear graph for the L_{27} orthogonal array in order to avoid confusion between factors. The assignment was performed as follows: Columns 1, 2, and 5 have been assigned to factors A, B, and C, respectively (according Table 1). Factor D is assigned to column 9. This configuration also allows the parameters A, B, and C to be set in a full factorial DOE, which allows for a detailed study on its influence. The final column assignation is shown in Table 2.

Additionally, two different infill patterns were introduced in the case study in order to explore their effects on the mechanical behavior (Figure 1C,D). This factor determines the pattern taken by the extruder to deposit the material inside the part, which could be beneficial in some cases [57]. Rectilinear and honeycomb patterns were used, as the results can be compared to those obtained by Gomez-Gras et al. from a similar experimental study performed with PLA specimens, in the same conditions and using the same machine [41,67].

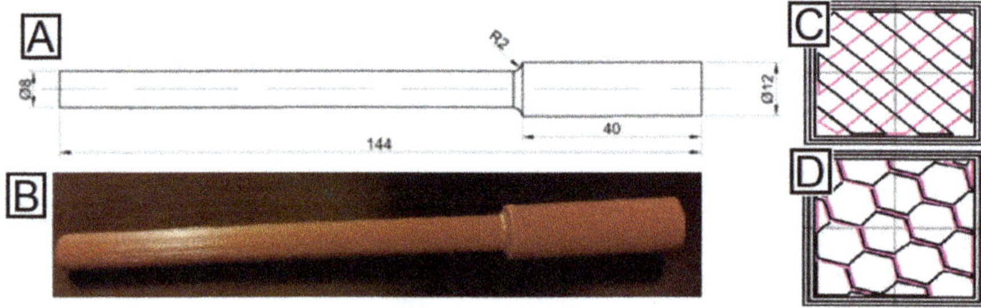

Figure 1. (**A**) Specimens used for the fatigue tests. (**B**) Overview of five specimens manufactured, all of them sharing the same manufacturing parameters. (**C**) Rectilinear infill pattern. (**D**) Honeycomb infill pattern (adapted from [41], with permission from Elsevier).

Table 2. L_{27} matrix column assignation along with signal and noise values for the life cycles of rectilinear and honeycomb infill patterns.

Test #	Factor				Rectilinear		Honeycomb	
	Layer Height (mm)	Nozzle Diameter (mm)	Infill Density (%)	Printing Speed (mm/s)	Signal (Num. of Cycles)	Noise (Num. of Cycles)	Signal (Num. of Cycles)	Noise (Num. of Cycles)
1	0.1	0.3	25	25	388	94	609	45
2	0.1	0.3	50	30	1961	955	1995	246
3	0.1	0.3	75	35	3549	2284	4395	389
4	0.1	0.4	25	30	512	124	378	50
5	0.1	0.4	50	35	569	20	1045	85
6	0.1	0.4	75	25	1330	236	2191	151
7	0.1	0.5	25	35	401	69	689	72
8	0.1	0.5	50	25	683	54	1078	283
9	0.1	0.5	75	30	2241	144	2592	201
10	0.2	0.3	25	30	1154	225	393	32
11	0.2	0.3	50	35	931	18	872	251
12	0.2	0.3	75	25	1720	235	3208	1116
13	0.2	0.4	25	35	484	55	929	168
14	0.2	0.4	50	25	1923	251	1933	187
15	0.2	0.4	75	30	2672	1033	6095	296
16	0.2	0.5	25	25	566	25	402	8
17	0.2	0.5	50	30	527	158	1021	245
18	0.2	0.5	75	35	756	117	1435	137
19	0.3	0.3	25	35	930	131	696	157
20	0.3	0.3	50	25	916	41	757	25
21	0.3	0.3	75	30	1764	741	2484	373
22	0.3	0.4	25	25	536	6	591	60
23	0.3	0.4	50	30	689	44	1044	102
24	0.3	0.4	75	35	1330	35	2222	36
25	0.3	0.5	25	30	2037	500	1362	170
26	0.3	0.5	50	35	819	60	2737	445
27	0.3	0.5	75	25	8262	324	6137	825

4. Test Samples Design and Manufacture

The test specimens were manufactured using a 2.85 mm ABS filament. There is no a specific standard focusing on fatigue testing for additive manufactured plastic parts. Therefore, special specimens have been designed, adapting their dimensions to the possibilities offered by the testing machine (Figure 1A,B). However, the design of the specimens is according to the ASTM D7774 standard [68], which regulates the test method for flexural fatigue properties of plastics.

The test samples were designed using SolidWorks®, then sliced using Slic3r, where the different building parameters were set according to the DOE. Finally, the parts were manufactured with a Pyramid dual extruder M® FFF machine (Oxfordshire, UK) oriented along the X axis. A total of 162 samples were manufactured—three repetitions for the 27 parameter set for the two infill patterns.

5. Fatigue Testing

The parts were tested using a GUNT WP 140 machine (Hamburg, Germany) (Figure 2), applying a rotational movement of 2800 min^{-1} and a force of 8 N. The load, applied in the direction perpendicular to the axis of rotation and along the longitudinal axis of the parts, generated a sinusoidal load in the fibers of the specimen. The geometry of the specimens causes failure in the critical section next to the diameter change, where the highest bending moment is being exerted.

A PCE-TC 3 thermographic camera (Palm Beach, FL, USA) was also installed to observe the changes in temperature of the specimen at the stress concentrator area. Its sensitivity is 0.15 °C and precision is of ± 2 °C. Both values are considered admissible for this kind of study, where the temperature can be considered as secondary to characterize the process.

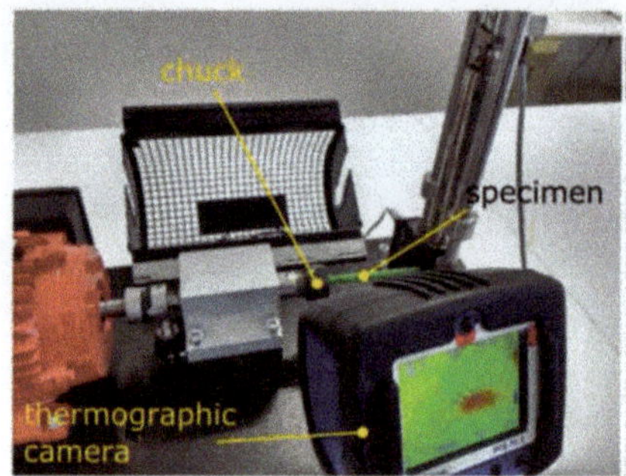

Figure 2. Experimental station (adapted from [41], with permission from Elsevier).

6. Statistical Analysis

To determinate the most influential factors in a DOE according to Taguchi's method, the signal-to-noise (S/N) ratio is used. Signal refers to the target magnitude (number of cycles) and noise represents the variability of that response. As the objective of the experimental plan was to find the parameters that maximize the number of cycles before failure, the aim of the statistical analysis is to maximize the signal and to minimize the noise, thus optimizing the S/N ratio. The ratio was calculated for each experiment using Equation (1), where η is the average S/N ratio, n is the number of experiments conducted at level i, and y_i is the measured value of the property.

$$\eta = -10 \cdot \log\left(\frac{1}{n} \sum_{i=1}^{n} \frac{1}{y_i^2}\right) \tag{1}$$

The optimization of the S/N ratio also defines the optimal factors by confirming whether there is a linear correlation between the signal and the S/N ratio, and the standard deviation and the S/N ratio.

To obtain the influence of each parameter and the interactions in the fatigue life, an analysis of variance (ANOVA) was performed on each parameter using the signal and the noise values. The parameters whose statistical influence was below 10% were not considered. The effect of the levels for each parameter and the interaction on the signal and noise were studied in order to find their influence on the response. The statistical result analysis shall deliver the printing parameters that lead to the highest fatigue lifespan for both of the infill patterns.

Wöhlers Curve

The optimal parameters found were used to manufacture a whole new set of parts that would be tested to different oscillating bending stress, so that a low-cycle fatigue study can be performed. The obtained results would lead to the determination of a Wöhler curve of the parameters set.

7. Results

In this section, the results obtained are presented in four subsections. First, the fatigue results acquired using Taguchi's DOE, and a fractography study are presented. Then, the comparison between the two infill patterns is shown, and finally, the resulting Wöhler curve is discussed.

7.1. Fatigue Results

The signal and noise response for each experiment are shown in Table 2. There was no correlation between the signal and the S/N ratio, or the noise and the S/N ratio. Therefore, a dual response approach was needed, so the factors that maximize the signal response and minimize the noise can be determined.

The results showed that the most influential factor in the signal was the infill density, for both of the infill patterns (42.2% for rectilinear and 72.4% for honeycomb), as it happened in the previous study done for PLA material [41]. The interaction between the layer height and the nozzle diameter was the next most influential in the number of cycles (18.8% for rectilinear and 17.3% for honeycomb). The other factors and interactions were declared non-influential, due to the fact that their influence was lower than 10%.

The noise results exhibited the same trend, with the infill density being the most influential factor (25.3% for rectilinear and 40% for honeycomb), followed by the interaction between the layer height and the nozzle diameter (20.0% for rectilinear and 15.8% for honeycomb). However, the nozzle diameter also exhibited a significant effect in both infills (17.4% for rectilinear and 13.4% for honeycomb). Again, the factors and interactions with an influence lower than 10% were ignored. The effect for each factor according to their level can be observed in Figure 3, where the evolution of all of the results are joined by discontinuous lines to guide the eye of the reader.

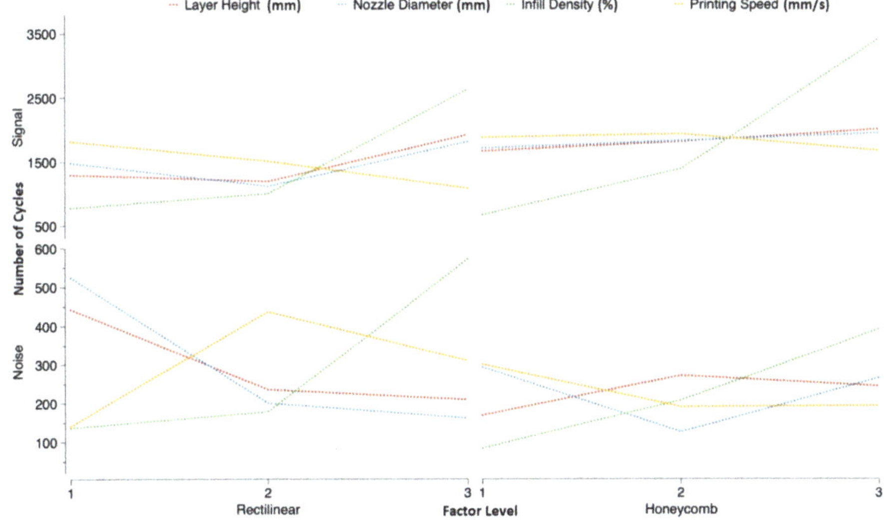

Figure 3. Factor effect on signal and noise for both infill patterns.

7.2. Optimal Factors for Rectilinear Infill Pattern

The results showed that the highest lifespan, using the rectilinear infill pattern, was obtained when layer height, nozzle diameter, and infill density were at their highest level. On the other hand, the lowest variance was obtained when the infill density was at the lowest level, and the layer height and nozzle diameter at their mid or highest level, due to the lower difference shown.

The interaction between the layer height and the nozzle diameter proved to be significant in the signal and the noise of the response. Since the significance of the infill density factor on the signal is higher than on the noise, its optimal level can be defined at 75%—level 3 (Figure 4).

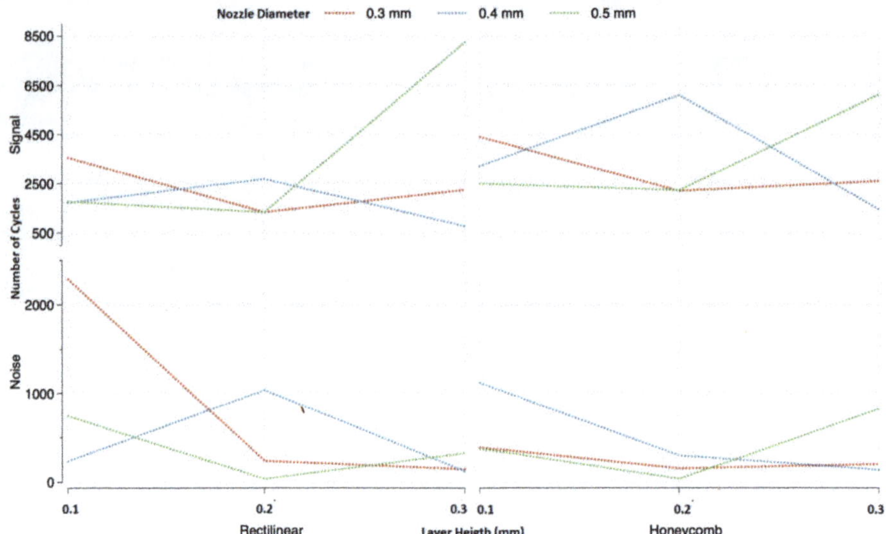

Figure 4. Interaction plots between nozzle diameter and layer height. Effect on signal and noise using rectilinear infill pattern on the left and honeycomb infill pattern on the right both at their highest level (75%).

It could be observed that the interaction between the layer height and nozzle diameter was complex, as the effects on the signal and the noise could not be separated from one another. The maximum signal is obtained when the nozzle diameter and layer height are selected at their highest level, observed in Figure 4. On the other hand, the effect of the layer height on the noise was minimized when the nozzle diameter was at its highest level, and its influence was almost as important as the interaction. Therefore, to minimize the variability of the signal, the nozzle diameter must be at its highest level.

The experiences that presented the best configurations of parameters for the rectilinear pattern were 9, 18, and 27 (Table 2). What those configurations have in common is that the nozzle diameter and the infill density are at their highest levels. It was observed that the best result for the rectilinear infill was experiment number 27, which corresponded to the three most influential factors at their highest level. This set of parameters presented an average life of 8262 cycles with just a 3.9% variance, which made this configuration the best one.

7.3. Optimal Factors for Honeycomb Infill Pattern

A similar analysis was performed for the honeycomb pattern results. The nozzle diameter and layer height maximized the lifespan at their middle and highest level and, like the rectilinear pattern, the infill density at its highest. The lowest noise was observed when the layer height was at its middle or highest level, nozzle diameter at its middle level, and infill density at its lowest (Figure 3).

The same situation using honeycomb happened as with a rectilinear pattern. The importance of the infill density in the signal was higher than in the noise, so, in order to maximize the cycles to failure, the infill pattern should be the highest—75% (Figure 4 right).

The same interaction was found to be significant using honeycomb, but in this case, there was no level for any of the factors that minimized the effect of the other. In order to maximize the signal, the nozzle diameter and the layer height needed to be at the same level. Minimizing the variance of the signal appeared to be more complicated; depending on the value of the nozzle diameter, the layer height could be at any of its levels. The lowest noise was found when the layer height was at its highest level and the nozzle diameter at its middle one, and vice versa. The best combinations of factors that would magnify the signal and minimize the noise should be when both factors are at levels 2 and 3.

The experiences that presented these combinations were 15, 18, 24, and 27. From these four, numbers 15 and 27 presented the highest life cycles, which were almost identical (Table 2). But experience 15 presented the lowest variance of the two (4.9% for experience 15 and 13.4% for experience 27). This difference in variance made the 15th experience the optimal one.

7.4. Fractography

Photographs of the broken specimens were taken after the fatigue tests. They were taken with a MOTIC SMC (Hong Kong, China) binocular loupe equipped with a MOTICAM 3 digital camera. The photographs showed singular aspects that describe the breaking mode found in the specimens tested. In all of the cases, the crack began around the area near the first or last printed layer, observed in Figure 5, on the left. This implied that the extruded filaments that were forming the curvature of the specimens acted as stress concentrators, so the cracks were formed there and then propagated inside the part.

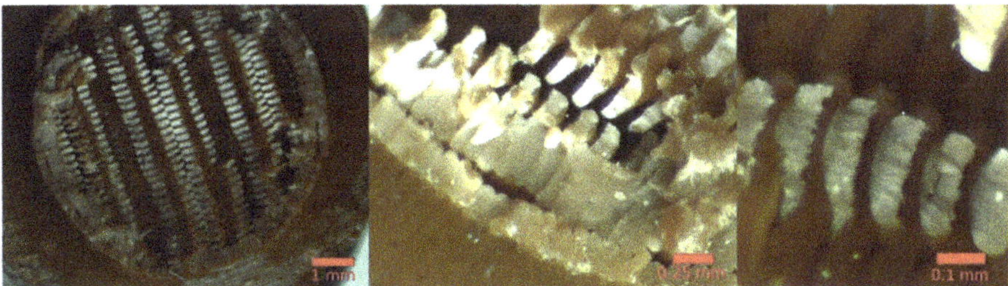

Figure 5. Image of the fractured area of the specimen.

In all of the cases, the type of break observed was ductile on the entire XY plane. The details of the fatigue marks are easily observed in Figure 5 in the middle and left, where the photographs from different specimens using different printing parameters are presented. However, it can be assessed that the type of break was the same in all of them. The propagation of the cracks as a combination of the bending and shear stress defined the failure mode of this type of material, as has already been discussed by other authors [64].

7.5. Infill Pattern Comparison

Figure 6 shows the comparison between the two infill patterns for all of the experiments. It can be observed that, depending on the factor levels, the difference in life cycles was significant and no relation was noticeable. However, the honeycomb configuration shows a better lifespan than the rectilinear in almost all of the configurations. Test number 27 showed the maximum life for both infill architectures, but the lifespan using the rectilinear pattern was 25% higher than using the honeycomb pattern.

On the other hand, there were two experiences that showed the highest lifespans using honeycomb pattern, numbers 15 and 27. Also, configuration number 15 showed that using the honeycomb pattern resulted in an over 50% of lifespan in comparison with the rectilinear configuration.

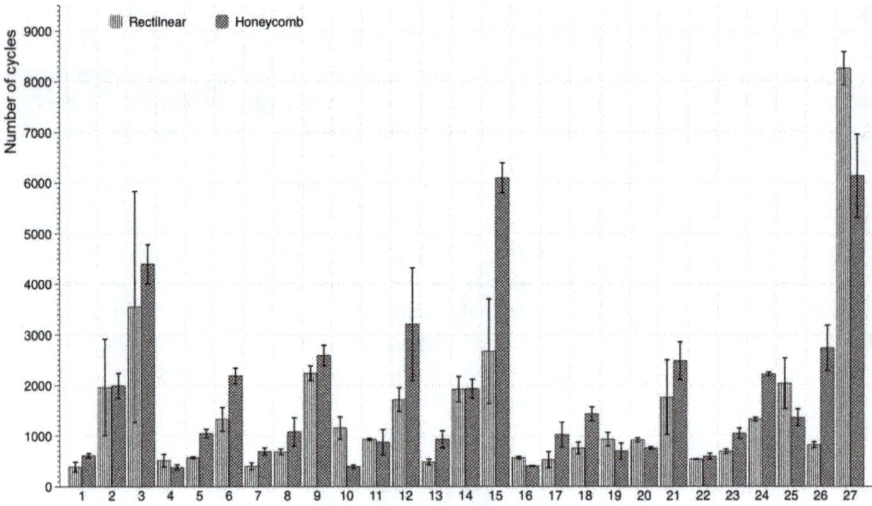

Figure 6. Lifespan comparison chart between rectilinear and honeycomb infill pattern.

7.6. Wöhler Curve

The analyzed results led to the conclusion that there was an optimal combination of parameters in the defined DOE, summarized in Table 3. This set of conditions was applied to print a second set of specimens, which were tested to different levels of bending stress, obtained by applying different forces at the specimen extreme point. Table 4 shows the eight different levels of force and the maximum bending stress to which the specimen were subjected in the stress concentrator area, calculated considering that the specimen can be modelled as a cylindrical cantilever.

Table 3. Optimal combination of factors and levels to maximize the expected cycles to failure.

Parameter	Value
Infill pattern	Honeycomb
Fill density	75%
Nozzle diameter	0.4 mm
Layer height	0.2 mm

Table 4. Forces applied for the Wöhler curve tests and maximum stress levels.

F (N)	M_{max} (N-mm)	σ_{max} (MPa)
8.0	832	28.7
8.5	884	30.5
9.0	936	32.3
9.5	988	34.0
10.0	1040	35.8
10.5	1092	37.6
11.0	1144	39.4
11.5	1196	41.2

With this data, different fatigue tests to construct the Wöhler curve were carried out at each of the indicated stress levels [38]. Following the protocol established by Wirsching, M.C. [69], and also that applied in our previous study [41], five repetitions were performed for each stress level, except for 28.7 MPa, as this stress was already tested for the results of the DOE analysis.

The least-square regression model was used to fit the linearized version of the potential Wöhler curve (Equation (2))

$$\log\left(2N_f\right) = -\frac{1}{b}\log\left(S_f\right) + \frac{1}{b}\log(S_a)$$

(2)

where $\log(S_a)$ is the independent variable, $\log\left(2N_f\right)$ is the dependent variable, the slope is $\frac{1}{b}$, and interception point is $\frac{-1}{b}\log\left(S_f\right)$. Thus, the S–N curve equation is Equation (3).

$$S_a = S_f\left(2N_f\right)^b$$

(3)

A potential curve, corresponding to Equation (3), was deduced from the testing, with a $R^2 = 0.9814$, and is represented in Figure 7. Furthermore, the model used in this figure is only valid for the low cycle fatigue domain.

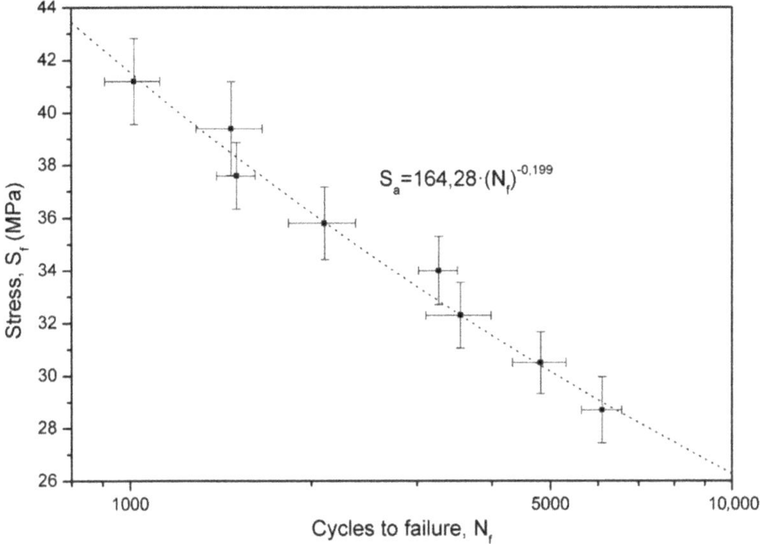

Figure 7. Wöhler curve for specimens manufactured with honeycomb infill, 75% infill density, 0.4 mm diameter nozzle, and 0.2 mm layer height.

8. Discussion

The results obtained showed that the infill density is the most important parameter affecting the live span of the ABS FFF produced parts. The other parameters studied do not have that much impact on the cycles to fail on their own, but instead on their interactions. It is also important that the influence of factors and interactions, for signal and noise, are the same and in the same order for the two infill patterns.

It is evident that when parts are more uniform or continuous, as the injected ones, their mechanical properties are better. Voids are always present when manufacturing parts using FFF technology, even if the parts are manufactured completely as solid. So comprehensively, the infill density has been found to be the most important factor affecting the life of a part—the more density, the more continuous, and the more life cycles the part can stand.

The interaction between the layer height and nozzle diameter has been found to be important. These two parameters also affect the continuity of the part. The higher they are, the more continuous the part is, as there are fewer interfaces inside the part. The nozzle diameter makes the extruded filament bigger, so the part is more continuous with lesser voids inside. Bigger layer heights cause the part to be manufactured with fewer layers, which is also more continuous. This result was also observed on the PLA specimens [41].

The printing speed does not affect the fatigue performance of the ABS FFF manufactured parts. This conclusion is reasonable, as the speed values that are tested in the experimental plan are not significantly different.

The difference between the two infill parameters in the cycles to failure is not evident. Overall, the honeycomb specimens are proven to have better results. However, this result varies according to the other parameters. For instance, if experiment number 27 is compared, the rectilinear pattern shows better results. This may be caused by the fact that the stress created by the load during the experiment is aligned with the layers, as specimens are printed along the X axis. The rectilinear pattern positions extruded filaments at 45° along the load, which causes an equal distribution of the stress along the plane, so the part is stronger. On the other hand, the honeycomb pattern does not transmit the stress the same way, or, at least, it is not proportional along the plane, making this pattern weaker in this case.

The evolution of the fatigue life versus stress amplitude of the selected printing conditions could be properly described by Wöhler's potential equation, as was also found in PLA [67]. This means that the selected range of the stress amplitudes corresponds to the same fatigue regime, elastic fatigue in this case, and no fatigue limit was observed.

9. Conclusions

In the present paper, the fatigue life cycles of the ABS parts manufactured with FFF technology using different building parameter configurations have been analyzed. Test samples have been built varying in layer height, nozzle diameter, infill density, printing speed, and infill pattern. The results obtained have confirmed the following:

1. The fatigue performance depends on the building parameters. This means that, by controlling the building parameters, the mechanical behavior of the FFF parts can also be controlled.
2. The infill density is the most important factor for the two infills structures studied. The fatigue life increases as the infill does. The infill strengthens the part causing an increased life. For any combination of building parameters, the higher the density inside the part, the higher the life span.
3. Selecting the right building parameters is not an easy task; as proven in this study, the selection of the right value of different parameters can increase the mechanical properties considerably, but some generalization can be extracted.
4. The improvement of the life of FFF parts is achieved when the parts are manufactured to be as continuous as possible, and also, when the direction of the extruded filaments or the infill pattern inside the part make the tension distribute equally.
5. This paper has also presented the S–N curve associated with the best 3D printing parameters. This curve can be adjusted by a simple Wöhler model, meaning that, at the tested stress levels, the ABS specimens are working inside the elastic region.
6. Further studies are needed to understand how the parameters studied, and others, affect the fatigue performance of FFF ABS produced parts. However, the obtained results in this study (and others with different materials) are expected to be similar for other FFF thermoplastics, not in value, but how the factors affect the life cycle.

10. Data Availability

The raw/processed data required to reproduce these findings cannot be shared at this time, as the data also forms part of an ongoing study.

Author Contributions: Conceptualization, J.A.T.R. and R.J.M.; Methodology, R.J.M. and M.D.E.; Software, J.L.F.; Validation, J.L.F., J.A.T.R. and R.J.M.; Formal Analysis, R.J.M.; Investigation, J.A.T.R. and M.D.E.; Resources, J.L.F.;

Data Curation, J.L.F. and R.J.M.; Writing-Original Draft Preparation, M.D.E.; Writing-Review & Editing, J.A.T.R. and R.J.M.; Visualization, J.A.T.R.; Supervision, M.D.E.; Project Administration, M.D.E., Funding Acquisition, M.D.E., J.A.T.R., R.J.M. and J.L.F.;

Funding: This research received no external funding.

Conflicts of Interest: The authors declare no conflict of interest.

References

1. Gordon, R. Trends in Commercial 3D Printing and Additive Manufacturing. *3D Print. Addit. Manuf.* **2015**, *2*, 89–90. [CrossRef]
2. Boschetto, A.; Bottini, L. Accuracy prediction in fused deposition modeling. *Int. J. Adv. Manuf. Technol.* **2014**, *73*, 913–928. [CrossRef]
3. Gurrala, P.K.; Regalla, S.P. DOE Based Parametric Study of Volumetric Change of FDM Parts. *Procedia Mater. Sci.* **2014**, *6*, 354–360. [CrossRef]
4. Xu, F.; Wong, Y.S.; Loh, H.T.; Fuh, J.Y.H.; Miyazawa, T. Optimal orientation with variable slicing in stereolithography. *Rapid Prototyp. J.* **1997**, *3*, 76–88. [CrossRef]
5. Pham, D.T.; Dimov, S.S.; Gault, R.S. Part orientation in stereolithography. *Int. J. Adv. Manuf. Technol.* **1999**, *15*, 674–682. [CrossRef]
6. Cheng, W.; Fuh, J.Y.H.; Nee, A.Y.C.; Wong, Y.S.; Loh, H.T.; Miyazawa, T. Multi-objective optimization of part-building orientation in stereolithography. *Rapid Prototyp. J.* **1995**, *1*, 12–23. [CrossRef]
7. Bártolo, P.J. *Stereolithography: Materials, Processes and Applications*; Springer Science & Business Media: New York, NY, USA, 2011, ISBN 9780387929033.
8. Adam, G.A.O.; Zimmer, D. On design for additive manufacturing: Evaluating geometrical limitations. *Rapid Prototyp. J.* **2015**, *21*, 662–670. [CrossRef]
9. Yuan, M.; Bourell, D. Orientation effects for laser sintered polyamide optically translucent parts. *Rapid Prototyp. J.* **2016**, *22*, 97–103. [CrossRef]
10. Durgun, I.; Ertan, R. Experimental investigation of FDM process for improvement of mechanical properties and production cost. *Rapid Prototyp. J.* **2014**, *20*, 228–235. [CrossRef]
11. Ajoku, U.; Saleh, N.; Hopkinson, N.; Hague, R.; Erasenthiran, P. Investigating mechanical anisotropy and end-of-vector effect in laser-sintered nylon parts. *Proc. Inst. Mech. Eng. Part B J. Eng. Manuf.* **2016**, *220*, 1077–1086. [CrossRef]
12. Es-Said, O.S.; Foyos, J.; Noorani, R.; Mendelson, M.; Marloth, R.; Pregger, B.A. Effect of Layer Orientation on Mechanical Properties of Rapid Prototyped Samples. *Mater. Manuf. Process.* **2000**, *15*, 107–122. [CrossRef]
13. Iyibilgin, O.; Leu, M.C.; Taylor, G.; Li, H.; Chandrashekhara, K. Investigation of Sparse Build Rapid Tooling by Fused Deposition Modeling. In Proceedings of the 25th International Solid Freedom Fabrication Symposium on Additive Manufacturing, Austin, TX, USA, 4–6 August 2014; pp. 542–556.
14. Takaichi, A.; Nakamoto, T.; Joko, N.; Nomura, N.; Tsutsumi, Y.; Migita, S.; Igarashi, Y. Microstructures and mechanical properties of Co-29Cr-6Mo alloy fabricated by selective laser melting process for dental applications. *J. Mech. Behav. Biomed. Mater.* **2013**, *21*, 67–76. [CrossRef] [PubMed]
15. Brandl, E.; Heckenberger, U.; Holzinger, V.; Buchbinder, D. Additive manufactured AlSi10Mg samples using Selective Laser Melting (SLM): Microstructure, high cycle fatigue, and fracture behavior. *Mater. Des.* **2012**, *34*, 159–169. [CrossRef]
16. Upadhyay, K.; Dwivedi, R.; Singh, A.K. Determination and Comparison of the Anisotropic Strengths of Fused Deposition Modeling P400 ABS. In *Advances in 3D Printing & Additive Manufacturing Technologies*; Springer: Singapore, 2017; pp. 9–28.
17. Kubalak, J.R.; Wicks, A.L.; Williams, C.B. Using multi-axis material extrusion to improve mechanical properties through surface reinforcement. *Virtual Phys. Prototyp.* **2018**, *13*, 32–38. [CrossRef]
18. Cantrell, J.T.; Rohde, S.; Damiani, D.; Gurnani, R.; DiSandro, L.; Anton, J.; Ifju, P.G. Experimental characterization of the mechanical properties of 3D-printed ABS and polycarbonate parts. *Rapid Prototyp. J.* **2017**, *23*, 811–824. [CrossRef]
19. Chacón, J.M.; Caminero, M.A.; García-Plaza, E.; Núñez, P.J. Additive manufacturing of PLA structures using fused deposition modelling: Effect of process parameters on mechanical properties and their optimal selection. *Mater. Des.* **2017**, *124*, 143–157. [CrossRef]

20. Olivier, D.; Travieso-Rodriguez, J.A.; Borros, S.; Reyes, G.; Jerez-Mesa, R. Influence of building orientation on the flexural strength of laminated object manufacturing specimens. *J. Mech. Sci. Technol.* **2017**, *31*, 133–139. [CrossRef]

21. Afrose, M.F.; Masood, S.H.; Iovenitti, P.; Nikzad, M.; Sbarski, I. Effects of part build orientations on fatigue behaviour of FDM-processed PLA material. *Prog. Addit. Manuf.* **2016**, *1*, 21–28. [CrossRef]

22. Croccolo, D.; De Agostinis, M.; Olmi, G. Experimental characterization and analytical modelling of the mechanical behaviour of fused deposition processed parts made of ABS-M30. *Comput. Mater. Sci.* **2013**, *79*, 506–518. [CrossRef]

23. Torrado, A.R.; Roberson, D.A. Failure Analysis and Anisotropy Evaluation of 3D-Printed Tensile Test Specimens of Different Geometries and Print Raster Patterns. *J. Fail. Anal. Prev.* **2016**, *16*, 154–164. [CrossRef]

24. Smith, W.C.; Dean, R.W. Structural characteristics of fused deposition modeling polycarbonate material. *Polym. Test.* **2013**, *32*, 1306–1312. [CrossRef]

25. Domingo-Espin, M.; Puigoriol-Forcada, J.M.; Garcia-Granada, A.A.; Lluma, J.; Borros, S.; Reyes, G. Mechanical property characterization and simulation of fused deposition modeling Polycarbonate parts. *Mater. Des.* **2015**, *83*, 670–677. [CrossRef]

26. Edwards, P.; Ramulu, M. Fatigue performance evaluation of selective laser melted Ti–6Al–4V. *Mater. Sci. Eng. A* **2014**, *598*, 327–337. [CrossRef]

27. Pandey, P.M.; Thrimurthulu, K.; Venkata Reddy, N. Optimal part deposition orientation in FDM by using a multicriteria genetic algorithm. *Int. J. Prod. Res.* **2004**, *42*, 4069–4089. [CrossRef]

28. Byun, H.S.; Lee, K.H. Determination of the optimal build direction for different rapid prototyping processes using multi-criterion decision making. *Robot. Comput. Integr. Manuf.* **2006**, *22*, 69–80. [CrossRef]

29. Paul, B.K.; Voorakarnam, V. Effect of Layer Thickness and Orientation Angle on Surface Roughness in Laminated Object Manufacturing. *J. Manuf. Process.* **2011**, *3*, 94–101. [CrossRef]

30. Pandey, P.M.; Venkata Reddy, N.; Dhande, S.G. Part deposition orientation studies in layered manufacturing. *J. Mater. Process. Technol.* **2007**, *185*, 125–131. [CrossRef]

31. Thrimurthulu, K.; Pandey, P.M.; Venkata Reddy, N. Optimum part deposition orientation in fused deposition modeling. *Int. J. Mach. Tools Manuf.* **2004**, *44*, 585–594. [CrossRef]

32. Xu, F.; Loh, H.T.; Wong, Y.S. Considerations and selection of optimal orientation for different rapid prototyping systems. *Rapid Prototyp. J.* **1999**, *5*, 54–60. [CrossRef]

33. Masood, S.H.; Rattanawong, W.; Iovenitti, P. Part Build Orientations Based on Volumetric Error in Fused Deposition Modelling. *Int. J. Adv. Manuf. Technol.* **2000**, *16*, 162–168. [CrossRef]

34. Frank, D.; Chandra, R.L.; Schmitt, R. An Investigation of Cause-and-Effect Relationships Within a 3D-Printing System and the Applicability of Optimum Printing Parameters from Experimental Models to Different Printing Jobs. *3D Print. Addit. Manuf.* **2015**, *2*, 131–139. [CrossRef]

35. Melenka, G.W.; Schofield, J.S.; Dawson, M.R.; Carey, J.P. Evaluation of dimensional accuracy and material properties of the MakerBot 3D desktop printer. *Rapid Prototyp. J.* **2015**, *21*, 618–627. [CrossRef]

36. Alexander, P.; Allen, S.; Dutta, D. Part orientation and build cost determination in layered manufacturing. *Comput. Des.* **1998**, *30*, 343–356. [CrossRef]

37. Puebla, K.; Arcaute, K.; Quintana, R.; Wicker, R.B. Effects of environmental conditions, aging, and build orientations on the mechanical properties of ASTM type I specimens manufactured via stereolithography. *Rapid Prototyp. J.* **2012**, *18*, 374–388. [CrossRef]

38. Moroni, L.; de Wijn, J.R.; van Blitterswijk, C.A. 3D fiber-deposited scaffolds for tissue engineering: Influence of pores geometry and architecture on dynamic mechanical properties. *Biomaterials* **2006**, *27*, 974–985. [CrossRef] [PubMed]

39. Majewski, C.; Hopkinson, N. Effect of section thickness and build orientation on tensile properties and material characteristics of laser sintered nylon-12 parts. *Rapid Prototyp. J.* **2011**, *17*, 176–180. [CrossRef]

40. Bellini, A.; Güçeri, S. Mechanical characterization of parts fabricated using fused deposition modeling. *Rapid Prototyp. J.* **2003**, *9*, 252–264. [CrossRef]

41. Gomez-Gras, G.; Jerez-Mesa, R.; Travieso-Rodriguez, J.A.; Lluma-Fuentes, J. Fatigue performance of fused filament fabrication PLA specimens. *Mater. Des.* **2018**, *140*, 278–285. [CrossRef]

42. Lee, C.S.; Kim, S.G.; Kim, H.J.; Ahn, S.H. Measurement of anisotropic compressive strength of rapid prototyping parts. *J. Mater. Process. Technol.* **2007**, *187*, 627–630. [CrossRef]

43. Ziemian, C.; Sharma, M. Anisotropic mechanical properties of ABS parts fabricated by fused deposition modelling. In *Mechanical Engineering*; In Tech: New York, NY, USA, 2012.

44. Ahn, S.H.; Montero, M.; Odell, D.; Roundy, S.; Wright, P.K. Anisotropic material properties of fused deposition modeling ABS. *Rapid Prototyp. J.* **2002**, *8*, 248–257. [CrossRef]

45. Domingo-Espin, M.; Borros, S.; Agullo, N.; Garcia-Granada, A.A.; Reyes, G. Influence of Building Parameters on the Dynamic Mechanical Properties of Polycarbonate Fused Deposition Modeling Parts. *3D Print. Addit. Manuf.* **2014**, *1*, 70–77. [CrossRef]

46. Tsouknidas, A.; Pantazopoulos, M.; Katsoulis, I.; Fasnakis, D.; Maropoulos, S.; Michailidis, N. Impact absorption capacity of 3D-printed components fabricated by fused deposition modelling. *Mater. Des.* **2016**, *102*, 41–44. [CrossRef]

47. Li, H.; Wang, T.T.; Sun, J.; Yu, Z. The effect of process parameters in fused deposition modelling on bonding degree and mechanical properties. *Rapid Prototyp. J.* **2017**, *24*, 80–92. [CrossRef]

48. Wu, W.; Geng, P.; Li, G.; Zhao, D.; Zhang, H.; Zhao, J. Influence of layer thickness and raster angle on the mechanical properties of 3D-printed PEEK and a comparative mechanical study between PEEK and ABS. *Materials* **2015**, *8*, 5834–5846. [CrossRef] [PubMed]

49. Akande, S.O.; Dalgarno, K.W.; Munguia, J. Low-Cost QA Benchmark for Fused Filament Fabrication. *3D Print. Addit. Manuf.* **2015**, *2*, 78–84. [CrossRef]

50. Tymrak, B.M.; Kreiger, M.; Pearce, J.M. Mechanical properties of components fabricated with open-source 3-D printers under realistic environmental conditions. *Mater. Des.* **2014**, *58*, 242–246. [CrossRef]

51. Casavola, C.; Cazzato, A.; Moramarco, V.; Pappalettere, C. Orthotropic mechanical properties of fused deposition modelling parts described by classical laminate theory. *Mater. Des.* **2016**, *90*, 453–458. [CrossRef]

52. Ziemian, S.; Okwara, M.; Ziemian, C.W. Tensile and fatigue behavior of layered acrylonitrile butadiene styrene. *Rapid Prototyp. J.* **2015**, *21*, 270–278. [CrossRef]

53. Huang, B.; Masood, S.H.; Nikzad, M.; Venugopal, P.R.; Arivazhagan, A. Dynamic Mechanical Properties of Fused Deposition Modelling Processed Polyphenylsulfone Material. *Am. J. Eng. Appl. Sci.* **2016**, *9*, 1–11. [CrossRef]

54. Hossain, M.; Ramos, J.; Espalin, D.; Perez, M.; Wicker, R. Improving Tensile Mechanical Properties of FDM-Manufactured Specimens via Modifying Build Parameters. In Proceedings of the International Solid Freeform Fabrication Symposium: An Additive Manufacturing Conference, Austin, TX, USA, 12 August 2013; pp. 380–392.

55. Baich, L.; Manogharan, G.; Marie, H. Study of infill print design on production cost-time of 3D printed ABS parts. *Int. J. Rapid Manuf.* **2015**, *5*, 308. [CrossRef]

56. Iyibilgin, O.; Yigit, C.; Leu, M.C. Experimental investigation of different cellular lattice structures manufactured by fused deposition modeling. In Proceedings of the Solid Freeform Fabrication Symposium, Austin, TX, USA, 12 August 2013; pp. 895–907.

57. Fernandez-Vicente, M.; Calle, W.; Ferrandiz, S.; Conejero, A. Effect of Infill Parameters on Tensile Mechanical Behavior in Desktop 3D Printing. *3D Print. Addit. Manuf.* **2016**, *3*, 183–192. [CrossRef]

58. Chan, K.S.; Koike, M.; Mason, R.L.; Okabe, T. Fatigue life of titanium alloys fabricated by additive layer manufacturing techniques for dental implants. *Metall. Mater. Trans. A* **2013**, *44*, 1010–1022. [CrossRef]

59. Riemer, A.; Leuders, S.; Thöne, M.; Richard, H.A.; Tröster, T.; Niendorf, T. On the fatigue crack growth behavior in 316L stainless steel manufactured by selective laser melting. *Eng. Fract. Mech.* **2014**, *120*, 15–25. [CrossRef]

60. Spierings, A.B. Fatigue performance of additive manufactured metallic parts. *Rapid Prototyp. J.* **2013**, *19*, 2, 88–94. [CrossRef]

61. Moore, J.P.; Williams, C.B. Fatigue properties of parts printed by PolyJet material jetting. *Rapid Prototyp. J.* **2015**, *21*, 675–685. [CrossRef]

62. Lee, J.; Huang, A. Fatigue analysis of FDM materials. *Rapid Prototyp. J.* **2013**, *19*, 291–299. [CrossRef]

63. Ziemian, C.; Ziemian, R.; Haile, K. Characterization of stiffness degradation caused by fatigue damage of additive manufactured parts. *Mater. Des.* **2016**, *109*, 209–218. [CrossRef]

64. Senatov, F.S.; Niaza, K.V.; Stepashkin, A.A.; Kaloshkin, S.D. Low-cycle fatigue behavior of 3d-printed PLA-based porous scaffolds. *Compos. Part B Eng.* **2016**, *97*, 193–200. [CrossRef]

65. Fischer, M.; Schöppner, V. Fatigue Behavior of FDM Parts Manufactured with Ultem 9085. *JOM* **2017**, *69*, 563–568. [CrossRef]

66. Puigoriol-Forcada, J.M.; Alsina, A.; Salazar-Martín, A.G.; Gomez-Gras, G.; Pérez, M.A. Flexural Fatigue Properties of Polycarbonate Fused-deposition Modelling Specimens. *Mater. Des.* **2018**, *155*, 414–421. [CrossRef]

67. Jerez-Mesa, R.; Travieso-Rodriguez, J.A.; Llumà-Fuentes, J.; Gomez-Gras, G.; Puig, D. Fatigue lifespan study of PLA parts obtained by additive manufacturing. *Procedia Manuf.* **2017**, *13*, 872–879. [CrossRef]

68. ASTM. *D7774-12: Standard Test Method for Flexural Fatigue Properties of Plastics, D20.10.24*; ASTM International: West Conshohocken, PA, USA, 2013.

69. Wirsching, P.H.; Light, M.C. Fatigue under wide band random stresses. *J. Struct. Div.* **1980**, *106–107*, 1593–1607.

Article

A Novel Route to Fabricate High-Performance 3D Printed Continuous Fiber-Reinforced Thermosetting Polymer Composites

Yueke Ming, Yugang Duan *, Ben Wang *, Hong Xiao and Xiaohui Zhang

State Key Lab for Manufacturing Systems Engineering, Xi'an Jiaotong University, Xi'an 710049, China; mingyueke@foxmail.com (Y.M.); xiaohongjxr@xjtu.edu.cn (H.X.); zhangxiaohui@mail.xjtu.edu.cn (X.Z.)
* Correspondence: ygduan@mail.xjtu.edu.cn (Y.D.); xawben@gmail.com (B.W.); Tel.: +86-029-8339-9516 (Y.D.); +86-029-8339-9516 (B.W.)

Received: 3 April 2019; Accepted: 24 April 2019; Published: 26 April 2019

Abstract: Recently, 3D printing of fiber-reinforced composites has gained significant research attention. However, commercial utilization is limited by the low fiber content and poor fiber–resin interface. Herein, a novel 3D printing process to fabricate continuous fiber-reinforced thermosetting polymer composites (CFRTPCs) is proposed. In brief, the proposed process is based on the viscosity–temperature characteristics of the thermosetting epoxy resin (E-20). First, the desired 3D printing filament was prepared by impregnating a 3K carbon fiber with a thermosetting matrix at 130 °C. The adhesion and support required during printing were then provided by melting the resin into a viscous state in the heating head and rapidly cooling after pulling out from the printing nozzle. Finally, a powder compression post-curing method was used to accomplish the cross-linking reaction and shape preservation. Furthermore, the 3D-printed CFRTPCs exhibited a tensile strength and tensile modulus of 1476.11 MPa and 100.28 GPa, respectively, a flexural strength and flexural modulus of 858.05 MPa and 71.95 GPa, respectively, and an interlaminar shear strength of 48.75 MPa. Owing to its high performance and low concentration of defects, the proposed printing technique shows promise in further utilization and industrialization of 3D printing for different applications.

Keywords: 3D printing; continuous carbon fiber; thermosetting epoxy resin; mechanical properties

---

1. Introduction

At present, the most commonly used polymer consumables in three-dimensional (3D) printing are thermoplastic filaments and powders, which exhibit weak load capacity, poor interlayer bonding, and low strength and hardness [1,2]. Usually, the high-performance fiber is used as reinforcement, and the polymeric resin is used as a matrix to improve the mechanical properties [3,4]. The fiber is used to enhance the corresponding strengths and moduli, whereas the resin is used to bond the fiber and uniformly distribute and transfer the external load [5,6].

As shown in Figure 1, 3D printing of fiber-reinforced composites has achieved remarkable milestones [7–9]. Initially, short fibers have been added into the thermoplastic filaments of the fused filament fabrication (FFF) process to attempt to improve performance. For instance, Tekinalp et al. [10] introduced short carbon fiber (SCF) into acrylonitrile butadiene styrene (ABS) and carried out 3D printing by using FFF equipment. All printed SCF/ABS composites exhibited a fiber orientation of 91.5%. Compared with the conventional compressive molded SCF/ABS composites, the tensile strength and tensile modulus of the 3D printed SCF/ABS composites, with a 40 wt % fiber content, increased to 65.0 MPa and 13.6 GPa, respectively. Furthermore, several research efforts have been made to replace the thermoplastic resin due to its defects with a thermosetting matrix [11]. For instance, Compton et al. [12]

synthesized a low-viscosity mixture by mixing several materials at a low temperature, where SCFs were used as a reinforcement, a thermosetting epoxy resin (EP) as a matrix, and an imidazole compound as a curing agent. The pre-formed samples were printed using FFF equipment. After curing, the printed SCF/EP composites with a fiber content of 35 wt % exhibited a tensile strength of 66.2 MPa. Later, continuous fiber reinforcements were used to further enhance the mechanical properties due to the ceilings of the short fiber reinforcement [13–15]. Yang et al. [16] utilized molten ABS to wrap the continuous carbon fiber (CCF) and carried out 3D printing by feeding, extruding, and cooling the mixture of the ABS filament. The printed CCF/ABS composites, with a 10 wt % fiber content, exhibited a tensile strength and flexural strength of 147 MPa and 127 MPa, respectively. However, the low interlaminar shear strength of 2.81 MPa reveals the vulnerable nature of the interlaminar performance due to the thermoplastic matrix. Nowadays, the continuous fiber-reinforced thermosetting polymer composites (CFRTPCs) are being used in 3D printing due to their strong intermolecular cross-linking [8]. Hao et al. [17] impregnated the continuous carbon fiber with epoxy resin (E-54). The pre-formed composites were then printed on the building platform and moved to a high-temperature chamber for curing. The as-printed CFRTPCs exhibited a tensile strength and flexural strength of 792.8 MPa and 202.0 MPa, respectively. However, Hao et al. did not discuss the detailed 3D printing issues about printing and post-curing methods.

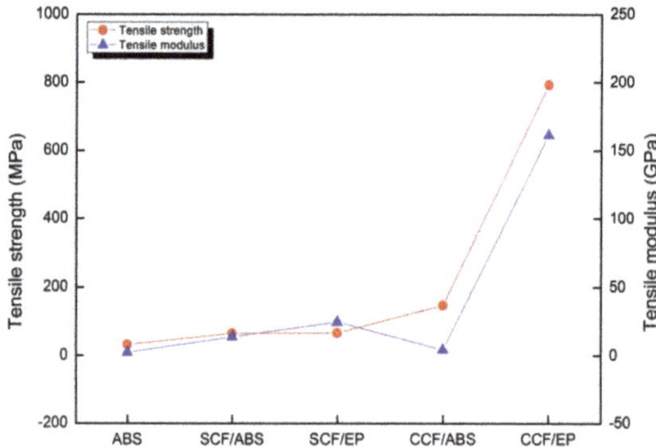

Figure 1. Performance evolution of 3D printing for different materials, including acrylonitrile butadiene styrene (ABS) [2], short carbon fiber (SCF)/ABS [10], SCF/epoxy resin (EP) [12], continuous carbon fiber (CCF)/ABS [13], and CCF/EP [17].

In this study, a novel 3D printing process is proposed to solve the challenges in printing and curing. The CFRTPC composites were fabricated using the relevant equipment. The fiber content, printing precision, fiber–resin interface, and internal voids were analyzed to characterize defects, deformations, and resin distribution during impregnating, printing, and curing modules. Finally, the mechanical properties of CFRTPCs were studied by tensile, flexural, and interlaminar shear testing.

2. Experimental

2.1. Materials

The thermosetting matrix system is comprised of an epoxy resin (E-20 (D.E.R. 671), 95 wt %, Dow, Pittsburg, CA, USA), and a curing agent (dicyandiamide (DICY), 5 wt %, Yongxin plasticization, Guangzhou, China). The dual interactions of amino and cyano groups in the DICY molecules allow an extended stable pot life with epoxy groups below 150 °C [18].

The 3K carbon fiber (Tenax®-J, HTS40, 200 TEX, Toho Tenax, Co., Ltd., Tokyo, Japan) was used as reinforcement. The tensile strength and tensile modulus of the 3K carbon fiber are 4400 MPa and 240 GPa, respectively.

2.2. 3D Printing Process

The high melting point (64–76 °C) allows E-20 to remain at a solid state at room temperature. The polymeric chains only possess intermolecular forces prior to the cross-linking reaction. The forces rapidly weaken and yield a low-viscosity liquid with increasing temperature, as shown in Figure 2. Therefore, the required adhesion and support during printing could be provided by using a solid-liquid resin transformation. Meanwhile, the impregnating and curing modules were separately designed to ensure fiber impregnation and to solve the problems during long-term post-curing. Hence, the whole process was separated into three modules: impregnating, printing, and curing. The low viscosity, at medium temperatures, promoted the fiber–resin interface during impregnating, whereas the high viscosity, at low temperatures, satisfied the printing requirements. Finally, a thermally initiated latent curing agent was used for curing at a higher temperature to avoid the premature cross-linking reactions between the first two modules.

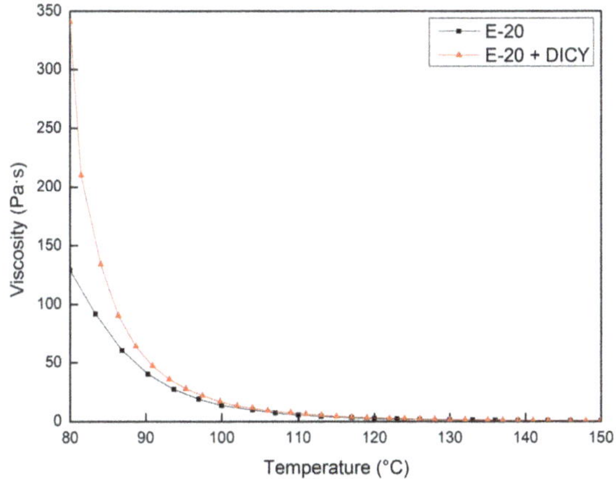

Figure 2. Viscosity–temperature curves of E-20 and the mixed matrix.

2.2.1. Impregnating

As shown in Figure 3a,b, the impregnation equipment was independently designed and built. The fiber was released from the supply coil and conveyed into the molten resin tank. According to the viscosity–temperature curve of the resin matrix (Figure 2) and the minimum reaction temperature (150 °C) of DICY, the impregnating temperature was set at 130 °C. Moreover, several yarn rollers were arranged in the resin tank to ensure the penetration of resin into the fiber bundles. The multiple deflections of the impregnating path and the control of the conveying tension resulted in efficient impregnation of the fiber. The width of the fiber bundle was extended from 1.5 to 3.5 mm. The infiltrated fiber bundle was then passed through a squeezing nozzle, which scraped off the excess resin and reshaped the extended bundle into a circle. By decreasing the temperature below the melting point (64–76 °C), the resin solidified together with the fiber to form a circular bar. Finally, through rewinding, the 3D printing filament of CFRTPCs was obtained, as shown in Figure 3c.

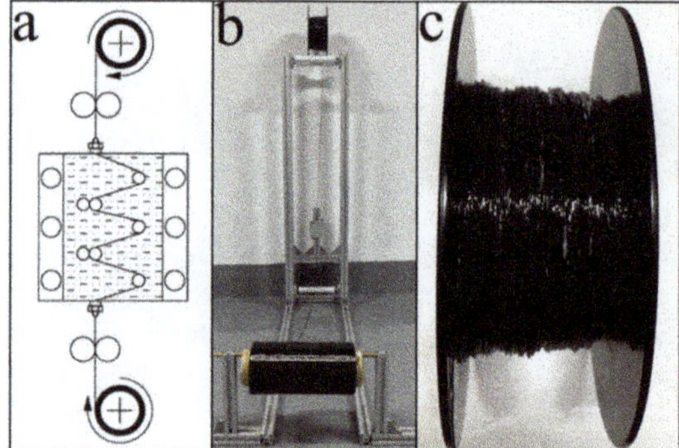

Figure 3. The as-designed impregnating equipment: (**a**) a schematic illustration of the impregnating process, (**b**) impregnating equipment, and (**c**) 3D printing filament of continuous fiber-reinforced thermosetting polymer composites (CFRTPCs).

2.2.2. Printing

The FFF-based printing equipment was designed and developed, as shown in Figure 4a,b. The 3D printing filament of CFRTPCs needed to pass through the heating head (130 °C) and printing nozzle in advance. The resin part of the filament was melted into a viscous state and resulted in a small molten resin tank inside the heating head. Hence, for further infiltration and subsequent printing, the filament was impregnated again due to the interactions of the heating head and the printing nozzle. After pulling out, the fiber was rapidly cooled down and attached to the platform or the former layer (Figure 4c). The heating head moved in the X-Y plane and along the cross-section contours and filling trajectories generated from the 3D model. After each layer's generation, the platform descended along the Z-direction by a distance equal to the layer thickness. These steps were repeated until the whole pre-formed sample was printed.

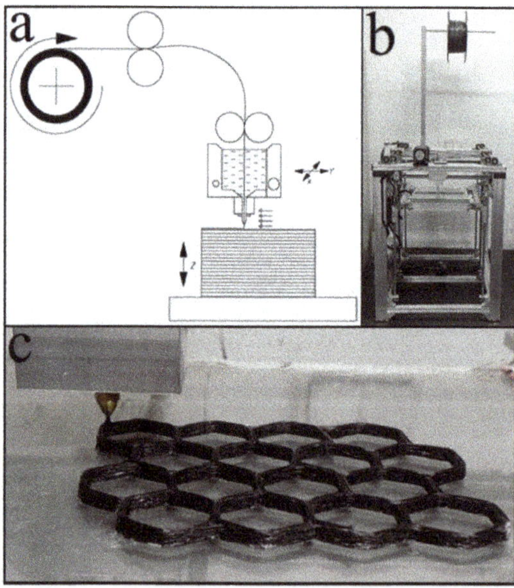

Figure 4. Working process of the printing module: (**a**) a schematic illustration of the printing process, (**b**) printing equipment, and (**c**) the working process of the printing nozzle.

2.2.3. Curing

It is worth noting that the resin melts and transforms into a viscous state due to the utilization of the thermally induced curing method. Once the constraints from the glassy resin are reduced, large deformations occur at the turning points in the planned path. In combination with a negative pressure condition, the pre-formed samples are prone to collapse.

Herein, a novel curing method is proposed to maintain the original shapes of the pre-formed CFRTPC samples and accommodate the complexity and variability of 3D printing technology (Figure 5). First, the pre-formed samples were completely buried in sodium chloride powders. Second, the pressure of −0.1 MPa was exerted by the external vacuum pump to promote the filling of internal voids and eliminate the trapped air. Then, the powders and pre-formed samples formed a dense entity to maintain the original shape. After that, the entire device was placed in an oven, infiltrating at 130 °C for 1 h and cured at 160 °C for 1 h. Finally, the CFRTPC samples were removed from the oven, rinsed with water, and dried to obtain the final products (Figure 5b).

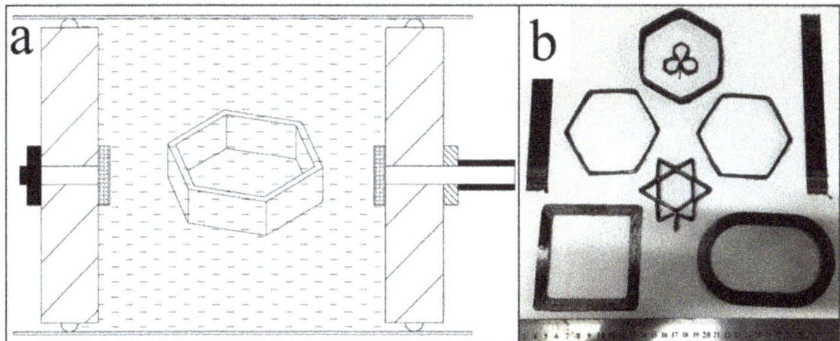

Figure 5. Curing module: (**a**) a schematic illustration of the curing process and (**b**) the cured CFRTPC samples.

2.3. Characterization

The fiber content of the 3D printing filament was calculated by thermal gravimetric analysis (TGA-DSC1, Mettler-Toledo GmbH, Greifensee, Schweiz). The surface of the impregnated filament and the 3D profile of the printed fiber bundle were scanned by laser scanning confocal microscopy (OLS4000, Olympus, Tokyo, Japan). The fiber–resin interface and resin distribution in the pre-formed and cured samples were observed with a scanning electron microscope (SEM, S-3000N, Hitachi, Tokyo, Japan) at an acceleration voltage of 30.0 kV. The corresponding internal structure and void distribution were obtained by the micron X-ray 3D imaging system (YXLON international GmbH, Hamburg, Germany) with an accelerating voltage of 80.0 kV and an image resolution of 8.0 μm.

Finally, the mechanical characterization was carried out using the electromechanical universal testing machine (MTS Systems, Co., Ltd, Shenzhen, China) to obtain the mechanical properties of the cured CFRTPCs. The experimental apparatus and operation specifications are shown in Figure 6. Moreover, five independently carried out measurements were averaged out for each property.

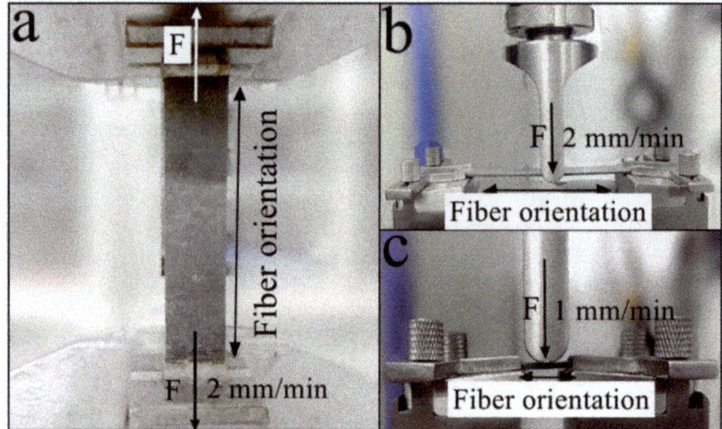

Figure 6. The specimen dimensions and operational specifications during (**a**) the tensile test, (**b**) the three-point bending test, and (**c**) the interlaminar shear test.

2.3.1. The Tensile Test

As shown in Figure 6a, the tensile test was conducted according to the ISO 527:1997 standard (plastics–determination of tensile properties). The dimensions of the tensile specimen were $250 \times 25 \times 2$ mm. The fiber was unidirectionally printed at $0°$, corresponding to the longest side of the specimen and force loading direction. The loading rate was 2 mm/min.

2.3.2. The Three-Point Bending Test

As shown in Figure 6b, the flexural test was conducted according to the ISO 14125:1998 standard (fiber-reinforced plastic composites–determination of flexural properties). The dimensions of the flexural specimen were $100 \times 15 \times 2$ mm. The fiber was unidirectionally printed at $0°$, which corresponds to the longest side of the specimen. The vertical force was applied at mid-point of the longitudinal direction, and the loading rate was 2 mm/min.

2.3.3. The Interlaminar Shear Test

As shown in Figure 6c, the interlaminar shear test was conducted according to the ISO 14130:1997 standard (fiber-reinforced plastic composites–determination of apparent interlaminar shear strength by short-beam method). The dimensions of the shear specimen were $20 \times 10 \times 2$ mm. The fiber was printed unidirectionally at $0°$, which corresponds to the longest side of the specimen. The vertical force was applied at the mid-point of the longitudinal direction, and the loading rate was 1 mm/min.

3. Results and Discussion

3.1. Fiber Content

The fiber content in CFRTPCs was mainly controlled by the impregnating module. After impregnating with the molten matrix, the continuous fiber bundle was extruded through a squeezing nozzle to scrape off the excess resin. The amount of the scraped resin can be controlled by adjusting the nozzle diameter.

As more resin was scraped off, the fiber content significantly increased with decreasing nozzle diameter, as shown in Figure 7. The highest fiber content of 71.05 wt % was achieved at a nozzle diameter of 0.6 mm. However, the damage to the fiber surface was exacerbated with the shrinkage of the nozzle diameter. The surface of the filament, impregnated by 1.0 mm and 0.6 mm nozzles, is presented in Figure 8. The smaller diameter and the sharp metal edges resulted in a large amount of fiber breakage and curling (Figure 8a). One should note that these defects inevitably influenced the mechanical properties of the obtained CFRTPCs by the subsequent printing and curing modules. The broken fibers were unable to carry and transfer stress during tensile and compressive processes, whereas the curled fibers caused void defects, which acted as crack sources and led to inferior interlayer performance. On the other hand, an excessively large nozzle diameter resulted in low fiber content and uneven resin distribution, which is also not desirable from an application viewpoint. Based on these observations, the squeezing nozzle diameter was set at 1.0 mm. The surface of the corresponding 3D printing filament, with a fiber content of 48.33 wt % and a diameter of 1.0 mm, is shown in Figure 8b.

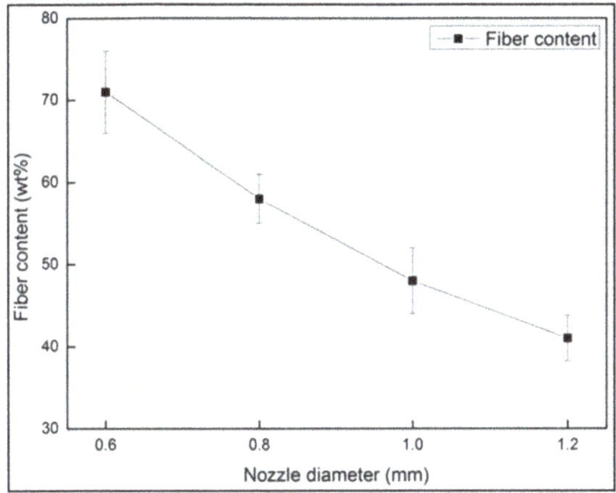

Figure 7. The influence of nozzle diameter on the fiber content in CFRTPCs.

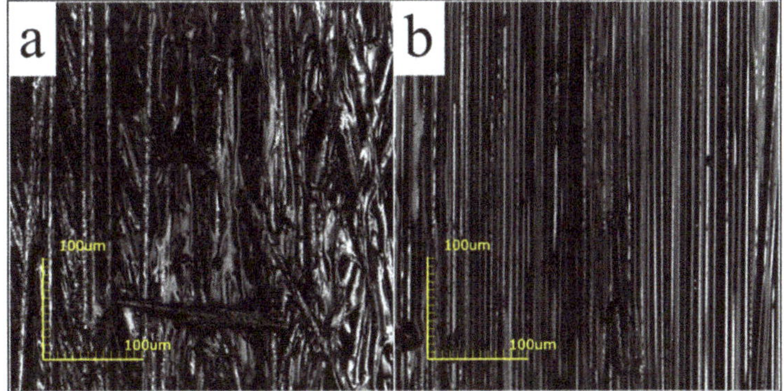

Figure 8. The surface of the filaments impregnated by (**a**) 0.6 mm and (**b**) 1.0 mm diameter nozzles.

3.2. Printing Precision

Based on the diameter of the impregnated filament, the diameter of the printing nozzle was determined to be 1.0 mm. During printing, the filament transforms into a viscous state inside the heating head (130 °C), which is soft and can be reshaped. After pulling out, the filament was flattened from a circular cross section into a rectangular shape by the printing nozzle, cooled down, and attached to the platform. The 3D profile of the printed fiber bundle is presented in Figure 9. The average width and thickness of the cross section are 1.8 mm and 0.4 mm, respectively.

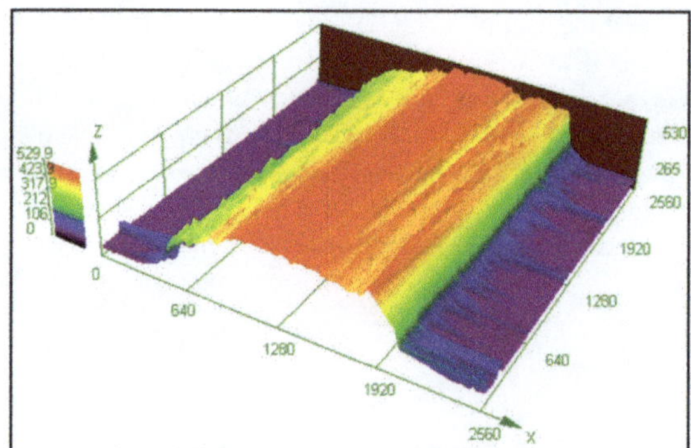

Figure 9. 3D profile of the printed fiber bundle with a width of 1.8 mm and a thickness of 0.4 mm.

A suitable printing speed, matching with the adhesion method of the resin's solid–liquid transition, renders an excellent printing quality and accuracy. Hence, a constant printing speed of 10 mm/s was selected, and repeated printing tests were carried out using lines, triangles, quads, and circles to measure the dimensional error from the planned paths. As shown in Figure 10a, the printing of the lines was slightly shorter due to the poor positioning accuracy of the large printing nozzle, which can be adjusted by code compensation. In addition, the fiber was dragged up from the corners, as shown in Figure 10b,c. The driving force during printing was mainly provided by the adhesion of the solid–liquid resin transition and the traction from the printed fiber bundle. The adhesion was always the same due to the constant temperature difference before and after printing. However, the traction had to be along the printing direction due to the vector characteristics, and it might cause a printing error when the curvature suddenly changes. Therefore, in the case of large angle deflections, it was necessary to reduce the printing speed and lower the nozzle height to increase the compaction force. Finally, during arc printing with constant curvature, as shown in Figure 10d, the fiber bundle rendered a uniform width with high precision.

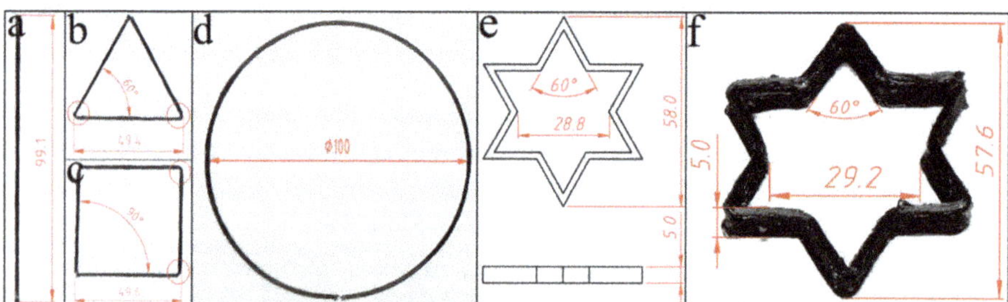

Figure 10. The printing tests and corresponding dimensions of (**a**) line, (**b**) triangle, (**c**) quad, and (**d**) circle and the comparison between the dimensional differences in the (**e**) original 3D model and (**f**) printed pre-formed sample.

Furthermore, a hexagonal star was selected to analyze the printing precision in the case of the Z-direction and complex structures. Comparing the printed pre-formed sample with the original 3D model (Figure 10e,f), the dimensional differences mainly appeared at the corner positions and contributed to the error in longitudinal and transverse directions. However, the other dimensions, such as height and angle, were well controlled because those were mainly determined by the settings of X–Y–Z motion distance and direction. The shape and size remained the same as the original 3D model. The overall dimensional error was less than 2%.

3.3. Interface and Voids

Considering the high-temperature stability, easy removal, and cost, sodium chloride (NaCl) was used as the final powder in the curing module. The diameter of the sodium chloride particles ranged from 10 to 50 μm. A flexible peel ply was used to wrap the pre-formed sample to prevent the embedment of these particles inside the sample during curing. The pressure difference due to the externally applied vacuum filled the gaps in the pre-formed sample and maintained the original shape. Finally, the high temperature activates the DICY and led to a cross-linking reaction.

Figure 11 shows that the proposed curing method effectively improves the fiber–resin interface. The tightly squeezed particles resulted in a compact structure. In the case of the pre-formed sample (Figure 11a,c), the surface of the fiber bundle was partially wrapped with resin, and dense voids were observed. One should note that the DICY, with a melting point of 208–211 °C, just scattered within the gaps of the fiber in the form of solid particles. Moreover, the void content in the pre-formed sample was ~10.05%. On the other hand, E-20 and DICY melted and formed a molten mixture with low viscosity (Figure 2) due to the infiltration at 130 °C for 1 h before curing. Under the influence of a negative pressure, the molten mixture moved and filled the inner voids. The hydrogen atoms on the amines in DICY then initiated an open ring reaction with the epoxy group during subsequent curing at 160 °C for 1 h. In addition, the nitrile group reacted with the hydroxyl group to produce amides, which further reacted with the epoxy group [18,19]. After diffusion, E-20 was continuously consumed until the cross-linking reaction was completed. Therefore, in the case of cured samples, the surface of the fiber was fully and evenly covered with resin (Figure 11b) and the internal voids were also well filled (Figure 11d). The void content was reduced from 10.05 to 2.53%. The degree of cross-linking reaction was over 99%, as measured by differential scanning calorimetry.

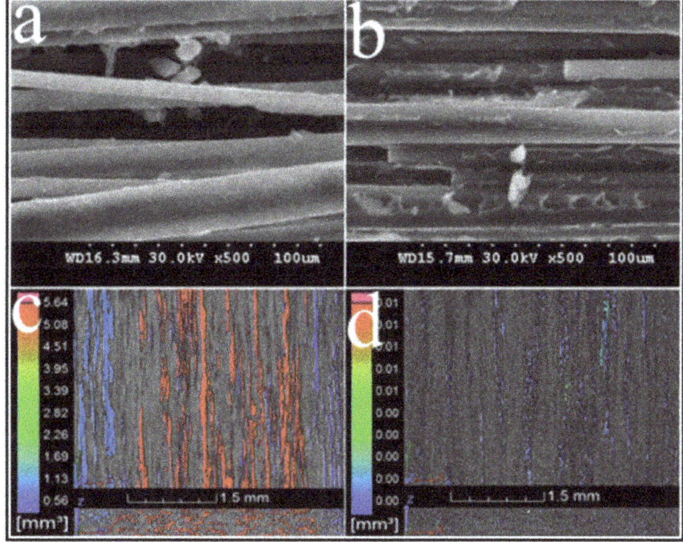

Figure 11. Fiber–resin interface and internal voids of the (**a,c**) pre-formed and (**b,d**) cured samples, respectively.

3.4. Mechanical Characterization

3.4.1. Tensile Test Results

Figure 12a shows that the tensile strength and tensile modulus of the cured CFRTPC samples were 1476.11 MPa and 100.28 GPa, respectively. The comparison of different materials (Figure 12b) reveals that the obtained tensile strength was twice as high as that of the 3D-printed CCF/E-54 (792.8 MPa) [17] and ~70% of the unidirectional (UD) composites (laminate, 2171.85 MPa) [20].

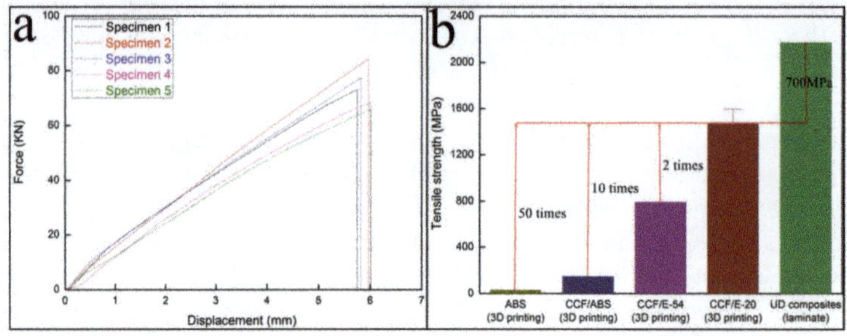

Figure 12. Tensile characterization: (**a**) the force–displacement curves and (**b**) tensile strengths of different materials.

3.4.2. Three-Point Bending Test Results

As shown in Figure 13a, the flexural strength and flexural modulus of the cured CFRTPC samples were 858.05 MPa and 71.95 GPa, respectively. The comparison of different materials (Figure 13b) reveals that the obtained flexural strength was ~4 times higher than that of the 3D-printed CCF/E-54 (202.0 MPa) [17] and close to the UD composites (laminate, 1703.01 MPa) [20].

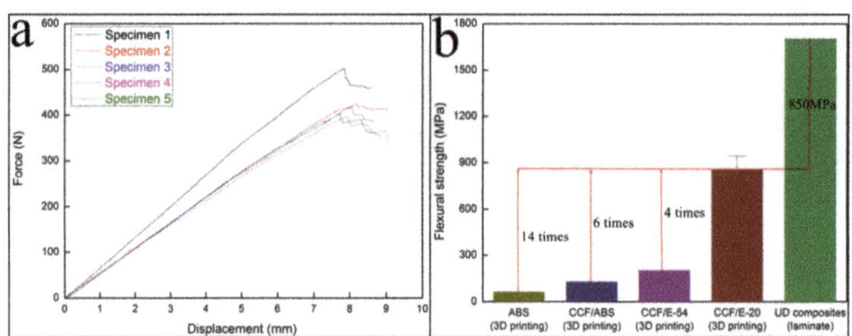

Figure 13. Flexural characterization: (**a**) the force-displacement curves and (**b**) flexural strengths of different materials.

3.4.3. Interlaminar Shear Test Results

The interlaminar shear strength of the cured CFRTPC samples was increased to be 48.75 MPa, but it was less than half of that of the UD composites (106.87 MPa, laminate) [20].

These results indicate that the mechanical properties, such as the tensile and flexural strengths and moduli, of the 3D-printed CFRTPCs were remarkably improved. One should note that a higher fiber content was obtained due to the introduction of the impregnating module. Moreover, the fiber and resin were evenly distributed, and an extensive cross-linking between the polymer chains of the thermosetting matrix was achieved due to the infiltration and curing processes. It is worth mentioning that a firm polymeric network can effectively carry and transfer the external load. However, the interlaminar shear strength of the 3D-printed CFRTPCs was quite low. Owing to the absence of any subsequent rolling or compaction process after the printing of each layer, the layer thickness was mainly determined by the Z-axis distance, which resulted in insufficient connectivity between layers. Hence, weak interlaminar bonding is one of the most vulnerable aspects of 3D-printed CFRTPCs.

3.5. Fracture Analysis

3.5.1. Tensile Fracture Mode

As shown in Figure 14a, the tensile fracture mode of the 3D-printed CCF/E-20 sample was fiber splitting. The whole sample was split into several strips along the printing direction. The multiple parts after the fracture indicate that there were voids and cracking defects inside the sample. Meanwhile, the same direction of fiber splitting also suggests that the bonding between the fiber bundles was not enough along the printing direction. The scratches, caused by the deflection of the fiber bundle at the exit of the printing nozzle, resulted in a higher amount of fiber damage. Therefore, when the load exceeded the bearing capacity limit, the sample split into multiple strips along the printing direction. One should note that the observed fracture mode was better than the fiber extraction from the resin of 3D-printed CCF/ABS [16], but it was still not as good as the regular brittle fracture of UD composites (laminate) [21], which suggests that the tensile properties can be further enhanced.

Figure 14. Fracture modes of the 3D-printed CCF/E-20 after (**a**) the tensile test, (**b**) the three-point bending test, and (**c**) the interlaminar shear test.

3.5.2. Flexural Fracture Mode

As shown in Figure 14b, the flexural fracture mode of 3D-printed CCF/E-20 sample was brittle rupture. The fractured part of the 3D-printed CCF/E-20 sample was divided into two regions, referred to as the flat compression fracture zone and the coarse tensile fracture zone, due to the dual influence of tensile and compressive stresses during bending. In the perpendicular direction, the fiber was tightly bonded to the resin and formed a strong fiber–resin interface, which implies that the external load can be effectively transferred and distributed.

3.5.3. Shear Fracture Mode

In the case of interlaminar shear analysis, the 3D-printed CCF/E-20 sample exhibited a multi-layer shear failure, as shown in Figure 14c. The test specimen was subjected to a diagonal triangular stress distribution, which resulted in a stress gradient between the layers [22]. Unlike 3D-printed

CCF/ABS [16], the fiber did not peel off from the resin, but the presence of multiple shearing surfaces and a low interlaminar shear strength of 48.75 MPa reveal that the interlayer performance of 3D-printed CCF/E-20 was not satisfactory.

4. Conclusions

In summary, a novel 3D printing technology for CFRTPCs was investigated in detail. The main conclusions are summarized below:

1. The whole fabrication process was separated into three independent modules: impregnating, printing, and curing. The required experimental equipment for each module was independently designed and developed.
2. A high fiber content of 48.33 wt % was achieved through impregnation. The dimensional precision was obtained by optimizing the printing process. The overall dimensional error was less than 2%. After curing, the fiber–resin interface was significantly improved, and the void content was reduced from 10.05 to 2.53%.
3. The mechanical properties of the 3D-printed CFRTPCs were evaluated by tensile, flexural, and interlaminar shear testing. The results reveal that the tensile strength and tensile modulus were 1476.11 MPa and 100.28 GPa, respectively; the flexural strength and flexural modulus were 858.05 MPa and 71.95 GPa, respectively; and the interlaminar shear strength was 48.75 MPa. Finally, the fracture analysis demonstrated the brittle nature of CFRTPCs and the inferior interlayer performance.

Author Contributions: Conceptualization: Y.M., Y.D., and B.W.; formal analysis: Y.M., B.W., and X.Z.; investigation: Y.M.; software: H.X.; supervision: Y.D.; writing—original draft: Y.M.; writing—review & editing: Y.D., B.W., H.X., and X.Z.

Funding: This research was funded by the National Key Research and Development Program of China (NO.2016YFB1100902), the National Natural Science Foundation of China (Grant No.51875440), and the China Postdoctoral Science Foundation (2018M640979).

Conflicts of Interest: The authors declare no conflict of interest.

References

1. Yu, X.; Tao, C.; Ru, J. Applications, Property Characterization of Polylactic Acid Product Formed by 3D Printing. *Modern Plast. Process. Appl.* **2016**, *28*, 51–53. [CrossRef]
2. Qiao, W.; Xu, H.; Chao, M.; Zhai, L.; Zhang, X.; Hao, H.; Zhang, Y.; Zhen, L. Research on Properties of ABS Filament Used in 3D Printing. *Eng. Plast. Appl.* **2016**, *44*, 18–23. [CrossRef]
3. Wang, X.; Jiang, M.; Zhou, Z.; Gou, J.; Hui, D. 3D printing of polymer matrix composites: A review and prospective. *Compos. Part B Eng.* **2017**, *110*, 442–458. [CrossRef]
4. Ngo, T.D.; Kashani, A.; Imbalzano, G.; Nguyen, K.T.; Hui, D. Additive manufacturing (3D printing): A review of materials, methods, applications and challenges. *Compos. Part B Eng.* **2018**, *143*, 172–196. [CrossRef]
5. Ning, F.; Cong, W.; Qiu, J.; Wei, J.; Wang, S. Additive manufacturing of carbon fiber reinforced thermoplastic composites using fused deposition modeling. *Compos. Part B Eng.* **2015**, *80*, 369–378. [CrossRef]
6. Blok, L.; Longana, M.; Yu, H.; Woods, B. An investigation into 3D printing of fibre reinforced thermoplastic composites. *Addit. Manuf.* **2018**, *22*, 176–186. [CrossRef]
7. Yang, C.; Wang, B.; Li, D.; Tian, X. Modelling and characterisation for the responsive performance of CF/PLA and CF/PEEK smart materials fabricated by 4D printing. *Virtual Phys. Prototyp.* **2017**, *12*, 69–76. [CrossRef]
8. Goh, G.D.; Yap, Y.L.; Agarwala, S.; Yeong, W.Y. Recent Progress in Additive Manufacturing of Fiber Reinforced Polymer Composite. *Adv. Mater. Technol.* **2019**, *4*, 1800271. [CrossRef]
9. Mohan, N.; Senthil, P.; Vinodh, S.; Jayanth, N. A review on composite materials and process parameters optimisation for the fused deposition modelling process. *Virtual Phys. Prototyp.* **2017**, *12*, 1–13. [CrossRef]
10. Tekinalp, H.L.; Kunc, V.; Velez-Garcia, G.M.; Duty, C.E.; Love, L.J.; Naskar, A.K.; Blue, C.A.; Ozcan, S. Highly oriented carbon fiber–polymer composites via additive manufacturing. *Compos. Sci. Technol.* **2014**, *105*, 144–150. [CrossRef]

11. Goh, G.; Dikshit, V.; Nagalingam, A.; Goh, G.; Agarwala, S.; Sing, S.; Wei, J.; Yeong, W. Characterization of mechanical properties and fracture mode of additively manufactured carbon fiber and glass fiber reinforced thermoplastics. *Mater. Des.* **2018**, *137*, 79–89. [CrossRef]

12. Compton, B.G.; Lewis, J.A.J.A.M. 3D-printing of lightweight cellular composites. *Adv. Mater.* **2015**, *26*, 5930–5935. [CrossRef] [PubMed]

13. Van Der Klift, F.; Koga, Y.; Todoroki, A.; Ueda, M.; Hirano, Y.; Matsuzaki, R. 3D Printing of Continuous Carbon Fibre Reinforced Thermo-Plastic (CFRTP) Tensile Test Specimens. *Open J. Compos. Mater.* **2016**, *6*, 18–27. [CrossRef]

14. Justo, J.; Távara, L.; García-Guzmán, L.; París, F. Characterization of 3D printed long fibre reinforced composites. *Compos. Struct.* **2018**, *185*, 537–548. [CrossRef]

15. Caminero, M.; Chacón, J.; García-Moreno, I.; Rodriguez, G. Impact damage resistance of 3D printed continuous fibre reinforced thermoplastic composites using fused deposition modelling. *Compos. Part B Eng.* **2018**, *148*, 93–103. [CrossRef]

16. Yang, C.; Tian, X.; Liu, T.; Cao, Y.; Li, D. 3D printing for continuous fiber reinforced thermoplastic composites: mechanism and performance. *Rapid Prototyp. J.* **2017**, *23*, 209–215. [CrossRef]

17. Hao, W.; Liu, Y.; Zhou, H.; Chen, H.; Fang, D. Preparation and characterization of 3D printed continuous carbon fiber reinforced thermosetting composites. *Polym. Test.* **2018**, *65*, 29–34. [CrossRef]

18. May, C.A.; Tanaka, Y. *Epoxy Resins: Chemistry and Technology*; Marcel Dekker: New York, NY, USA, 1988; Volume 3, pp. 285–464.

19. Gu, X.Y.; Zhang, J.Y. Study on the Curing Mechanism of Epoxy/DICY System by FT-IR. *Polym. Mater. Sci. Eng.* **2006**, *22*, 182–184. [CrossRef]

20. Hexcel Company: Datasheet resources. Available online: https://www.hexcel.com/user_area/content_media/raw/HexPly_85517_us_DataSheet.pdf (accessed on 14 June 2018).

21. Greenhalgh, E.S. *Failure Analysis and Fractography of Polymer Composites*; Woodhead: Cambridge, UK, 2010; pp. 108–116.

22. Cui, W.C.; Wisnom, M.R.; Jones, M. Failure mechanisms in three and four point short beam bending tests of unidirectional glass/epoxy. *J. Strain Anal. Eng.* **1992**, *27*, 235–243. [CrossRef]

Article

Experiments on the Ultrasonic Bonding Additive Manufacturing of Metallic Glass and Crystalline Metal Composite

Guiwei Li [1,2], Ji Zhao [1,3], Jerry Ying Hsi Fuh [2], Wenzheng Wu [1,*], Jili Jiang [1], Tianqi Wang [1] and Shuai Chang [2]

[1] School of Mechanical and Aerospace Engineering, Jilin University, Changchun 130025, China; ligw15@mails.jlu.edu.cn (G.L.); jzhao@jlu.edu.cn (J.Z.); jiangjl16@mails.jlu.edu.cn (J.J.); wangtq18@mails.jlu.edu.cn (T.W.)

[2] Department of Mechanical Engineering, National University of Singapore, Singapore 117576, Singapore; jerry.fuh@nus.edu.sg (J.Y.H.F.); msecs@nus.edu.sg (S.C.)

[3] School of Mechanical Engineering and Automation, Northeastern University, Shenyang 110004, China

[*] Correspondence: wzwu@jlu.edu.cn; Tel.: +86-138-9481-4109

Received: 24 August 2019; Accepted: 13 September 2019; Published: 14 September 2019

Abstract: Ultrasonic vibrations were applied to weld Ni-based metallic glass ribbons with Al and Cu ribbons to manufacture high-performance metallic glass and crystalline metal composites with accumulating formation characteristics. The effects of ultrasonic vibration energy on the interfaces of the composite samples were studied. The ultrasonic vibrations enabled solid-state bonding of metallic glass and crystalline metals. No intermetallic compound formed at the interfaces, and the metallic glass did not crystallize. The hardness and modulus of the composites were between the respective values of the metallic glass and the crystalline metals. The ultrasonic bonding additive manufacturing can combine the properties of metallic glass and crystalline metals and broaden the application fields of metallic materials.

Keywords: metallic glasses; composite materials; interfaces; additive manufacturing; ultrasonic bonding; 3D printing

---

1. Introduction

Metallic glass (MG), also known as amorphous alloy or liquid metal, is produced via modern rapid-solidification metallurgy [1,2]. The internal atoms are arranged in a short-range-ordered, long-range-disordered amorphous structure due to the rapid cooling of liquid melt [3]. It has the excellent mechanical, physical, and chemical properties of metals and glass and has broad application prospects in the automotive, aerospace, medical, communication, and industrial automation fields [4–12]. However, these materials have glassy interior microstructures, making them brittle, and their critical forming sizes make them difficult to manufacture bulk blanks, limiting their applications [13]. Recently, some researchers have employed traditional additive manufacturing technologies to manufacture bulk metallic glass, which still cannot improve its mechanical properties [14–22]. Cu, Al, and other conventional crystalline metals, in contrast, form crystals because of the ordering of their internal atoms. They have high plasticity because of the crystal slip deformation and twinning deformation under stress, but the mechanical properties, such as strength and hardness, are much lower than those of glassy metals [23,24]. To synthesize the advantages of metallic glass and crystalline metals, they can be combined with additive manufacturing to manufacture bulk composites. This has been explored by various researchers, as described below.

Li et al. made Fe-based metallic glass and crystalline Cu composite parts by selective laser melting [25]. Kim et al. employed electron beam welding to bond Zr-based metallic glass and stainless steel, but the surface of the Zr-based metallic glass crystallized easily [26]. Li et al. explored the joint effect of Zr-based metallic glass and crystalline metal by using laser-foil-printing additive manufacturing, and the results illustrated that Zr-based metallic glass can be welded to Zr 702 alloy [27,28]. Feng et al. investigated the fracture mechanism of Zr-based metallic glass and crystalline Cu composites processed by explosive welding [29]. Wang employed laser impact welding to bond Fe-based metallic glass and crystalline Cu and found the interface hardness to be much higher than that of crystalline Cu [30].

The above studies mainly focused on the manufacture of bulk metallic glass composites with a high-energy beam, such as laser beams. These methods were very complex, required high-quality raw materials, and often crystallized the metallic glass. An alternative method that might avoid such issues is ultrasonic bonding additive manufacturing. Ultrasonic additive manufacturing is a hybrid additive manufacturing technique, which combines the capabilities of ultrasonic bonding and CNC milling [31,32]. Based on ultrasonic bonding, this method is simpler, has lower quality requirements for raw materials, increases the sample temperature less, and can bond various kinds of materials [33]. In this study, we used a custom ultrasonic bonding system to bond Ni-based metallic glass ribbons with Al and Cu crystalline ribbons, and we analyzed the interfaces of metallic glass and crystalline metal composites processed by ultrasonic bonding additive manufacturing.

2. Material and Methods

Figure 1 shows the principle of ultrasonic bonding additive manufacturing. First, a Ni-based metallic glass ribbon and a crystalline metal ribbon were placed on the fixture, and ultrasonic bonding parameters were set to activate the ultrasonic bonding system. Then, the horn was brought into contact with the ribbon and pressed down to perform ultrasonic consolidation. After ultrasonic consolidation, the specimen remained under pressure from the horn for a short time to prevent the specimen from warping. This consolidation combined the two metal ribbons into one piece, which was then combined with the next metal ribbon until the entire part was completed. This additive manufacturing technology has great potential for manufacturing high-performance bulk metallic glass composites and functional graded materials.

$Ni_{82.2}Cr_7B_3Si_{4.8}Fe_3$ (wt %) metallic glass ribbons (Miai Metal Material Co. LTD, Kunshan, China), Al ribbons, and Cu ribbons with cross-sectional dimensions of 1.7×0.04 mm, 1.7×0.1 mm, and 1.7×0.1 mm, respectively, were used, considering that Ni-based metallic glass has a good welding capacity [34]. A custom ultrasonic bonding system (Dongguan Jieshi Ultrasonic Automation Co., Ltd., Dongguan, China) was used to bond metallic glass and crystalline metal, using a frequency of 35 kHz and a power of 800 W. Figure 2a shows the cross-sectional morphology of a three-layer Al/Ni-based (MG) composite sample processed by the ultrasonic bonding system. Figure 2b,c show the EDS mapping analysis of the Al and Ni elements in Figure 2a, respectively. Figure 2e,f show the EDS line analysis of the Al and Ni elements along the pink line in Figure 2d, respectively. These two elements diffused slightly at the interfaces. Figure 2 illustrates that metallic glass can be bonded with crystalline metal, and the ultrasonic bonding process can be employed to manufacture bulk metallic glass composites additively with layer-by-layer accumulating formation characteristics. The effects of ultrasonic vibration energy on the quality of the interfaces in the Ni-based metallic glass composites were studied.

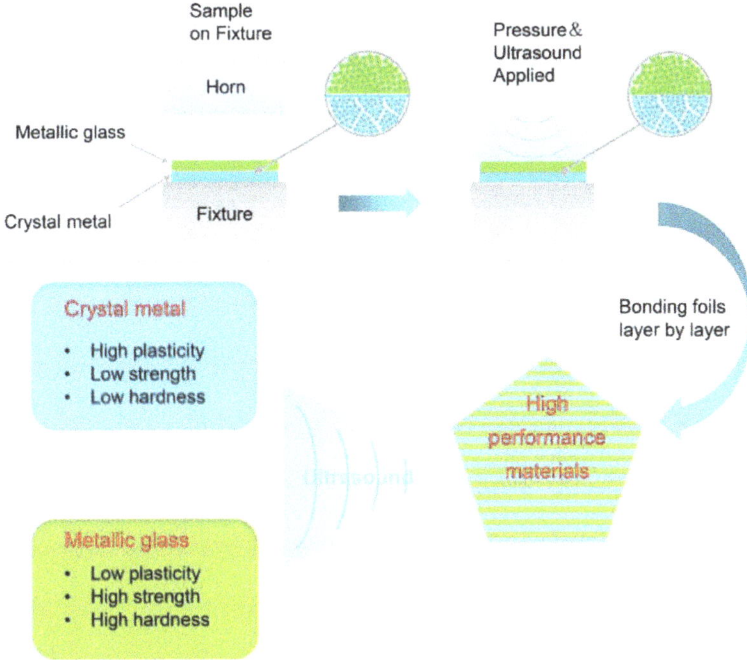

Figure 1. Schematic of ultrasonic bonding additive manufacturing.

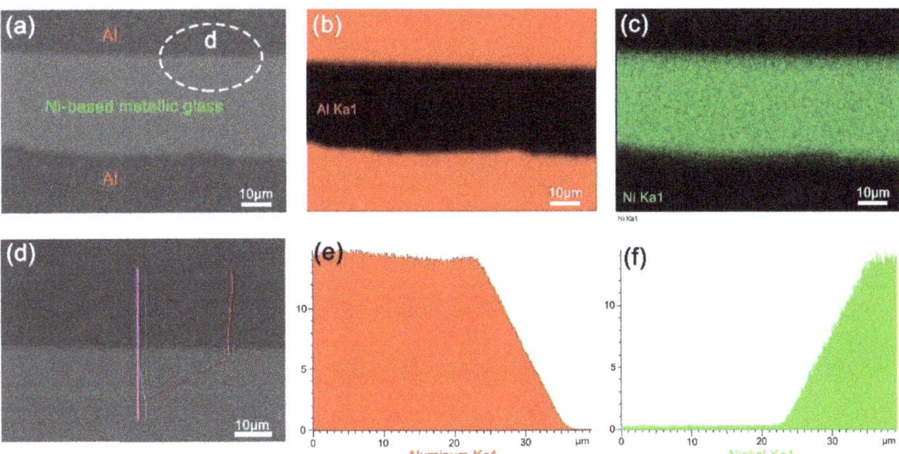

Figure 2. The cross-sectional SEM image of Al/Ni-based (MG) composites (**a**), EDS mapping analysis of Al (**b**) and Ni (**c**) elements, and EDS line analysis of Al (**e**) and Ni (**f**) elements along the pink line of the cross-sectional SEM image (**d**).

During ultrasonic consolidation, the energy inputted via the ultrasonic bonding system into the interior of the consolidated sample is defined as $Q = Pt$, where Q is the input energy, P is the power, and t is the ultrasonic bonding time that the ultrasonic wave acted on the ribbons. The time between the ultrasonic emission and the start of the ultrasonic bonding system is the delay time, and the time that the horn continues to press on the ribbons after the ultrasonic emission is the hold time. The bonding time *t* is the main parameter which affects the quality of consolidation between a layer of metallic glass and a layer of crystalline metal. Table 1 shows the experimental scheme.

Table 1. Ultrasonic bonding experiment parameters.

Material	Fixed Factors			Control Factors		
	Factor	Level	Unit	Factor	Value	Unit
Ni-based (MG) and Al	Pressure	0.18	MPa	Bonding time	60	ms
	Delay time	40	ms		80	
	Hold time	50	ms		160	
Ni-based (MG) and Cu	Pressure	0.18	MPa	Bonding time	40	ms
	Delay time	40	ms		60	
	Hold time	50	ms		140	

The cross-sectional morphologies of the consolidated samples were observed by SEM (ZEISS-EVO18, Carl Zeiss NTS, Oberkochen, Germany). The sample cross section was first polished with 8000 mesh sandpaper and then polished on a polisher (Kejing Automation Equipment Co. LTD, Shenyang, China) with a 50 nm SiO_2 polishing liquid. The bonding interface of the torn sample was observed using a digital microscope (Keyence Singapore PTE LTD, Singapore, Singapore). The phase composition of the bonding interface of the sample was tested by XRD (X′ Pert PRO MPD, PANalytical BV, Almelo, Netherlands) using Cu Kα radiation at λ = 1.54 Å, operated at 40 kV and 40 mA. The interior hardness and modulus of the consolidated samples were tested by nanoindentation (Nano Indenter G200, Agilent, Oak Ridge, USA) using a peak holding time of 3 s and a surface approach velocity of 10 nm/s.

3. Results and Discussion

3.1. SEM of the Cross-Sectional Morphology

Figure 3 shows the cross-sectional morphologies of the metallic glass and crystalline metal composites at various bonding times. After polishing, the surfaces of the metallic glass were relatively smooth, but the surfaces of the crystalline metals showed some obvious scratches. This result occurred mainly because the atomic arrangement inside the metallic glass had short-range order, long-range disorder, and better wear resistance than the crystalline metal. The metallic glass is much harder than Al or Cu, and under the same polishing force, slightly more of the crystalline metal was removed than the Ni-based metallic glass.

Figure 3a–c show a clear boundary between the Ni-based metallic glass and Al but no gaps at the interfaces. The consolidation interfaces could remain relatively flat with the increase of bonding times, indicating that the metallic glass and Al can be bonded well over a wide range of bonding times. Figure 3d,e show that the bonding interfaces between the Ni-based metallic glass and Cu are relatively flush and there are no gaps at the interfaces. However, when the bonding time was 140 ms, the joint interface appeared corrugated, and some areas showed penetration of the Ni-based metallic glass into the Cu. Thus, Ni-based metallic glass can be bonded well to Cu only over a limited range of ultrasonic vibration energies. This may be connected with the ductility of the crystalline metals. Al is more malleable than Cu and tends to form a stable plastic flow when forming a welding joint. Since the metallic glass is harder than the Cu, the excessive input energy from the ultrasonic bonding system into the bonding sample may have resulted in an unstable plastic flow of raw materials, which made the metallic glass penetrate into Cu.

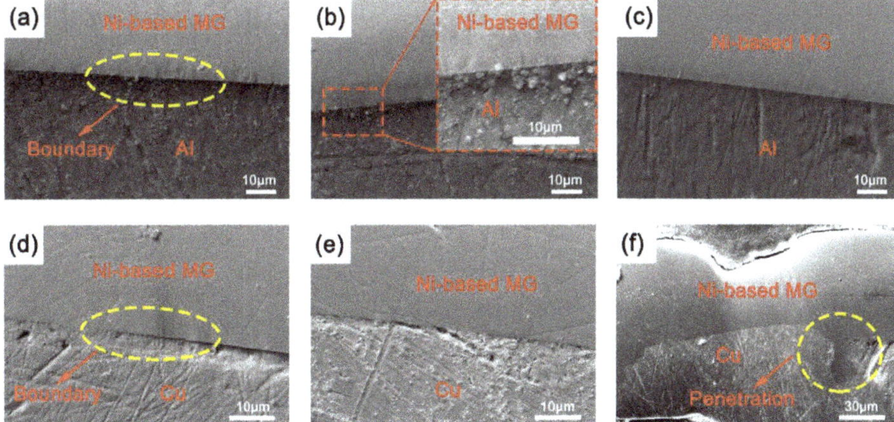

Figure 3. Cross-sectional morphologies of Al/Ni-based (MG) composite samples (**a–c**) at bonding times of 60, 80, and 160 ms, as well as cross-sectional morphologies of Cu/Ni-based (MG) composite samples (**d–f**) at bonding times of 40, 60, and 140 ms.

3.2. Phase Analysis at the Joint Interface

Figure 4a,b show a topographical view of the junction of the two materials after fracturing the consolidated samples under a digital microscope. The joint boundary shows that under ultrasonic vibrations, the metallic glass and crystalline metals can be bonded well in a solid state. Because the metallic glass has higher strength than the crystalline metals, the crystalline metals remained on the metallic glass ribbons after fracturing. The phases in the metallic glass, crystalline metals, and bonding interfaces were also analyzed by XRD. As shown in Figure 4c, the XRD pattern of the Ni-based metallic glass showed only two diffuse peaks. In contrast, Al showed five distinct, sharp diffraction peaks corresponding to its (111), (200), (220), (311), and (222) planes. The diffraction peaks at the joint interfaces of the Al/Ni-based (MG) composite samples were the superposition of the peaks for the Ni-based metallic glass and the Al, and no other crystalline peaks appeared, indicating no new substances formed at the joint interfaces. As shown in Figure 4d, the XRD pattern of the crystalline Cu had three distinct, sharp peaks corresponding to the (111), (200), and (220) planes. The diffraction peaks at the joint interfaces of the Cu/Ni-based (MG) composite samples were the superposition of the Ni-based metallic glass and the Cu diffraction peaks, and no other crystalline peaks appeared, indicating that no new substances formed at the joint interfaces and the metallic glass did not crystallize.

Table 2 shows the values of the force to tear the composite samples apart. When the bonding times were relatively short, the moderate input energy from the ultrasonic bonding system into the composite samples made the raw materials bond better, and the samples required a larger force to tear them apart. However, when the input energy was relatively excessive, the bonded samples required a smaller force to tear the raw materials apart. Figure 5a,b show the topographical view of the junction of the Cu/Ni-based (MG) composite samples after tearing apart at a bonding time of 140 ms. The crystalline metal remained on the metallic glass ribbon, and the fracturing boundary was relatively neat. This may be connected with the plastic flow of the raw materials during bonding, the excessive input energy resulting in more plastic flow of raw materials, and the area affected by ultrasonic vibrations being clearer. Then, the two materials were easier to be torn apart along the boundary of the ultrasonic-vibration-affected zone.

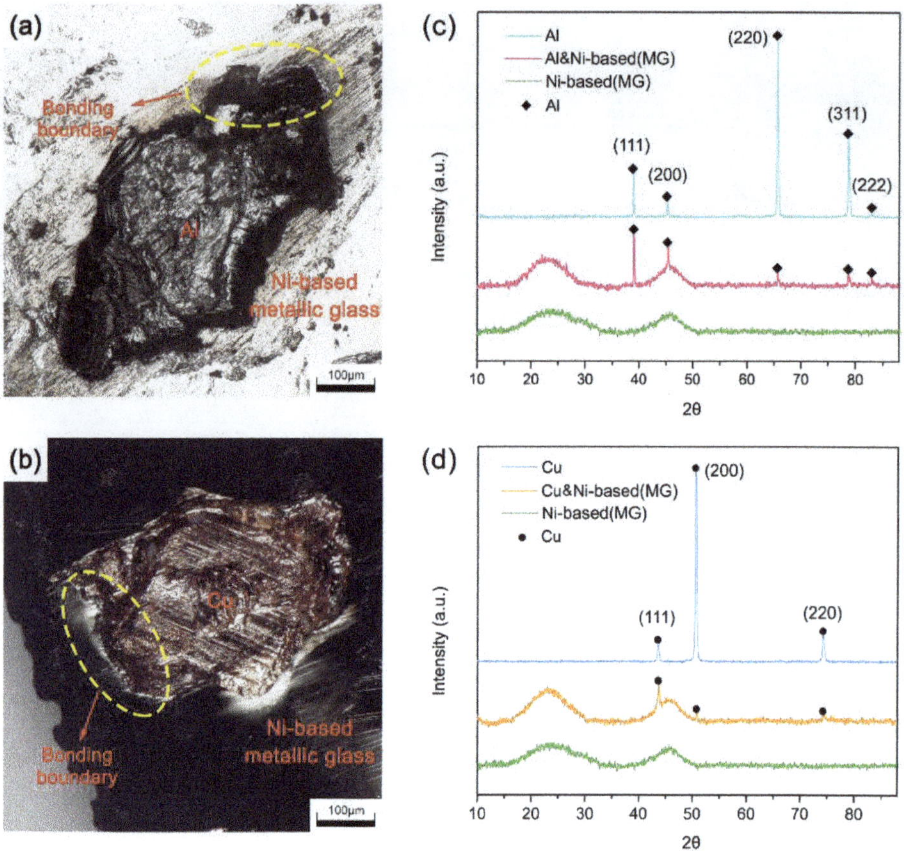

Figure 4. The joint surface morphology (**a**) of the Al/Ni-based (MG) composite sample at a bonding time of 80 ms, and the XRD results (**c**) of the joint surface and relative raw materials. The joint surface morphology (**b**) of the Cu/Ni-based (MG) composite sample at a bonding time of 60 ms, and the XRD results (**d**) of the joint surface and relative raw materials.

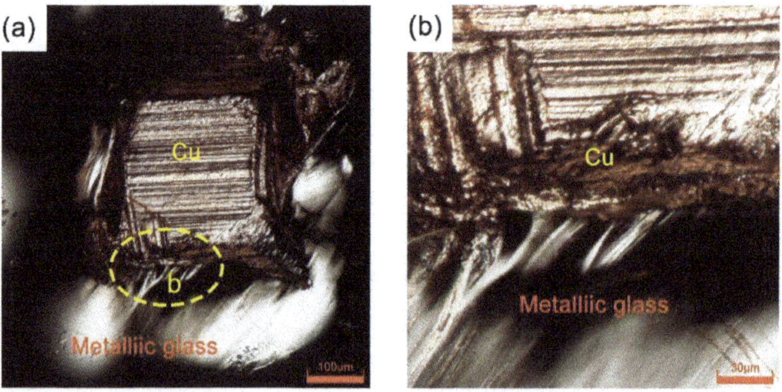

Figure 5. The joint surface morphology of the Cu/Ni-based (MG) composite sample at a bonding time of 140 ms (**a**), and a larger view of the boundary (**b**).

Table 2. The force to tear the samples apart.

Materials	Ni-based (MG) and Al			Ni-based (MG) and Cu		
Bonding time (ms)	60	80	160	40	60	140
Force (N)	11.71 ± 3.24	12.61 ± 1.22	7.23 ± 1.97	10.82 ± 0.90	10.71 ± 1.61	5.85 ± 1.50

3.3. Hardness and Modulus Inside the Consolidated Samples

To analyze how the ultrasonic vibration energy affected the mechanical properties of the consolidated metallic glass and crystalline metal composites, we tested the hardness and modulus of the Ni-based metallic glass, the crystalline metals, and consolidated specimens by nanoindentation. As shown in Figure 6a,b, during the nanoindentation test, three sampling points were selected in the middle of the raw materials' cross sections and the joint interfaces of the composite samples, respectively. The depths of indentations were all 1600 nm to determine the critical compressive loads. The indentation curves in Figure 6a,b correspond to the hardness and modulus of the sampling points on the different sections selected closest to the average value of the section. The indentation curves of the consolidated samples were located between the indentation curves of the metallic glass and the crystalline metals. The compressive loads that the composite specimens could withstand were between the limited loads of the two materials. Also, the unloading curves of the consolidated samples had significant inflexion, mainly because the crystalline metals have higher plasticity than the metallic glass. For the same displacement into the surface, the indenter produced more permanent plastic deformation in the metallic glass than in the crystalline metals; thus, when the consolidated sample was unloaded, the indenter detached from the metallic glass before the crystalline metals. This may also be connected with the phase transformations and residual deformation of the raw materials [35,36]. In the process of loading, it was easier for the crystalline metal than the metallic glass to form a phase transition, and the residual deformation of the crystalline metal was smaller than that of the metallic glass after unloading.

Figure 6c,d show the hardness (H) and elastic modulus (E) of the cross sections of the Ni-based metallic glass, crystalline metals, and consolidated samples, respectively. The hardness and elastic modulus of the consolidated specimens were between the respective values of the two materials. The hardness values of Al, Cu, and Ni-based metallic glass were 0.75 ± 0.11, 1.26 ± 0.11, and 9.41 ± 1.59 GPa, respectively. Further, the hardness values of Al/Ni-based (MG) composites for bonding times of 80 and 160 ms as well as Cu/Ni-based (MG) composites for bonding times of 60 and 140 ms were 2.49 ± 0.13, 2.79 ± 0.44, 1.84 ± 0.27, and 2.84 ± 0.33 GPa, respectively. The hardness of the ultrasonically bonded samples was significantly different from that of the Ni-based metallic glass and crystalline metal materials, as revealed by Tukey's test ($p < 0.05$). The modulus values of Al, Cu, and Ni-based metallic glass were 67.97 ± 4.20, 84.15 ± 12.07, and 129.16 ± 9.52 GPa, respectively. Further, the modulus values of Al/Ni-based (MG) composites for bonding times of 80 and 160 ms as well as Cu/Ni-based (MG) composites for bonding times of 60 and 140 ms were 89.68 ± 4.82, 93.53 ± 5.35, 106.58 ± 8.19, and 117.35 ± 10.47 GPa, respectively. The elastic modulus values of the Ni-based metallic glass and Al consolidated samples were significantly different from that of the two materials, as revealed by Tukey's test ($p < 0.05$), but the elastic modulus values of the Ni-based metallic glass and Cu consolidated samples were not obviously different from that of the two materials. This result mainly occurred because of the slight difference in elastic modulus between Cu and the Ni-based metallic glass. As shown in Figure 3f, when the ultrasonic bonding system inputted too much energy into the interior of the consolidated sample, this caused penetration, which increased the error in the elastic modulus at the consolidation interface of the Cu/Ni-based (MG) composite samples relative to the Al/Ni-based (MG) composite samples. As the ultrasonic vibration energy inputted into the consolidated sample increased, the elastic modulus and hardness at the consolidation interfaces increased. This result may be connected with the plastic flow of the raw materials at the interface junction during bonding. The higher inputted ultrasonic energy may have caused the two materials to diffuse more deeply, which slightly increased the hardness and modulus. Table 3 shows the deformation relative to yielding (H/E) and the resistance to plastic indentation ratios (H^3/E^2) calculated based on nanoindentation results. H/E and H^3/E^2 ratios are important and valuable parameters for predicting the resistance of samples to plastic deformation [35,37]. A higher ratio means better sample resistance to plastic deformation. Table 3 illustrates that ultrasonic bonding additive manufacturing can combine the durability of metallic glass and crystalline metals.

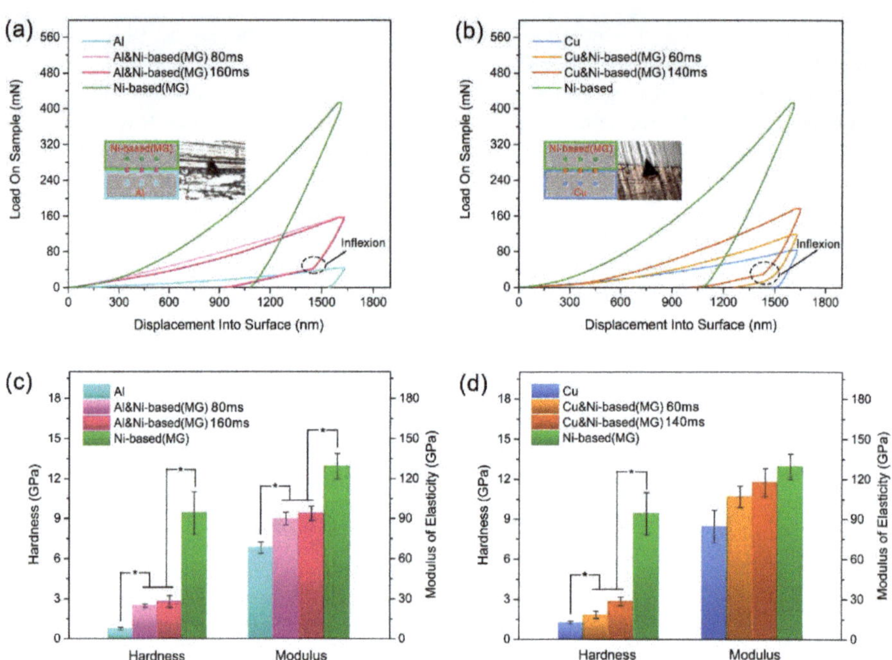

Figure 6. Nanoindentation curves (**a**) and the hardness and modulus (**c**) of the cross section of Al/Ni-based (MG) composite specimens at bonding times of 80 and 160 ms. Nanoindentation curves (**b**) and the hardness and modulus (**d**) of the cross section of the Cu/Ni-based (MG) composite samples at bonding times of 60 and 140 ms. Data presented as mean ± standard deviation, * $p < 0.05$.

Table 3. The deformation relative to yielding (H/E) and resistance to the plastic indentation (H^3/E^2).

Samples	H/E	H^3/E^2 (GPa)
Al	0.01111 ± 0.00185	0.00010 ± 0.00004
Ni-based (MG) and Al (80 ms)	0.02776 ± 0.00005	0.00192 ± 0.00009
Ni-based (MG) and Al (160 ms)	0.02977 ± 0.00368	0.00256 ± 0.00106
Ni-based (MG)	0.06726 ± 0.00279	0.03948 ± 0.00588
Ni-based (MG) and Cu (60 ms)	0.01722 ± 0.00125	0.00056 ± 0.00016
Ni-based (MG) and Cu (140 ms)	0.02413 ± 0.00099	0.00166 ± 0.00030
Cu	0.01527 ± 0.00356	0.00031 ± 0.00018

4. Conclusions

Al/Ni-based (MG) and Cu/Ni-based (MG) composites were manufactured additively via ultrasonic vibrations. The range of the inputted ultrasonic vibration energy to bond Ni-based metallic glass and Al well was wider than that of Ni-based metallic glass and Cu. No intermetallic compounds formed at the junction of the metallic glass composite samples, and the Ni-based metallic glass did not crystallize after formation. The hardness and modulus of the interior of the composite specimens produced by ultrasonic additive manufacturing were between the respective values of the two materials. The mechanical properties of the metallic glass and crystalline metals were fused by ultrasonic vibrations. Ultrasonic bonding can also be combined with traditional machining or laser cutting to perform layer-by-layer cumulative formation. We believe that this technique will allow 3D printing of bulk, complex, high-performance structures of metallic glass composite parts and promote the development and application of metallic glass in industrial fields and others.

Author Contributions: G.L. conceived the study and established the collaboration with W.W. and J.Y.H.F. W.W. proposed the key idea of this paper. J.J. and T.W. carried out the experiments. G.L. analyzed the data, wrote and modified the Manuscript. S.C. improved the figures. J.Z. assisted with discussing the idea and results. J.Y.H.F. reviewed the manuscript. All authors did a favor for the final manuscript.

Funding: This research was supported by the National Natural Science Foundation of China (No. 51675226), the Key Scientific and Technological Research Project of Jilin Province (No. 20180201055GX), the Project of International Science and Technology Cooperation of Jilin Province (No. 20170414043GH), Industrial Innovation Project of Jilin Province (No. 2019C038-1), Key Laboratory of CNC Equipment Reliability, Ministry of Education and the China Scholarship Council (CSC).

Conflicts of Interest: The authors declare no conflict of interest.

References

1. Ma, E.; Ding, J. Tailoring structural inhomogeneities in metallic glasses to enable tensile ductility at room temperature. *Mater. Today* **2016**, *19*, 568–579. [CrossRef]
2. Zhang, F.; Wu, J.; Jiang, W.; Hu, Q.; Zhang, B. New and Efficient Electrocatalyst for Hydrogen Production from Water Splitting: Inexpensive, Robust Metallic Glassy Ribbons Based on Iron and Cobalt. *ACS Appl. Mater. Interfaces* **2017**, *9*, 31340–31344. [CrossRef] [PubMed]
3. Sheng, H.W.; Luo, W.K.; Alamgir, F.M.; Bai, J.M.; Ma, E. Atomic packing and short-to-medium-range order in metallic glasses. *Nature* **2006**, *439*, 419–425. [CrossRef] [PubMed]
4. Trexler, M.M.; Thadhani, N.N. Mechanical properties of bulk metallic glasses. *Prog. Mater. Sci.* **2010**, *55*, 759–839. [CrossRef]
5. Meagher, P.; O'Cearbhaill, E.D.; Byrne, J.H.; Browne, D.J. Bulk Metallic Glasses for Implantable Medical Devices and Surgical Tools. *Adv. Mater.* **2016**, *28*, 5755–5762. [CrossRef] [PubMed]
6. Sekol, R.C.; Kumar, G.; Carmo, M.; Gittleson, F.; Hardesty-Dyck, N.; Mukherjee, S.; Schroers, J.; Taylor, A.D. Bulk metallic glass micro fuel cell. *Small* **2013**, *9*, 2081–2085. [CrossRef]
7. Li, H.F.; Zheng, Y.F. Recent advances in bulk metallic glasses for biomedical applications. *Acta Biomater* **2016**, *36*, 1–20. [CrossRef] [PubMed]
8. Wang, W.H. Bulk Metallic Glasses with Functional Physical Properties. *Adv. Mater.* **2009**, *21*, 4524–4544. [CrossRef]
9. Chen, W.; Zhou, H.; Liu, Z.; Ketkaew, J.; Shao, L.; Li, N.; Gong, P.; Samela, W.; Gao, H.; Schroers, J. Test sample geometry for fracture toughness measurements of bulk metallic glasses. *Acta Mater.* **2018**, *145*, 477–487. [CrossRef]
10. Li, H.X.; Lu, Z.C.; Wang, S.L.; Wu, Y.; Lu, Z.P. Fe-based bulk metallic glasses: Glass formation, fabrication, properties and applications. *Prog. Mater. Sci.* **2019**, *103*, 235–318. [CrossRef]
11. Dudina, D.V.; Georgarakis, K.; Aljerf, M.; Li, Y.; Braccini, M.; Yavari, A.R.; Inoue, A. Cu-based metallic glass particle additions to significantly improve overall compressive properties of an Al alloy. *Compos. Part. A Appl. Sci. Manuf.* **2010**, *41*, 1551–1557. [CrossRef]
12. Chuai, D.; Liu, X.; Yu, R.; Ye, J.; Shi, Y. Enhanced microwave absorption properties of flake-shaped FePCB metallic glass/graphene composites. *Compos. Part. A Appl. Sci. Manuf.* **2016**, *89*, 33–39. [CrossRef]

13. Kruzic, J.J. Bulk Metallic Glasses as Structural Materials: A Review. *Adv. Eng. Mater.* **2016**, *18*, 1308–1331. [CrossRef]

14. Żrodowski, Ł.; Wysocki, B.; Wróblewski, R.; Krawczyńska, A.; Adamczyk-Cieślak, B.; Zdunek, J.; Błyskun, P.; Ferenc, J.; Leonowicz, M.; Święszkowski, W. New approach to amorphization of alloys with low glass forming ability via selective laser melting. *J. Alloy. Compd.* **2019**, *771*, 769–776. [CrossRef]

15. Pauly, S.; Löber, L.; Petters, R.; Stoica, M.; Scudino, S.; Kühn, U.; Eckert, J. Processing metallic glasses by selective laser melting. *Mater. Today* **2013**, *16*, 37–41. [CrossRef]

16. Mahbooba, Z.; Thorsson, L.; Unosson, M.; Skoglund, P.; West, H.; Horn, T.; Rock, C.; Vogli, E.; Harrysson, O. Additive manufacturing of an iron-based bulk metallic glass larger than the critical casting thickness. *Appl. Mater. Today* **2018**, *11*, 264–269. [CrossRef]

17. Guo, S.; Wang, M.; Zhao, Z.; Zhang, Y.Y.; Lin, X.; Huang, W.D. Molecular dynamics simulation on the micro-structural evolution in heat-affected zone during the preparation of bulk metallic glasses with selective laser melting. *J. Alloy. Compd.* **2017**, *697*, 443–449. [CrossRef]

18. Li, X. Additive Manufacturing of Advanced Multi-Component Alloys: Bulk Metallic Glasses and High Entropy Alloys. *Adv. Eng. Mater.* **2018**, *20*, 1700874. [CrossRef]

19. Gibson, M.A.; Mykulowycz, N.M.; Shim, J.; Fontana, R.; Schmitt, P.; Roberts, A.; Ketkaew, J.; Shao, L.; Chen, W.; Bordeenithikasem, P.; et al. 3D printing metals like thermoplastics: Fused filament fabrication of metallic glasses. *Mater. Today* **2018**, *21*, 697–702. [CrossRef]

20. Ouyang, D.; Xing, W.; Li, N.; Li, Y.; Liu, L. Structural evolutions in 3D-printed Fe-based metallic glass fabricated by selective laser melting. *Addit. Manuf.* **2018**, *23*, 246–252. [CrossRef]

21. Bordeenithikasem, P.; Stolpe, M.; Elsen, A.; Hofmann, D.C. Glass forming ability, flexural strength, and wear properties of additively manufactured Zr-based bulk metallic glasses produced through laser powder bed fusion. *Addit. Manuf.* **2018**, *21*, 312–317. [CrossRef]

22. Bordeenithikasem, P.; Shen, Y.; Tsai, H.-L.; Hofmann, D.C. Enhanced mechanical properties of additively manufactured bulk metallic glasses produced through laser foil printing from continuous sheetmetal feedstock. *Addit. Manuf.* **2018**, *19*, 95–103. [CrossRef]

23. Ashby, M.; Greer, A. Metallic glasses as structural materials. *Scr. Mater.* **2006**, *54*, 321–326. [CrossRef]

24. Hufnagel, T.C.; Schuh, C.A.; Falk, M.L. Deformation of metallic glasses: Recent developments in theory, simulations, and experiments. *Acta Mater.* **2016**, *109*, 375–393. [CrossRef]

25. Li, N.; Zhang, J.; Xing, W.; Ouyang, D.; Liu, L. 3D printing of Fe-based bulk metallic glass composites with combined high strength and fracture toughness. *Mater. Des.* **2018**, *143*, 285–296. [CrossRef]

26. Kim, J.; Kawamura, Y. Dissimilar welding of Zr41Be23Ti14Cu12Ni10 bulk metallic glass and stainless steel. *Scr. Mater.* **2011**, *65*, 1033–1036. [CrossRef]

27. Li, Y.; Shen, Y.; Chen, C.; Leu, M.C.; Tsai, H.-L. Building metallic glass structures on crystalline metal substrates by laser-foil-printing additive manufacturing. *J. Mater. Process. Technol.* **2017**, *248*, 249–261. [CrossRef]

28. Li, Y.; Shen, Y.; Leu, M.C.; Tsai, H.-L. Building Zr-based metallic glass part on Ti-6Al-4V substrate by laser-foil-printing additive manufacturing. *Acta Mater.* **2018**, *144*, 810–821. [CrossRef]

29. Feng, J.; Chen, P.; Zhou, Q. Investigation on Explosive Welding of Zr53Cu35Al12 Bulk Metallic Glass with Crystalline Copper. *J. Mater. Eng. Perform.* **2018**, *27*, 2932–2937. [CrossRef]

30. Wang, X.; Luo, Y.; Huang, T.; Liu, H. Experimental Investigation on Laser Impact Welding of Fe-Based Amorphous Alloys to Crystalline Copper. *Materials* **2017**, *10*, 523. [CrossRef]

31. Ward, A.A.; Zhang, Y.; Cordero, Z.C. Junction growth in ultrasonic spot welding and ultrasonic additive manufacturing. *Acta Mater.* **2018**, *158*, 393–406. [CrossRef]

32. Levy, A.; Miriyev, A.; Sridharan, N.; Han, T.; Tuval, E.; Babu, S.S.; Dapino, M.J.; Frage, N. Ultrasonic additive manufacturing of steel: Method, post-processing treatments and properties. *J. Mater. Process. Technol.* **2018**, *256*, 183–189. [CrossRef]

33. Lin, J.-Y.; Nambu, S.; Koseki, T. Evolution of bonding interface during ultrasonic welding between steel and aluminium alloy. *Sci. Technol. Weld. Join.* **2018**, *24*, 83–91. [CrossRef]

34. Wu, W.; Jiang, J.; Li, G.; Fuh, J.Y.H.; Jiang, H.; Gou, P.; Zhang, L.; Liu, W.; Zhao, J. Ultrasonic additive manufacturing of bulk Ni-based metallic glass. *J. Non-Cryst. Solids* **2019**, *506*, 1–5. [CrossRef]

35. Warcholinski, B.; Gilewicz, A.; Kuprin, A.S.; Tolmachova, G.N.; Ovcharenko, V.D.; Kuznetsova, T.A.; Zubar, T.I.; Khudoley, A.L.; Chizhik, S.A. Mechanical properties of Cr-O-N coatings deposited by cathodic arc evaporation. *Vacuum* **2018**, *156*, 97–107. [CrossRef]
36. Warcholinski, B.; Gilewicz, A.; Kuznetsova, T.A.; Zubar, T.I.; Chizhik, S.A.; Abetkovskaia, S.O.; Lapitskaya, V.A. Mechanical properties of Mo(C)N coatings deposited using cathodic arc evaporation. *Surf. Coat. Technol.* **2017**, *319*, 117–128. [CrossRef]
37. Zavaleyev, V.; Walkowicz, J.; Kuznetsova, T.; Zubar, T. The dependence of the structure and mechanical properties of thin ta-C coatings deposited using electromagnetic venetian blind plasma filter on their thickness. *Thin Solid Film.* **2017**, *638*, 153–158. [CrossRef]

Communication

Additive and Substractive Surface Structuring by Femtosecond Laser Induced Material Ejection and Redistribution

Xxx Sedao [1,2,*], Matthieu Lenci [3], Anton Rudenko [1], Alina Pascale-Hamri [2], Jean-Philippe Colombier [1] and Cyril Mauclair [1,2]

[1] Laboratoire Hubert Curien, UMR 5516 CNRS, Université de Lyon, Université Jean Monnet, 42000 Saint-Etienne, France; anton.rudenko@univ-st-etienne.fr (A.R.); jean.philippe.colombier@univ-st-etienne.fr (J.-P.C.); cyril.mauclair@univ-st-etienne.fr (C.M.)

[2] GIE Manutech-USD, 20 rue Benoit Lauras, 42000 Saint-Etienne, France; alina.hamri@manutech-usd.fr

[3] Mines Saint-Etienne, Univ Lyon, CNRS, UMR 5307 LGF, Centre SMS, F - 42023 Saint-Etienne, France; lenci@emse.fr

* Correspondence: xxx.sedao@univ-st-etienne.fr; Tel.: +33-469-663-206

Received: 8 November 2018; Accepted: 30 November 2018; Published: 4 December 2018

Abstract: A novel additive surface structuring process is devised, which involves localized, intense femtosecond laser irradiation. The irradiation induces a phase explosion of the material being irradiated, and a subsequent ejection of the ablative species that are used as additive building blocks. The ejected species are deposited and accumulated in the vicinity of the ablation site. This redistribution of the material can be repeated and controlled by raster scanning and multiple pulse irradiation. The deposition and accumulation cause the formation of µm-scale three-dimensional structures that surpass the initial surface level. The above-mentioned ablation, deposition, and accumulation all together constitute the proposed additive surface structuring process. In addition, the geometry of the three-dimensional structures can be further modified, if desirable, by a subsequent substractive ablation process. Microstructural analysis reveals a quasi-seamless conjugation between the surface where the structures grow and the structures additively grown by this method, and hence indicates the mechanic robustness of these structures. As a proof of concept, a sub-mm sized re-entrant structure and pillars are fabricated on aluminum substrate by this method. Single units as well as arrayed structures with arbitrary pattern lattice geometry are easily implemented in this additive surface structuring scheme. Engineered surface with desired functionalities can be realized by using this means, i.e., a surface with arrayed pillars being rendered with superhydrophobicity.

Keywords: ultrafast laser; femtosecond; ablation; scanning; additive surface structuring; hydrophobicity

---

1. Introduction

Laser based additive manufacturing (AM) processes offer unprecedented new opportunities in design and manufacture [1]. In the vast majority of laser additive manufacturing (LAM) processes, high power, continuous, or long-pulsed lasers are commonly used as high-efficiency thermal sources to heat up metallic powders/bulk, and cause melting/cohesion. Ultrafast laser pulses, due to their low onset of thermal effect, on the one hand, are extensively used in the high-precision substractive processing sector, such as surface structuring [2,3], bulk inscription [4,5], and thin-film scribing [6]; however, on the other hand, they find themselves available for a fairly limited amount of applications in LAM. The few existing applications include ultrafast laser additive manufacturing/surface structuring for high melting temperature metals [7,8], laser-induced forward transfer (LIFT) for stacking up miniature building blocks together toward high precision, μm-scale additive processes [9,10], and improving surface finishing of metal parts produced by conventional LAM [11,12]. In this communication, we would like to broaden the spectrum of ultrafast LAM by exploring a concept of ultrafast laser additive surface structuring (LASS): by taking advantage of the material spallation/phase explosion induced by intense ultrafast laser irradiation, the energetically ejected material from the phase explosion can be transferred and accumulated to form a predefined structure. This approach is powder-free, and the structure building material is directly acquired from the substrate where the additive features are to be grown. Moreover, "finishing touches" can be made to the as-built structures by using the very same ultrafast laser, in order to improve the fineness of the surface finishing of the additively grown features. In the following, the concept is explained firstly. Then, as a proof of concept, two types of structures are fabricated using the proposed ultrafast LASS approach. The aspect of added functions at the surface with these ultrafast laser grown structures is also discussed.

2. Materials and Methods

A schematic drawing of the laser setup and raster scan strategy is depicted in Figure 1a. For the sake of clarity, the material reported in the Results section is aluminum. In order to verify whether the proposed ultrafast LASS is generic, a few other materials were also chosen for the same test, such as copper and titanium alloy Ti6Al4V, and comparable results were obtained from these materials, too. Nonetheless, a similar test on stainless steel 316L did not yield satisfactory results, suggesting a certain material dependency and limitation of this process. The targets were prepared by conventional metallography procedures with a surface roughness Ra = 20 nm. The micromachining was carried out using a fiber fs laser system (Tangerine HP, Amplitude Systems, Pessac, France). The laser has a central wavelength of 1030 nm with a pulse duration of 300 fs, and a tunable repetition rate from single shot to two MHz. The linearly polarized laser pulses were attenuated, sent through a Galvano scanner, and focused through a 100-mm telecentric f-theta lens. The focused laser spot exhibits a Gaussian profile and the spot diameter (at $1/e^2$) measures $2\omega_0 = 22$ μm. A raster scanning strategy was used. The overlap ratio of 0.91 between successive laser pulses and successive scan tracks was kept constant for all of the experiments. This relatively high overlap ratio was intended for a homogeneous energy deposition [13]. The laser fluence quoted in this paper is the peak fluence $F = 2\varepsilon/\pi\,\omega_0{}^2$, with ε being the laser pulse energy.

Figure 1. Ultrafast laser additive surface structuring (LASS) explained: (**a**) Laser scan path is indicated by the red arrows: the laser beam movement starts from the far side and advances linearly toward the near side to form a scan track. The scan tracks start from the right side and end on the left. The numbers 1 and 2 represent the beginning and the end of the process, respectively; (**b**) scanning electron microscope (SEM) micrograph showing the site of ablation, and additively grown part. A mechanically prepared cross-section is given as inset. (**c**) A Focused Ion Beam (FIB) milled cross-section of the additively fabricated part. The platinum layer appearing in the micrograph is a protective coating deposited prior to the milling process.

The topographical analysis was performed using an optical microscope (OM, Zeiss, Oberkochen, Germany) and scanning electron microscope (SEM, FEI Europe B.V., Eindhoven, Netherlands). Cross-section samples for OM were prepared by wheel-saw halving, cold-setting resin mounting, and mechanical polishing. Cross-section samples for SEM inspection were made using a dual beam system SEM and Focused Ion Beam (FIB, FEI Helios Nanolab 600i dualbeam workstation). Energy-dispersive X-ray spectroscopy (EDX, Brucker, Bremen, Germany) was used to evaluate the surface chemistry change. A Vikers microindentation (Cetim, Saint-Etienne, France) was also performed to access hardness, with a loading force of two N. All of the hardness values quoted in the text were averaged values from five individual measurements. For the wetting test, a three-μL water droplet was deposited on the surface by a microsyringe, and the side-on view was captured by a dedicated camera. The contact angle was deduced from the registered images using software Digidrop (version 13.06.3.12GB, GBX, Romans sur Isère, France).

3. Results

The lower-right part of the schematic drawing shown in Figure 1a can be used to illustrate the concept of the LASS by intense ultrafast laser irradiation (roughly defined as fluence >5~10 times the ablation threshold). Then, the material experiences a photo excitation, electron–phonon non-equilibration-enhanced electron heat conduction before equilibrium with the lattice, and then undergoes spallation and phase explosion [14]. Various species, depending on the laser fluence, consisting of anything from liquid layers and/or large droplets to a mixture of vapor-phase atom clusters and droplets etc., are ejected as a consequence of stress release. Under certain circumstances (a more dedicated discussion is in the following paragraphs), the ejected species land on the adjacent site, adhere to the vicinity of ablation site, and form an accumulation. Such a material redistribution is somehow similar to the work of Temmler, et al. [15] (although the mechanism governing the material displacement is totally different). By raster scanning a small area on a substrate, the accumulation continues on and eventually forms a three-dimensional structure by itself, which will be termed as additive structures (AS), as indicated by the white bump in Figure 1a. An AS on aluminum sample was fabricated in this way, and its SEM micrograph is given in Figure 1b. The material to form the AS was sourced from the ablation groove, as marked in Figure 1b, at a laser fluence of 18 J/cm^2, and 500-kHz repetition rate. The lateral dimension of the AS is about half of the groove width, and the longitudinal dimension is approximately 50 μm, which is better seen from the mechanical cross-section presented in the inset.

In order to characterize the AS, we performed microstructural and chemical analysis. The SEM micrograph in Figure 1c shows a FIB prepared cross-section from the substrate–AS junction region. Laminated structures are observed on the AS, the formation of which is due to ablative and accumulative processes. The cross-section elaborates also that the cohesion between the AS and substrate is at the microstructural level. A borderline is evident at the junction, but no major voids nor pores are located there, which implies a robust conjugation between the substrate and the AS. The AS is relatively dense. There are some sub-μm sized voids in the bulk region of the AS, but they are sparse. Nonetheless, it is evident that some micropores are located at the outermost surface region and the lower outer surface region. Those micropores located in the outermost surface region could be potentially eliminated by a substractive ablation process step, which is discussed later in this communication. An EDX spectrum was acquired from different locations across the entire AS, and compared with the spectrum obtained from the substrate. In the AS, a small increase of oxygen and nitrogen is deduced from their characteristic peaks' intensity in the spectra. This chemical composition change may be due to an oxidation/nitration or air trapping (due to the presence of the sub-μm voids) within the AS during the process.

As mentioned in Section 2, the AS can be achieved on aluminum but not stainless steel; a typical comparison can be viewed in Figure 2a,b, with the cross-sections of these two materials treated at a fluence of 18 J/cm^2 and a 250-kHz repetition rate. The laser processing is shown to be strongly material-dependent in the ~100 kHz repetition rate regime [16,17], resulting either in material accumulation on the side of the ablation groove in the aluminum case, or to significant roughness development in the case of stainless steel. The laser plasma hydrodynamics and ablation particles shielding may be taken account of in these observations.

The interaction of ultrafast laser at fluence exceeding the ablation threshold with metal targets is associated with a large variety of physical phenomena, taking place from femtosecond up to microsecond timescales [14,18]. At 250 kHz (a pulse-to-pulse interval of four microseconds), the pulse-by-pulse thermal accumulation alone does not play a major role, as the irradiated surfaces have enough time to cool down during the interval. Besides thermal accumulation, the laser ablation by high-intensity pulses is accompanied by plasma plume expansion away from the metal target [18]. Figure 2c depicts the initiation of ablation simulated by two-dimensional (2D) compressible Navier–Stokes equations [2,19], taking account of Gaussian pulse energy deposition on the surface, lattice heating, and hydrodynamic movement, which causes strong pressure gradients and material removal. The lowest density corresponds to hot laser-induced plasma, the highest density corresponds to unaffected ablation area, and the intermediate density corresponds to a liquid state. The central part of the plume leaves the surface and expands longitudinally above the surface; however, the corners of the plasma plume remain less hot, may re-solidify before the arrival of the next pulse, and then form a redistributed aluminum layer above the initial level. In this case, the consequent pulse-by-pulse ablation with a scan from left to right leads to a pronounced microstructure on the left side of the laser-processed area, as displayed in Figure 2a. Other possible mechanisms of material re-deposition include Marangoni melt flow and recoil vapor pressure [15,20]. However, these effects fail to explain why the accumulated material is observed above the ablated area, but not outside it.

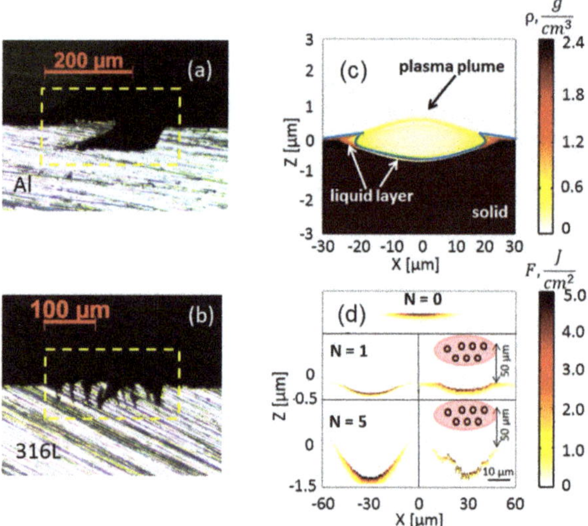

Figure 2. Cross-section images of laser processed (**a**) aluminum, and (**b**) stainless steel 316L, at a fluence of 18 J/cm^2 and pulse repetition rate of 250 kHz. The dashed line boxes are visual guides for highlighting the features inside the ablation grooves; (**c**) Density snapshot showing plasma plume expansion from aluminum surface 50 picoseconds after irradiation; (**d**) Multi-pulse simulations of energy deposition on stainless steel surface. Top row N = 0, initial surface; middle row N = one pulse; and bottom row N = five pulses. In the middle and bottom rows, unaffected (left) /affected (right) laser structuring by the presence of nanoparticles of r = 50 nm and concentration of C = 50 μm^{-2}.

Another characteristic feature of the ultrashort laser ablation of metal targets is the formation of nanoparticles of sizes up to r = 100 nm in plasma plume [21–25]. The species are known to be moving much slower than the plasma plume, with the velocities slightly exceeding 100 m/s [22,25]. The presence of slow moving nanoparticles generated by a laser pulse may influence the following laser pulse interaction with the surface. To investigate the effects of light absorption on the surface, with and without transmission through nanoparticles, we use an ablation model, including Maxwell equations [26] coupled with electron-ion heat transfer equations [19]. The spherical nanoparticles with a radius r = 50 nm and a concentration of C = 50 μm^{-2} are distributed randomly in an ellipsoid-centered region at distance of 50 μm above the initial surface level, with a radius Rz = 10 μm in the propagation direction, and Rx = 20 μm in the transverse direction. The nanoparticles are assumed to be created by the first laser pulse. The numerical results show the surface profile after N = one and N = five pulse irradiation with and without taking account of the energy loss in transmission through nanoparticles in Figure 2d. Although only a slight difference is seen after the first pulse irradiation, the inhomogeneous distribution of the energy due to the presence of nanoparticles seriously degrades the ablation quality after five consecutive pulses: the ablation profile is now asymmetric, with roughness features up to a few hundred nm. It is worth mentioning that owing to the low reflectivity of stainless steel (R = 0.56) compared to that of aluminum (R = 0.96), the greater transmission of laser pulse through steel nanoparticles would cause a stronger surface roughness. The effects of the interaction between the nanoparticles and laser-induced plasma created by the next pulse may as well lead to uncontrolled material deposition inside the crater. Both phenomena may have an impact on the quality and/or surface morphology [22]. The difference in ablated volume and depth is revealed by simulations, taking into account the presence of nanoparticles in Figure 2d already after five pulses of irradiation. Thermal effects might be at the origin of further ablation crater degradation, as the diffusivity of stainless steel $D = k_i/\rho C_i \approx 5 \times 10^{-6}$ m^2/s (k_i is the lattice thermal conductivity, C_i is the lattice heat capacity, and ϱ is the lattice density), which is six times lower than the diffusivity of aluminum $D \approx 3 \times 10^{-5}$ m^2/s. For aluminum, the material accumulates only at the corners of the laser-processed area, leaving the central part inside the ablation crater free from micro-debris [16], and facilitating the proposed technique of additive surface structuring.

The application of such AS for surface functionalization is multifold. One of the simplest is implemented by building two basic AS blocks against one to another, and is reported in Figure 3a. The inset is a mechanical cut cross-section revealing the negatively inclining feature of the sidewalls of the AS. Such geometry is otherwise termed the re-entrant shape, and is capable of rendering a surface liquid repellent property [27]. In order to test this functionality, a large surface of a few mm^2 area was fabricated with repeating "re-entrant" pattern periodically (Figure 3b). The surface was hydrophilic and anisotropic right after the LASS process, but turned to isotropic and hydrophobic after a thermal treatment at 150 °C for two hours: a droplet could sit on the AS surface without spreading, as shown in the inset. Such surface morphology, once capped with a low surface energy coating, might also exhibit oleophobic characteristics [28]; experiments on coating preparation and oleophobicity test are currently underway. It is also worth noting that upscaling such ultrafast LASS to an even larger surface, in the order of cm^2 towards m^2, is fairly straightforward nowadays, as it is based on standard ultrafast laser beam shaping, scanning, and processing tools [3].

Furthermore, an additional merit offered by the ultrafast LASS process is worth illustrating. Given that the ultrafast laser is also a high-precision micromachining tool, additional laser processing steps can be added after the AS is fabricated, and the surface finishing of the AS can be improved by using the very same laser source. For instance, the edges of AS can be sharpened by ultrafast laser ablation, as shown in Figure 4. The four sub-mm sized pillars were fabricated by the LASS process (by circulating the laser beam from the center to outer region), and they had a mushroom-shaped geometry right after the process (a side-on view of the mushroom-shaped AS is given as an inset in Figure 4a). The cap part of the "mushroom" is trimmed to a pillar shape, as shown in Figure 4a, after an additional ultrafast laser substractive scan along the contour of the AS at a reduced laser fluence (7 J/cm^2). The sharp edge is better viewed in Figure 4b, where the sample is tilted by a nearly 90° angle. The pillars in this particular case are 250 μm in diameter and 100-μm high. Pillars with height (the elevation from the substrate surface to the peak of the pillars) up to approximately 300 μm and small diameters can also be made by this strategy; the side-on view is given in the Figure 4b inset. The large surface production of the AS, which is similar to those in Figure 4a, is also easily implemented, and an example is presented in Figure 4c. The inset in Figure 4c is a wetting test on the surface with a large array of sub-mm sized pillars, where a superhydrophobicity contact angle of 150 °C is observed (after a thermal treatment at 150 °C for two hours). Mechanical property-wise, the microindentation test revealed a decreased hardness at the surface of the AS, a Vickers hardness value at two N load $HV_{2N} = 230$ at the AS, compared to a Vickers $HV_{2N} = 252$ of the substrate. The reduction of the hardness is thought to be related to the presence of the micropores and reduced density in the AS compared to bulk aluminum. It is obvious that the substractive process step can be applied to remove the previously mentioned micropores lodged in the outermost surface region. If this surface porous layer were removed, the microhardness would be further improved. A more comprehensive study on this matter and process optimization will surely be meaningful toward further exploring the value of this additive and subtractive surface structuring approach.

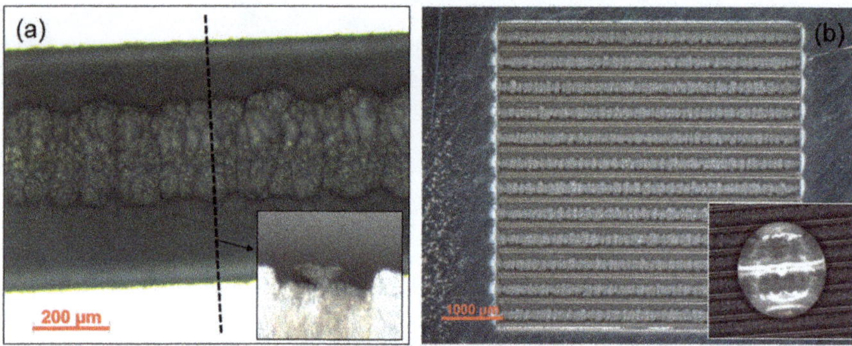

Figure 3. (**a**) Top view of the re-entrant structure realized by building the additive structures (AS) block from two sides. The dash line and the arrow are the visual indication of a cross-section. The inset is a cross-section of this re-entrant structure; (**b**) The structure unit shown in (**a**) is easily implemented on a large surface scale. This surface exhibits isotropic hydrophobicity (lower bottom inset).

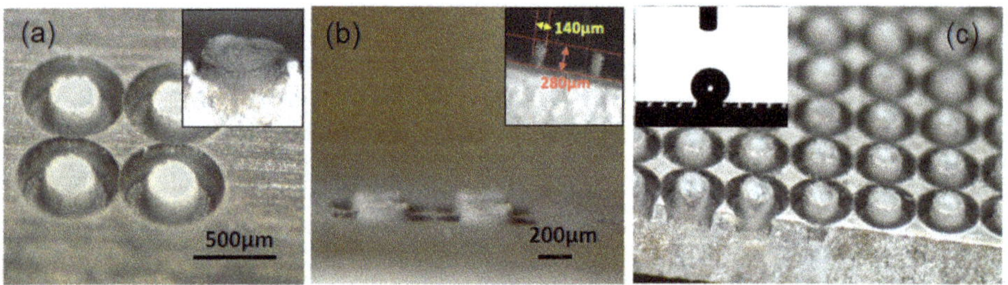

Figure 4. (**a**) Four sub-mm sized pillars made by the ultrafast laser additive surface structuring (LASS) process, plus ultrafast laser substractive ablation at the contour of the pillar blocks to sharpen the edges. The inset shows a single AS unit without its edge being sharpened; (**b**) Tilted view of the pillars. The pillars can be grown higher, at the cost of pillar diameter, as shown in the inset; (**c**) Pillars similar to those in (**a**) are fabricated on a large scale for achieving a superhydrophobicity at the surface, as demonstrated in the inset.

4. Discussion

This report proposes a novel ultrafast LASS approach that is based on intense femtosecond laser irradiation, ablation, and subsequent material ejection. The inhomogeneous distribution of hot plasma induced by intense laser irradiation contributes to the accumulation of ablated species near the ablation area. The accumulation leads to the formation of the additive manufactured structures. As a proof of concept, one-dimensional re-entrant surface structures are fabricated using the approach. The advantage of ultrafast laser refining the additively manufactured structures is also demonstrated. These femtosecond laser additively manufactured geometries, once fabricated on a large scale, can render the underlying surfaces with desired properties such as anisotropic wetting and superhydrophobicity. Although the above results were obtained from aluminum substrates, further study suggests that the method is quite generic and applicable to certain other metals and/or alloys. Supplementary tests have confirmed the applicability of this process mode to copper and titanium alloy Ti6Al4V, at adapted laser fluences. Nonetheless, a similar test on stainless steel 316L did not yield satisfactory results. Numerical analysis suggests that the presence of laser-induced nanoparticles, optical properties such as reflectance and transmission, and material properties such as diffusivity may play a role in the observed discrepancies in stainless steel with respect to the other metal/alloys. It is clear that a better understanding is still required toward the further development of the process. In the meantime, the mechanical strength of these structures needs to be characterized quantitatively. Surface functionalization, especially potentials in adding superomniphobic (hydrophobic and oleophobic at the same time) properties to an engineered surface, will be explored.

Author Contributions: Conceptualization, X.S. and C.M.; FIB, EDX and SEM, M.L.; Simulation, A.R. and J.-P.C.; Wetting characterization and indentation test, A.P.-H.

Funding: The authors thank French ANR for the Equipex MANUTECH-USD support (Investissements d'Avenir ANR-10-EQPX-36-01). This work was carried out for LABEX MANUTECH-SISE financed project ICRAM. This work was also supported by the IMOTEP project within the program 'Investissements d'Avenir' operated by ADEME.

References

1. Gu, D.D.; Meiners, W.; Wissenbach, K.; Poprawe, R. Laser additive manufacturing of metallic components: materials, processes and mechanisms. *Int. Mater. Rev.* **2012**, *57*, 133–164. [CrossRef]

2. Sedao, X.; Abou Saleh, A.; Rudenko, A.; Douillard, T.; Esnouf, C.; Reynaud, S.; Maurice, C.; Pigeon, F.; Garrelie, F.; Colombier, J.-P. Self-Arranged Periodic Nanovoids by Ultrafast Laser-Induced Near-Field Enhancement. *ACS Photonics* **2018**, *5*, 1418–1426. [CrossRef]

3. Saint-Pierre, D.; Granier, J.; Egaud, G.; Baubeau, E.; Mauclair, C. Fast Uniform Micro Structuring of DLC Surfaces Using Multiple Ultrashort Laser Spots through Spatial Beam Shaping. *Phys. Procedia* **2016**, *83*, 1178–1183. [CrossRef]

4. Liu, Z.; Siegel, J.; Garcia-Lechuga, M.; Epicier, T.; Lefkir, Y.; Reynaud, S.; Bugnet, M.; Vocanson, F.; Solis, J.; Vitrant, G.; et al. Three-Dimensional Self-Organization in Nanocomposite Layered Systems by Ultrafast Laser Pulses. *ACS Nano* **2017**, *11*, 5031–5040. [CrossRef] [PubMed]

5. Mauclair, C.; Cheng, G.; Huot, N.; Audouard, E.; Rosenfeld, A.; Hertel, I.V.; Stoian, R. Dynamic ultrafast laser spatial tailoring for parallel micromachining of photonic devices in transparent materials. *Opt. Express* **2009**, *17*, 3531–3542. [CrossRef] [PubMed]

6. Bonse, J.; Mann, G.; Krüger, J.; Marcinkowski, M.; Eberstein, M. Femtosecond laser-induced removal of silicon nitride layers from doped and textured silicon wafers used in photovoltaics. *Thin Solid Films* **2013**, *542*, 420–425. [CrossRef]

7. Nie, B.; Yang, L.; Huang, H.; Bai, S.; Wan, P.; Liu, J. Femtosecond laser additive manufacturing of iron and tungsten parts. *Appl. Phys. A* **2015**, *119*, 1075–1080. [CrossRef]

8. Nie, B.; Huang, H.; Bai, S.; Liu, J. Femtosecond laser melting and resolidifying of high-temperature powder materials. *Appl. Phys. A* **2015**, *118*, 37–41. [CrossRef]

9. Delaporte, P.; Alloncle, A.-P. Laser-induced forward transfer: A high resolution additive manufacturing technology. *Opt. Laser Technol.* **2016**, *78*, 33–41. [CrossRef]

10. Piqué, A.; Serra, P. *Laser Printing of Functional Materials: 3D Microfabrication, Electronics and Biomedicine*; Wiley: Hoboken, NJ, USA, 2018.

11. Mingareev, I.; Bonhoff, T.; El-Sherif, A.F.; Meiners, W.; Kelbassa, I.; Biermann, T.; Richardson, M. Femtosecond laser post-processing of metal parts produced by laser additive manufacturing. *J. Laser Appl.* **2013**, *25*, 052009. [CrossRef]

12. Baubeau, E.; Missemer, F.; Bertrand, P. System and Method for Additively Manufacturing by Laser Melting of a Powder Bed 2017. U.S. Patent 15,762,482, 27 September 2018.

13. Eichstädt, J.; Huis, A.J. Analysis of Irradiation Processes for Laser-Induced Periodic Surface Structures. *Phys. Procedia* **2013**, *41*, 650–660. [CrossRef]

14. Zhigilei, L.V.; Lin, Z.; Ivanov, D.S. Atomistic Modeling of Short Pulse Laser Ablation of Metals: Connections between Melting, Spallation, and Phase Explosion†. *J. Phys. Chem. C* **2009**, *113*, 11892–11906. [CrossRef]

15. Temmler, A.; Küpper, M.; Walochnik, M.A.; Lanfermann, A.; Schmickler, T.; Bach, A.; Greifenberg, T.; Oreshkin, O.; Willenborg, E.; Wissenbach, K.; et al. Surface structuring by laser remelting of metals. *J. Laser Appl.* **2017**, *29*, 012015. [CrossRef]

16. Sedao, X.; Mathieu, L.; Rudenko, A.; Pascale-Hamri, A.; Faure, N.; Colombier, J.-P.; Mauclair, C. Influence of Pulse Repetition Rate on Morphology and Material Removal Efficiency of Ultrafast Laser Ablated Metallic Surfaces. *Opt. Laser Eng.*. Submitted.

17. Zuhlke, C.A.; Anderson, T.P.; Alexander, D.R. Formation of multiscale surface structures on nickel via above surface growth and below surface growth mechanisms using femtosecond laser pulses. *Opt. Express* **2013**, *21*, 8460–8473. [CrossRef] [PubMed]

18. Anisimov, S.I.; Luk'yanchuk, B.S. Selected problems of laser ablation theory. *Phys.-Usp.* **2002**, *45*, 293. [CrossRef]

19. Colombier, J.P.; Combis, P.; Audouard, E.; Stoian, R. Guiding heat in laser ablation of metals on ultrafast timescales: an adaptive modeling approach on aluminum. *New J. Phys.* **2012**, *14*, 013039. [CrossRef]

20. Volkov, A.N.; Zhigilei, L.V. Melt dynamics and melt-through time in continuous wave laser heating of metal films: Contributions of the recoil vapor pressure and Marangoni effects. *Int. J. Heat Mass Trans.* **2017**, *112*, 300–317. [CrossRef]

21. Bourquard, F.; Loir, A.-S.; Donnet, C.; Garrelie, F. In situ diagnostic of the size distribution of nanoparticles generated by ultrashort pulsed laser ablation in vacuum. *Appl. Phys. Lett.* **2014**, *104*, 104101. [CrossRef]

22. Noël, S.; Hermann, J.; Itina, T. Investigation of nanoparticle generation during femtosecond laser ablation of metals. *Appl. Surf. Sci.* **2007**, *253*, 6310–6315. [CrossRef]

23. Amoruso, S.; Ausanio, G.; Bruzzese, R.; Vitiello, M.; Wang, X. Femtosecond laser pulse irradiation of solid targets as a general route to nanoparticle formation in a vacuum. *Phys. Rev. B* **2005**, *71*, 033406. [CrossRef]

24. Guillermin, M.; Colombier, J.P.; Valette, S.; Audouard, E.; Garrelie, F.; Stoian, R. Optical emission and nanoparticle generation in Al plasmas using ultrashort laser pulses temporally optimized by real-time spectroscopic feedback. *Phys. Rev. B* **2010**, *82*, 035430. [CrossRef]

25. Tsakiris, N.; Anoop, K.K.; Ausanio, G.; Gill-Comeau, M.; Bruzzese, R.; Amoruso, S.; Lewis, L.J. Ultrashort laser ablation of bulk copper targets: Dynamics and size distribution of the generated nanoparticles. *J. Appl. Phys.* **2014**, *115*, 243301. [CrossRef]

26. Rudenko, A.; Mauclair, C.; Garrelie, F.; Stoian, R.; Colombier, J.P. Light absorption by surface nanoholes and nanobumps. *Appl. Surf. Sci.* **2019**, *470*, 228–233. [CrossRef]

27. Dufour, R.; Perry, G.; Harnois, M.; Coffinier, Y.; Thomy, V.; Senez, V.; Boukherroub, R. From micro to nano reentrant structures: hysteresis on superomniphobic surfaces. *Colloid Polym Sci* **2013**, *291*, 409–415. [CrossRef]

28. Li, J.; Qin, Q.H.; Shah, A.; Ras, R.H.A.; Tian, X.; Jokinen, V. Oil droplet self-transportation on oleophobic surfaces. *Sci. Adv.* **2016**, *2*, e1600148. [CrossRef]

 *materials*

Article

Effects of the Quenching Rate on the Microstructure, Mechanical Properties and Paint Bake-Hardening Response of Al–Mg–Si Automotive Sheets

Guanjun Gao, Yong Li *, Zhaodong Wang, Hongshuang Di, Jiadong Li and Guangming Xu

State Key Laboratory of Rolling and Automation, Northeastern University, Shenyang 110819, China; wwwgaoguanjun@126.com (G.G.); zhaodongwang@263.net (Z.W.); dhshuang@mail.neu.edu.cn (H.D.); lijd@ral.nue.edu.cn (J.L.); xu_gm@epm.neu.edu.cn (G.X.)
* Correspondence: liyong20190525@126.com; Tel.: +86-158-0405-5285

Received: 20 September 2019; Accepted: 28 October 2019; Published: 31 October 2019

Abstract: The quenching rate of Al–Mg–Si alloys during solution treatment is an important parameter for the automotive industry. In this work, the effect of the different quenching rates on the microstructures, mechanical properties, and paint bake-hardening response of Al–Mg–Si sheets was studied. Large dimples form on the fracture surface of a sample at a quenching rate of 0.01 °C/s. When the quenching rate increased to 58.9 °C/s, the dimples became smaller. The recrystallized grains and textures were slightly affected by quenching rates beyond 1.9 °C/s. Thus, higher r values of the samples were achieved with slower quenching rates. Furthermore, only the Al(FeMn)SiCr insoluble phases were observed in samples with a rapid quenching rate. Sufficient solute atoms and vacancies resulted in the improvement of the precipitation kinetics and paint bake-hardening capacity for Al–Mg–Si sheets at rapid rates. With a decrease in the quenching rate, the formation of the rod-like coarse β' phases consumed many solute atoms and vacancies, leading to the deterioration of the paint bake-hardening capacity. This study provides a critical reference on quenching rates for industrial practices, so that good mechanical properties can be achieved using precision control of the quenching process.

Keywords: Al–Mg–Si alloy; quenching rate; microstructures; mechanical properties; paint bake-hardening; precipitates

1. Introduction

In recent years, Al–Mg–Si alloys have become the preferred material for use as automotive outer body sheets due to their high strength-to-weight ratio, good formability, high corrosion resistance, and attractive hardening potential after artificial aging [1–4]. For heat-treatable Al–Mg–Si alloys, precipitation hardening is the main strengthening mechanism utilized in processing. The precipitation sequence of Al–Mg–Si alloys is as follows [5–8]: SSSS → clusters/GP zones → pre-β'' → β'' → β' → β, Si, where SSSS represents a supersaturated solid solution. The main strengthening precipitates are highly coherent GP zones with a large amount of needle-shaped β'' phase after the paint bake-hardening treatment (generally at 185 °C for 20–30 min) [5,9]. At peak aging period, the coarsened β'' phase is semicoherent with the matrix [8]. The post β' phases are semicoherent rods and produce coarser precipitate microstructures with a lower strength contribution [10,11]. The equilibrium β phase, which is not coherent with the matrix, forms as plates with dimensions of several micrometers composed of Mg_2Si [12]. The abovementioned precipitation structures and phases are closely related to both the alloy composition and the heat treatment process.

The heat treatment of Al–Mg–Si alloys involves solution annealing, quenching, and aging. The solute atoms redissolve and remain in the SSSS form during solution annealing and quenching. Then, SSSS transforms into strengthening phases after aging [13]. The paint bake-hardening response is obviously affected by the quenching rate. Therefore, the quenching rate is an important parameter during the heat treatment process of automotive sheets. In order to achieve an improved paint bake-hardening response, sufficient supersaturated solute atoms and vacancies are necessary after quenching. Obviously, a faster than critical quenching rate effectively suppresses precipitation during quenching. The alloy composition also affects the microstructure and properties at the same quenching rate. Excessive Si promotes high strength and high conductivity, while excessive Mg plays a negative role in the strength and the conductivity [14]. With a large Mg/Si ratio, the alloy possesses a negative natural aging effect on the maximum hardness during artificial aging, while the alloy with a small Mg/Si ratio shows a negligible negative NA effect [15]. Dispersoids containing Mn and/or Cr have been commonly applied for grain size control during homogenization or annealing treatment. The hardness can increase by inducing Mn and/or Cr dispersoids formed during solution treatment [16,17].

Fan et al. [18] found that the rapidly quenched alloy showed a better combination of properties compared to the slowly quenched material in both naturally and artificially aged tempers. Coarse Si and Mg_2Si precipitates formed during slow quenching affected fracture behavior. Using a series of AlMgSi(Cu) alloys quenched in various fluids, Garric et al. [19] studied the impact of the quenching rate on the microstructure and mechanical properties obtained after aging. The results showed that the usual 10 °C/s could be refined depending on the properties of interest given the applications. However, only three or four quenching rates were studied in their work, and the effect of higher or lower quenching rate has yet to be fully explored. Here, 10 different cooling rates were used during quenching. The main aims were to optimize the quenching rate parameters and to investigate its effects on the microstructures, mechanical properties, and paint bake-hardening response of Al–Mg–Si automotive sheets.

2. Materials and Methods

In this study, a commercial cold-rolled AA6016 alloy sheet with a composition of Al—1.10%, Si—0.55%, Mg—0.14%, and Fe—0.10% Cr (wt.%) was provided by MINGTAI Aluminum (Henan, China). Individual sheets, with dimensions of 200 mm × 200 mm × 1 mm, were solution heat-treated at 560 °C for 5 min in an air circulation furnace. After heat preservation, the sheets were directly quenched at 11 different quenching rates to room temperature (RT). The different quenching rates from 58.9 °C/s to 0.01 °C/s were measured using three thermocouples placed on the sheets. The highest quenching rate (58.9 °C/s) was achieved by water cooling, and the other rates were achieved by controlling the temperature and pressure by cooling air. The lowest cooling rate (0.01 °C/s) was achieved by cooling with a 560 °C air furnace. The temperature for each quenching rate was measured from 560 °C to 230 °C, because the range of 230–560 °C was the quenching-sensitive region for this AA6016 alloy.

The as-quenched sheets were immediately pre-aged at 80 °C for 5 h. Subsequently, all sheets were subjected to paint bake-hardening at 185 °C after storage at RT for 168 h. One set of the as-quenched sheets was isothermally heated at 400 °C for 2 h as a control. A schematic representation of the heat treatment procedure is presented in Figure 1, and the measured cooling curves for the as-prepared sheets are presented in Figure 2.

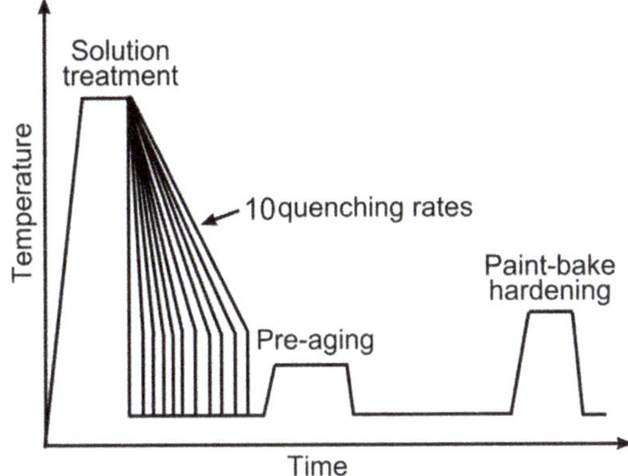

Figure 1. Representation of the heat treatment procedure.

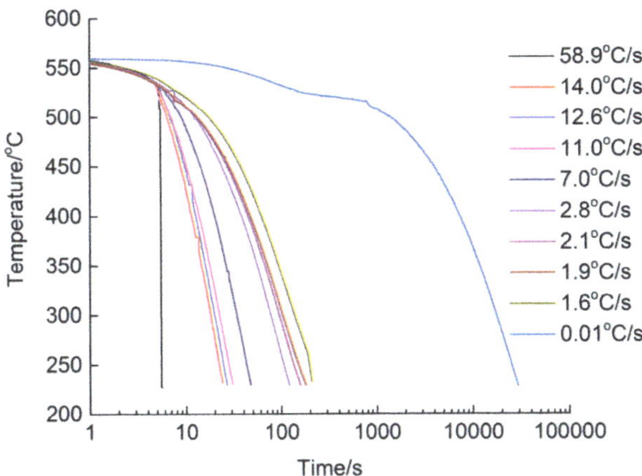

Figure 2. Ten cooling curves of different quenching rates from 58.9 °C/s to 0.01 °C/s.

After pre-aging and paint bake-hardening, the sheets were analyzed using the Vickers hardness and tensile tests. The Vickers hardness test was performed using a KB3000BVRZ-SA Vickers hardness tester (Leica Microsystems Company, Germany) with a load of 49 N and a dwell time of 10 s. In order to reduce the testing error, seven indentations were measured, and final values were obtained after removal of maximum and minimum values. The tensile test was performed in accordance with the China Standard (GB/T 228–2002) using an INSTRON-4206 electronic universal tester (Instron Corporation, USA) at a load speed of 3 mm/min. The properties of sheets in three directions of 0°, 45°, and 90° with respect to the rolling direction were tested before and after paint bake-hardening.

The microstructure of the sheets was characterized using an Imager M2m ZEISS optical microscope (ZEISS Company, Germany). The samples were first mechanically ground, followed by electro-polishing in 10 vol % perchloric acid in an alcohol solution at 25 V for 30 s. The surface and fracture morphology were then observed using a ZEISS ULTRA 55 field emission scanning electron microscope (SEM) (ZEISS Company, Germany) at a 15 kV operating voltage, equipped with an X-ray energy dispersive spectrometer (EDS) system. The recrystallization texture of the samples after solution treatment was characterized by electron backscattered diffraction (EBSD) analysis on a ZEISS ULTRA 55 SEM. The step size of the EBSD analysis was set to 2.5 μm. EBSD samples were prepared by electrolytic polishing using a solution of 10% perchloric acid and 90% ethanol at 35 V and 20–40 s.

The size and distribution of particles and precipitates in the samples were investigated using a Tecnai G^2 F20 transmission electron microscope (TEM) (FEI Company, USA) at a 200 kV operating voltage. The TEM samples were first mechanically ground to approximately 70 μm and then electrolytically polished using a Struers TenuPol-5 jet-polisher (Struers Company, USA) at an operating voltage of 18 V, in an electrolyte solution of 30% by vol nitric acid in methanol (stored between −25 °C and −30 °C). All TEM images were obtained along the <001> Al zone axis.

Differential scanning calorimetry (DSC) samples with a thickness of 0.5 mm, a diameter of 5 mm, and a weight of 25 mg were cleaned using an ultrasonic cleaner. They were then tested using a Q100 model DSC unit (TA Instruments, USA) under an argon atmosphere, and the test temperature ranged from 30 °C to 400 °C with a heating rate of 10 °C/min. Pure aluminum samples were used as reference materials, and the baseline measurements were performed measuring pure aluminum samples.

3. Results

3.1. Microstructures

Figure 3 shows the microstructures of samples with different quenching rates or annealing treatment at 400 °C for 2 h. No particles were present in the matrix of the sample with a quenching rate of 58.9 °C/s (the black dots in Figure 3a,b were impurities after electro-polishing). With a decrease in the quenching rate, an increasing number of particles were observed in the samples. In particular, the enrichment of the particles was more obvious at the grain boundaries, as shown in Figure 3d,e. Coarse irregular particles formed both in the grains and at grain boundaries when a sample was cooled at the lowest quenching rate (0.01 °C/s). These coarse particles were larger than 20 μm. Meanwhile, some particles below the size of 10 μm were scattered around the coarser ones (Figure 3f). For a sample with annealing treatment at 400 °C for 2 h after water quenching, no coarse particles larger than 20 μm were observed; only a high density of micron particles uniformly distributed in the matrix, as shown in Figure 3g.

Figure 4 shows SEM micrographs of the particles distributed in the matrix. We found that the size and density of particles increased with decreasing quenching rates. A closer look at the micron particles indicated that these particles were square, spherical, or rod-shaped. According to our EDS analysis (Figure 4f,g), the coarse irregular particles larger than 20 μm were Si phase, while some square particles were β phase (arrows in Figure 4). Thus, we found that an equilibrium β phase was obtained by the annealing treatment method we had adopted here.

Figure 5 shows the fracture morphology of samples at different quenching rates before and after paint bake-hardening at 185 °C for 20 min. A large number of small and deep dimples were uniformly distributed in the 58.9 °C/s quenching rate sample before paint bake-hardening. Conversely, the size and density of the dimples became larger and lower with a decrease in the quenching rate. After paint bake-hardening, a sharp decline in the count occurred for the fast-quenched sample. Some areas of the fracture in this sample presented no dimples, and only river-like streaks were observed. Meanwhile, in the case of the sample with a quenching rate of 0.01 °C/s, the morphology of the fracture changed little. This result indicated that paint bake-hardening had no effect on the morphology of fractures in samples with small quench rates.

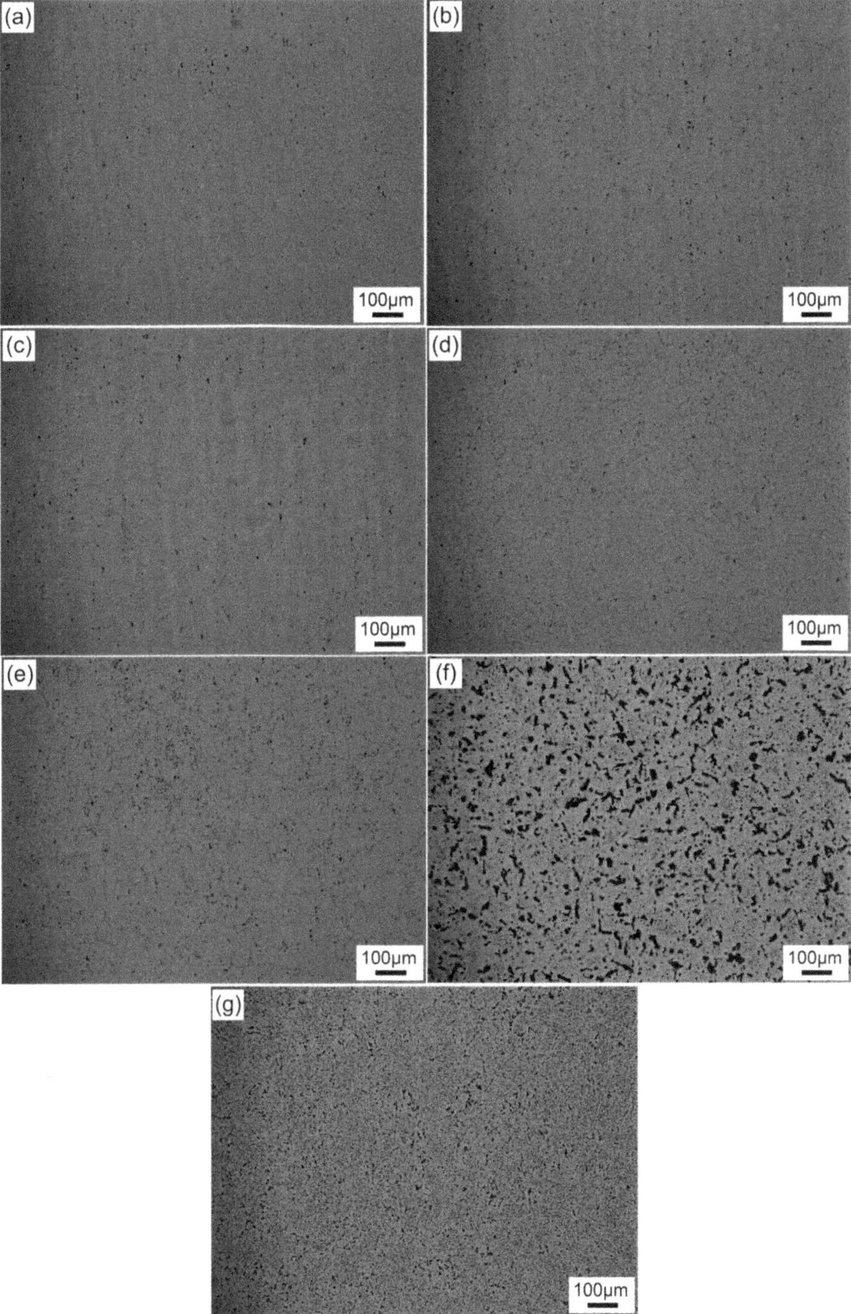

Figure 3. Microstructures of samples with the different quenching rates or annealing treatment at 400 °C for 2 h after pre-aging treatment: (**a**) 58.9 °C/s, (**b**) 14.0 °C/s, (**c**) 7.0 °C/s, (**d**) 2.8 °C/s, (**e**) 1.6 °C/s, (**f**) 0.01 °C/s, and (**g**) annealing at 400 °C for 2 h after water quenching.

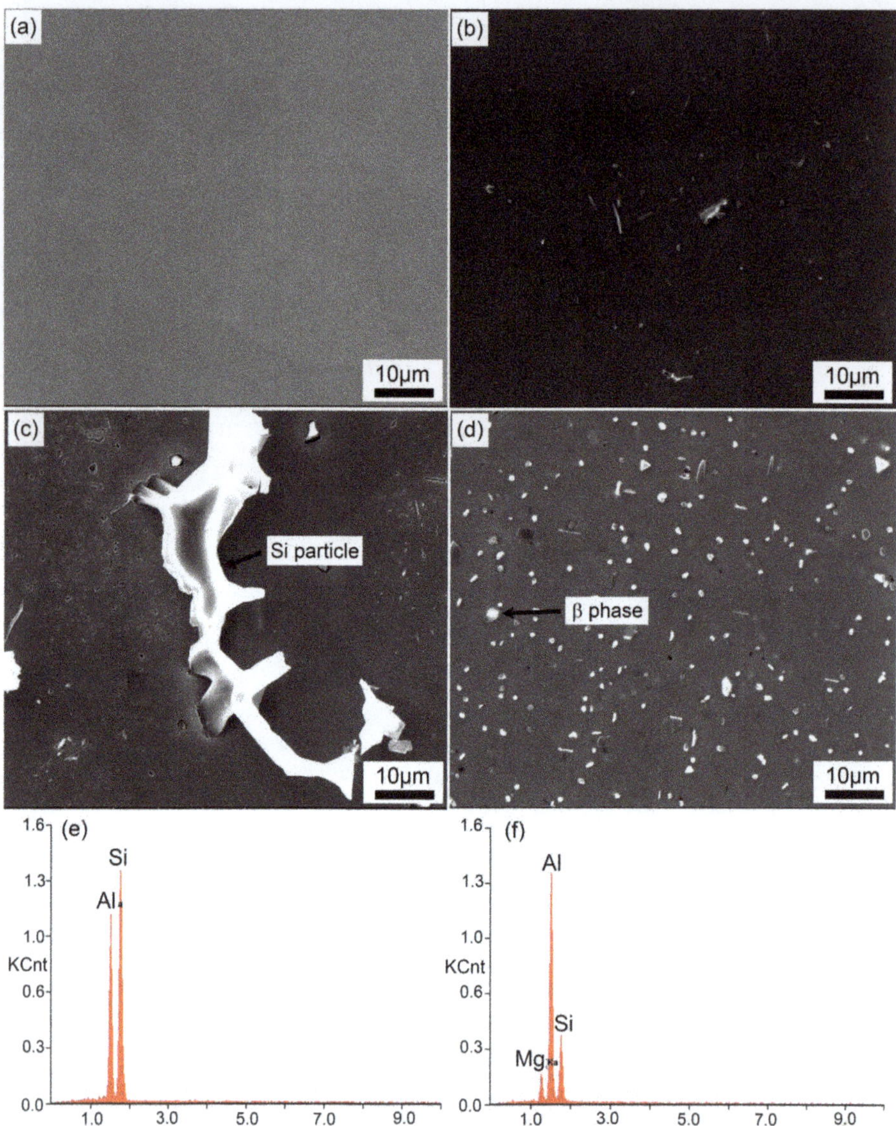

Figure 4. Scanning electron microscope (SEM) micrographs and energy dispersive spectrometer (EDS) analysis of particles distributed in samples with different quenching rates after pre-aging treatment: (**a**) 58.9 °C/s, (**b**) 2.8 °C/s, (**c**) 0.01 °C/s, (**d**) annealing treatment, (**e**) EDS spectra of Si phase, and (**f**) EDS spectra of β phase.

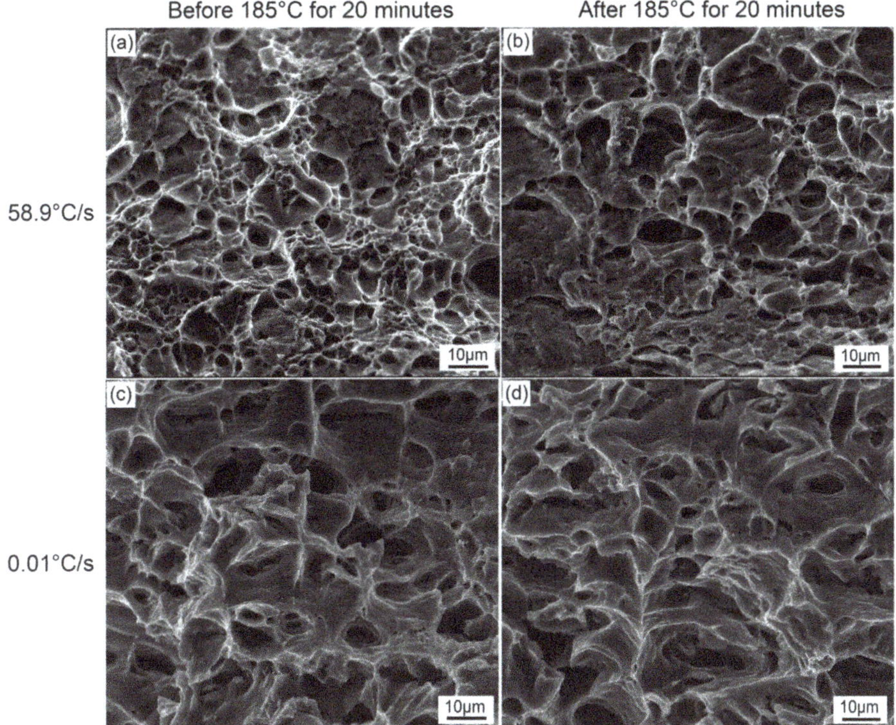

Figure 5. Fracture morphology of samples with different quenching rates before and after paint bake-hardening at 185 °C for 20 min: (**a**) and (**c**), 58.9 °C/s; (**b**) and (**d**), 0.01 °C/s.

3.2. Mechanical Property Characterization

Figure 6 shows the engineering stress–strain curves of samples with different quenching rates before and after paint bake-hardening at 185 °C for 20 min and 60 min. These results revealed that the yield stress of a sample with a quenching rate of 58.9 °C/s before paint bake-hardening was ~130 MPa. Additionally, the yield stress of samples decreased with decreasing quenching rate. When the sample was cooled at a quenching rate of 0.01 °C/s, the yield stress decreased to ~40 MPa. After paint bake-hardening, the strength of the sample increased significantly, and a decrease in elongation occurred. However, the paint bake-hardening treatment had no effect on the strength and elongation of the sample when the quenching rate was decreased to 0.01 °C/s.

In general, the standard for automobile sheets of aluminum requires the r value (10% deformation) to be higher than 0.6 [20]. The average r value was calculated using

$$\bar{r} = \frac{r_{0°} + 2r_{45°} + r_{90°}}{4} \tag{1}$$

where $r_{0°}$, $r_{45°}$, and $r_{90°}$ are the r values in three different directions [21]. Figure 7 shows the average r value of samples with the T4P state (a state of being placed at RT after pre-aging). After pre-aging at 80 °C for 5 h, the average r values of samples with different quenching rates were all higher than 0.6, which suggested that a good deep drawability had developed through these treatments. The r value remained a constant value of ~0.6 with a quenching rate higher than 7.0 °C/s and heat treatment at 400 °C for 2 h, and this value increased with decreasing quenching rate. The r value eventually reached a value of ~0.7 when the quenching rate was dropped to 0.01 °C/s.

In order to study the average *r* value of these samples, texture analysis was performed on EBSD maps with a scanning size of 2.5 μm. More than 500 grains were included for statistically validity. Figure 8 shows inverse pole figure (IPF) maps displaying the recrystallized grains and texture of samples with different quenching rates after solution treatment. The grains have been completely recrystallized with a solution treatment at 560 °C for 5 min. The average grain size calculated by the linear intercept method was ~22.1, 21.6, and 22.4 μm for the samples with quenching rates of 58.9 °C/s, 7.0 °C/s, and 1.9 °C/s, respectively.

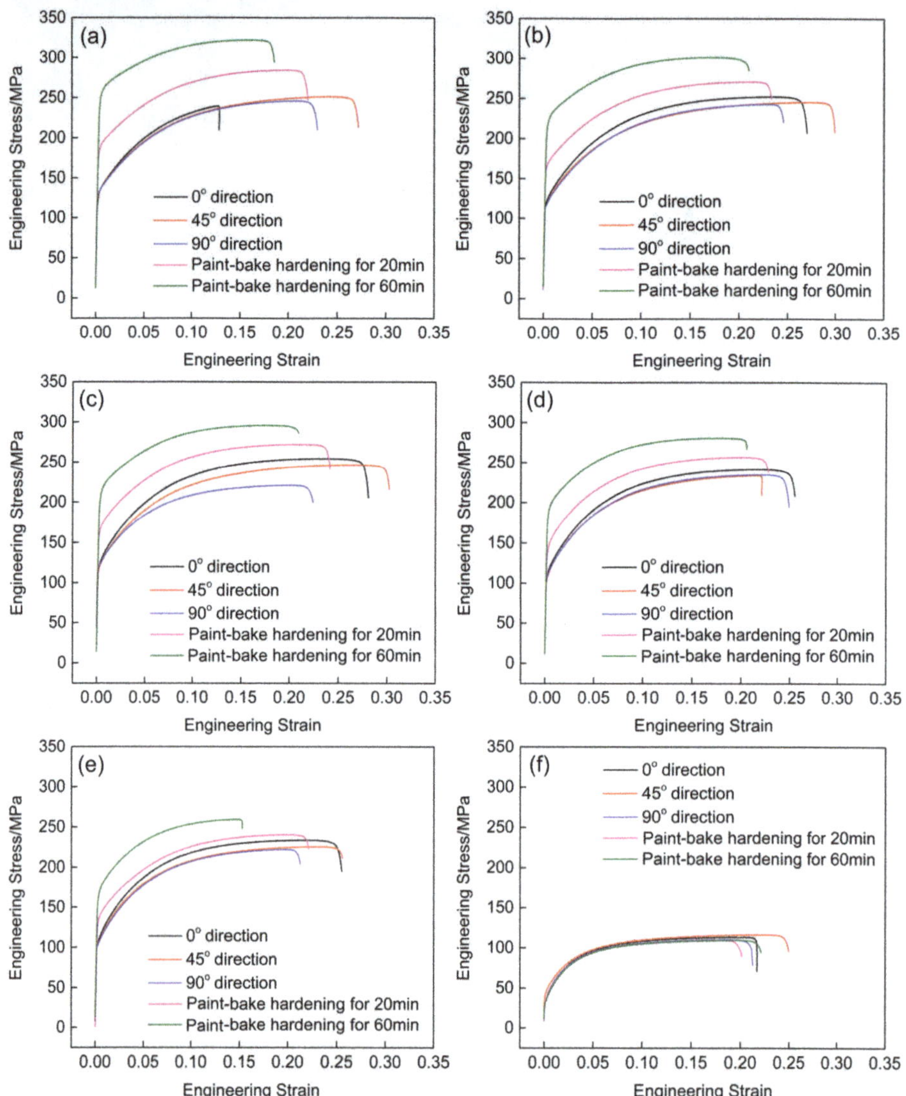

Figure 6. Engineering stress–strain curves of samples with different quenching rates after pre-aging treatment: (**a**) 58.9 °C/s, (**b**) 14.0 °C/s, (**c**) 7.0 °C/s, (**d**) 2.8 °C/s, (**e**) 1.6 °C/s, and (**f**) 0.01 °C/s.

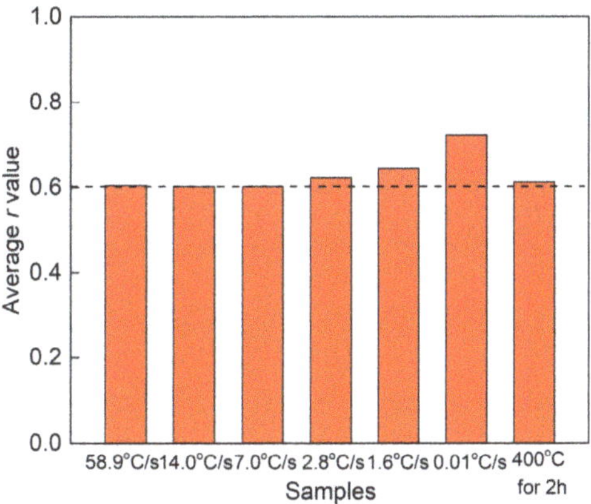

Figure 7. Average *r* value of samples in the T4P state.

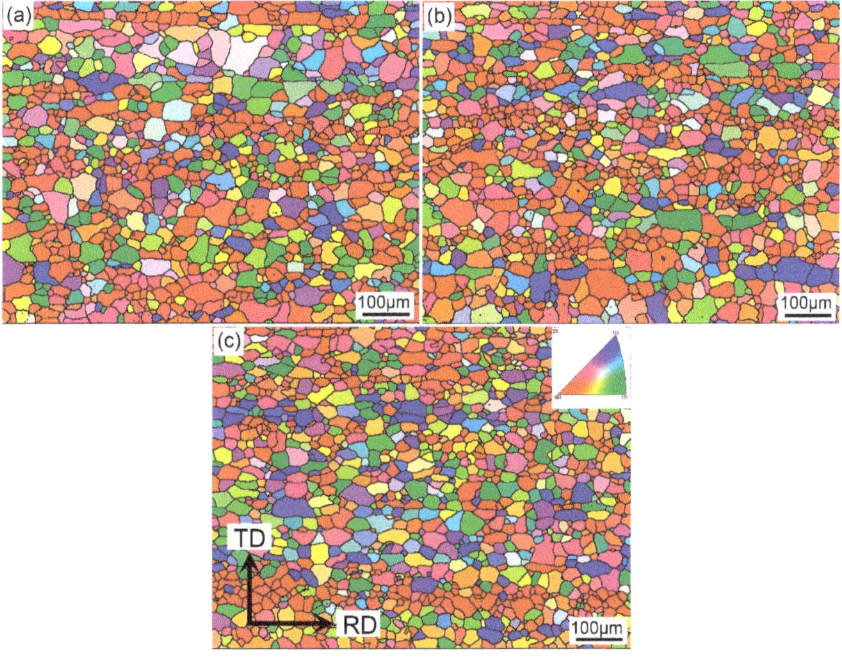

Figure 8. Inverse pole figure (IPF) maps showing the recrystallized grains and texture of the samples with different quenching rates after solution treatment: (**a**) 58.9 °C/s, (**b**) 7.0 °C/s, and (**c**) 1.9 °C/s.

Figure 9 shows the final texture of solutionized samples with different quenching rates. We found that their texture components were quite similar. In samples with quenching rates of 58.9 °C/s, 7.0 °C/s, and 1.9 °C/s, the most obvious texture was a recrystallized Cube {001} <100> orientation, which had the highest volume fraction and intensity. In addition, all the samples included Cube$_{ND}$ {001} <310> and P {011} <122> textures that were beneficial to the as-developed deep drawability [22]. The volume fraction and intensity of the texture components changed slightly with the decrease in quenching rate.

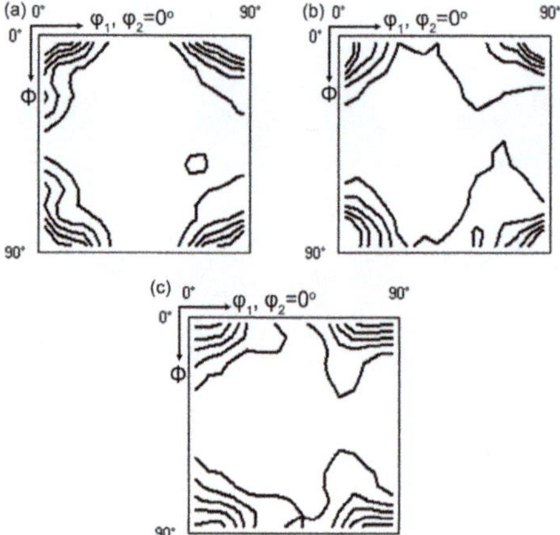

Figure 9. Texture of solutionized samples with different quenching rates: (**a**) 58.9 °C/s, (**b**) 7.0 °C/s, and (**c**) 1.9 °C/s.

3.3. Paint Bake-Hardening

The capacity for paint bake-hardening is an important performance index of automotive body sheets [1,5,23]. After paint bake-hardening treatment, the strength of an alloy sheet is improved to meet the requirements for concave resistance. Figure 10 shows the paint bake-hardening response of samples with different quenching rates. The yield stress of a sample at a quenching rate of 58.9 °C/s increased to 55 MPa and 122 MPa after paint bake-hardening at 185 °C for 20 min and 60 min, respectively. This demonstrated a good capacity for a rapid aging response. However, the strength increment decreased with a decrease in the quenching rate after the paint bake-hardening treatment. When the quenching rate dropped to 0.01 °C/s, no increase in the yield stress occurred (Figure 10a). The peak-aged Vickers hardness values of the samples are presented in Figure 10b. The hardness increment, as well as the yield stress, decreased with decreasing quenching rate. Moreover, the peak-aged hardness did not seriously decrease with a decrease in the quenching rate from 58.9 °C/s to 7.0 °C/s. In addition, no increase in the hardness of a sample with a quenching rate of 0.01 °C/s occurred, even after employing a paint bake-hardening treatment at 185 °C for 4 h. The sample heat treated at 400 °C for 2 h also showed a poor hardening response. As a result, the sample with an extremely slow quenching rate had lost its age-hardening capacity.

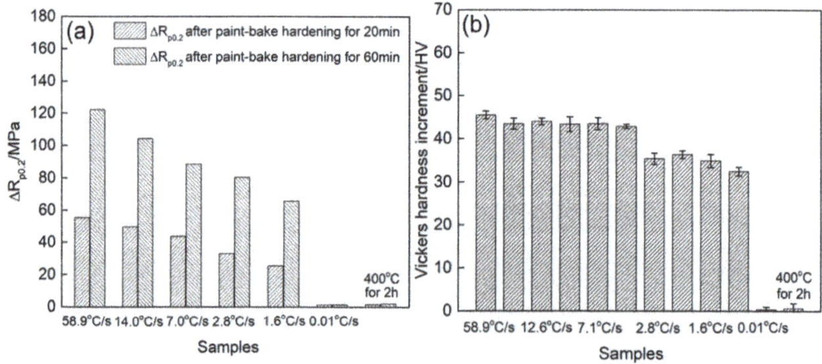

Figure 10. Paint bake-hardening response of samples with different quenching rates: (**a**) Yield stress increment after paint bake-hardening at 185 °C for 20 min and 60 min, and (**b**) Vickers hardness increment after paint bake-hardening at 185 °C for 4 h.

3.4. Precipitation Observation

To further investigate the paint bake-hardening response of the samples with different quenching rates, the precipitates in the matrix were observed using TEM. Figure 11 shows the TEM bright-field images of the samples with different quenching rates after paint bake-hardening at 185 °C for 4 h. In the sample with the quenching rate of 58.9 °C/s, only spherical particles of 100–200 nm size were observed, and uniformly distributed in the matrix, while in the sample with the quenching rate of 7.0 °C/s, a small amount of rod-like precipitates appeared. The length of the precipitates was ~500 nm, and the width was less than 50 nm. Obviously, a certain orientation relationship existed between the rod-like precipitates and the matrix. With the decrease of quenching rate, the density and size of these rod-like precipitates increased significantly. These rod-like precipitates co-existed with the spherical particles in the matrix. When the quenching rate dropped to 1.6 °C/s, these rod-like phases were seriously coarsened and lengthened. The width of the phases was ~200 nm, and the length reached ~3000 nm.

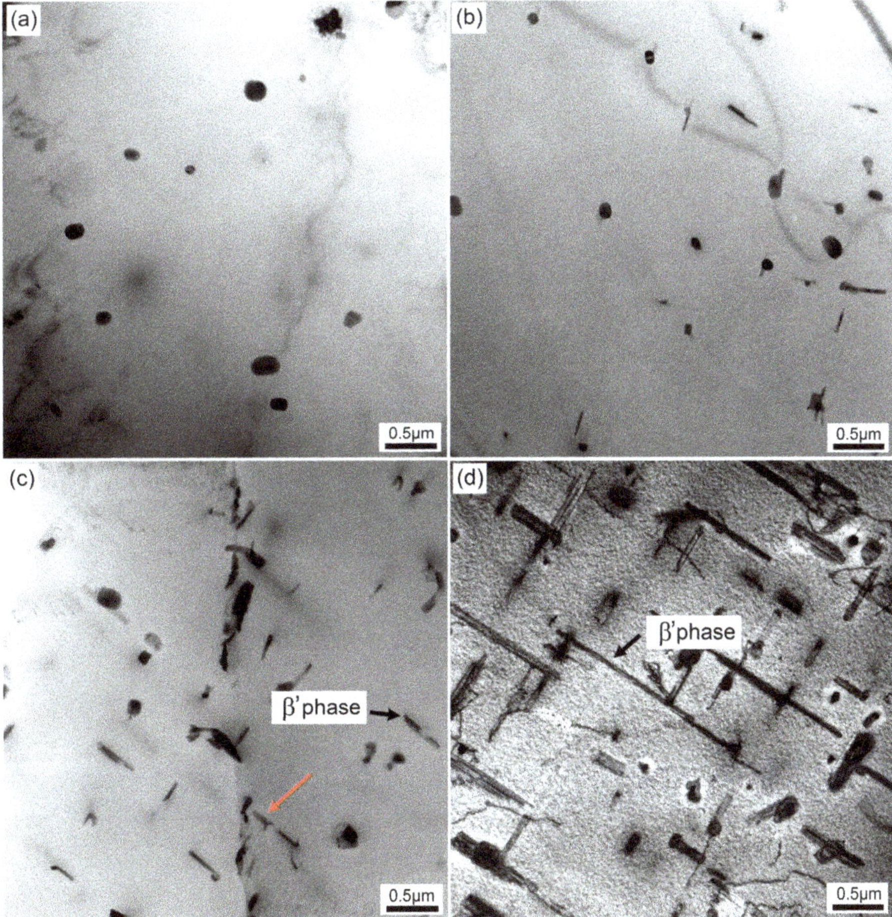

Figure 11. TEM bright-field images of samples with different quenching rates after paint bake-hardening at 185 °C for 4 h: (**a**) 58.9 °C/s, (**b**) 7.0 °C/s, (**c**) 2.8 °C/s, and (**d**) 1.6 °C/s.

Figure 12 shows enlarged TEM bright-field images of samples with different quenching rates. A large amount of fine dot-like and needle-like precipitates was observed in the matrix. Previous studies [24] have indicated that these dot-like precipitates were the needle-like ones viewed in another direction. In samples with different quenching rates, the density and size of the precipitates in the matrix significantly dropped (from ~40 nm to ~15 nm) when the quenching rate decreased from 58.9 °C/s to 7.0 °C/s. As the quenching rate continued to drop to 2.8 °C/s, the size of the needle-like precipitates decreased in width. In a sample with a quenching rate of 1.6 °C/s, the density of the precipitates further decreased in the matrix.

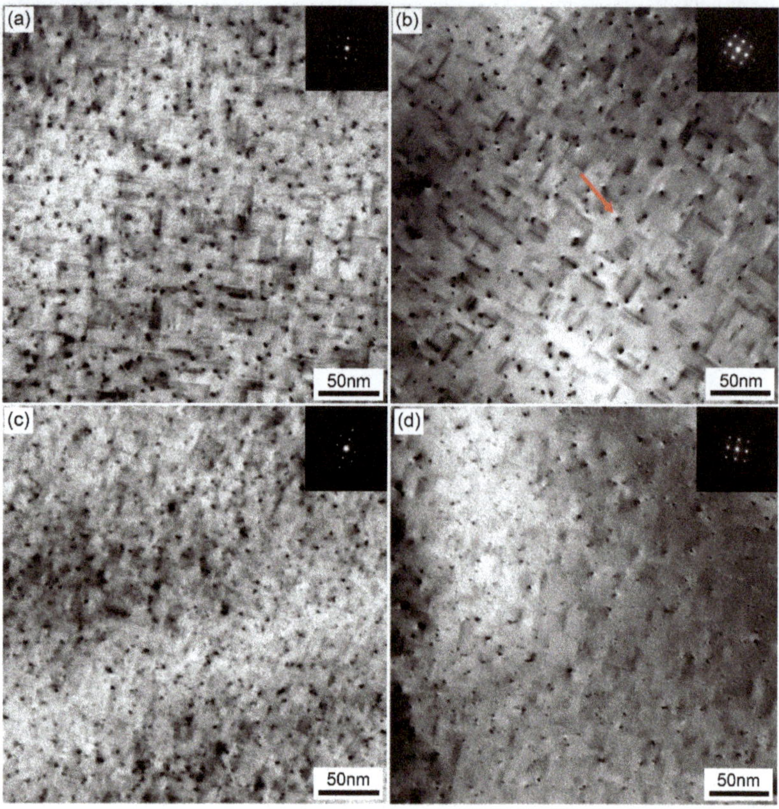

Figure 12. Enlarged TEM bright-field images of samples with different quenching rates after paint bake-hardening at 185 °C for 4 h: (**a**) 58.9 °C/s, (**b**) 7.0 °C/s, (**c**) 2.8 °C/s, and (**d**) 1.6 °C/s.

In order to further study the effect of these precipitates in samples with different quenching rates on the paint bake-hardening response, HRTEM analysis and the corresponding FFT (Fast Fourier Transform) patterns based on the large rod-like and fine needle-like precipitates in Figures 11 and 12 were analyzed, as shown in Figure 13. According to the corresponding FFT patterns, the rod-shaped precipitate was identified as a late phase β'. The orientation relationships between these rod-like precipitates and the matrix were $(001)//(001)_{Al}$, $[010]//[100]_{Al}$, and $[100]//[\bar{3}50]_{Al}$. This observation was consistent with previous studies by Vissers et al. [11]. Additionally, the orientation relationships between the fine needle-like precipitates and the matrix were $(010)//(00\bar{1})_{Al}$, $[001]//[130]_{Al}$, and $[100]//[3\bar{2}0]_{Al}$. Based on the HRTEM and corresponding FFT patterns, the precipitates observed were identified as being β'' phase, which was the main hardening precipitate for a Al–Mg–Si alloy system [8,25,26].

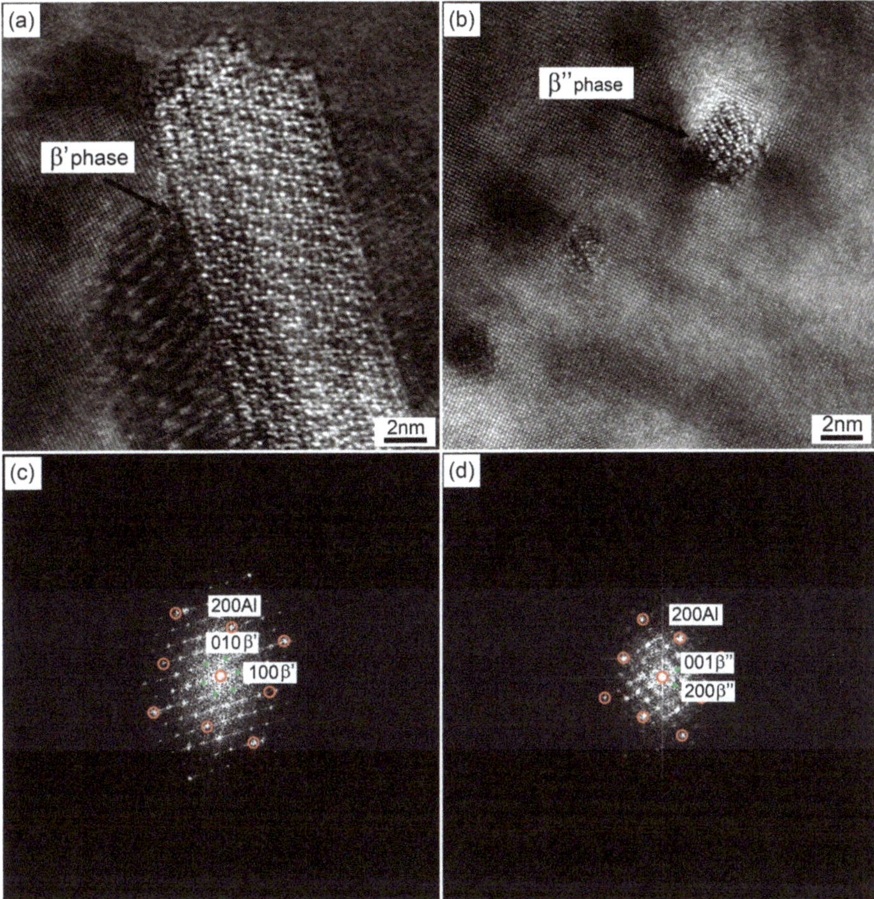

Figure 13. (**a,b**), HRTEM images of the large rod-like and fine needle-like precipitates in Figures 11 and 12 (red arrows marked); and (**c,d**), the corresponding FFT (Fast Fourier Transform) patterns from samples in (**a,b**).

4. Discussion

It is an often noted fact that dispersed particles and strengthening precipitates (β" and β' precipitate) are common in Al–Mg–Si alloys. The size and density of these precipitates are greatly influenced by quenching rates [27,28]. These dispersed particles change slightly during subsequent heat treatments [29]. When a sample was cooled to RT with a very slow quenching rate, large-sized Si particles formed in the matrix. During the fracture process of this alloy, most cracks germinated in the vicinity of these coarse particles and became favorable sites for crack nucleation [30], resulting in the large dimples in Figure 5b,d. However, the strength of the sample was very low at a quenching rate of 0.01 °C/s, regardless of the paint bake-hardening treatment (Figure 6). The improvement in plasticity caused by this decrease in strength was the main reason for the formation of the large dimples. In contrast, no coarse particles formed in the matrix of a sample with a quenching rate of 58.9 °C/s after pre-aging treatment, and the strength was relatively higher. Thus, the dimple size of the fracture surface decreased. After paint bake-hardening treatment, precipitates rapidly grew in a one-dimensional direction, and the strength of the sample increased, while the plasticity decreased [5,31]. This resulted in a further decrease in the size of dimples, and in some areas, no dimples existed.

During solution treatment, the rolled microstructure transformed into recrystallized grains, and the soluble phase redissolved into the matrix. This recrystallization was prior to the redissolution of the soluble phase. If the alloy was kept at a high temperature for a long time, grain boundary migration occurred. With the grain boundary migration, the larger grains swallowed up adjacent smaller grains, resulting in an increase in the average grain size [32,33]. In this study, when the sample was cooled to RT with a very slow quenching rate of 1.9 °C/s, the driving force provided by external energy at high temperature was not enough to cause grain boundary migration or abnormal grain growth. The grain size of a sample with a quenching rate of 1.9 °C/s was similar to that of 58.9 °C/s. Furthermore, the volume fraction and intensity of main recrystallization texture changed slightly with decreasing quenching rate (Figures 8 and 9), while the average r value increased with decreasing quenching rate. Generally, the r value of alloy sheets is closely related to recrystallized grain orientations [21,34,35]. Retaining a rolling texture such as S {123} <634> or B {011} <211> leads to a higher r value. In contrast, a recrystallized cube texture {001} <100> results in a lower r value. According to our experimental results, although the recrystallized textures of samples with the different quenching rates were similar, the r values increased with decreasing quenching rate. This may have been due to the coarse Si phase formed with this small quenching rate, which hindered the strain in the thickness direction more effectively than in the width direction. Therefore, the coarse Si phase was the main contributor to the increase in the r value.

It has been reported previously that the precipitation sequence for Al–Mg–Si alloys is SSSS → clusters/GP zones → pre-β″ → β″ → β′ → β, Si. In this study, no coarse rod-like β′ precipitates formed in the matrix, and only spherical particles were observed in a sample with a quenching rate of 58.9 °C/s (Figure 11a). These spherical particles were confirmed to be an Al(FeMn)SiCr insoluble phase in previous research [13]. Although the samples were treated by paint bake-hardening at 185 °C for 4 h, only nanometer-scale needle-like β″ precipitates formed, rather than rod-like β′ precipitates. With a decrease in the quenching rate, the residence time of samples at higher temperature was prolonged. Sufficient energy promoted the rapid nucleation and growth of precipitates, leading to the formation of micron-sized β′ precipitates. As the quenching rate continued to decrease, the density and size of the β′ precipitates further increased (Figure 11d). When a sample was cooled with an extremely small quenching rate (0.01 °C/s), all solute atoms were separated from the matrix. A large amount of coarse Si phase accompanied by β′ phase formed, which was observed by optical microscopy (Figure 3e). The consumption of supersaturated solute atoms and vacancies led to the decrease of cluster density during pre-aging treatment, which decreased the strength of the alloy [36].

In the practice of automotive manufacturing, the general paint bake-hardening treatment is to hold an alloy at 185 °C for 20–30 min. During a paint bake-hardening treatment, precipitates rapidly grow in a one-dimensional direction, and an elastic distortion region forms around them. This effectively hinders the dislocation movement, resulting in the occurrence of a rapid aging response [5,8]. When the time of paint bake-hardening treatment at 185 °C is extended to 4–10 h, most precipitates transform into β″ phase. In the peak-aged state, the hardness of the alloy reaches its maximum. In this study, DSC analysis of samples with different quenching rates was performed to investigate this precipitation behavior, as shown in Figure 14. We observed that the temperature of peak II related to β″ phase increased with decreasing quenching rate. For a sample with a quenching rate of 0.01 °C/s, no endothermic or exothermic reaction occurred during DSC heating. Consequently, a higher density of β″ phase formed in the rapidly quenched sample when holding at a temperature for the same amount of time.

As shown in Figure 12, the density and size of the β″ precipitates decreased with quenching rate decrease after a paint bake-hardening treatment at 185 °C for 4 h. For a sample with a quenching rate of 58.9 °C/s, all solute atoms dissolved into the matrix to form SSSS. When a sample was rapidly quenched to RT, there were many supersaturated solute atoms and vacancies in the matrix. During the subsequent paint bake-hardening treatment, sufficient solute atoms and vacancies provided the raw materials and diffusion channels for the formation of β″ strengthening phases. Not only that, with solute atoms and vacancies, more nucleation sites were provided for the formation of β″ precipitates [37–39]. The abovementioned factors ensured that the strength and hardness of a sample obviously increased after paint bake-hardening treatment (Figure 10). Additionally, the diffusion distance was short due to sufficient vacancies. The formation of the β″ precipitates needed less energy. Hence, this obviously decreased the precipitation temperature of β″ phase structures (Figure 14).

When the quenching rate of the sample decreased, the hold time at high temperature was prolonged. A large number of solute atoms and vacancies were consumed to form a coarse β′ phase structure (Figure 13). This led to a decrease in the density and size of β″ precipitates after a paint bake-hardening treatment [36,40]. Compared with β′ phase, the strengthening effect of β″ precipitates was more remarkable [5,41]. Therefore, the strength and hardness increment decreased with decreasing quenching rate. Furthermore, the consumption of solute atoms and vacancies increased the diffusion distance, and the nucleation and growth of β″ precipitates needed more energy. Thus, an exothermic peak related to the β″ phase appeared at higher temperatures in the DSC curves (Figure 14). For a sample with an extremely slow quenching rate (0.01 °C/s), the solute atoms and vacancies were almost consumed, leading to no precipitation reaction during the DSC heating process.

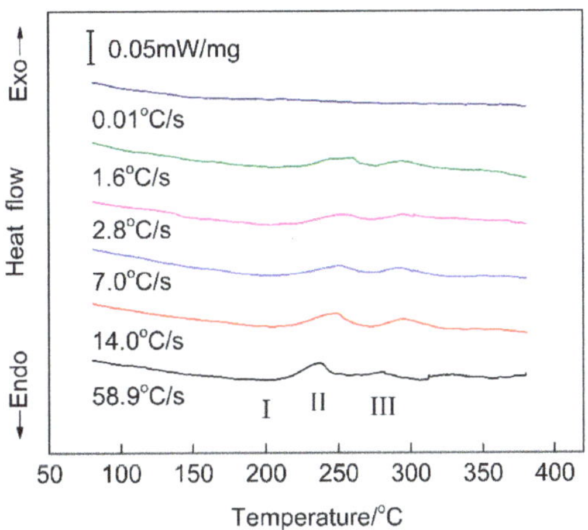

Figure 14. Differential scanning calorimetry (DSC) curves for samples with different quenching rates.

5. Conclusions

The effects of different quenching rates on the microstructure, mechanical properties, and paint bake-hardening response of Al—1.10% and Si—0.55% Mg automotive sheets were investigated in this study. The conclusions are as follows:

1. For a sample with a quenching rate of 0.01 °C/s, large dimples formed on the fracture surface due to the coarse Si phase and good plasticity. With a quenching rate of 58.9 °C/s, these dimples became smaller. The plasticity of a sample rapidly deteriorated, and the size of dimples further decreased after paint bake-hardening treatment;

2. The recrystallized grains and textures were slightly affected by the quenching rate beyond 1.9 °C/s. While higher r values of samples with low quenching rates were achieved, this may have been related to the hindrance of coarse particles due to the strain in thickness direction during the deformation of the sample;

3. For a sample with a rapid quenching rate, only the Al(FeMn)SiCr insoluble phase was observed. With a decrease in the quenching rate, the holding time at higher temperature was prolonged. This led to sufficient energy in the system, which promoted the nucleation and growth of precipitates, leading to the formation of rod-like coarse β′ phase structures;

4. Sufficient solute atoms and vacancies were formed in a sample with a rapid quenching rate, resulting in the improvement of the precipitation kinetics and paint bake-hardening capacity. With a decrease in the quenching rate, the solute atoms and vacancies were consumed. Consequently, the precipitation kinetics of the β'' phase decreased, and the paint bake-hardening capacity deteriorated.

Author Contributions: Conceptualization, G.G. and Y.L.; Software, J.L.; Validation, G.G. and Y.L.; Investigation, all authors; Data curation, G.G., Y.L., H.D. and G.X.; Writing—original draft preparation, G.G. and Y.L.; Writing—review and editing, G.G. and Y.L.; Project administration, Y.L. and Z.W.; Funding acquisition, Y.L. and Z.W.

Funding: This project was supported by the National Natural Science Foundation of China, grant number 51790485.

Acknowledgments: The authors acknowledge financial support from the Project 51790485 supported by National Natural Science Foundation of China. We would like to thank LetPub (www.letpub.com) for providing linguistic assistance during the preparation of this manuscript.

Conflicts of Interest: The authors declare no conflict of interest.

References

1. Miller, W.S.; Zhuang, L.; Bottema, J.; Wittebrood, A.J.; Smet, P.D.; Haszler, A.; Vieregge, A. Recent development in aluminum alloys for the automotive industry. *Mater. Sci. Eng. A* **2000**, *280*, 37–49. [CrossRef]
2. Hirsch, J.; Al-Samman, T. Superior light metals by texture engineering: Optimized aluminum and magnesium alloys for automotive applications. *Acta Mater.* **2013**, *61*, 818–843. [CrossRef]
3. Ding, X.P.; Cui, H.; Zhang, J.X.; Li, H.X.; Guo, M.X.; Lin, Z.; Zhuang, L.Z.; Zhang, J.S. The effect of Zn on the age hardening response in an Al-Mg-Si alloy. *Mater. Des.* **2015**, *65*, 1229–1235. [CrossRef]
4. Pogatscher, S.; Antrekowitsch, H.; Leitner, H.; Ebner, T.; Uggowitzer, P.J. Mechanisms controlling the artificial aging of Al-Mg-Si Alloys. *Acta Mater.* **2011**, *59*, 3352–3363. [CrossRef]
5. Chen, J.H.; Costan, E.; van Huis, M.A.; Xu, Q.; Zandbergen, H.W. Atomic Pillar-Based Nanoprecipitates Strengthen AlMgSi Alloys. *Science* **2006**, *312*, 416–419. [CrossRef]
6. Marioara, C.D.; Nordmark, H.S.; Andersen, J.; Holmestad, R. Post-β'' phases and their influence on microstructure and hardness in 6xxx Al-Mg-Si alloys. *J. Mater. Sci.* **2006**, *41*, 471–478. [CrossRef]
7. Wang, X.; Esmaeili, S.; Lloyd, D.J. The sequence of precipitation in the Al-Mg-Si-Cu alloy AA6111. *Metall. Mater. Trans. A* **2006**, *37*, 2691–2699. [CrossRef]
8. Zandbergen, H.W.; Andersen, S.J.; Jansen, J. Structure Determination of Mg5Si6 Particles in Al by Dynamic Electron Diffraction Studies. *Science* **1997**, *277*, 1221–1225. [CrossRef]
9. Ding, L.; He, Y.; Wen, Z.; Zhao, P.; Jia, Z.; Liu, Q. Optimization of the pre-aging treatment for an AA6022 alloy at various temperatures and holding times. *J. Alloys Compd.* **2015**, *647*, 238–244. [CrossRef]
10. Andersen, S.J.; Marioara, C.D.; Vissers, R.; Frøseth, A.; Zandbergen, H.W. The structural relation between precipitates in Al-Mg-Si alloys, the Al-matrix and diamond silicon, with emphasis on the trigonal phase U1-MgAl2Si2. *Mater. Sci. Eng. A* **2007**, *444*, 157–169. [CrossRef]
11. Vissers, R.; van Huis, M.A.; Jansen, J.; Zandbergen, H.W.; Marioara, C.D.; Andersen, S.J. The crystal structure of the β' phase in Al-Mg-Si alloys. *Acta Mater.* **2007**, *55*, 3815–3823. [CrossRef]
12. Takeda, M.; Ohkubo, F.; Shirai, T.; Takeda, M. Stability of metastable phases and microstructures in the ageing process of Al-Mg-Si ternary alloys. *J. Mater. Sci.* **1998**, *33*, 2385–2390. [CrossRef]
13. Gao, G.; He, C.; Li, Y.; Li, J.; Wang, Z.; Misra, R.D.K. Influence of different solution methods on microstructure, precipitation behavior and mechanical properties of Al-Mg-Si alloy. *Trans. Nonferrous Met. Soc.* **2018**, *28*, 839–847. [CrossRef]
14. Xu, X.; Yang, Z.; Ye, Y.; Wang, G.; He, X. Effects of various Mg/Si ratios on microstructure and performance property of Al-Mg-Si alloy cables. *Mater. Charact.* **2016**, *119*, 114–119. [CrossRef]
15. Tao, G.; Liu, C.; Chen, J.; Lai, Y.; Ma, P.; Liu, L. The influence of Mg/Si ratio on the negative natural aging effect in Al-Mg-Si-Cu alloys. *Mater. Sci. Eng. A* **2015**, *642*, 241–248. [CrossRef]
16. Zhan, H.; Hu, B. Analyzing the microstructural evolution and hardening response of an Al-Si-Mg casting alloy with Cr addition. *Mater. Charact.* **2018**, *142*, 602–612. [CrossRef]

17. Qian, X.; Parson, N.; Chen, X. Effects of Mn addition and related Mn-containing dispersoids on the hot deformation behavior of 6082 aluminum alloys. *Mater. Sci. Eng. A* **2019**, *764*, 138–253. [CrossRef]

18. Fan, Z.; Lei, X.; Wang, L.; Yang, X.; Sanders, R. Influence of quenching rate and aging on bendability of AA6016 sheet. *Mater. Sci. Eng. A* **2018**, *730*, 317–327. [CrossRef]

19. Garric, V.; Colas, K.; Donnadieu, P.; Renou, G.; Urvoy, S.; Kapusta, B. Correlation between quenching rate, mechanical properties and microstructure in thick sections of Al Mg Si(Cu) alloys. *Mater. Sci. Eng. A* **2019**, *753*, 253–261. [CrossRef]

20. Prillhofer, R.; Rank, G.; Berneder, J.; Antrekowitsch, H.; Uggowitzer, P.; Pogatscher, S. Property Criteria for Automotive Al-Mg-Si Sheet Alloys. *Materials* **2014**, *7*, 5047–5068. [CrossRef]

21. Wang, X.; Guo, M.; Cao, L.; Luo, J.; Zhang, J.; Zhuang, L. Effect of heating rate on mechanical property, microstructure and texture evolution of Al-Mg-Si-Cu alloy during solution treatment. *Mater. Sci. Eng. A* **2015**, *621*, 8–17. [CrossRef]

22. Wang, X.; Guo, M.; Zhang, Y.; Xing, H.; Li, Y.; Luo, J.; Zhang, J.; Zhuang, L. The dependence of microstructure, texture evolution and mechanical properties of Al-Mg-Si-Cu alloy sheet on final cold rolling deformation. *J. Alloys Compd.* **2016**, *657*, 906–916. [CrossRef]

23. Li, H.; Yan, Z.; Cao, L. Bake hardening behavior and precipitation kinetic of a novel Al-Mg-Si-Cu aluminum alloy for lightweight automotive body. *Mater. Sci. Eng. A* **2018**, *728*, 88–94. [CrossRef]

24. Yang, W.; Wang, M.; Zhang, R.; Zhang, Q.; Sheng, X. The diffraction patterns from β″ precipitates in 12 orientations in Al-Mg-Si alloy. *Scr. Mater.* **2010**, *62*, 705–708. [CrossRef]

25. Guo, M.X.; Zhang, Y.D.; Yuan, B.; Sha, G.; Zhang, J.S.; Zhuang, L.Z.; Lavernia, E.J. Influence of aging pathways on the evolution of heterogeneous solute-rich features in peak-aged Al-Mg-Si-Cu alloy with a high Mg/Si ratio. *Phil. Mag. Lett.* **2019**, *99*, 49–56. [CrossRef]

26. Guo, M.X.; Sha, G.; Cao, L.Y.; Liu, W.Q.; Zhang, J.S.; Zhuang, L.Z. Enhanced bake-hardening response of an Al-Mg-Si-Cu alloy with Zn addition. *Mater. Chem. Phys.* **2015**, *162*, 15–19. [CrossRef]

27. Zhu, S.; Li, Z.; Yan, L.; Li, X.; Huang, S.; Yan, H.; Zhang, Y.; Xiong, B. Effects of Zn addition on the age hardening behavior and precipitation evolution of an Al-Mg-Si-Cu alloy. *Mater. Charact.* **2018**, *145*, 258–267. [CrossRef]

28. Fallah, V.; Langelier, B.; Ofori-Opoku, N.; Raeisinia, B.; Provatas, N.; Esmaeili, S. Cluster evolution mechanisms during aging in Al-Mg-Si alloys. *Acta Mater.* **2016**, *103*, 290–300. [CrossRef]

29. Brito, C.; Vida, T.; Freitas, E.; Cheung, N.; Spinelli, J.E.; Garcia, A. Cellular/dendritic arrays and intermetallic phases affecting corrosion and mechanical resistances of an Al-Mg-Si alloy. *J. Alloys Compd.* **2016**, *673*, 220–230. [CrossRef]

30. Li, Y.; Qiu, S.; Zhu, Z.; Han, D.; Chen, J.; Chen, H. Intergranular crack during fatigue in Al-Mg-Si aluminum alloy thin extrusions. *Int. J. Fatigue* **2017**, *100*, 105–112. [CrossRef]

31. Ding, L.; Jia, Z.; Nie, J.; Weng, Y.; Cao, L.; Chen, H.; Wu, X.; Liu, Q. The structural and compositional evolution of precipitates in Al-Mg-Si-Cu alloy. *Acta Mater.* **2018**, *145*, 437–450. [CrossRef]

32. Zhang, C.; Wang, C.; Guo, R.; Zhao, G.; Chen, L.; Sun, W.; Wang, X. Investigation of dynamic recrystallization and modeling of microstructure evolution of an Al-Mg-Si aluminum alloy during high-temperature deformation. *J. Alloys Compd.* **2019**, *773*, 59–70. [CrossRef]

33. Yao, X.; Zheng, Y.F.; Quadir, M.Z.; Kong, C.; Liang, J.M.; Chen, Y.H.; Munroe, P.; Zhang, D.L. Grain growth and recrystallization behaviors of an ultrafine grained Al–0.6wt%Mg–0.4wt%Si–5vol.%SiC nanocomposite during heat treatment and extrusion. *J. Alloys Compd.* **2018**, *745*, 519–524. [CrossRef]

34. Liu, Y.S.; Kang, S.B.; Ko, H.S. Texture and plastic anisotropy of Al–Mg–0.3Cu–1.0Zn alloys. *Scr. Mater.* **1997**, *37*, 411–417. [CrossRef]

35. Engler, O.; Hirsch, J. Texture control by thermomechanical processing of AA6xxx Al-Mg-Si sheet alloys for automotive applications—A review. *Mater. Sci. Eng. A* **2002**, *336*, 249–262. [CrossRef]

36. Serizawa, A.; Hirosawa, S.; Sato, T. Three-Dimensional Atom Probe Characterization of Nanoclusters Responsible for Multistep Aging Behavior of an Al-Mg-Si Alloy. *Metall. Mater. Trans. A* **2008**, *39*, 243–251. [CrossRef]

37. Birol, Y. Preaging to improve bake hardening in a twin-roll cast Al-Mg-Si alloy. *Mater. Sci. Eng. A* **2005**, *391*, 175–180. [CrossRef]

38. Birol, Y. Pre-straining to improve the bake hardening response of a twin-roll cast Al-Mg-Si alloy. *Scr. Mater.* **2005**, *52*, 169–173. [CrossRef]

39. Birol, Y.; Karlik, M. The interaction of natural ageing with straining in a twin-roll cast AlMgSi automotive sheet. *Scr. Mater.* **2006**, *55*, 625–628. [CrossRef]
40. Aruga, Y.; Kozuka, M.; Takaki, Y.; Sato, T. Formation and reversion of clusters during natural aging and subsequent artificial aging in an Al-Mg-Si alloy. *Mater. Sci. Eng. A* **2015**, *631*, 86–96. [CrossRef]
41. Jin, S.; Ngai, T.; Zhang, G.; Zhai, T.; Jia, S.; Li, L. Precipitation strengthening mechanisms during natural ageing and subsequent artificial aging in an Al-Mg-Si-Cu alloy. *Mater. Sci. Eng. A* **2018**, *724*, 53–59. [CrossRef]

 *materials* 

Article

Effects of Deformation Parameters on Microstructural Evolution of 2219 Aluminum Alloy during Intermediate Thermo-Mechanical Treatment Process

Lei Liu [1], Yunxin Wu [2,*] and Hai Gong [2]

[1]   Light Alloy Research Institute, Central South University, Changsha 410083, China; Larill@csu.edu.cn
[2]   State Key Laboratory of High Performance Complex Manufacturing, Central South University, Changsha 410083, China; csu2019@163.com
*   Correspondence: csu1991@163.com; Tel.: +86-188-9035-5864

Received: 30 July 2018; Accepted: 15 August 2018; Published: 22 August 2018

Abstract: To explore the effective way of grain refinement for 2219 aluminum alloy, the approach of 'thermal compression tests + solid solution treatment experiments' was applied to simulate the process of intermediate thermo-mechanical treatment. The effects of deformation parameters (i.e., temperature, strain, and strain rate) on microstructural evolution were also studied. The results show that the main softening mechanism of 2219 aluminum alloy during warm deformation process is dynamic recovery, during which the distribution of $CuAl_2$ phase changes and the substructure content increases. Moreover, the storage energy is found to be decreased with the increase in temperature and/or the decrease in strain rate. In addition, complete static recrystallization occurs and substructures almost disappear during the solid solution treatment process. The average grain size obtained decreases with the decrease in deforming temperature, the increase in strain rate, and/or the increase in strain. The grain refinement mechanism is related to the amount of storage energy and the distribution of precipitated particles in the whole process of intermediate thermal-mechanical treatment. The previously existing dispersed fine precipitates are all redissolved into the matrix, however, the remaining precipitates exist mainly by the form of polymerization.

Keywords: 2219 aluminum alloy; intermediate thermo-mechanical treatment; storage energy; $CuAl_2$ phase; grain refinement

---

1. Introduction

2219 aluminum alloy, which has the advantages of high strength and resistance to stress corrosion, good weldability, and service performance, was considered as the third generation of space materials and has been widely used in aerospace industry [1]. Its good mechanical properties are closely related to microstructural characteristics and depend largely on their chemical composition, processing history, and heat treatment process [2]. During the manufacturing process, strength and plasticity targets are two of the most important performance parameters to be achieved. For 2219 aluminum alloy that can be strengthened by heat treatment, solid solution strengthening and precipitation strengthening are effective means to improve its strength [3–5]. Concerning the improvement of ductility, the grain refinement approach may be often applied [6]. Also, grain refinement is the only method that can effectively improve both the ductility and the strength of the materials [7,8]. So, factories attach great importance to controlling the grain size, especially for those metals and alloys without phase change [9]. Therefore, the theory and technology of grain refinement for 2219 aluminum alloy are of great significance to be investigated.

Thermo-mechanical treatment (TMT), an advanced and coupled metallurgical technology, has an unparalleled advantage in refining grain and improving plasticity and the comprehensive properties of metals and alloys [10,11]. It is mainly through coupling the dislocations and other defects produced by deformation with the morphology and distribution of the precipitated phase during heat treatment that the grain refinement is achieved, thus improving the strength and ductility of the material [12,13]. According to the study of E.D. Russo [14,15], TMT is defined as the intermediate thermo-mechanical treatment (ITMT) when the precipitated phase only acts as auxiliary particles, such as the core of the recrystallized nucleation or the obstructor of grain boundary migration (GBM) during the recrystallization process. Many scholars [16–37] have made extensive efforts on the process of ITMT. Some scholars [16,17] focused on the traditional combination of cold deformation (CD) and recrystallization annealing treatment, also known as thermal mechanical processing (TMP). J. Waldman [18] proposed a new FA-ITMT process with 'subsection cooling and thermal insulation homogenization treatment + CD + solid solution treatment (SST)'. B.R. Ward [19] improved the process to enlarge its application. J. Wert [20] developed a new superplastic pretreatment (SPPT) process and the grain size of the 7050-aluminum alloy was refined to 10 µm. Although the technological measures adopted before 'CD + SST' are different, all the purposes are to precipitate dispersed second-phase particles. Some studies [21,22] focused on the effect of precipitation on the recrystallization process (i.e., recrystallization kinetics, texture evolution, grain size, etc.). With regard to the effect of ITMT parameters on the microstructure evolution, S. Primig [23] found that the grain size of the recrystallization solution treatment (RST) process increases with the increasing of heating rate, J. Waldman [10] found that the grain refinement can be achieved by increasing the RST temperature, and H. Yoshida [24] found that RST period has little effect on the grain refinement. The deformation parameters have a great influence on the grain size and orientation and the distribution of the precipitated phase, so it is an important part of the ITMT process. However, the study on the effects of deformation parameters were rarely reported. In addition, most of the studies on ITMT processes are focused on the optimization and improvement of 7000 serials aluminum alloy [25–28], Al-Li alloy [29], and 6000 series aluminum alloys [30,31], Ni-rich Ti-51.5 at.% Ni shape memory alloy [32], Ti-28Nb-35.4Zr alloy [33], direct-quenched low-carbon strip steel [34], commercial austenitic stainless steel [35–37], and so on. However, the grain refinement effect of the ITMT process for 2219 aluminum alloy is rarely reported. Therefore, it is urgent to study the effects of deformation parameters on microstructural evolution of 2219 aluminum alloy during the ITMT process.

In this paper, the ITMT process of 2219 aluminum alloy was simulated by warm compression + solution treatment experiment. The evolution of grain size and the second phase on 2219 aluminum alloy during warm deformation and solution treatment were studied. In addition, the grain refinement mechanism of 2219 aluminum alloy and the influence of the deformation parameters were discussed in detail. Finally, the optimal deformation parameters of ITMT were summarized.

2. Materials and Methods

To explore the influence of deformation parameters (temperature: T, strain: ε, strain rate: $\dot{\varepsilon}$) on the microstructural evolution during the process of ITMT (warm deformation (WD) + SST), the experimental scheme was adopted as shown in Figure 1.

The warm deformation stage was conducted on the Gleeble-3500 thermal simulator unit (Dynamic Systems Inc., New York, NY, USA). Different deformation parameters were arranged with the temperature range of 210–300 °C, the strain rate range of 0.01–5 s^{-1}, and the strain range of 0–0.9. The samples were water quenched immediately after the deformation stage to retain the deformed microstructure. Subsequently, the samples were heated to 540 °C in a quenching furnace and kept for 4 h during the solution treatment stage, and the water quenched was also adopted to preserve the solution-treated microstructure.

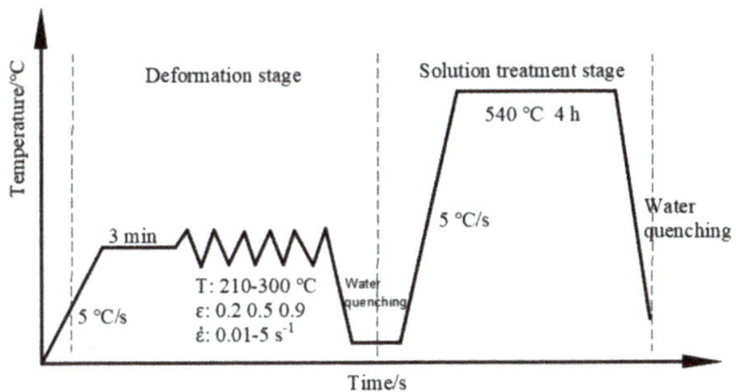

Figure 1. Experimental scheme of intermediate thermo-mechanical treatment process.

The samples were taken from 2219 aluminum alloy plate after hot rolling, and their chemical composition is shown in Table 1. All the samples were machined to cylinder shape with the height of 15 mm and the diameter of 10 mm. During the warm compression stage, some lubricants were added to the ends of the samples to reduce friction with the dies. After the experiment, the samples were cut along the deformation direction for microstructural observation by scanning electron microscope (SEM, Oxford Instruments Inc., Oxford, UK) and electron back-scattered diffraction (EBSD, Helios Nanolab 600i, FEI Company, Hillsboro, OS, USA) method. The different phases existing in present samples under different TMT stages were detected by an X-Ray diffractometer (XRD, Advance D8, Bruker Beijing Scientific Technology Co., Ltd., Beijing, China). The samples for SEM were ground and mechanically polished, and the samples for EBSD were ground and mechanically and electrolytically polished. The electrolyte used was the mixture of nitric acid and methanol solution with volume ratio of 3:7. The samples were electrolyzed under the voltage of 22 V and kept for 55 s. The results were processed by Channel 5 software (HKL Technology, Inc., Danbury, CT, USA), in which the high angle grain boundaries (HAGB: misorientation >15°) were expressed in coarse black solid lines, and the low angle grain boundaries (LAGB: misorientation >2°) were expressed in fine white solid lines.

Table 1. Chemical composition of the 2219 aluminum alloy studied (mass fraction, %).

Cu	Mn	Si	Zr	Fe	Mg	Zn	V	Ti	Al
5.8–6.8	0.2–0.4	≤0.2	0.1~0.25	≤0.3	≤0.02	0.10	0.05~0.15	0.02~0.1	Bal

3. Results and Discussion

3.1. Initial Micrographs Analysis of the Undeformed Sample

The microstructures of the samples in the initial state are shown in Figure 2. As shown in Figure 2a, the initial grain of the samples presents a typical elongated state after hot rolling, with the length >500 μm and the width >100 μm, which may make the material exhibit anisotropic properties. Some fine recrystallized grains are found at local grain boundaries, which indicates dynamic recrystallization initiated during hot rolling. A large number of substructures (i.e., fine white solid lines) can be found inside the grain. This is because many dislocations accumulate and intertwine in local areas as a result of the dislocation density increasing and their interaction effects during the previous large plastic deformation, thus forming an uneven distribution and making a grain into many small crystal blocks (i.e., subgrains) with slightly different misorientations. Moreover, the inverse pole figure (IPF) shows that (001) and (101) grains are dominant and (111) grains are few. By the misorientation analysis (Figure 2b), LAGB content (77.6%) dominates, far exceeding that of HAGB (22.4%). In addition, as shown in Figure 2c, the distribution of the second phase in the solid solution matrix is approximately dispersed, and its particles are approximately spherical with the size of about 10 µm. However, a few parts of the precipitated phase are connected to lumps or chains. The second phase can be determined to be $CuAl_2$ (θ) phase by energy disperse spectroscopy (EDS, Oxford Instruments Inc., Oxford, UK) and XRD analysis.

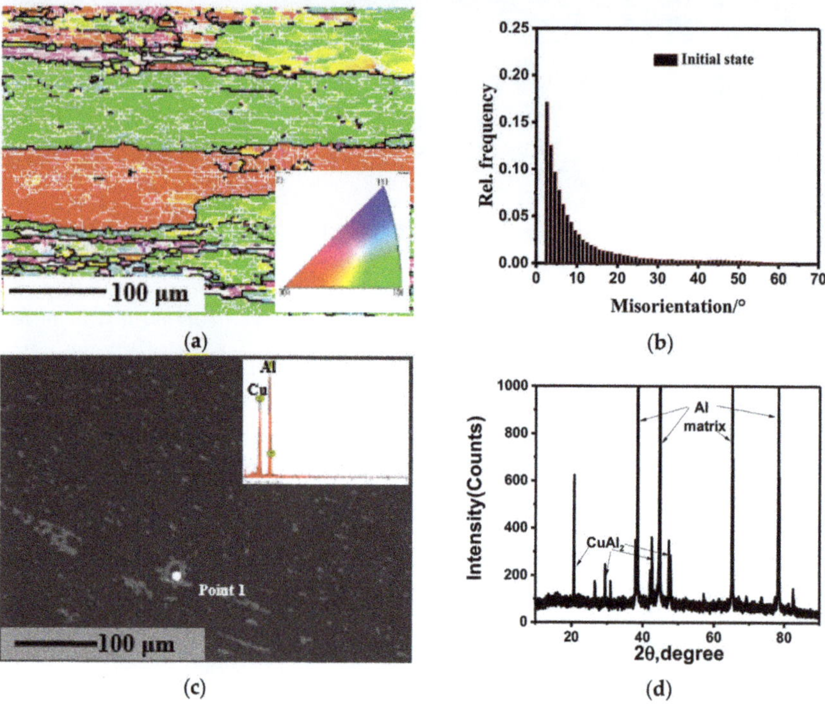

Figure 2. Initial microstructure of 2219 aluminum alloy used in the experiment: (**a**) EBSD micrograph; (**b**) the frequency distribution of misorientation; (**c**) SEM micrograph and EDS analysis of Point 1; (**d**) XRD analysis.

3.2. Effects of Deformation Parameters on the Deformed Microstructure during WD Process

3.2.1. Effects on Flow Behavior

The true stress–strain curves of 2219 aluminum alloy under different warm deformation conditions are shown in Figure 3. It can be clearly seen that the flow stress is very sensitive to the deformation parameters and increases with the decrease in temperature and/or the increase in strain rate. On one hand, the slip system is limited and the dislocation accumulation and entanglement are serious at lower temperature, which makes the flow stress higher. With the increase of temperature, the diffusion of the vacancy and the slip and climbing of dislocation become easier, thus the dynamic softening effect is enhanced and the flow stress is reduced [38]. On the other hand, the lower strain rate provides sufficient time for dislocation movement, making the dynamic recovery (DRV) more thorough, and the softening effect becomes more obvious [39,40].

In addition, each true stress–strain curve can be divided into two stages: Short work hardening stage and long stable flowing stage. This characteristic marks obvious dynamic recovery process. In the stage of work hardening, the flow stress increases rapidly with the increase of strain. It is due to the rapid increment and accumulation of dislocation, which makes the resistance of the deformation increase sharply, and the dynamic softening effect is too weak to make up for the work hardening effect [41]. In the stage of stable flowing, the flow stress is almost invariable, which is due to the counteraction of the work hardening effect and dynamic softening effect in the deformation process [42]. That is to say, the proliferation and reduction of dislocations have reached a kind of equilibrium. In particular, the flow stresses of the 2219 aluminum alloy decrease slightly under the conditions of 270 °C and 300 °C-0.01 s^{-1}, which may characterize the occurrence of partial dynamic recrystallization.

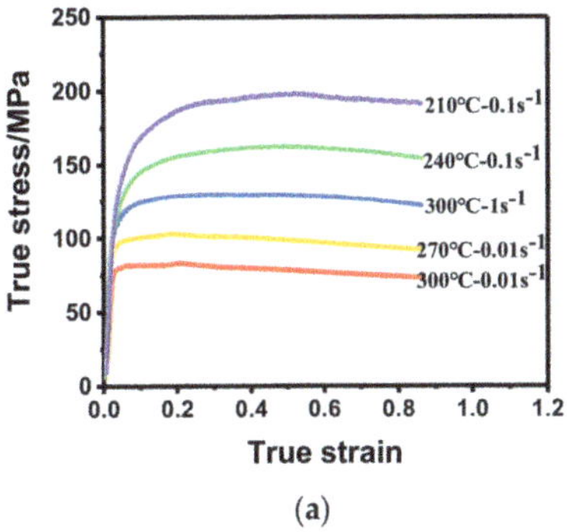

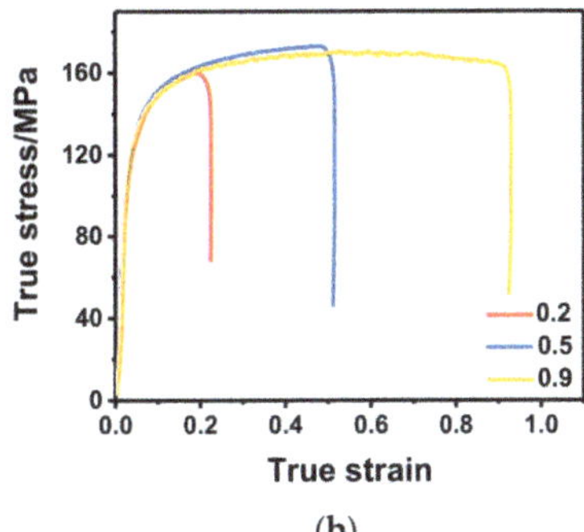

Figure 3. True stress–strain curves under: (**a**) different temperatures and strain rates; (**b**) the strain of 0.2, 0.5, and 0.9.

According to the analysis above, the evolution of increment and disappearance of dislocation determine the magnitude of flow stress, and the mathematical relationship between them can be expressed as follows [43]:

$$\sigma = \alpha \mu b \sqrt{\rho} \tag{1}$$

where α is the material constant, μ is the shear modulus, b is the Burgers vector, ρ is the dislocation density, and σ is the flow stress (MPa). Therefore, the flow stress is proportional to the root of dislocation density.

In addition, the relationship between storage energy and dislocation density can be expressed as follows:

$$E_{st} = c\mu\rho b^2 \tag{2}$$

where E_{st} is the storage energy (kJ) and c is the material parameter. It can be seen that the storage energy is proportional to the dislocation density. According to Equations (1) and (2), the relationship between storage energy and flow stress can be obtained and shown as follows:

$$E_{st} = c_1 \mu^{-1} \sigma^2 \tag{3}$$

where $c_1 = c/\alpha^2$. It can be seen from Equation (3) that the storage energy is proportional to the square of the flow stress. That is to say, the storage energy also decreases with the increase in temperature and the decrease in strain rate.

3.2.2. Effects on Subgrain Evolution

The EBSD micrographs of 2219 aluminum alloy after different deformation conditions are shown in Figure 4. It is observed that the microstructures obtained show obvious characteristics of DRV, which is in accordance with the true stress–strain curves of Figure 3. The cutting surface is a random diameter plane when the deformed sample is cut by wire electrode discharge machining (WEDM). Therefore, the grains exhibit various forms as a result of the different planes selected. But all the grains have different degrees of fragmentation and deformation because all of them are compressed. According to the IPF, it can also be seen that the grains have different dominant orientations in different cutting planes. In addition, many LAGBs are observed inside the grains, which shows that sufficient storage energy is introduced and the driving force for recrystallization can be provided with thermal energy in the subsequent annealing process [44]. According to the probability density function of the dislocations (Figure 5a) and the volume ratio of LAGB and HAGB (Table 2), it can be seen that the distribution of the misorientation remains almost the same and LAGBs occupy the dominant content. However, the quantitative content of LAGBs is changed, and it is closely related to the deformation parameters.

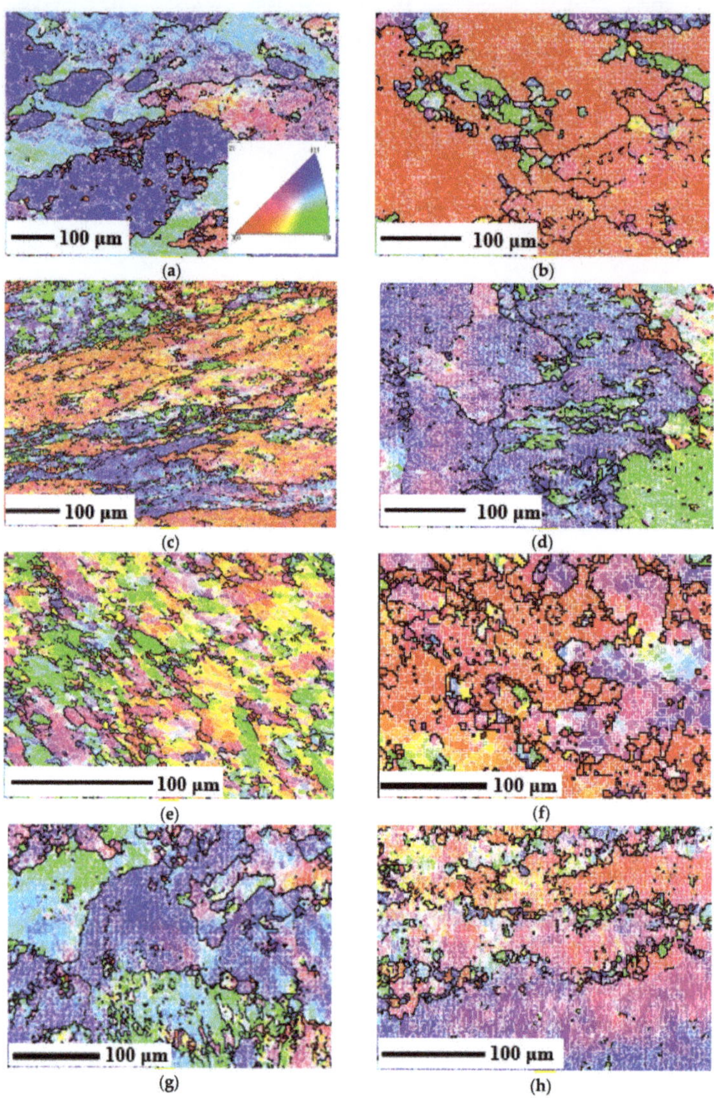

Figure 4. EBSD micrographs of the deformed samples under the condition of: (**a**) 240 °C-0.3 s^{-1}-0.2; (**b**) 240 °C-0.3 s^{-1}-0.5; (**c**) 240 °C-0.3 s^{-1}-0.9; (**d**) 210 °C-0.01 s^{-1}-0.9; (**e**) 270 °C-0.01 s^{-1}-0.9; (**f**) 300 °C-0.01 s^{-1}-0.9; (**g**) 300 °C-0.1 s^{-1}-0.9; (**h**) 300 °C-1 s^{-1}-0.9.

From Figure 4a–c, it can be seen that the area of HAGBs increases with the increase of strain due to grain deformation. In addition, the content ratio of LAGB and HAGB increases from 79.0% to 84.3%.

This shows that the density of dislocation cell and the content of LAGB increase with the increase of strain, so more energy can be stored in the dislocations on the subgrain boundaries. Therefore, the storage energy gradually increases with the increase in strain [45].

From Figure 4d–f, it can be seen that with the increase of the deformation temperature, not only the structures of initial grains are changed, but also some new fine recrystallized grains (Figure 4e) and the grown equiaxed recrystallization grains (Figure 4f) are formed near the grain boundaries, which is due to the continuous dynamic recrystallization (CDRX) occurring by the continuous lattice rotation near the grain boundaries [46,47]. This is consistent with the decreasing trend of the corresponding flow curves at large strains in Figure 3. Meanwhile, according to Figure 5b and Table 2, the content ratio of the LAGBs against HAGBs under the deformed conditions of 270 °C-0.01 s^{-1} (77.4%) and 300 °C-0.01 s^{-1} (76.8%) are lower than the initial state (77.6%). It is proved again that DRX of 2219 aluminum alloy has occurred at this condition. It is also observed that only a small portion of the recrystallized grains are found in local areas and the initial deformed grains occupy the dominant contents, which indicates the main dynamic softening mechanism is DRV. Moreover, the driving force for DRV increases with the increase in temperature and dislocation density [48]. Therefore, DRV proceeds fiercely and the dislocation density becomes lower as the temperature increases, so storage energy will be reduced.

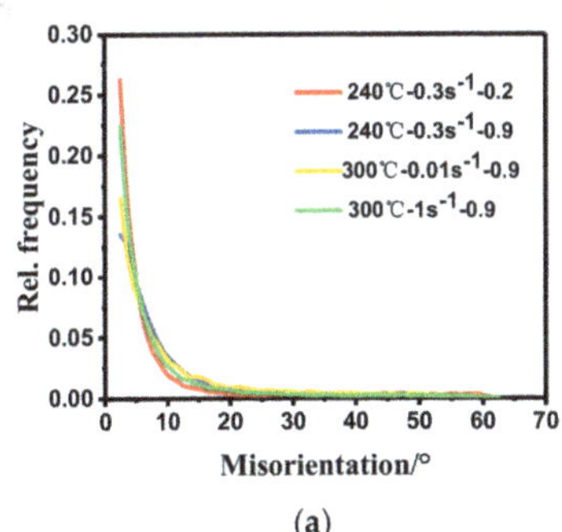
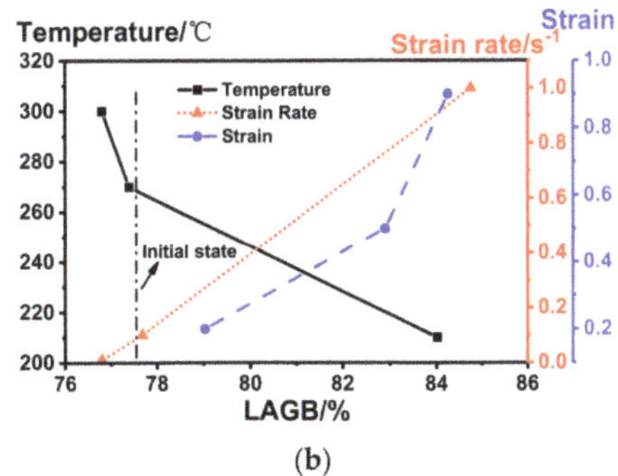

Figure 5. (a) Frequency distribution of misorientation under different deforming conditions; (b) the relationship between the content of LAGBs and deforming temperature, strain rate, and strain.

Table 2. The content of LAGBs and HAGBs under different deformation conditions.

Deformation Parameters	Variables		LAGBs	HAGBs
Initial State	-		77.6%	22.4%
240 °C-0.3 s^{-1}	strain	0.2	79.0%	21.0%
		0.5	82.9%	17.10%
		0.9	84.3%	15.7%
0.01 s^{-1}-0.9	temperature/°C	210	84.0%	16.0%
		270	77.4%	22.6%
		300	76.8%	23.2%
300 °C-0.9	strain rate/s^{-1}	0.01	76.8%	23.2%
		0.1	77.7%	22.3%
		1	84.8%	15.2%

From Figures 4f–h and 5b, and Table 2, it is observed that the content ratio of LAGBs against HAGBs increases from 76.8% to 84.8% with the increase in strain rate at the deformed temperature of 300 °C, which shows the increasing contents of substructures. This is because the increase in strain rate makes the deformation time shortened and DRV process incomplete. Therefore, the dislocation density increases greatly and the storage energy can be increased with the increase in strain rate.

3.2.3. Effects on the Distribution of Precipitated Phase

The SEM micrographs of the deformed samples are shown in Figure 6. It can be seen that the distribution of the second phase is relatively dispersed and small parts of the second phase are united together, which is similar to the initial state (Figure 1c). This indicates that the second phase of 2219 aluminum alloy cannot be broken apart during warm deformation. By EDS and XRD analysis in Figure 6e,f, the precipitated phase is still $CuAl_2$ (θ) phase. It can be observed that deformation can affect the shape and size of the precipitated phase [49,50], but the content of the precipitated phase is only determined by the solubility under different temperatures in terms of thermodynamics.

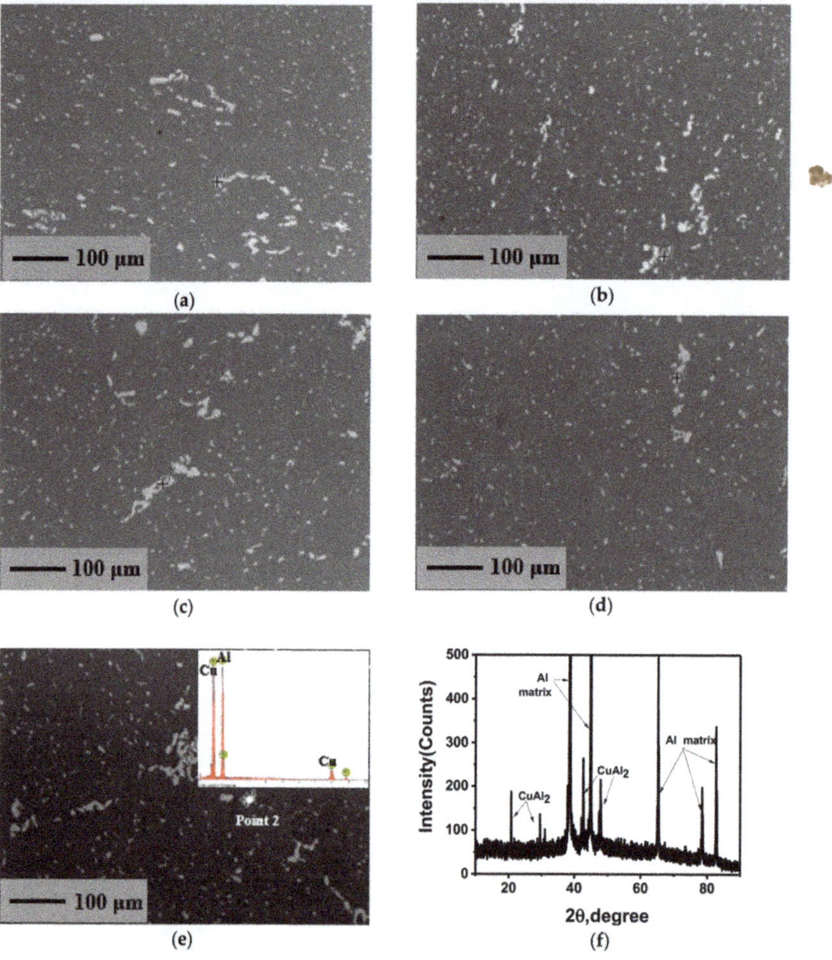

Figure 6. SEM micrograph of the deformed samples under the condition of: (**a**) 240 °C-0.3 s^{-1}-0.2; (**b**) 240 °C-0.3 s^{-1}-0.9; (**c**) 210 °C-0.01 s^{-1}-0.9; (**d**) 300 °C-0.01 s^{-1}-0.9; (**e**) 300 °C-0.1 s^{-1}-0.9 and EDS analysis of Point 2; (**f**) XRD analysis of the deformed samples.

The precipitates in 2219 aluminum alloy play an important role in the process of ITMT, and the relationship between precipitates and substructures may be the main mechanism of grain refinement [51]. Studies [52,53] show that a large number of dislocations and local deformable zones are formed around the large precipitated particles, and the deformed energy is introduced. Thus, the recrystallized core will be formed during the subsequent SST process (i.e., particles stimulate nucleation, PSN). The dispersed small particles apply resistance to the migration of grain boundaries or subgrain boundaries to limit the growth of recrystallized grains during the SST process [52], and finally refined grains will be obtained [53].

As shown in Figure 6a,b, the condensation of the precipitated phase is slightly relieved and the distribution is more diffuse as the strain increases. This is due to the crush of large lump precipitates and the movement of particles as the strain increases. In addition, the formation and spheroidization of the second phase can be promoted by the introduction of crystal defects (vacancies and dislocation, etc.) during warm deformation. This can effectively impede the movement of dislocations and promote the formation of dislocation walls, and eventually form a large number of polygonal substructures through DRV [54]. From Figure 6c,d, it can be seen that the content of θ phase is slightly reduced, the condensation phenomenon is slightly relieved, and the distribution is more diffuse as the temperature increases. This is because the solubility of $CuAl_2$ increases with the increase in temperature, which makes a part of the precipitates resolve to the matrix and thus reduces the content of the dispersed phase. In addition, solute atoms diffuse and migrate more readily and the alloy has a better plastic mobility as the temperature increases, thus making the distribution more uniform. From Figure 6d,e, it is known that large precipitate lumps become serious and the distribution is relatively nonuniform as the strain rate increases. This is due to the shorter deformation time and poor fluidity at the high strain rate.

3.3. Effects of Deformation Parameters on the Microstructures during SST Process

3.3.1. Effects on Grain Refinement

The EBSD microstructures of the samples after SST are shown in Figure 7. It can be clearly seen that (001), (101), and (111) grains are randomly arranged, and there are no deformed microstructures, which indicates that complete static recrystallization (SRX) has occurred. The elongated (or squashed) and broken grains are reproduced and grown to new uniform and fine equiaxed grains because of the increase of the atomic diffusion capacity when the deformed metal is solution treated at a high temperature. In addition, all the substructures have almost disappeared and the frequency distribution of the misorientations is shown in Figure 8a. It can be seen that the distribution curves of misorientations are similar to each other and the content of LAGBs is less than 8%. However, the grain size under different deformation parameters is quite different. The expectation (EX) and the deviation coefficient (EX/s, s: Standard deviation) of the grain size by mathematical statistics are shown in Table 3. It can be seen that all the samples are refined to different degrees compared with the initial state, which indicates that the deformation parameters have a significant effect on grain refinement during the recrystallization process. Moreover, the grain size during the SRX process can be determined by the following formula:

$$d = 2R = 2\int_0^{t_R} v\,dt = 2\sqrt[4]{\frac{3v}{\pi N}} \tag{4}$$

where d is the diameter of grains, N is the nucleation rate, and v is the growth rate. It can be seen that the grain size decreases with the increase in nucleation rate and/or the decrease in growth rate.

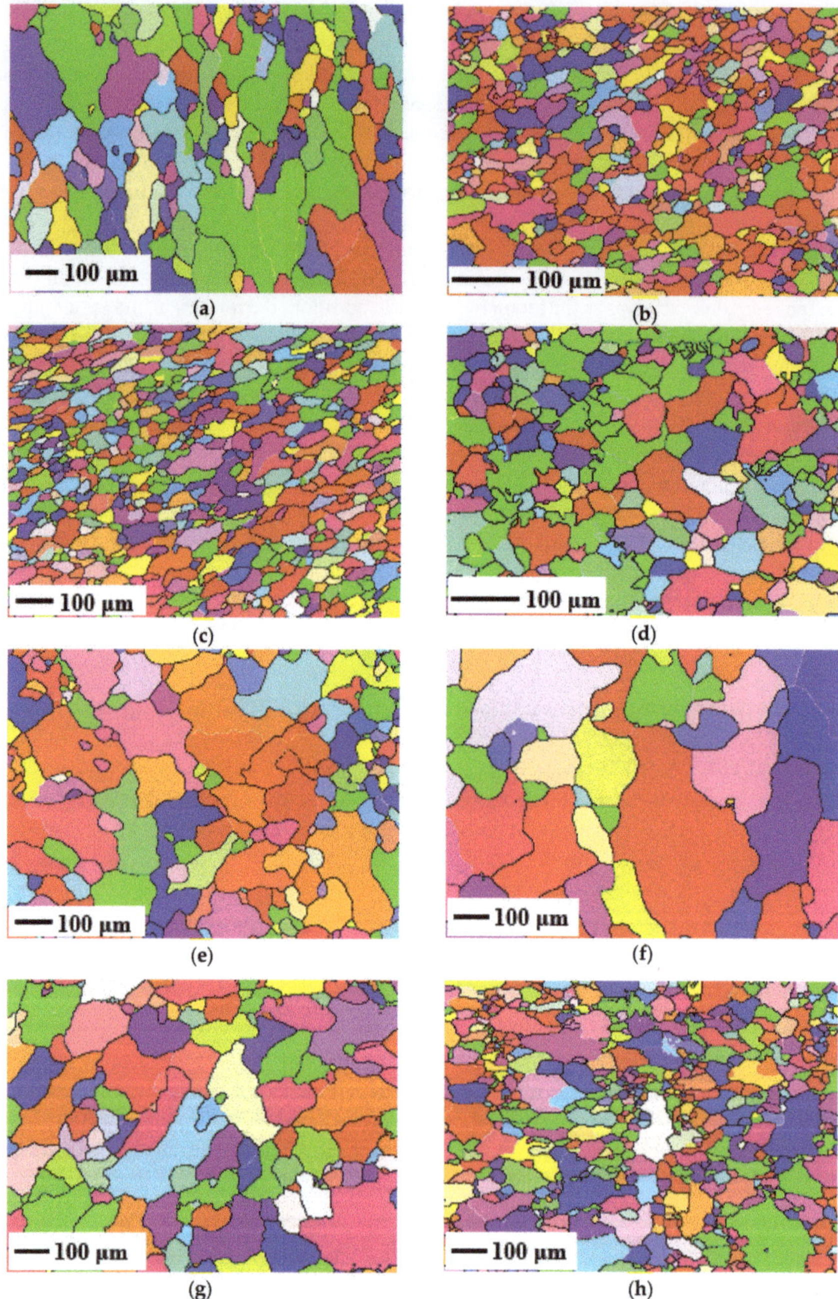

Figure 7. EBSD micrograph of the solution-treated samples under the condition of: (**a**) 240 °C-0.3 s^{-1}-0.2; (**b**) 240 °C-0.3 s^{-1}-0.5; (**c**) 240 °C-0.3 s^{-1}-0.9; (**d**) 210 °C-0.01 s^{-1}-0.9; (**e**) 270 °C-0.01 s^{-1}-0.9; (**f**) 300 °C-0.01 s^{-1}-0.9; (**g**) 300 °C-0.1 s^{-1}-0.9; (**h**) 300 °C-1 s^{-1}-0.9.

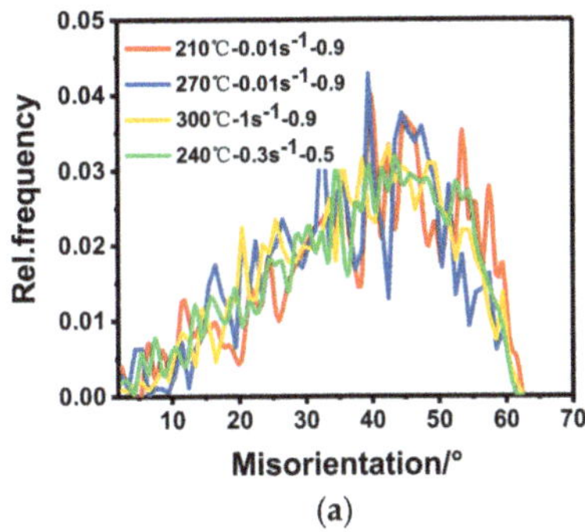
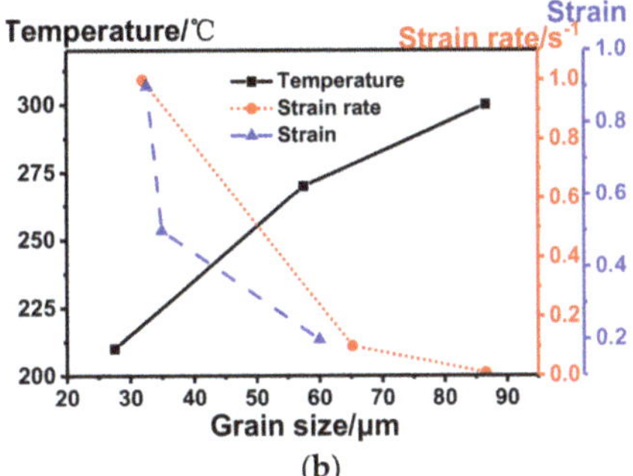

(a)

(b)

Figure 8. (**a**) Frequency distribution of misorientation and grain size under different conditions; (**b**) the relationship between grain size and temperature, strain rate and strain, respectively.

Table 3. Grain size deviation under different deformation conditions.

Deformation Parameters	Variables	Average Value of Grain Size, Expectation/μm	Coefficient of Variation(s/EX)	Misorientation Fraction(>15°)
240 °C-0.3 s^{-1}	0.2	60	0.72	92.9%
	0.5	35	0.57	94.9%
	0.9	33	0.55	93.4%
0.01 s^{-1}-0.9	210 °C	27	0.58	93%
	270 °C	58	0.91	95.5%
	300 °C	87	0.92	93.0%
300 °C-0.9	0.01 s^{-1}	87	0.92	93.0%
	0.1 s^{-1}	65	0.78	96.3%
	1 s^{-1}	32	0.73	93.5%

From Figure 7a–c and Table 3, the grain size is refined from 60 μm to 33 μm after SST, and the size deviation coefficient is becoming smaller as the amount of deformation increases. This is due to the fact that when the amount of deformation is small, only a few grains are deformed and the deformation distribution inside the metal is quite uneven, so the amount of grain nucleation is relatively less and the new grains can grow up quickly as a result of the different deviations in grain sizes. Thus, the coarse recrystallized grains are obtained. As the amount of deformation increases, more crystal defects of the alloy and more unstable substructures will be introduced, thus more storage energy will be obtained, which motivates the development of SRX and the increase of the nucleation rate during the SST process. Moreover, the refined and uniformly distributed precipitates can effectively hinder the atoms' diffusion and grain boundary migration, which can limit the growth of recrystallized grains. Therefore, the grain size decreases with the increase in strain. It is also found that the grain size under the strain of 0.5 and 0.9 is little changed. The main reason is that the smaller precipitation phases cannot become the core of the recrystallization, and the nucleation density cannot be increased with the continuous increase in the strain, so the grains cannot be further refined [55].

As shown in Figure 7d–f and Table 3, the average grain size increases gradually with the range of 27 μm to 87 μm, and the deviation coefficient increases from 0.58 to 0.92 with the increase in deforming temperature. This is related to the amount of storage energy. According to the previous discussion, the storage energy decreases with the increase in deforming temperature, so the nucleation rate of SRX decreases with the increase in deforming temperature during the SST process. Moreover, the amount of the precipitated phase decreases with the increase in deforming temperature, which will lead to the decrease in the nucleation rate and weaken the pinning effect. Therefore, the average grain size will increase with the increase in deforming temperature.

As shown in Figure 7f–h and Table 3, the recrystallized grain size decreases gradually from 87 μm to 32 μm with the increase in strain rate and the deviation coefficient is also reduced from 0.92 to 0.73. This is also related to the effect of storage energy and the precipitated phase. According to the previous discussion, the storage energy increases with the increase in strain rate, so the nucleation rate of SRX also increases with the increase in strain rate. In addition, although the nucleation rate is high at high strain rate, the distribution of the precipitated phase is even worse with the increase in the strain rate, which results in the inhomogeneity of the growth and the final uneven distribution of grain size.

Generally, the relationship between yield stress and average grain size of alloys can be characterized by the Hall–Petch formula [8,56]:

$$\sigma_s = \sigma_0 + k_y d^{-1/2} \tag{5}$$

where σ_s is the yield stress of the material, σ_0 and k_y are material constants, and d is the average grain size. It can be seen that yield stress increases with the decrease in grain size.

According to the study of Wert [57], the coefficient k_y in the formula is about 0.12, which means that the effect of recrystallized grain size on yield stress of 2219 aluminum alloy is very small. Thus, the yield stress change caused by grain refinement from 87 μm to 27 μm can be calculated as $\Delta\sigma_s = k_y \times \Delta d^{(-1/2)} = 0.01$ MPa, which indicates that the strengthening effect caused by grain refinement is negligible, and the precipitation strengthening is the main strengthening mechanism of 2219 aluminum alloy. This is consistent with the existing conclusions [58,59].

3.3.2. Effects on Precipitated Phase

The SEM micrograph, EDS analysis, and XRD analysis of the samples after SST are shown in Figure 9. It can be seen that the precipitated phase still exists after SST, and it is determined to be $CuAl_2$ phase by EDS and XRD analysis. This is because the content of Cu exceeds the solubility in Al at this temperature. In addition, the previously existing dispersed fine precipitates are all redissolved into the matrix to obtain the supersaturated solid solution for the subsequent aging hardening. However, the second phase remaining is mainly characterized by the form of polymerization, that is to say, large $CuAl_2$ lump precipitates cannot be redissolved in the matrix during the SST process, which will have a great impact on the final mechanical performance. Internal cracks can be easily introduced due to the high strength and brittleness of the precipitates polymerization in subsequent processing, thus resulting in defect damage of the component. Therefore, the condensation of $CuAl_2$ phase should be avoided and its distribution should be made dispersed in the process of casting and forging for 2219 aluminum alloy, otherwise, the large $CuAl_2$ lump precipitates will be inherited in the final products due to the inability of subsequent processing to break down its polymerization state.

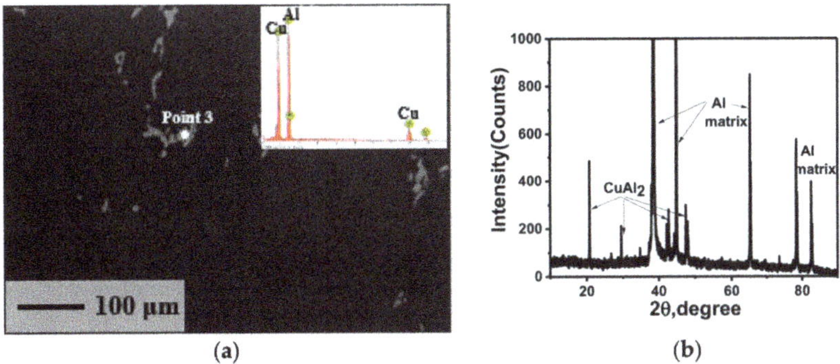

Figure 9. (a) The SEM micrograph of the solution-treated samples under the condition of 300 °C-0.01 s^{-1}-0.9 with EDS analysis of Point 3; (b) XRD analysis of the sample.

4. Conclusions

In this study, the approach of "WD + SST" was put forward to investigate the grain refinement of 2219 aluminum alloy, and the thermal compression tests and solid solution treatment experiments were adopted to simulate the process. The effects of deformation parameters on the law of microstructure evolution at various stages during the intermediate thermal-mechanical treatment process were also studied. The following conclusions can be drawn:

1. During the warm deformation process of 2219 aluminum alloy, the flow stress is very sensitive to temperature, strain rate, and strain. The storage energy is found to be proportional to the square of the flow stress, and it decreases with the increase in temperature and/or the decrease in strain rate, and it rises first and then keeps a relatively stable state with the increase in strain. In addition, the main softening mechanism is determined to be dynamic recovery. Under relatively high temperature (270 °C, 300 °C) and lower strain rate (0.01 s^{-1}), incomplete continuous dynamic recrystallization can also occur.

2. During the warm deformation process, the grain morphology changes and the substructure content increases. Moreover, the proportion of the low angle grain boundaries increases with the decrease in deforming temperature, the increase in strain rate, and/or the increase in strain. In addition, the distribution of $CuAl_2$ phase is more dispersed with the increase in deforming temperature, the decrease in strain rate, and/or the increase in strain. However, some $CuAl_2$ phase particles are still polymerized.

3. During the solid solution treatment process of 2219 aluminum alloy, complete static recrystallization occurred and substructure almost disappeared. The average grain size obtained decreased with the decrease in deforming temperature, the increase in strain rate, and/or the increase in strain. The grain refinement mechanism is related to the amount of storage energy and the distribution of precipitated particles in the whole process of intermediate thermal-mechanical treatment. The previously existing dispersed fine precipitates are all redissolved into the matrix, however, the precipitates remaining exist mainly by the form of polymerization.

4. According to the experimental results, the optimum deformation parameters for industrial processing of 2219 aluminum alloy are as follows: T < 240 °C, ε > 0.5, and $\dot{\varepsilon}$ > 1 s^{-1}, which can get better grain-refining effects at the same time.

Author Contributions: Conceptualization, L.L. and Y.W.; Methodology, L.L.; Formal Analysis, L.L. and H.G.; Writing-Original Draft Preparation, L.L.; Writing-Review & Editing, Y.W. and H.G.

Funding: The research was funded by the National Natural Science Foundation of China (Grant Number: 51405520, U1637601 and 51327902), the Fundamental Research Funds for the Central Universities of Central South

University (Grant Number: 502221801) and the Project of State Key Laboratory of High Performance Complex Manufacturing, Central South University (Grant Number: ZZYJKT2017-06).

Conflicts of Interest: The authors declare no conflict of interest.

References

1. An, L.H.; Cai, Y.; Liu, W.; Yuan, S.J.; Zhu, S.Q. Effect of pre-deformation on microstructure and mechanical properties of 2219 aluminum alloy sheet by thermomechanical treatment. *Trans. Nonferr. Met. Soc. China* **2012**, *22*, s370–s375. [CrossRef]

2. Venkatesh, B.D.; Chen, D.L.; Bhole, S.D. Effect of heat treatment on mechanical properties of Ti-6Al-4V ELI alloy. *Mater. Sci. Eng. A* **2009**, *506*, 117–124. [CrossRef]

3. Zhang, J.; Chen, B.; Zhang, B. Effect of initial microstructure on the hot compression deformation behavior of a 2219 aluminum alloy. *Mater. Des.* **2012**, *34*, 15–21. [CrossRef]

4. Qu, W.Q.; Song, M.Y.; Yao, J.S.; Zhao, H.Y. Effect of Temperature and Heat Treatment Status on the Ductile Fracture Toughness of 2219 Aluminum Alloy. *Mater. Sci. Forum* **2011**, *689*, 302–307. [CrossRef]

5. Xia, C.; Zhang, W.; Kang, Z.; Jia, Y.; Wu, Y. High strength and high electrical conductivity Cu-Cr system alloys manufactured by hot rolling–quenching process and thermomechanical treatments. *Mater. Sci. Eng. A* **2012**, *538*, 295–301. [CrossRef]

6. Song, R.; Ponge, D.; Raabe, D.; Speer, J.G.; Matlock, D.K. Overview of processing, microstructure and mechanical properties of ultrafine grained bcc steels. *Mater. Sci. Eng. A* **2006**, *441*, 1–17. [CrossRef]

7. Liu, S.; Zhong, Q.; Zhang, Y.; Liu, W.; Zhang, X. Investigation of quench sensitivity of high strength Al-Zn-Mg-Cu alloys by time-temperature-properties diagrams. *Mater. Des.* **2010**, *31*, 3116–3120. [CrossRef]

8. Petch, N.J. The Cleavage Strength of Polycrystals. *J. Iron. Steel. Inst.* **1953**, *174*, 25–28.

9. Moallemi, M.; Kermanpur, A.; Najafizadeh, A.; Rezaee, A.; Baghbadorani, H.S. Formation of nano/ultrafine grain structure in a 201 stainless steel through the repetitive martensite thermomechanical treatment. *Mater. Lett.* **2012**, *89*, 22–24. [CrossRef]

10. Waldman, J.; Sulinski, H.; Markus, H. The effect of ingot processing treatments on the grain size and properties of Al alloy 7075. *Metall. Trans.* **1974**, *5*, 573–584. [CrossRef]

11. Popov, N.N.; Prokoshkin, S.D.; Sidorkin, M.Y.; Sysoeva, T.I.; Borovkov, D.V. Effect of thermomechanical treatment on the structure and functional properties of a 45Ti-45Ni-10Nb alloy. *Russ. Metall.* **2007**, *2007*, 59–64. [CrossRef]

12. Elmay, W.; Prima, F.; Gloriant, T.; Bolle, B.; Zhong, Y. Effects of thermomechanical process on the microstructure and mechanical properties of a fully martensitic titanium-based biomedical alloy. *J. Mech. Behav. Biomed.* **2013**, *18*, 47–56. [CrossRef] [PubMed]

13. Kim, H.Y.; Ohmatsu, Y.; Kim, J.I.; Hosoda, H.; Miyazaki, S. Mechanical Properties and Shape Memory Behavior of Ti-Mo-Ga Alloys. *Mater. Sci. Eng. A* **2006**, *438–440*, 839–843. [CrossRef]

14. Russo, E.D.; Conserva, M.; Gatto, F.; Markus, H. Thermo-mechanical treatments on high strength Al-Zn-Mg(-Cu) alloys. *Metall. Trans.* **1973**, *4*, 1133–1144. [CrossRef]

15. Russo, E.D.; Conserva, M.; Buratti, M.; Gatto, F. A new thermo-mechanical procedure for improving the ductility and toughness of Al-Zn-Mg-Cu alloys in the transverse directions. *Mater. Sci. Eng.* **1974**, *14*, 23–36. [CrossRef]

16. Lademo, O.G.; Pedersen, K.O.; Berstad, T.; Furu, T.; Hopperstad, O.S. An experimental and numerical study on the formability of textured AlZnMg alloys. *Eur. J. Mech. A-Solid* **2008**, *27*, 116–140. [CrossRef]

17. Tajally, M.; Emadoddin, E. Mechanical and anisotropic behaviors of 7075 aluminum alloy sheets. *Mater. Des.* **2011**, *32*, 1594–1599. [CrossRef]

18. Waldman, J.; Sulinski, H.V.; Markus, H. Processes for the Fabrication of 7000 Series Aluminum Alloys. U.S. Patent 3,847,681, 12 November 1974.

19. Ward, B.R.; Agrawal, S.P.; Ashton, R.F. Method of Producing Superplastic Aluminum Sheet. U.S. Patent 4,486,244, 4 December 1984.

20. Wert, J.A.; Paton, N.E.; Hamilton, C.H.; Mahoney, M.W. Grain refinement in 7075 aluminum by thermomechanical processing. *Metall. Trans. A* **1981**, *12*, 1267–1276. [CrossRef]

21. Vandermeer, R.A. Microstructural descriptors and the effects of nuclei clustering on recrystallization path kinetics. *Acta Mater.* **2005**, *53*, 1449–1457. [CrossRef]

22. Poole, W.J.; Militzer, M.; Wells, M.A. Modelling recovery and recrystallization during annealing of AA 5754 aluminum alloy. *Met. Sci. J.* **2003**, *19*, 1361–1368.

23. Primig, S.; Leitner, H.; Knabl, W.; Lorich, A.; Clemens, H. Influence of the heating rate on the recrystallization behavior of molybdenum. *Mater. Sci. Eng. A* **2012**, *535*, 316–324. [CrossRef]

24. Yoshida, H. Effect of pre-heat treatment on the superplasticity of a 7475 alloy sheet. *J. Jpn. Inst. Light. Met.* **1991**, *41*, 446–452. [CrossRef]

25. Kaibyshev, R.; Sakai, T.; Musin, F.; Nikulin, I.; Miura, H. Superplastic behavior of a 7055 aluminum alloy. *Scripta Mater.* **2001**, *45*, 1373–1380. [CrossRef]

26. Binlung, O.U.; Yang, J.; Yang, C. Effects of Step-Quench and Aging on Mechanical Properties and Resistance to Stress Corrosion Cracking of 7050 Aluminum Alloy. *Mater. T. Jim.* **2007**, *41*, 783–789.

27. Tsai, T.C.; Chuang, T.H. Role of grain size on the stress corrosion cracking of 7475 aluminum alloys. *Mater. Sci. Eng. A* **1997**, *225*, 135–144. [CrossRef]

28. Zuo, J.; Hou, L.; Shi, J.; Cui, H.; Zhuang, L. The mechanism of grain refinement and plasticity enhancement by an improved thermomechanical treatment of 7055 Al alloy. *Mater. Sci. Eng. A* **2017**, *702*, 42–52. [CrossRef]

29. Ye, L.Y.; Zhang, X.M.; Liu, Y.W.; Tang, J.G.; Zheng, D.W. Effect of two-step aging on recrystallized grain size of Al-Mg-Li alloy. *Mater. Sci. Technol.* **2014**, *27*, 125–131. [CrossRef]

30. Troeger, L.P.; Jr, E.A.S. Microstructural and mechanical characterization of a super-plastic 6xxx aluminum alloy. *Mater. Sci. Eng. A* **2000**, *277*, 102–113. [CrossRef]

31. Esmaeili, S.; Lloyd, D.J.; Jin, H. A thermomechanical process for grain refinement in precipitation hardening AA6xxx aluminum alloys. *Mater. Lett.* **2011**, *65*, 1028–1030. [CrossRef]

32. Aboutalebi, M.R.; Karimzadeh, M.; Salehi, M.T.; Abbaasi, S.M.; Morakabati, M. Influences of Aging and Thermomechanical Treatments on the Martensitic Transformation and Superelasticity of Highly Ni-rich Ti-51.5 at.% Ni Shape Memory Alloy. *Thermochim. Acta* **2015**, *616*, 14–19. [CrossRef]

33. Ozan, S.; Lin, J.; Li, Y.; Zhang, Y.; Munir, K. Deformation mechanism and mechanical properties of a thermomechanically processed β Ti-28Nb-35.4Zr alloy. *J. Mech. Behav. Biomed.* **2018**, *78*, 224–234. [CrossRef] [PubMed]

34. Saastamoinen, A.; Kaijalainen, A.; Porter, D.; Suikkanen, P. The effect of thermomechanical treatment and tempering on the subsurface microstructure and bendability of direct-quenched low-carbon strip steel. *Mater. Charact.* **2017**, *134*, 172–181. [CrossRef]

35. Misra, R.D.K.; Kumar, B.R.; Somani, M.; Karjalainen, P. Deformation processes during tensile straining of ultrafine/nanograined structures formed by reversion in metastable austenitic steels. *Scripta Mater.* **2008**, *59*, 79–82. [CrossRef]

36. Eskandari, M.; Kermanpur, A.; Najafizadeh, A. Formation of Nanocrystalline Structure in 301 Stainless Steel Produced by Martensite Treatment. *Metall. Mater. Trans. A* **2009**, *40*, 2241–2249. [CrossRef]

37. Forouzan, F.; Najafizadeh, A.; Kermanpur, A. Production of nano/submicron grained AISI 304L stainless steel through the martensite reversion process. *Mater. Sci. Eng. A* **2010**, *527*, 7334–7339. [CrossRef]

38. Chen, F.; Cui, Z.; Chen, S. Recrystallization of 30Cr2Ni4MoV Ultra-super-critical Rotor Steel during Dot Deformation Part 3 Metadynamic Recrystallization. *Mater. Sci. Eng. A* **2011**, *528*, 5073–5080. [CrossRef]

39. Lin, Y.C.; Chen, M.S.; Zhong, J. Microstructural evolution in 42CrMo steel during compression at elevated temperatures. *Mater. Lett.* **2008**, *62*, 2132–2135. [CrossRef]

40. Momeni, A.; Dehghani, K. Prediction of dynamic recrystallization kinetics and grain size for 410 martensitic stainless steel during hot deformation. *Met. Mater. Int.* **2010**, *16*, 843–849. [CrossRef]

41. Lin, Y.C.; Chen, X.M.; Wen, D.X.; Chen, M.S. A physically-based constitutive model for a typical nickel-based superalloy. *Comp. Mater. Sci.* **2014**, *83*, 282–289.

42. Khelfaoui, F.; Guénin, G. Influence of the recovery and recrystallization processes on the martensitic transformation of cold worked equiatomic Ti–Ni alloy. *Mater. Sci. Eng. A* **2003**, *355*, 292–298. [CrossRef]

43. Honeycombe, R.W.K.; Bhadeshia, H. *Pobert William Kerr. Steels: Microstructure and Properties*; Edward Arnold: London, UK, 1981.

44. Peng, X.N.; Guo, H.Z.; Shi, Z.F.; Qin, C.; Zhao, Z.L. Microstructure characterization and mechanical properties of TC4-DT titanium alloy after thermomechanical treatment. *T. Nonferr. Metal. Soc.* **2014**, *24*, 682–689. [CrossRef]

45. Wert, J.A.; Austin, L.K. Modeling of thermo-mechanical processing of heat-treatable aluminum alloys. *Metall. Trans. A* **1988**, *19*, 617–625. [CrossRef]

46. Nie, W.; Shang, C. Microstructure and toughness of the simulated welding heat affected zone in X100 pipeline steel with high deformation resistance. *Acta Metall. Sin.* **2012**, *48*, 797–806. [CrossRef]

47. Sha, G.; Wang, Y.B.; Liao, X.Z.; Duan, Z.C.; Ringer, S.P. Influence of equal-channel angular pressing on precipitation in an Al-Zn-Mg-Cu alloy. *Acta Mater.* **2009**, *57*, 3123–3132. [CrossRef]

48. Deschamps, A.; Fribourg, G.; Brechet, Y.; Chemin, J.L.; Hutchinson, C.R. In situ evaluation of dynamic precipitation during plastic straining of an Al-Zn-Mg-Cu alloy. *Acta Mater.* **2012**, *60*, 1905–1916. [CrossRef]

49. Lang, Y.; Zhou, G.; Hou, L.; Zhang, J.; Zhuang, L. Significantly enhanced the ductility of the fine-grained Al-Zn-Mg-Cu alloy by strain-induced precipitation. *Mater. Des.* **2015**, *88*, 625–631. [CrossRef]

50. Kita, K.; Kobayashi, K.; Monzen, R. Dispersions of Particles and Heat Resistance in a Cu-Fe-P Alloy (Special Issue on Composite Materials). *J. Soc. Mater. Sci.* **2000**, *49*, 482–487. [CrossRef]

51. Huo, W.; Hou, L.; Lang, Y.; Cui, H.; Zhuang, L. Improved thermo-mechanical processing for effective grain refinement of high-strength AA 7050 Al alloy. *Mater. Sci. Eng. A* **2015**, *626*, 86–93. [CrossRef]

52. Zuo, J.; Hou, L.; Shi, J.; Cui, H.; Zhuang, L. Precipitates and the evolution of grain structures during double-step rolling of high-strength aluminum alloy and related properties. *Acta Metall. Sin.* **2016**, *52*, 1105–1114.

53. Yang, H.S.; Mukherjee, A.K.; Roberts, W.T. Effect of cooling rate in over-ageing and other thermomechanical process parameters on grain refinement in an AA7475 aluminum alloy. *J. Mater. Sci.* **1992**, *27*, 2515–2524. [CrossRef]

54. Hall, E.O. The Deformation and Ageing of Mild Steel: II Characteristics of the Lüders Deformation. *Proc. Phys. Soc.* **1951**, *64*, 747–753. [CrossRef]

55. Wert, J.A. *Strength of Metals and Alloys*; Pergamon Press: Oxford, UK, 1980.

56. Ma, K.; Wen, H.; Hu, T.; Topping, T.D.; Isheim, D. Mechanical behavior and strengthening mechanisms in ultrafine grain precipitation-strengthened aluminum alloy. *Acta Mater.* **2014**, *62*, 141–155. [CrossRef]

57. Yang, Y.; Yu, K.; Li, Y.; Zhao, D.; Liu, X. Evolution of nickel-rich phases in Al-Si-Cu-Ni-Mg piston alloys with different Cu additions. *Mater. Des.* **2012**, *33*, 220–225. [CrossRef]

58. Gourdet, S.; Montheillet, F. An experimental study of the recrystallization mechanism during hot deformation of aluminum. *Mater. Sci. Eng. A* **2000**, *283*, 274–288. [CrossRef]

59. Sitdikov, O.; Sakai, T.; Miura, H.; Hama, C. Temperature effect on fine-grained structure formation in high-strength Al alloy 7475 during hot severe deformation. *Mater. Sci. Eng. A* **2009**, *516*, 180–188. [CrossRef]

Article

A Physically Based Constitutive Model and Continuous Dynamic Recrystallization Behavior Analysis of 2219 Aluminum Alloy during Hot Deformation Process

Lei Liu [1], Yunxin Wu [2,*], Hai Gong [2], Shuang Li [2] and A. S. Ahmad [2]

[1] Light Alloy Research Institute, Central South University, Changsha 410083, China; Larill@csu.edu.cn
[2] State Key Laboratory of High Performance Complex Manufacturing, Central South University, Changsha 410083, China; csu2019@163.com (H.G.); cmeell@163.com (S.L.); csu_cmee@163.com (A.S.A.)
* Correspondence: csu1991@163.com; Tel.: +86-188-9035-5864

Received: 3 August 2018; Accepted: 13 August 2018; Published: 15 August 2018

Abstract: The isothermal compression tests of the 2219 Al alloy were conducted at the temperature and the strain rate ranges of 623–773 K and 0.01–10 s^{-1}, respectively, and the deformed microstructures were observed. The flow curves of the 2219 Al alloy obtained show that flow stress decreases with the increase in temperature and/or the decrease in strain rate. The physically based constitutive model is applied to describe the flow behavior during hot deformation. In this model, Young's modulus and lattice diffusion coefficient are temperature-dependent, and the creep exponent is regarded as a variable. The predicted values calculated by the constitutive model are in good agreement with the experimental results. In addition, it is confirmed that the main softening mechanism of the 2219 Al alloy during hot deformation is dynamic recovery and incomplete continuous dynamic recrystallization (CDRX) by the analysis of electron backscattered diffraction (EBSD) micrographs. Moreover, CDRX can readily occur under the condition of high temperatures, low strain rates, and large strains. Meanwhile, the recrystallization grain size will also be larger.

Keywords: 2219 aluminum alloy; constitutive model; microstructural evolution; continuous dynamic recrystallization; hot deformation

1. Introduction

The 2219 Al alloy has long been used in the manufacture of various aerospace components (i.e., oxidizer and fuel tanks) due to its high strength, high fracture toughness, and reliable weldability [1]. During the formative process of the components, the hot working method is of great significance in obtaining the specific shapes and required properties [2]. Moreover, the characterization of flow behaviors in the hot working process largely determines the accuracy of the simulation by finite element method (FEM) [3]. It has been confirmed that the main softening mechanism in hot deformation process involves dynamic recrystallization (DRX) and dynamic recovery (DRV) [4], but the microstructural evolution is often very difficult to illustrate for its complexity [5–7]. Therefore, it is extremely important and necessary to investigate both of the flow behaviors and microstructural evolution on 2219 Al alloy during hot deformation [8–10].

Over the last decades, many efforts [11–19] have been made on the establishment of constitutive models to characterize flow behaviors accurately for various metals and alloys. A mathematical relationship between flow stress and deformation parameters (i.e., temperature, strain rate, strain) is usually used to represent the constitutive model [2]. Firstly, the constitutive model was constructed on the classical Hollomon equation [11] and Ludwik equation [12]. Later, Johnson and Cook [13] proposed the famous Johnson-Cook model, in which work hardening, strain rate, and temperature are taken into account. In addition, Sellars and McTegart [14] proposed another widely applied constitutive model by relating the temperature-compensated strain-rate parameter (i.e., Z parameter) to the flow stress. However, the constitutive model is obviously phenomenological and all the material parameters do not have obvious physical meaning. Recently, according to Mirzadeh et al. [15], a physically based constitutive model was proposed when the glide and climb of dislocations is the main deformation mechanism, in which only two material parameters remain and both of them have physical and metallurgical meaning. To characterize the whole flow curve under different strains, the material parameters are used to be expressed as functions of strain (i.e., strain compensation). This method has been successfully applied for steels [16,17], magnesium alloys [18], commercial-purity aluminum [9], and AA2030 aluminum alloy [19].

During the hot deformation process, DRX often takes place for most metals and alloys. Moreover, different kinds of thermomechanical processing (TMP) may lead to different DRX phenomena, such as discontinuous DRX (DDRX), continuous DRX (CDRX), and geometric DRX (GDRX) [20]. In order to recognize the differences of the three DRX types, many efforts [21–37] have been made on the observation and revelation of microstructural evolution during the hot forming process. DDRX usually occurs with new strain-free grains nucleation and the grains growth in low-stacking-fault energy (SFE) metals and alloys [21]. The investigation of DDRX behavior mainly focuses on the mechanism of nucleation and numerical models, which has been widely applied in different metals and alloys, such as 304 stainless steel [22,23], as-extruded 7075 aluminum alloy [24], and 3Cr2NiMnMo steel [25]. Recently, based on the published researches by Perdrix [26] and Montheillet et al. [27,28], CDRX occurs with high-angle grain boundaries (HAGBs) formed by the progressive lattice rotation of subgrains near grain boundaries during hot deformation. This mechanism has been verified in Mg alloys [29] and Al–Zn alloys [30]. Another transformation mechanism of low-angle grain boundaries (LAGBs) into HAGBs is by the progressive misorientation angles increases. This has also been reported in steel [31] and Al alloys [32]. Based on the mechanisms above, two numerical models of CDRX, the Gourdet-Montheillet model [33] and Toth et al. model [34], are proposed to characterize this process. Apart from DDRX and CDRX, GDRX is a field with less study, but is still observed in some metals and alloys during a large amount of hot deformation, such as commercial-purity Al [35] and Mg–Zn–Zr alloys [36,37]. However, the microstructural evolution with constitutive analysis for 2219 Al alloy under the hot deformation process is rarely reported.

In this study, the flow behavior of the 2219 Al alloy was studied by isothermal compression test, and an effective constitutive model based on physical mechanism was developed. The microstructure evolution during hot deformation was also investigated. Finally, the influence of different TMP conditions on CDRX behavior was discussed.

2. Materials and Methods

Isothermal-compression tests were conducted to investigate the flow behaviors and microstructural evolution of aluminum 2219 alloy during hot deformation. The chemical compositions (wt.%) of 2219 Al alloy used in this experiment are shown in Table 1.

Table 1. The chemical compositions of 2219 Al alloy used in this test (wt.%).

Si	Fe	Cu	Mn	Mg	Zn	V	Ti	Zr	Al
0.20	0.30	5.8~6.8	0.20~0.40	0.02	0.10	0.05~0.15	0.02~0.10	0.10~0.25	Bal

The samples were cut from wrought billet and machined into a cylindrical shape with a diameter of 10 mm and height of 15 mm by a wire-cutting electrodischarge machine (WEDM, DK7745 by Zhengda (Zhengda Science and Technology Co., Limited, Taizhou, China)). Then, the samples were solution treated at 813 K for 2 h in an electric resistance furnace (SG-XS1200 by Sager (Sager Furnace

Co., Limited, Shanghai, China)), followed by water quenching. Hot compression tests were conducted using a Gleeble-3500 thermosimulator (Dynamic Systems Inc., New York, NY, USA.). According to the traditionally used forming temperature for hot working and actual industrial processing, the deforming temperature: $T = (0.7–0.85) T_m$ was adopted in this paper. Four different compression temperatures (773 K, 723 K, 673 K, 623 K) and four different strain rates (0.01 s^{-1}, 0.1 s^{-1}, 1 s^{-1}, 10 s^{-1}) were applied in the tests; the final true strain is up to 0.9. Each temperature was tested only once under four different strain rates, respectively. Tantalum foil and boron nitride were used to reduce the friction between anvils and the specimen. Each sample was heated to the specified forming temperature at a heating rate of 5 K/s, and then held for 3 min to eliminate the temperature gradient. The true stress-strain data were automatically collected from the computer system during the compression process. After the tests, all the samples were immediately water quenched to preserve the deformed microstructure. Then, the compressed samples were sectioned parallel to the compression axis for subsequent microstructural analysis by the electron backscattered diffraction (EBSD) technique. After the samples were grinded, they were then electropolished in a solution of 30% nitric acid and 70% methanol at 18 V for 50 s. HKL Channel5 software was used to analyze the EBSD data. In all the EBSD images, the LAGBs (orientation angle: 2°–15°) were marked by thin white lines, and the HAGBs (orientation angle: >15°) were marked by the thick black lines. Figure 1 shows the EBSD microstructure of 2219 Al alloy before compression. Coarse equiaxed recrystallized grains and deformed grains with many substructures were both detected.

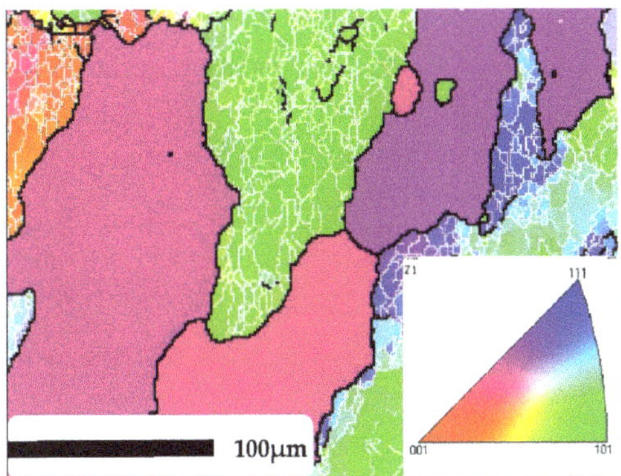

Figure 1. The electron backscattered diffraction (EBSD) micrograph of 2219 Al alloy before deformation.

3. Results and Discussion

3.1. Flow Behaviors

The true stress-strain curves of the 2219 Al alloy at different temperatures and strain rates are illustrated in Figure 2. Although lubricants were used to reduce friction and isothermal conditions were controlled by the equipment itself, the temperature rise caused by deformation heat and small friction existed cannot be ignored. According to the friction correction criterion proposed by Roebuck [38], the range of B (i.e., barreling coefficient) obtained in this experiment was: 1.02 < B < 1.09, which means the measured stresses didn't need to be corrected. In addition, the deformation heat produced as a result of a high strain rate would have a great influence on the flow stress [39], therefore the flow curves under the strain rates of 1 s^{-1} and 10 s^{-1} were modified by temperature compensation according to Humphreys F.J [40]. It can be clearly seen from Figure 2 that the flow stress increased rapidly to a peak value and then slightly fell to a relatively steady value, which shows an obvious characteristic of a single-peak type. Thus, each flow stress curve can be divided into three parts by strains: work-hardening part, softening part, and steady part [41,42]. The mechanism of each part is mainly controlled by the competition between work hardening and DRV and/or DRX during the hot deformation process [43]. For the work-hardening part, the rapid accumulation and proliferation of dislocations leads to obvious work hardening, but DRV develops too slowly. The synthesis result is the rapid increase of flow stress at this stage. For the softening part, high dislocation density largely promotes the development of DRV and the formation of DRX, which has exceeded the effect of work hardening. That is to say, the flow stress will slightly decrease. For the steady part, the multiplication and annihilation of dislocations approximately reached an offsetting state due to the effect of continuous deformation and DRV and/or DRX, respectively, thus, the flow stresses almost remained relatively stable values.

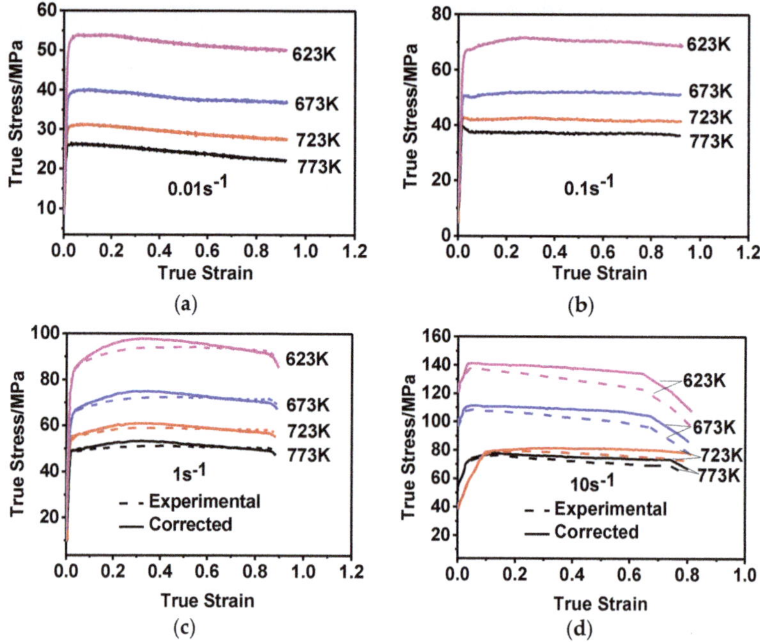

Figure 2. The true strain-stress curves obtained at various strain rates and temperatures of (**a**) 0.01 s^{-1}; (**b**) 0.1 s^{-1}; (**c**) 1 s^{-1}; (**d**) 10 s^{-1}.

In addition, it is observed from Figure 2 that flow stress exhibited strong dependence on temperature and strain rate. Moreover, flow stress increased with the increase in strain rate and decrease in temperature. This phenomenon indicates that the effect of work hardening is more significant at low temperatures and/or high strain rates. Conversely, DRV and/or DRX developed readily at high temperatures and/or lower strain rates, such as 773 K/0.01 s^{-1}, 773 K/0.1 s^{-1}, 723 K/0.01 s^{-1}. This is because hot compression is a thermal-activation process [5]. Therefore, there will be a larger driving force for the development of DRV and/or DRX due to the easier movement and migration of dislocation and grain boundary [44] at higher temperatures. On the other hand, it will develop more sufficiently to consume more dislocation density at lower strain rate. Thus, either of higher forming temperature and lower strain rate can promote the development of DRV and/or DRX, and finally result in the decrease of the flow stress.

3.2. Physically Based Constitutive Modeling

The activation energy means the energy required to overcome the barriers during the deformation process [45]. However, the value of activation energy on 2219 Al alloy (133 KJ/mol) calculated by Arrhenius model exhibits obvious deviation from that of the self-diffusion activation energy on Al alloy (142 KJ/mol) [46]. Some studies [15,46–49] indicate that the reason for the deviation may be the dependence of Young's modulus and self-diffusion coefficient on temperature. Therefore, a physically constitutive model was proposed by considering Young's modulus (E) and self-diffusion coefficient (D) as functions of temperature. Thus, the mathematical relationship is expressed as follows [47]:

$$\dot{\varepsilon}/D(T) = B[sinh(\alpha\sigma/E(T))]^5, \tag{1}$$

where, $\dot{\varepsilon}$ (s^{-1}) and σ (MPa) are the strain rate and the flow stress, respectively, B and α are the material constants, and the index 5 represents the ideal value of the creep exponent. However, microstructural evolution (i.e., DRX) generally has a certain influence on the value of creep exponent, so it is reasonable to be considered as a variable (i.e., n) [39].

Therefore, Equation (1) can be rewritten as:

$$\dot{\varepsilon}/D(T) = B[sinh(\alpha\sigma/E(T))]^n. \tag{2}$$

The relationships between self-diffusion activation energy and temperature, Young's modulus and temperature can be expressed as [15,47–49]:

$$D(T) = D_0 exp(-Q_{sd}/RT), \tag{3}$$

$$E(T) = 2G(1+v), \tag{4}$$

where $D(T)$ is the lattice diffusion coefficient of aluminum alloy, and D_0 is the pre-exponent coefficient of the lattice diffusion (1.7×10^{-4} m^2/s), Q_{sd} is the activation energy of lattice diffusion (142 KJ/mol), R is the universal gas constant (8.31 J·mol^{-1} K^{-1}), and T is the forming temperature (K). v is Poisson's ratio (0.33), and G is the shear modulus; its relation with temperature can be expressed as [46]:

$$G/G_0 = 1 + [\eta(T - 300)/T_M], \tag{5}$$

where G_0 is the shear modulus at 300 K (2.54×10^4 MPa), T_M is the melting temperature of 2219 Al alloy (916 K), and η indicates the temperature dependence of shear modulus (–0.50). According to the parameters above, $D(T)$ and $E(T)$ can be expressed as follows:

$$D(T) = 1.7 \times 10^{-4} exp(-142000/8.31T), \tag{6}$$

$$E(T) = 6.7564 \times 10^4 [1 - 0.5(T - 300/916)]. \tag{7}$$

In order to obtain the three unknown parameters (α, n, B) in Equation (2), the integrated physically based constitutive equations of the power, exponential and hyperbolic sine laws are summarized and expressed as follows [2]:

$$\dot{\varepsilon}/D(T) = \begin{cases} B_1[\sigma/E(T)]^{n_1} \\ B_2 exp[\beta\sigma/E(T)] \\ B[sinh(\alpha\sigma/E(T))]^n \end{cases}, \tag{8}$$

where B_1, B_2, n_1 are material parameters and $\alpha = \beta/n_1$.

Based on the power and exponential functions of Equation (8), the material parameters n_1 and β can be obtained from the gradient of the graph of $\ln[\dot{\varepsilon}/D(T)]$ against $\ln[\sigma/E(T)]$ and $\ln[\dot{\varepsilon}/D(T)]$ against $\sigma/E(T)$. These plots are shown in Figure 3 (i.e., taking the strain of 0.5 as an example).

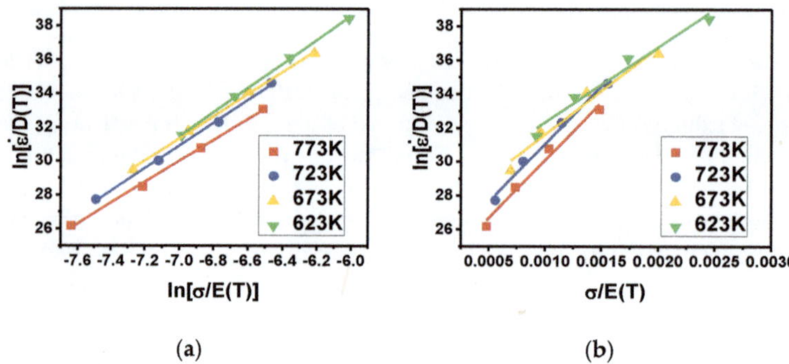

Figure 3. Relationships between material parameters: (**a**) $\ln\left[\dot{\varepsilon}/D(T)\right]$ against $\ln[\sigma/E(T)]$; (**b**) $\ln\left[\dot{\varepsilon}/D(T)\right]$ against $\sigma/E(T)$.

Therefore, the parameter α can be obtained by evaluating $\alpha = \beta/n_1$ at different temperatures and strains; the values are shown in Figure 4a. It can be clearly seen that α is dependent on both temperature and strain and increases with the increase in temperature and/or strain. So α can be expressed as a function of temperature and strain, and its 3D nonlinear surface fitting (Parabola 2D Method) is shown in Figure 4b. Generally, the accuracy of the model's fitting can be evaluated in terms of correlation coefficient (R^2), which is expressed as follows [39]:

$$R^2 = \frac{\sum_{i=1}^{N}(E_i - E)(P_i - P)}{\sqrt{\sum_{i=1}^{N}(E_i - E)^2 \sum_{i=1}^{N}(P_i - P)^2}},\tag{9}$$

where E_i and P_i are the experimental and predicted values of each point, respectively, E and P are the average experimental and the predicted values, and N is the total number of the data sample used. According to the data analysis, α has very good fitting with correlation coefficient $R^2 = 0.972$; therefore, α can be mathematically expressed as follows:

$$\alpha(\varepsilon, T) = -5088.79315 - 108.92456\varepsilon + 13.83184T + 319.01959\varepsilon^2 - 0.00757T^2.\tag{10}$$

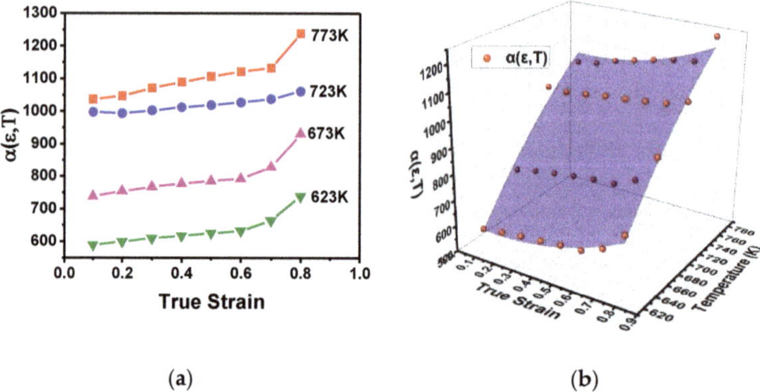

Figure 4. The variation of α: (**a**) at different strains and temperatures; (**b**) 3D illustration and its nonlinear surface fitting.

According to the hyperbolic sine function of Equation (8), the material parameters n and $\ln B$ can be obtained from the slope and intercept of the graph of $\ln\left[\dot{\varepsilon}/D(T)\right]$ against $\ln\{sinh[\alpha\sigma/E(T)]\}$. The plot under the strain of 0.5 is shown in Figure 5. It can be clearly seen that the fitting lines at different temperatures have an approximate slope (n), but the intercepts are completely different. That means n is only a function of strain, while B is a function of temperature and strain.

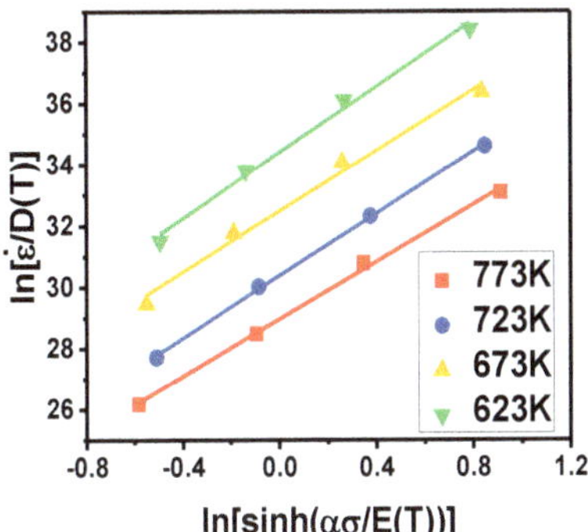

Figure 5. Relationship between $\ln\left[\dot{\varepsilon}/D(T)\right]$ against $\ln\{sinh[\alpha\sigma/E(T)]\}$ for different parameters.

The variation of $\ln B$ under different temperatures and strains is represented in Figure 6a, and the 3D nonlinear surface fitting (Poly 2D) is shown in Figure 6b. According to the data analysis, the correlation coefficient of $\ln B$ is $R^2 = 0.992$; therefore, $\ln B$ is expressed mathematically as follows:

$$\ln B(\varepsilon, T) = 82.33815 - 4.1088\varepsilon - 0.10644T - 0.93899\varepsilon^2 + 4.8375 \times 10^{-5}T^2 + 0.00588\varepsilon T. \tag{11}$$

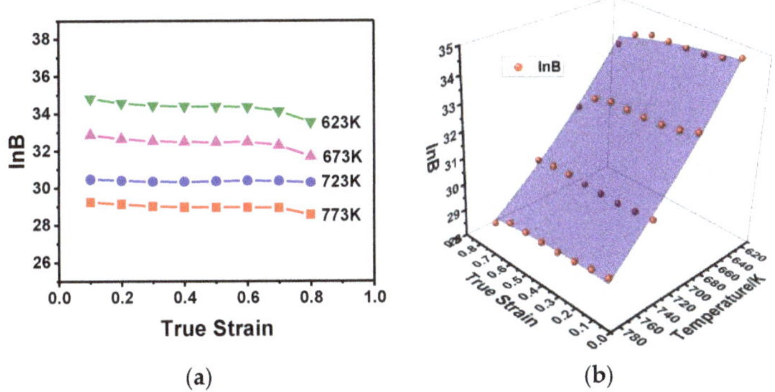

Figure 6. The variation of $\ln B$: (**a**) at different strains and temperatures; (**b**) 3D illustration and its nonlinear surface fitting.

The values of n at the different strains are plotted in Figure 7 and fitted by a 5th order polynomial. It can be seen that n is close to 5 and the slight deviation can be attributed to the microstructural evolution. As analyzed from the data, correlation coefficient of n is $R^2 = 0.999$. Therefore, n is expressed mathematically as follows:

$$n(\varepsilon) = 5.18187 + 1.83318\varepsilon - 17.24749\varepsilon^2 + 52.16674\varepsilon^3 - 72.65807\varepsilon^4 + 39.98397\varepsilon^5. \tag{12}$$

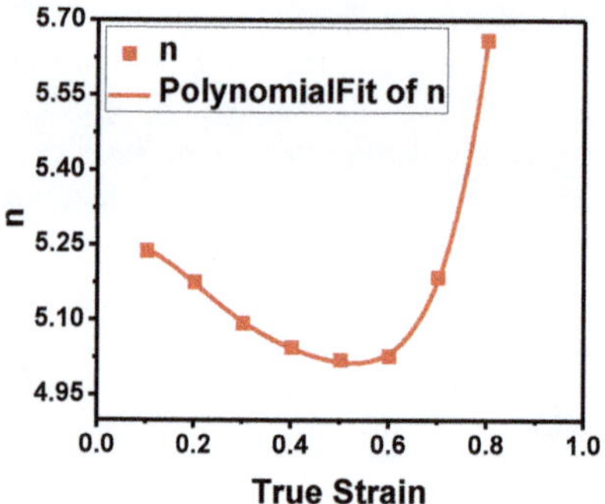

Figure 7. Relationship between true strain and n by polynomial fitting.

The physically based constitutive model of 2219 Al alloy was established based on the governing equations above; from the integrated expression, shown in Equation (13), the flow stress of 2219 Al alloy in the temperature range of 623 K to 773 K, strain rate range of 0.01 s^{-1} to 10^{-1}, and the strain range of 0~0.8 can be predicted.

$$\begin{cases} \sigma = E(T)/\alpha(\varepsilon,T)\text{arcsinh}\{exp[\ln[\dot{\varepsilon}/D(T)] - lnB(\varepsilon,T)]/n(\varepsilon)\} \\ D(T) = 1.7 \times 10^{-4}exp(-142000/8.31T) \\ E(T) = 6.7564 \times 10^{4}[1 - 0.5(T - 300)/916] \\ lnB(\varepsilon,T) = 82.33815 - 4.1088\varepsilon - 0.10644T - 0.93899\varepsilon^2 + 4.8375 \times 10^{-5}T^2 + 0.00588\varepsilon T \\ \alpha(\varepsilon,T) = -5088.79315 - 108.92456\varepsilon + 13.83184T + 319.01959\varepsilon^2 - 0.00757T^2 \\ n(\varepsilon) = 5.18187 + 1.83318\varepsilon - 17.24749\varepsilon^2 + 52.16674\varepsilon^3 - 72.65807\varepsilon^4 + 39.98397\varepsilon^5 \end{cases} \quad (13)$$

3.3. Verification of the Model

In order to verify the physically based constitutive model, the experimental and predicted stresses under different deformation conditions are compared, as shown in Figure 8. It is obvious that the predicted flow stresses obtained from the physically based model are in good agreement with the experimental flow stresses for both modeling and prediction sets. To quantify the accuracy of the model, the correlation coefficient (R^2) and average absolute relative error (AARE) were determined; the AARE is expressed as follows [39]:

$$AARE(\%) = 1/N \sum_{i=1}^{N}|(E_i - P_i)/E_i| \times 100\%, \quad (14)$$

where E_i and P_i are the experimental and predicted values of each point, and N is the total number of the data sample used.

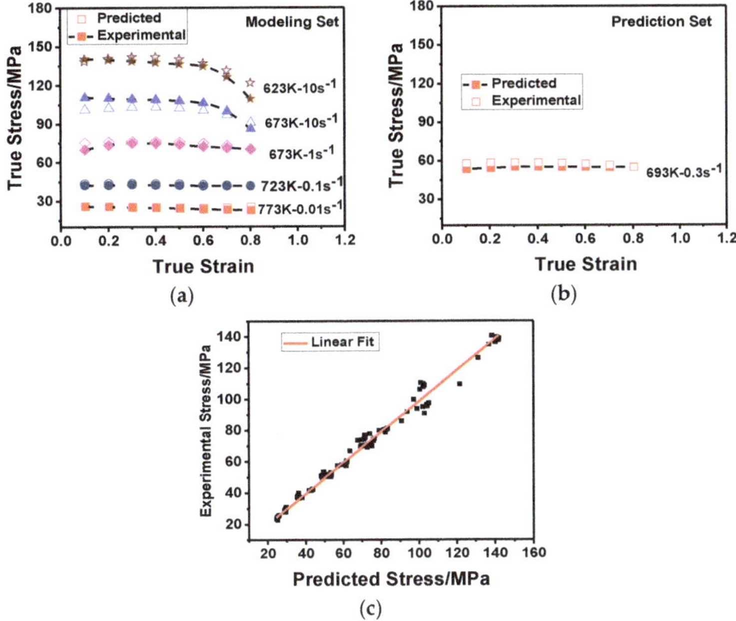

Figure 8. Correlations between the experimental and predicted flow stresses of the physically based constitutive model: (**a**) under modeling set; (**b**) under prediction set; (**c**) by linear fit.

The correlation between the experimental and predicted flow stresses is $R^2 = 0.985$ and the AARE = 3.88%. Conclusively, the physically based constitutive model can be used to predict the flow stress of 2219 Al alloy during hot deformation with high accuracy.

3.4. Microstructural Evolution

3.4.1. The Formation of CDRX

Microstructure observation was applied to investigate the formation and development of DRV and DRX during hot deformation [4]. Figure 9 shows the EBSD micrograph of the 2219 Al alloy after compression at a temperature of 673 K and strain rate of 10 s⁻¹. Comparing with the samples before compression (Figure 1), it is observed that the grains were flattened in the direction of compression and a large amount of substructures with LAGBs were generated in the original grains. The deformed microstructure showed obvious characteristics of DRV. The SFE of the metals and alloys had a significant effect on the softening mechanisms (DRV and/or DRX). For a 2219 Al alloy with high SFE, it is more difficult to dissociate the perfect dislocation into two partials, but perfect dislocation glide, climb, and cross slip can occur easily. During hot deformation, rapid DRV takes place readily, and the stored energy is decreased by rearrangement and annihilation of dislocations, which generally retards the development of DDRX. However, it can be seen from Figure 9a that many fine grains with a size less than 10 μm appeared near the grain boundaries and a few incomplete fine grains (marked with black dotted circles) composed of HAGBs; partial LAGBs were also observed. It is reasonable to deduce that these incomplete fine grains will develop to whole recrystallized grains with continued deformation; that is to say, these partial LAGBs will transform into HAGBs [50]. Therefore, the special formation mechanism of new recrystallized grains is proved to be CDRX rather than the traditional DDRX on 2219 Al alloy during hot deformation [51,52].

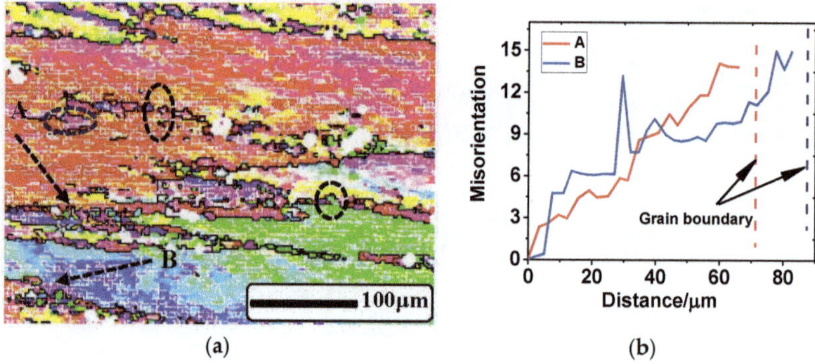

(a) (b)

Figure 9. EBSD measurement showing continuous dynamic recrystallization (CDRX) in 2219Al alloy. (a) Micrograph of the deformed sample under the condition of 673 K and 10 s^{-1}; (b) the cumulative misorientation along the dotted line from A and B to the grain boundary.

Figure 9b exhibits the cumulative misorientation of the two dotted lines (shown in Figure 9a) from the interior of the grain (i.e., A and B) to the grain boundaries, respectively. An obvious increase of misorientation from the interior to the boundary was observed. Thus, as deformation continued, the misorientation of LAGBs near a grain boundary may gradually increase and the LAGBs will transform into HAGBs once the misorientation reached the critical value (15°) [53]. The development mechanism of substructures with LAGBs near the grain boundaries can be ascribed to the progressive lattice rotation [54]. Grain boundary sliding (GBS) was more likely to occur at specifically oriented boundaries, and then these grain boundaries got serrated due to the migration of the HAGBs as shown in Figure 9a. The small serrations on partial boundaries may be eliminated by GBS, but the large serrations or mantles may be gradually formed in the nonsliding boundaries. As GBS can still operate on the mantles, the remaining parts have to sustain the shear effects and distortion begins to occur. As a result, local lattices near the mantles have to rotate and asymmetric bulges are generated. Then, subgrains with intermediate grain boundaries are formed. New recrystallized grains will be formed as long as the partial LAGBs of these subgrains transform into HAGBs [55]. The results of the formation of CDRX are consistent with the study of Yin [50], Shimizu [52], and Beer [53].

3.4.2. The Influence of TMP on CDRX

The EBSD micrographs of the compressed samples under different deformed conditions are illustrated in Figure 10 to investigate the influence of TMP on CDRX. Among them, Figure 10a–c shows the microstructures under the conditions of 673 K, 10 s^{-1}, and the strains of 0.2, 0.5, and 0.9, respectively. It can be seen clearly that, as the deformation continued, the original grains were gradually elongated, grain boundaries became more and more jagged, and the amount of small, recrystallized grains near the grain boundaries increased progressively. Thus, the increase in the strain can promote the formation of CDRX. However, the percentage of recrystallized grains is still very small; even at the strain of 0.9, the original grains and their internal substructures occupied the majority, which also proves that the predominant softening mechanism of the 2219 Al alloy during hot deformation was DRV.

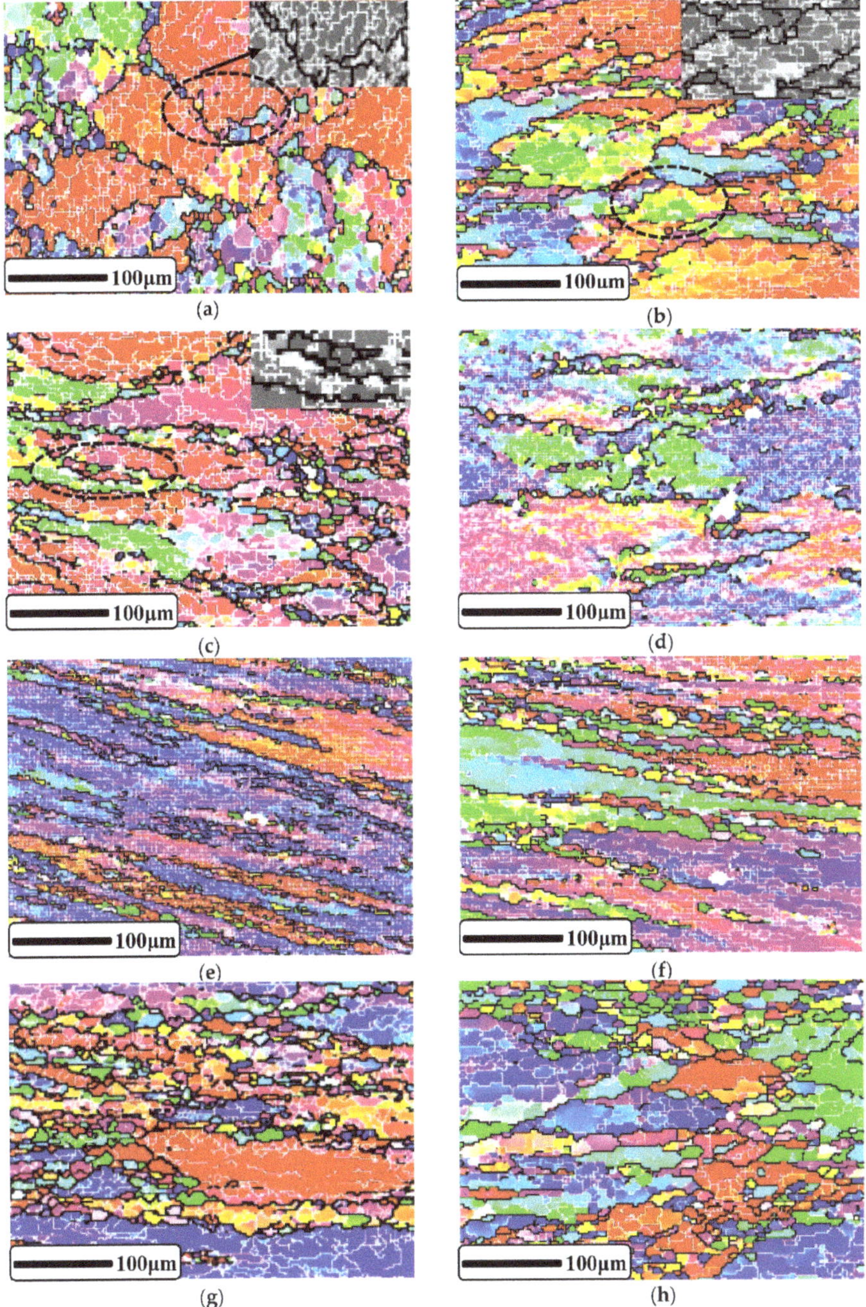

Figure 10. The EBSD micrographs of the deformed samples under conditions of: (**a**) 673 K and 0.1 s^{-1} with the strain of 0.2; (**b**) 673 K and 0.1 s^{-1} with the strain of 0.5; (**c**) 673 K and 0.1 s^{-1} with the strain of 0.9; (**d**) 623 K and 1 s^{-1} with the strain of 0.9; (**e**) 673 K and 1 s^{-1} with the strain of 0.9; (**f**) 773 K and 1 s^{-1} with the strain of 0.9; (**g**) 773 K and 0.1 s^{-1} with the strain of 0.9; (**h**) 773 K and 0.01 s^{-1} with the strain of 0.9.

In addition, Figure 10d–f shows the microstructures deformed at a strain rate of $1\ \text{s}^{-1}$ and strain of 0.9 at different temperatures of 623 K, 673 K, and 773 K, respectively. It can be seen that, as the temperature increased, the amount of recrystallized grains near the grain boundaries gradually increased, and also the grain size gradually increased. This is due to the fact that the mobility of the grain boundaries will increase with the increase in temperature. Moreover, DRX is a thermal-activation process; therefore, there will be a greater driving force to promote the growth of the new grains at high temperature. So, the increase of forming temperature can promote the development of CDRX and growth of new grains. Finally, Figure 10f–h show the microstructures deformed at the temperature of 773 K, strain of 0.9, and different strain rates of $1\ \text{s}^{-1}$, $0.1\ \text{s}^{-1}$, and $0.01\ \text{s}^{-1}$, respectively. It can also be observed that as the strain rate increased, the amount of recrystallized grains decreased, and the growth of new fine grains was greatly restricted. This is because the reduction of forming time restrains the growth of recrystallized grains at high strain rate. Therefore, the increase in strain rates limits the development of CDRX and the growth of its recrystallized grains. The results of the influence of TMP on CDRX obtained are consistent with the study of Wang [51] on AA7050 aluminum alloy.

4. Conclusions

In this study, the flow curves under different deformation conditions of 2219 Al alloy were obtained based on thermal-compression tests. The physically based constitutive model was established to describe its flow behavior. The microstructure evolution behavior of the 2219 Al alloy in thermal compression was also studied. The following conclusions can be drawn:

1. The flow stress of the 2219 Al alloy is very sensitive to temperatures and strain rates, and its value decreases with the increase in temperatures and/or the decrease in strain rates.
2. The physically based constitutive model of 2219 Al alloy established is proved to have good predictive performance, which can be used to accurately describe the flow behavior of the 2219 Al alloy in the temperature range of 623 K to 773 K, strain rate range of $0.01\ \text{s}^{-1}$ to 10^{-1}, and the strain range of 0~0.8.
3. It has been proved that the main microstructure evolution of 2219 Al alloy under hot deformation is DRV and incomplete CDRX. Moreover, CDRX can occur readily at high temperatures, low strain rates and high strains; meanwhile, the recrystallized grains size will also be larger.

Author Contributions: Conceptualization, L.L.; methodology, Y.W. and H.G.; software, S.L.; validation, L.L. and S.L.; formal analysis, L.L. and H.G.; writing—original draft preparation, L.L.; writing—review and editing, Y.W. and A.S.A.

Funding: The research was funded by the National Natural Science Foundation of China (Grant Number: 51405520, U1637601 and 51327902), the Fundamental Research Funds for the Central Universities of Central South University (Grant Number: 502221801), and the Project of State Key Laboratory of High Performance Complex Manufacturing, Central South University (Grant Number: ZZYJKT2017-06).

Acknowledgments: The authors are very grateful for the experimental materials provided by the Collaborative Innovation Center of Nonferrous Metals, the thermal compression instruments provided by Suzhou Research Institute for Nonferrous Metal Co., Ltd, and the methodological guidance given by all colleagues in B405 office.

Conflicts of Interest: The authors declare no conflict of interest.

References

1. Kaibyshev, R.; Sitdikov, O.; Mazurina, I.; Lesuer, D.R. Deformation behavior of a 2219 Al alloy. *Mater. Sci. Eng. A* **2002**, *334*, 104–113. [CrossRef]
2. Mirzadeh, H. Constitutive modeling and prediction of hot deformation flow stress under dynamic recrystallization conditions. *Mech. Mater.* **2015**, *85*, 66–79. [CrossRef]
3. Toros, S.; Ozturk, F. Flow curve prediction of Al–Mg alloys under warm forming conditions at various strain rates by ANN. *Appl. Soft Comput.* **2011**, *11*, 1891–1898. [CrossRef]

4. Huang, C.Q.; Jie, D.; Wang, S.X.; Liu, L.L. A physical-based constitutive model to describe the strain-hardening and dynamic recovery behaviors of 5754 aluminum alloy. *Mater. Sci. Eng. A* **2017**, *699*, 106–113. [CrossRef]

5. Lin, Y.C.; Chen, X.M. A critical review of experimental results and constitutive descriptions for metals and alloys in hot working. *Mater. Des.* **2011**, *32*, 1733–1759. [CrossRef]

6. Salehi, M.S.; Serajzadeh, S. A neural network model for prediction of static recrystallization kinetics under non-isothermal conditions. *Comp. Mater. Sci.* **2010**, *49*, 773–781. [CrossRef]

7. Sun, Z.; Liu, L.; Yang, H. Microstructure evolution of different loading zones during TA15 alloy multi-cycle isothermal local forging. *Mater. Sci. Eng. A* **2011**, *528*, 5112–5121. [CrossRef]

8. Mcqueen, H.J.; Ryan, N.D. Constitutive analysis in hot working. *Mater. Sci. Eng. A* **2002**, *322*, 43–63. [CrossRef]

9. Ashtiani, H.R.R.; Parsa, M.H.; Bisadi, H. Constitutive equations for elevated temperature flow behavior of commercial purity aluminum. *Mater. Sci. Eng. A* **2012**, *545*, 61–67. [CrossRef]

10. Changizian, P.; Zarei-Hanzaki, A.; Roostaei, A.A. The high temperature flow behavior modeling of AZ81 magnesium alloy considering strain effects. *Mater. Des.* **2012**, *39*, 384–389. [CrossRef]

11. Hollomon, J.H.; Member, J. Tensile Deformation. *Met. Technol.* **1945**, *12*, 268–290.

12. Ludwig, P. *Element der Technologischen Mechanik//Elemente der Technologischen Mechanik*; Springer: Berlin/Heidelberg, Germany, 1909.

13. Johnson, G.R.; Cook, W.H. A constitutive model and data for metals subjected to large strains, high strain rates and high temperatures. *Eng. Fract. Mech.* **1983**, *21*, 541–548.

14. Sellars, C.M.; Mctegart, W.J. On the mechanism of hot deformation. *Acta Metall. Mater.* **1966**, *14*, 1136–1138. [CrossRef]

15. Mirzadeh, H.; Cabrera, J.M.; Najafizadeh, A. Constitutive relationships for hot deformation of austenite. *Acta Mater.* **2011**, *59*, 6441–6448. [CrossRef]

16. Mirzadeh, H.; Cabrera, J.M.; Najafizadeh, A. Modeling and Prediction of Hot Deformation Flow Curves. *Metall. Mater. Trans. A* **2012**, *43*, 108–123. [CrossRef]

17. Lin, Y.C.; Chen, M.S.; Zhong, J. Constitutive modeling for elevated temperature flow behavior of 42CrMo steel. *Comp. Mater. Sci.* **2008**, *42*, 470–477. [CrossRef]

18. Yu, D.H. Modeling high-temperature tensile deformation behavior of AZ31B magnesium alloy considering strain effects. *Mater. Des.* **2013**, *51*, 323–330. [CrossRef]

19. Ashtiani, H.R.R.; Shahsavari, P. Strain-dependent constitutive equations to predict high temperature flow behavior of AA2030 aluminum alloy. *Mech. Mater.* **2016**, *100*, 209–218. [CrossRef]

20. Huang, K.; Logé, R.E. A review of dynamic recrystallization phenomena in metallic materials. *Mater. Des.* **2016**, *111*, 548–574. [CrossRef]

21. Chen, F.; Cui, Z.; Chen, S. Recrystallization of 30Cr2Ni4MoV ultra-super-critical rotor steel during hot deformation. Part I: Dynamic recrystallization. *Mater. Sci. Eng. A* **2011**, *528*, 5073–5080. [CrossRef]

22. Stewart, G.R.; Jonas, J.J.; Montheillet, F. Kinetics and Critical Conditions for the Initiation of Dynamic Recrystallization in 304 Stainless Steel. *Iron Steel I Jpn.* **2004**, *44*, 1581–1589. [CrossRef]

23. Marchattiwar, A.; Sarkar, A.; Chakravartty, J.K.; Kashyap, B.P. Dynamic Recrystallization during Hot Deformation of 304 Austenitic Stainless Steel. *J. Mater. Eng. Perform.* **2013**, *22*, 2168–2175. [CrossRef]

24. Quan, G.Z.; Mao, Y.P.; Li, G.S.; Lv, W.Q.; Wang, Y. A characterization for the dynamic recrystallization kinetics of as-extruded 7075 aluminum alloy based on true stress–strain curves. *Comp. Mater. Sci.* **2012**, *55*, 65–72. [CrossRef]

25. Li, X.; Wu, X.C.; Zhang, X.X.; Li, M.Y. Dynamic Recrystallization of Hot Deformed 3Cr2NiMnMo Steel: Modeling and Numerical Simulation. *J. Iron Steel Res. Int.* **2013**, *20*, 98–104. [CrossRef]

26. Perdrix, C.; Perrin, M.Y.; Montheillet, F. Mechanical behavior and structure development of aluminum during high amplitude hot deformation/comportment mecanique et evolution structural de l'aluminium au cours d'une deformation a chaud de grande amplitude. *Metal* **1981**, *78*, 309–320.

27. Montheillet, F.; Gilormini, P.; Jonas, J.J. Relation between axial stresses and texture development during torsion testing: A simplified theory. *Acta Metall. Mater.* **1985**, *33*, 705–717. [CrossRef]

28. Gourdet, S.; Montheillet, F. An experimental study of the recrystallization mechanism during hot deformation of aluminium. *Mater. Sci. Eng. A* **2000**, *283*, 274–288. [CrossRef]

29. Ion, S.E.; Humphreys, F.J.; White, S.H. Dynamic recrystallisation and the development of microstructure during the high temperature deformation of magnesium. *Acta Metall. Mater.* **1982**, *30*, 1909–1919. [CrossRef]

30. Gardner, K.J.; Grimes, R. Recrystallization during hot deformation of aluminium alloys. *Met. Sci.* **1979**, *13*, 216–222. [CrossRef]

31. Yanushkevich, Z.; Belyakov, A.; Kaibyshev, R. Microstructural evolution of a 304-type austenitic stainless steel during rolling at temperatures of 773–1273 K. *Acta Mater.* **2015**, *82*, 244–254. [CrossRef]

32. Kaibyshev, R.; Shipilova, K.; Musin, F.; Motohashi, Y. Continuous dynamic recrystallization in an Al–Li–Mg–Sc alloy during equal-channel angular extrusion. *Mater. Sci. Eng. A* **2005**, *396*, 341–351. [CrossRef]

33. Gourdet, S.; Montheillet, F. A model of continuous dynamic recrystallization. *Acta Mater.* **2003**, *51*, 2685–2699. [CrossRef]

34. Tóth, L.S.; Estrin, Y.; Lapovok, R.; Gu, C. A model of grain fragmentation based on lattice curvature. *Acta Mater.* **2010**, *58*, 1782–1794. [CrossRef]

35. Finegan, B.A.; Mcfarlane, H.J. Evolution of flow stress in aluminium during ultra-high straining at elevated temperatures. Part II. *Philos. Mag. A* **1989**, *60*, 473–485.

36. Mcqueen, H.J.; Myshlyaev, M.M.; Mwembela, A. Microstructural Evolution and Strength in Hot Working of ZK60 and other Mg Alloys. *Can. Metall. Q.* **2013**, *42*, 97–112. [CrossRef]

37. Myshlyaev, M.M.; Mcqueen, H.J.; Mwembela, A.; Konopleva, E. Twinning, dynamic recovery and recrystallization in hot worked Mg–Al–Zn alloy. *Mater. Sci. Eng. A* **2002**, *337*, 121–133. [CrossRef]

38. Roebuck, B.; Lord, J.D.; Brooks, M.; Loveday, M.S.; Sellars, C.M. Measurement of flow stress in hot axisymmetric compression tests. *High Temp. Technol.* **1997**, *23*, 59–83. [CrossRef]

39. Wei, H.L.; Liu, G.Q.; Zhang, M.H. Physically based constitutive analysis to predict flow stress of medium carbon and vanadium microalloyed steels. *Mater. Sci. Eng. A* **2014**, *602*, 127–133. [CrossRef]

40. Humphreys, F.J.; Hatherly, M. *Recrystallization and Related Annealing Phenomena*, 2nd ed.; Pergamon Press: Oxford, UK, 1995.

41. Lin, Y.C.; Chen, M.S.; Zhong, J. Prediction of 42CrMo steel flow stress at high temperature and strain rate. *Mech. Res. Commun.* **2008**, *35*, 142–150. [CrossRef]

42. Wu, K.; Liu, G.Q.; Hu, B.F.; Wang, C.F.; Zhang, Y.W. Effect of processing parameters on hot compressive deformation behavior of a new Ni–Cr–Co based P/M superalloy. *Mater. Sci. Eng. A* **2011**, *528*, 4620–4629. [CrossRef]

43. Solhjoo, S. Determination of critical strain for initiation of dynamic recrystallization. *Mater. Des.* **2010**, *31*, 1360–1364. [CrossRef]

44. Lin, Y.C.; Chen, M.S.; Zhong, J. Microstructural evolution in 42CrMo steel during compression at elevated temperatures. *Mater. Lett.* **2008**, *62*, 2132–2135. [CrossRef]

45. Reyes-Calderón, F.; Mejía, I.; Cabrera, J.M. Hot deformation activation energy (QHW) of austenitic Fe–22Mn–1.5Al–1.5Si–0.4C TWIP steels microalloyed with Nb, V, and Ti. *Mater. Sci. Eng. A* **2013**, *562*, 46–52. [CrossRef]

46. Frost, H.J.; Ashby, M.F. *Deformation-Mechanism Maps: The Plasticity and Creep of Metals and Ceramics*; Pergamon Press: Oxford, UK, 1982.

47. Cabrera, J.M.; Omar, A.A.; Prado, J.M.; Jonas, J.J. Modeling the flow behavior of a medium carbon microalloyed steel under hot working conditions. *Metall. Mater. Trans. A* **1997**, *28*, 2233–2244. [CrossRef]

48. Cabrera, J.M.; Jonas, J.J.; Prado, J.M. Flow behaviour of medium carbon microalloyed steel under hot working conditions. *Met. Sci.* **2013**, *12*, 579–585. [CrossRef]

49. Thomas, A.; El-Wahabi, M.; Cabrera, J.M.; Prado, J.M. High temperature deformation of Inconel 718. *J. Mater. Process. Technol.* **2006**, *177*, 469–472. [CrossRef]

50. Yin, X.Q.; Park, C.H.; Li, Y.F.; Ye, W.J.; Zuo, Y.T. Mechanism of continuous dynamic recrystallization in a 50Ti-47Ni-3Fe shape memory alloy during hot compressive deformation. *J. Alloy Compd.* **2017**, *693*, 426–431. [CrossRef]

51. Wang, S.; Luo, J.R.; Hou, L.G.; Zhang, J.S.; Zhuang, L.Z. Physically based constitutive analysis and microstructural evolution of AA7050 aluminum alloy during hot compression. *Mater. Des.* **2016**, *107*, 277–289. [CrossRef]

52. Shimizu, I. Theories and applicability of grain size piezometers: The role of dynamic recrystallization mechanisms. *J. Struct. Geol.* **2008**, *30*, 899–917. [CrossRef]

53. Beer, A.G.; Barnett, M.R. Microstructural Development during Hot Working of Mg-3Al-1Zn. *Metall. Mater. Trans. A* **2007**, *38*, 1856–1867. [CrossRef]
54. Lin, Y.C.; Wu, X.Y.; Chen, X.M.; Chen, J.; Wen, D.X. EBSD study of a hot deformed nickel-based superalloy. *J. Alloy Compd.* **2015**, *640*, 101–113. [CrossRef]
55. Drury, M.R.; Humphreys, F.J. The development of microstructure in Al-5% Mg during high temperature deformation. *Acta Metall. Mater.* **1986**, *34*, 2259–2271. [CrossRef]

Article

Stability Research Considering Non-Linear Change in the Machining of Titanium Thin-Walled Parts

Haining Gao [1,2,*] and Xianli Liu [2]

[1]   College of Mechanical and Energy Engineering, HuangHuai University, Zhumadian 463000, China
[2]   School of Mechanical & Power Engineering, Harbin University of Science and Technology,
      Harbin 50080, China
*   Correspondence: hngao@hrbust.edu.cn; Tel.: +86-136-5466-2869

Received: 24 May 2019; Accepted: 26 June 2019; Published: 28 June 2019

Abstract: Aiming to solve the problem whereby the damping process effect is significant and difficult to measure during low-speed machining of titanium alloy thin-walled parts, the ploughing coefficient of the flank face is obtained based on the frequency-domain decomposition (FDD) of the measured vibration signal and the energy balance principle, and then the process-damping prediction model is obtained. Aiming to solve the problem of non-linear change of dynamic characteristics of a workpiece caused by the material removal effect in the machining of titanium alloy thin-walled parts, a prediction model of dynamic characteristics of a workpiece is established based on the structural dynamic modification method. Meanwhile, the effect of material removal on the process-damping coefficient is studied, and the internal relationship between the process-damping coefficient and the dynamic characteristics of the workpiece is revealed. The stability lobe diagram is obtained by the full discretization in the titanium alloy milling process. The correctness of the model and stability prediction is verified by experiments under different working conditions. It is found that the coupling characteristics of process-damping and workpiece dynamic characteristics control the stability of the milling process. The research results can provide theoretical support for accurate characterization and process optimization of titanium alloy thin-walled workpiece milling.

Keywords: stability lobe diagram; milling; process-damping; dynamic characteristics; thin-walled weak rigidity parts

---

1. Introduction

Titanium alloy thin-walled parts have found an increasingly wide use in the aviation manufacture industry because they can meet the requirements of high performance and high stability. Due to the low stiffness and high material removal rate, chattering occurs very easily in the machining process, which reduces the surface precision of the workpiece and the service life of cutters and machine tools, limiting production efficiency. It has become an urgent problem to be solved in the aviation manufacturing industry.

Titanium alloys are mostly machined at low speed. In the process of low-speed machining, the periodic friction and extrusion between the tool flank and the machined surface will produce a ploughing effect, which improves the processing stability [1,2]. Sisson et al [3] and Peters et al [4] studied the process-damping phenomenon in a low-speed cutting process with an experimental method. It was found that the ultimate cutting depth obtained by considering process-damping has been effectively improved, especially in the low-speed range. Wu et al [5] constructed a ploughing force model which can accurately characterize the process-damping. It was assumed that ploughing force was a linear function of the total volume of the cutter–worker interference and ploughing coefficient. Subsequently, researchers have conducted in-depth research on the accurate acquisition of the total volume of cutter–worker interference and ploughing coefficient. Lee et al [6] solved the relative vibration of tool–workpiece based on an iterative algorithm, and then obtained the total volume of tool–workpiece interference. The total volume of cutter–workpiece interference was calculated based on a numerical algorithm [7]. Cao et al [8] proposed two integration methods to calculate the cutter–worker interference area. The identification of ploughing coefficient is as follows. Wu et al [5] calculated ploughing coefficient by the contact theory. Shawky et al [9] identified ploughing coefficient by using the relationship between measured normal cutting force and static interference volume function. Tun et al [10] proposed a method for identifying process-damping coefficients directly by experimental stability limits. Ahmadi et al [11] presented a method to obtain the average process-damping coefficient of a dynamic cutting system by steady cutting experiments. Wan et al [12] provided a new method to identify process-damping based on the model proposed by Ahmadi and a modal analysis method. In addition, Jin et al [13] modeled process-damping through finite-element simulation of the vibration tool. Feng et al [14] found that the cutting vibration is the key factor affecting process-damping. However, there is no consensus on the analysis of the process-damping effect.

The inherent properties of the workpiece (modal mass, stiffness, and damping ratio) change with the material removal, especially for the processing of weak rigid parts. This material removal effect makes it more difficult to accurately model the dynamic characteristics of the process system. The finite-element method (FEM) and the structural modification method (SDM) are often used to study material quality removal effect. Adetoro et al [15], Song et al [16], and Campa et al [17] used a two-dimensional finite-element model to study the variation of workpiece modal parameters with the processing location and material removal. Combining with the stability solution method (frequency-domain method and semi-discrete method), the stability lobe diagrams under the corresponding working conditions were obtained. Ding et al [18] proposed a 3D FEM model to predict the modal parameters of each machining step. FEM can effectively raise the prediction accuracy when the cutting step is reduced, but it also reduces the processing efficiency. Besides, FEM cannot simulate the condition of small radial depth of cut. The structural modification method was proposed by Özgüven et al [19] in 1990. The main idea is to calculate the frequency-response function of the modified system by using the frequency-response function of the original system and modifying the dynamic structure matrix of the system. Salih et al [20] and Budak et al [21] obtained milling stability lobe diagrams with material removal based on structure-modification method and frequency-domain method. Song et al [22] developed a new method for dynamic modification of equal mass structure to predict the variation of inherent properties of workpiece with material removal.

Accurate acquisition of material mass-removal effect and process-damping effect is an effective way to enhance the stability prediction accuracy in the machining of titanium alloy. With the removal of material, the vibration resistance of the workpiece decreases, and the cutting vibration amplitude increases with the same cutting parameters. The process-damping coefficient is affected by the amplitude of vibration [10]. The interaction mechanism between material removal effect and process-damping has not been revealed in the existing literature. Meanwhile, tool stiffness in the cross-feed direction is lower than that of the workpiece-clamping system. The traditional two-degree-of-freedom dynamic equation of workpiece or cutting system is not applicable under this condition. In this paper, a new general dynamic model of the dynamic characteristics of the workpiece and process-damping is established. The milling stability is obtained by discrete method, and the effects of workpiece dynamic characteristics and process-damping on the stability are investigated. The research results can provide theoretical support for accurate characterization and process optimization of titanium alloy thin-walled workpiece milling.

2. Defining a Dynamic Model for Process-Damping

The workpiece has weak rigidity in the normal (Y) direction, and the tool has weak rigidity in the feed (X) direction in in the machining of titanium alloy, as shown in Figure 1. Therefore, a two-degree-of-freedom dynamic model is founded in this work, which considered the vibration characteristics of the workpiece and the tool. The shear force is produced on the rake face due to shear effect, and the ploughing force is produced on the flank face due to friction and extrusion effect in the process of cutting material. The shear force and ploughing force can be broken into two sections—the static milling force and the ploughing milling force. Ignoring the modal coupling effect of the process system, the dynamic equation of cutting process is established, as shown in Equation (1).

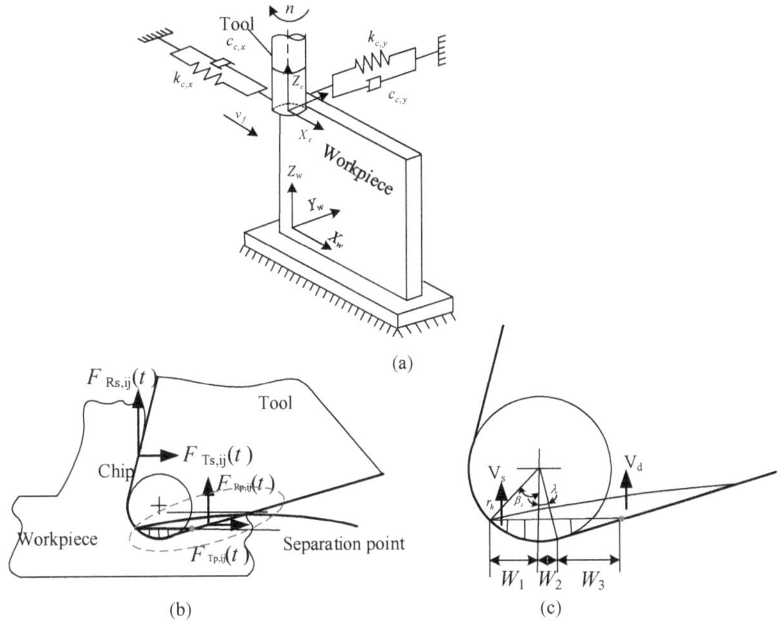

Figure 1. Dynamic model of milling system. (**a**) Three-dimensional model, (**b**) Schematic of chip generation, (**c**) Area of the indented volum.

$$\begin{cases} [m_{xx}]\{\ddot{x}(t)\} + [c_{xx}]\{\dot{x}(t)\} + [k_{xx}]\{x(t)\} = \{F_x(t)\} \\ [m_{yy}]\{\ddot{y}(t)\} + [c_{yy}]\{\dot{y}(t)\} + [k_{yy}]\{y(t)\} = \{F_y(t)\} \end{cases} \tag{1}$$

$[m_{xx}]$, $[c_{xx}]$, $[k_{xx}]$ are the mass, structural damping, and stiffness matrices of the cutting system, respectively. $[m_{yy}]$, $[c_{yy}]$, $[k_{yy}]$ are the mass, structural damping, and stiffness matrices of the workpiece, respectively. $\{F_i(t)\}$ is the cutting force in the X and Y directions. $\{i(t)\}$ is the response caused by the cutting force. $i = x \, or \, y$.

The calculation formula of the cutting force in Y direction is shown in Equation (2).

$$\{F_y(t)\} = F_{ys,st}(t) + F_{ys,dy}(t) + F_{yp,st}(t) + F_{yp,dy}(t) \tag{2}$$

$F_{ys,st}(t)$ and $F_{yp,st}(t)$ are the static cutting forces produced by shearing and friction on rank and flank face. $F_{ys,dy}(t)$ and $F_{yp,dy}(t)$ are the dynamic cutting force produced by shearing and friction on rank and flank face.

Static cutting force has no influence on the regeneration effect of machining process, so it is often ignored in the process of dynamic modeling.

The dynamic equation of processing is obtained as shown in Equation (3).

$$[m_{yy}]\{\ddot{y}(t)\} + [c_{yy}]\{\dot{y}(t)\} + [k_{yy}]\{y(t)\} = F_{ys,dy}(t) + F_{yp,dy}(t) \tag{3}$$

$$\begin{cases} F_{ys,dy}(t) = g(\phi_{i,j}) \sum_{j=0}^{N-1} (\sin(\phi_{i,j})F_{tsj}(t) - \cos(\phi_{i,j})F_{rsj}(t)) \\ F_{tsj}(t) = K_{Ts}a_p h(\phi_{i,j}) \\ F_{rsj}(t) = K_{Rs}a_p h(\phi_{i,j}) \\ h(\phi_{i,j}) = \Delta x \sin(\phi_{i,j}) + \Delta y \cos(\phi_{i,j}) \end{cases} \tag{4}$$

$$\begin{cases} F_{yp,dy}(t) = g(\phi_{i,j}) \sum_{j=0}^{N-1} (\sin(\phi_{i,j})F_{tpj}(t) - \cos(\phi_{i,j})F_{rpj}(t)) \\ F_{tpj}(t) = K_{Tf}V_{d,j} \\ F_{rpj}(t) = K_{Rf}V_{d,j} \\ V_{d,j} = \frac{l_w^2}{2v_c}(\dot{x}\sin(\phi_{i,j}) + \dot{y}\cos(\phi_{i,j})) \end{cases} \tag{5}$$

$g(\phi_{i,j}(t))$ is a window function for judging the processing state. Its expression is shown in Equation (6). K_{Ts} and K_{Rs} are the shear force coefficients on the rake face. K_{Tf} and K_{Rf} are the ploughing force coefficients on the flank face.

$$g(\phi_{i,j}(t)) = \begin{cases} 1, & if\,\phi_{st} \leq \theta_{i,j} \leq \phi_{ex} \\ 0, & otherwise \end{cases} \tag{6}$$

l_w is the boundary length between static indented and dynamic indented volumes.

$$l_w = W_1 + W_2 + W_3 \tag{7}$$

The angle of cutting rotation for the cutter tooth is $\phi_{i,j}(t)$. The expression of the tool rotation angle considering the tool helix angle, is shown in Equation (8).

$$\phi_{i,j} = \frac{2\pi\Omega t}{60} + \frac{2\pi(j-1)}{N} - \frac{dz\tan\beta}{R} \tag{8}$$

where Ω is the spindle speed, N is the number of cutter teeth, β is the helix angle of the cutting tool, dz is the length of the axial unit.

Similarly, the dynamic shear force $F_{xs,dy}(t)$ and the dynamic ploughing force $F_{xp,dy}(t)$ in the X direction are obtained. We can get the dynamic equations with equivalent process-damping as follows.

$$\begin{bmatrix} m_{xx} & 0 \\ 0 & m_{yy} \end{bmatrix}\begin{Bmatrix} \ddot{x} \\ \ddot{y} \end{Bmatrix} + \begin{bmatrix} k_{xx} & 0 \\ 0 & k_{yy} \end{bmatrix}\begin{Bmatrix} x \\ y \end{Bmatrix} + \left(\begin{bmatrix} c_{xx} & 0 \\ 0 & c_{yy} \end{bmatrix} + \begin{bmatrix} c_{px} & 0 \\ 0 & c_{py} \end{bmatrix}\right)\begin{Bmatrix} \dot{x} \\ \dot{y} \end{Bmatrix} \\ = \tfrac{1}{2}a_p\begin{bmatrix} a_{xx} & a_{xy} \\ a_{yx} & a_{yy} \end{bmatrix}\begin{Bmatrix} \Delta x \\ \Delta y \end{Bmatrix} \tag{9}$$

$$\begin{Bmatrix} c_{px} \\ c_{py} \end{Bmatrix} = \frac{l_w^2}{2v_c}\begin{bmatrix} -\cos(\phi_{i,j}) & -\sin(\phi_{i,j}) \\ \sin(\phi_{i,j}) & -\cos(\phi_{i,j}) \end{bmatrix}\begin{bmatrix} K_{Tf}\sin(\phi_{i,j}) & K_{Tf}\cos(\phi_{i,j}) \\ K_{Rf}\sin(\phi_{i,j}) & K_{Rf}\cos(\phi_{i,j}) \end{bmatrix} \tag{10}$$

$$\begin{cases} a_{xx} = \sum_{j=0}^{N-1} -g(\phi_{i,j})[\sin(2\phi_{i,j})K_{Ts} + (1 - \cos(2\phi_{i,j}))K_{Rs}] \\ a_{xy} = \sum_{j=0}^{N-1} -g(\phi_{i,j})[(\cos(2\phi_{i,j}) + 1)K_{Ts} + \sin(2\phi_{i,j})K_{Rs}] \\ a_{yx} = \sum_{j=0}^{N-1} g(\phi_{i,j})[(1 - \cos(2\phi_{i,j}))K_{Ts} + \sin(2\phi_{i,j})K_{Rs}] \\ a_{yy} = \sum_{j=0}^{N-1} g(\phi_{i,j})[\sin(2\phi_{i,j})K_{Ts} - (\cos(2\phi_{i,j}) + 1)K_{Rs}] \end{cases} \tag{11}$$

where c_{px} and c_{py} are the process-damping coefficients in machining process.

3. Prediction of Part Dynamics

The change of dynamic characteristics caused by material removal effect can be regarded as structural modification. Combining the original frequency-response function of the workpiece with the change value of the frequency-response function caused by material removal, the inherent properties of the workpiece after material removal can be obtained. The response of process system composed of workpiece and fixture under the cutting force in frequency domain can be expressed as:

$$\{y(\omega)\} = [[k_{yy}] - \omega^2[m_{yy}] + \tau[c_{yy}]]^{-1}\{F_y(\omega)\} \tag{12}$$

where τ is the unit imaginary number.

The receptance matrix, $[\alpha]$ is defined by:

$$\alpha = [[k_{yy}] - \omega^2[m_{yy}] + \tau[c_{yy}]]^{-1} \tag{13}$$

The change of inherent properties of workpiece caused by material removal effect can be expressed as $[\Delta M]$, $[\Delta C]$ and $[\Delta K]$. The dynamic structure-modification matrix is shown in Equation (14).

$$[D] = [\Delta k_{yy}] - \omega^2[\Delta m_{yy}] + \tau[\Delta c_{yy}] \tag{14}$$

The receptance matrix is obtained by simultaneous Equation (13) and Equation (14).

$$[\gamma] = [[[k_{yy}] + [\Delta k_{yy}]] - \omega^2[[m_{yy}] + [\Delta m_{yy}]] + \tau[[c_{yy}] + [\Delta c_{yy}]]]^{-1} \tag{15}$$

The modified receptance matrix of workpiece structure can be expressed as follows.

$$\left[\gamma_{ab} \vdots \gamma_{ab}\right] = \left[\alpha_{ab} \vdots 0\right]\left[[I] - [D]\begin{bmatrix} \gamma_{bb} & \gamma_{bc} \\ \gamma_{cb} & \gamma_{cc} \end{bmatrix}\right] \tag{16}$$

where α_{ij} and γ_{ij} are submatrices of $[\alpha]$ and $[\gamma]$ respectively.

4. Solving the Process-Damping Coefficient

Side milling is a discontinuous cutting style. Because of the inconvenience of the experimental acquisition equipment and the periodic change of the cutting signal, the process-damping coefficient cannot be obtained directly in milling experiments. In this paper, four eddy current displacement sensors are used to obtain the vibration signal of the workpiece and the cutting system during the machining process, as shown in Figure 2. Then the frequency-domain decomposition method is used to analyze the vibration signal of the process system, and the total damping coefficient of the process system is obtained. Finally, the ploughing coefficients of the process system are obtained.

The workpiece deformation $y(t)$ in the Y direction can be expressed by the mode shape U and the mode displacement $q(t)$.

$$\begin{cases} y(t) = Uq(t) \\ y(t) = \{y_1(t) \quad y_2(t)\}^T \\ q(t) = \{q_1(t) \quad q_2(t)\}^T \\ U = \left[\left\{\begin{matrix} u_1 \\ u_2 \end{matrix}\right\}_1 \quad \left\{\begin{matrix} u_1 \\ u_2 \end{matrix}\right\}_1\right] = \left[\{U\}_1 \quad \{U\}_2 \right] \end{cases} \tag{17}$$

Converting the power spectrum of the time domain deformation of experimental points y1, y2 into the frequency domain, the formula is as follows.

$$S_{yy,[2\times2]} = Y(j\omega) \cdot Y^{*T}(j\omega) \tag{18}$$

$Y(j\omega)$ is the Fourier spectrum of workpiece deformation, $*$ and T represents the conjugate of a complex number and transpose of a matrix.

Equation (17) is introduced into Equation (18), and then Equation (18) is converted into Equation (19).

$$S_{yy,[2\times2]} = UQ(j\omega)Q^{*T}(j\omega)U^{*T} \tag{19}$$

$Q(j\omega)$ is the Fourier spectrum of $q(t)$.

$Q(j\omega)Q^{*T}(j\omega)$ is the 2*2 matrix. Due to the orthogonality of the modal, the asymmetry element is 0. In addition, because the modal of the system can be well separated (as shown in Figure 3), the Equation (19) can be simplified as follows.

$$S_{yy,[2\times2]} \approx U_1 Q_1 Q_1^{*T} U_1^{*T} \tag{20}$$

The magnitude of direct frequency-response functions (FRF) at any point can be obtained by the ratio of the power spectrum of the dynamic displacement of the point to the power spectrum of the exciting force. Therefore, the frequency-response function of position 1 is calculated as shown in Equation (21).

$$\frac{S_{11}(j\omega)}{S_{ff}} = |H_{11}|^2 \tag{21}$$

where S_{ff} is a fixed value [23].

The power spectrum of the deflection at point y1 can be transformed into the following form.

$$S_{11}(j\omega) = u_1 Q_1 Q_1^{*T} u_1^{*T} \tag{22}$$

When the frequency range is the same, the frequency-response function can be expressed by modal parameters.

$$\begin{cases} H_{11} = \dfrac{1/m_{yy}}{\omega_{n,y}^2 - \omega^2 + 2\xi_r \omega \omega_{n,y} j} \\ |H_{11}|^2 = \dfrac{(1/m_{yy})^2}{(\omega_{n,y}^2 - \omega^2)^2 + (2\xi_r \omega \omega_{n,y})^2} \end{cases} \tag{23}$$

Equations (22) and (23) are substituted into Equation (21), the position sensitive detector (PSD) of mode coordinate q1 is obtained according to the modal parameters of the main mode.

$$Q_1 Q_1^* = \frac{S_{ff}/u_1 u_1^* m_{r,y}^2}{(\omega_{n,y} - \omega^2)^2 + (2\xi_r \omega \omega_{n,y})^2} \tag{24}$$

Inverse Fourier transform for Equation (24).

$$\begin{cases} \mathcal{F}^{-1}(Q_1 Q_1^*) = C_2 e^{-\xi_{r,y}\omega_{n,y}t} \cos(\omega_d t) \\ C_2 = \dfrac{S_{ff}/u_1 u_1^* m_{r,y}^2}{2\omega_{n,y}^3 \xi_{r,y} \sqrt{1-\xi_{r,y}^2}} \\ \omega_d = \omega_{n,y} \sqrt{1-\xi_{r,y}^2} \end{cases} \tag{25}$$

The extreme points occur at $t = i\pi/\omega_d$. The absolute value of $\mathcal{F}^{-1}$ can be expressed as a linear function. Substituting $t = i\pi/\omega_d$ in Equation (25), we get the peaks and valleys of $\mathcal{F}^{-1}$.

$$|P_{x,i}| = 2C_2^2 e^{\frac{-2i\pi\xi_{r,y}}{\sqrt{1-\xi_{r,y}^2}}} \tag{26}$$

Logarithmic operation is conducted to linearize Equation (26).

$$\begin{cases} 2\ln|P_{x,i}| = \delta_0 + \delta_1 i \\ \delta_0 = 2\ln C_2 \\ \delta_1 = -\dfrac{2\pi\xi_{r,y}}{\sqrt{1-\xi_{r,y}^2}} \end{cases} \tag{27}$$

The total damping ratio can be calculated by Equation (2).

$$\xi_{r,y} = \frac{-\delta_1}{\sqrt{\delta_1^2 + 4\pi^2}} \tag{28}$$

It is impractical to obtain the frequency-response function in machining process, so the autocorrelation of q1 is resolved by the vibration signal measured in cutting process.

The power spectrum matrix of the workpiece vibration measured at points y1, y2 can be expressed as follows.

$$\begin{aligned} S^e_{yy,[2\times2]}(j\omega_i) &= Y_e(j\omega_i) \cdot Y_e(j\omega_i)^{*T} \\ &= [\{V\}_1 \quad \{V\}_2] \begin{bmatrix} \kappa_1 & 0 \\ 0 & \kappa_2 \end{bmatrix} [\{V\}_1 \quad \{V\}_2]^T \\ &= \{V\}_1\kappa_1\{V\}_1^T + \{V\}_2\kappa_2\{V\}_2^T \end{aligned} \tag{29}$$

where ω_i is the frequency line. (κ_1, κ_2), $(\{V\}_1, \{V\}_2)$ are the eigenvalues and normalized eigenvectors of $S^e_{yy,[2\times2]}(j\omega_i)$ matrix, respectively. If $\kappa_1 > \kappa_2$ near the main mode, Equation (29) can be simplified as follows.

$$S^e_{yy,[2\times2]}(j\omega_i) \approx \{V\}_1\kappa_1\{V\}_1^T \tag{30}$$

The similarity between eigenvectors and modal shapes is determined on each frequency line using modal assurance criteria (MAC).

$$MAC(\omega_i) = \frac{\left|\{U\}_1^T \cdot \{V\}_1^T\right|}{(\{U\}_1^T \cdot \{V\}_1) \cdot (\{U\}_1^T \cdot \{V\}_1)} \tag{31}$$

When the value of $MAC(\omega_i)$ is close to 1, $Q_1 Q_1^* \approx \kappa_1$.

The difference between the total damping coefficient and the structural damping coefficient is used to calculate the process-damping coefficient.

$$c_{p,y} = 2m_{yy}\omega_{n,y}\xi_{r,y} - c_{s,y} \tag{32}$$

In the same way, the X-directional process-damping coefficient can be solved by Equation (33).

$$c_{p,x} = 2m_{xx}\omega_{n,x}\xi_{r,x} - c_{s,x} \tag{33}$$

The energy of the average process-damping effect is equal to the energy consumed by the dynamic plough force in the rotational period of the spindle, thus the ploughing force coefficients can be obtained.

$$\begin{cases} \int_0^T \left[\sum_{i,j} g(\phi_{i,j})\frac{l_w^2}{2v_c}\cos(\phi_{i,j})(-\cos(\phi_{i,j})K_{Tf} - \sin(\phi_{i,j})K_{Rf})dz_{i,j}\right]\dot{x}(t)dt = \int_0^T -c_{p,x}\dot{x}^2(t)dt \\ \int_0^T \left[\sum_{i,j} g(\phi_{i,j})\frac{l_w^2}{2v_c}\cos(\phi_{i,j})(\sin(\phi_{i,j})K_{Tf} - \cos(\phi_{i,j})K_{Rf})dz_{i,j}\right]\dot{y}(t)dt = \int_0^T -c_{p,y}\dot{y}^2(t)dt \end{cases} \tag{34}$$

5. Experimental

5.1. Experimental Setup

The experiment was carried out on a three-axis machining center manufactured by Dalian machine tool(Dalian machine tool group, Dalian, Liaoning, China). Its model is VDL-1000E. Solid carbide cutting tools with four teeth, diameter 10 and helix angle 30° are used for processing thin-walled titanium alloy parts. Its model is GM-4E-D10.0. The coating material is TiAlN. The extended length of the cutting tool is 110 mm. The geometric dimension of thin-walled titanium alloy parts is 200 × 200 × 5 mm. The workpiece is fixed on the vise. The vice is bolted to the workbench. The extended length of the workpiece is 100 mm. The milling mode is down milling and dry cutting. The rotary dynamometer produced by Kistler Company (Winterthur, Switzerland) was used to collect cutting force data. Its model is 5236B. Three-component acceleration sensors produced by PCB Sensor Company was used to collect the acceleration data of cutting process. Its model is 368F. The three-dimensional sensitivity is 2.462 mv/g, 2.534 mv/g and 2.487 mv/g, respectively. The initial modal parameters of the cutting system and workpiece are obtained by using the unidirectional acceleration sensor produced by PCB Sensor Company and the impact hammer. The vibration signals of the workpiece and the tool are measured using four ST-2-U-05-00-20-KH07 eddy current sensors. The experimental machining site is shown in Figure 2.

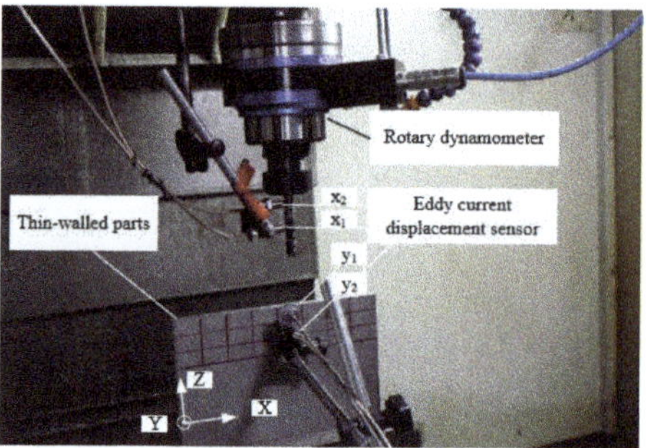

Figure 2. Experimental machining site.

5.2. The Acquisition of Dynamic Characteristics

Modal parameters of cutting system are usually obtained by hammer impact experiment. Generally, the real part and imaginary part of the frequency-response function are fitted by curve fitting method to obtain the modal parameters of the process system, as shown in Figure 3. Machining process is mainly controlled by first-order modal parameters. The initial first-order modal parameters of the unmachined workpiece and the tool are shown in Table 1.

Table 1. The initial first-order modal parameters.

Position	Natural Frequency (Hz)	Rigidity (N/m)	Damping Ratio (ξ_s)
Tools (X direction)	963	4.85×10^7	0.0591
Workpiece (Y direction)	652	8.54×10^6	0.0310

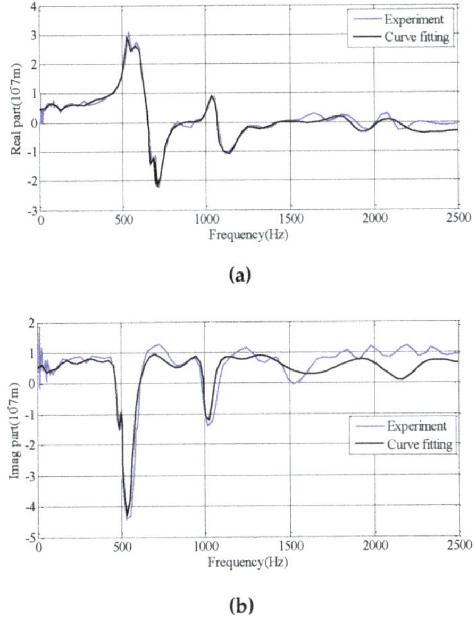

(a)

(b)

Figure 3. Fitting the measured frequency-response function.

The thickness of thin-walled titanium alloy parts is reduced from 5 mm to 0 mm by machining along the wall thickness. The process diagram is shown in Figure 4. The comparison results between the experimental and predicted values of the modal parameters of the workpiece at different machining positions is shown in Table 2.

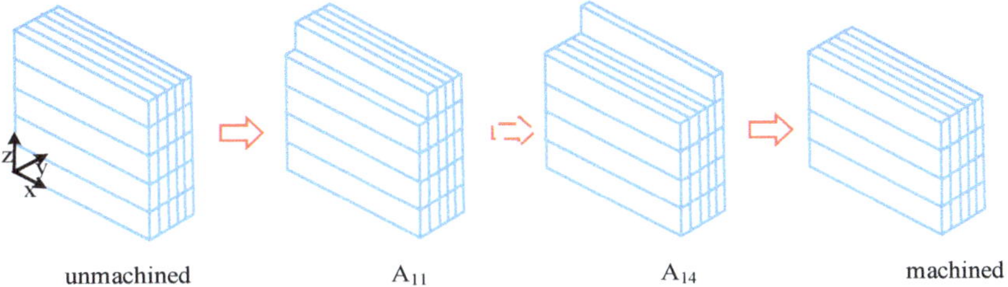

Figure 4. Schematic diagram of machining process.

Table 2 showed that with the removal of the material, the natural frequency of the workpiece increased, and the stiffness and damping ratio decreased gradually. The vibration resistance of workpiece is weakened. In addition, the changes of them displayed a non-linear feature. Therefore, if the damping ratio is assumed to be a fixed value or an equal proportion change, the prediction of dynamic force and stability will deviate.

By comparing the predicted value and measured value of the modal parameters of the workpiece, it can be seen that the maximum error of natural frequency, stiffness and damping ratio is 10.07%, 8.85% and 7.90%, respectively. The error is within acceptable range to verify the accuracy of the workpiece dynamic model.

Table 2. The modal parameters of the workpiece at different machining positions.

	Measured			Predicted		
Position	Natural Frequency (Hz)	Rigidity (N/m)	Damping Ratio ($\xi_{s,y}$)	Natural Frequency (Hz)	Rigidity (N/m)	Damping Ratio ($\xi_{s,y}$)
A11	675	8.47×10^6	0.0308	612	7.82×10^6	0.0287
A12	690	8.39×10^6	0.0304	633	7.73×10^6	0.0283
A13	718	8.30×10^6	0.0299	647	7.68×10^6	0.0279
A14	782	8.21×10^6	0.0291	725	7.63×10^6	0.0268
A15	864	8.11×10^6	0.0275	789	7.59×10^6	0.0259

5.3. Ploughing Force Coefficient

With $ap = 6$ mm, $ae = 0.5$ mm, $n = 900$ r/min, $f = 0.1$ mm/z, the steps to obtain the total damping ratio in the feed direction and normal direction of position A11 are shown in Figure 5.

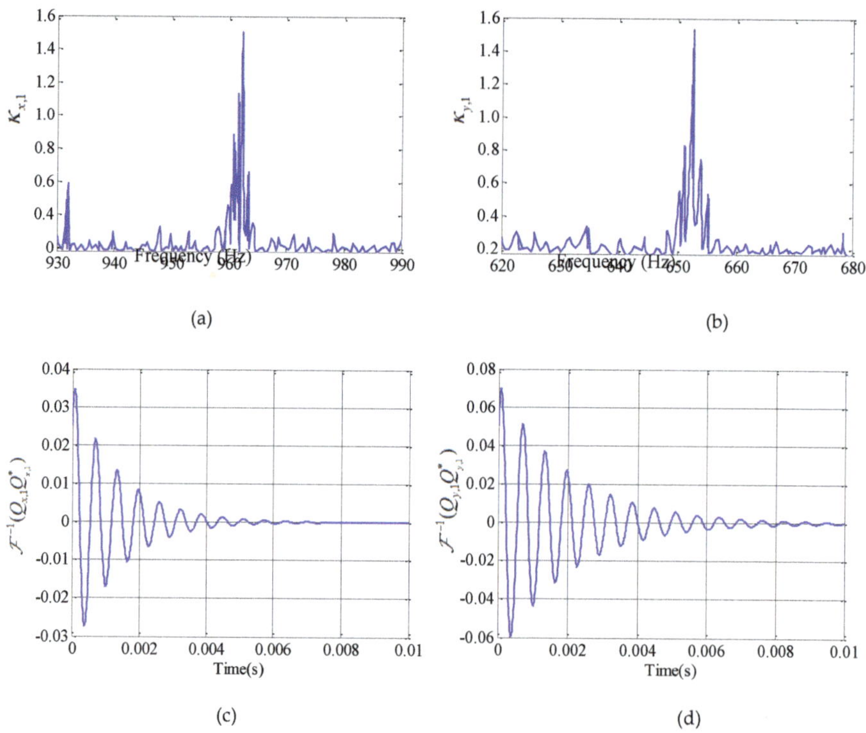

(a)

(b)

(c)

(d)

Figure 5. *Cont.*

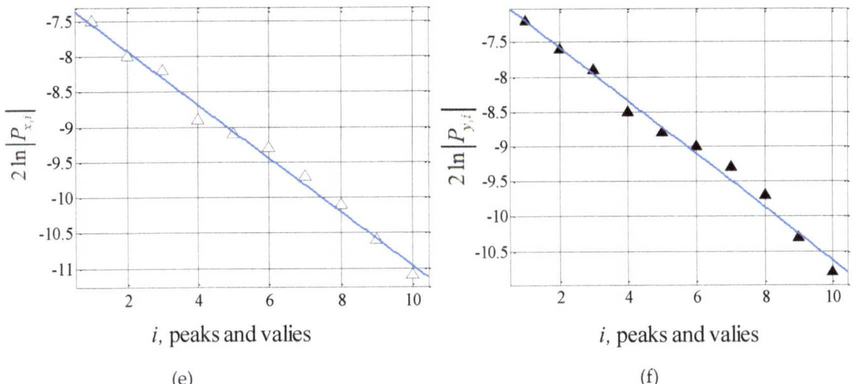

(e)

(f)

Figure 5. Identification of damping ratio in Z11. (**a**) Measured $\kappa_{x,1}$, (**b**) Measured $\kappa_{y,1}$, (**c**) $\mathcal{F}^{-1}(Q_{x,1}Q_{x,1}^*)$, (**d**) $\mathcal{F}^{-1}(Q_{y,1}Q_{y,1}^*)$, (**e**) Logarithmic decrement to identify damping, (**f**) Logarithmic decrement to identify damping in Y direction.

The overall damping ratios are estimated by substituting the slope of the fitted lines in Equation (27), $\xi_{r,x} = 0.0801, \xi_{r,y} = 0.0798$. The process-damping coefficient is evaluated as $c_{p,x} = 2m_x\omega_{n,x}(\xi_{r,x} - \xi_{s,x}) = 53.39(Ns/m)$, $c_{p,y} = 2m_y\omega_{n,y}(\xi_{r,y} - \xi_{s,y}) = 31.09(Ns/m)$. The ploughing force coefficients are obtained to be $K_{Tf} = 3.735 \times 10^{13}N/m^2$, $K_{Rf} = 1.208 \times 10^{13}N/m^2$.

By analyzing the cutting vibration obtained by the cutting process shown in Figure 4, the variation law of the vibration amplitude and the process-damping coefficient with the material removal effect is obtained, as shown in Figure 6.

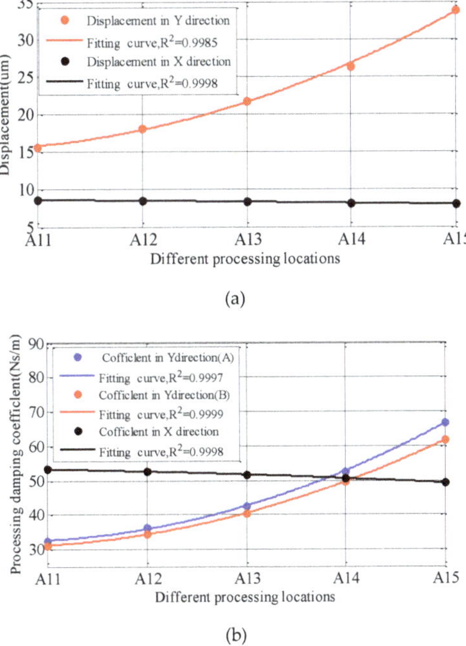

(a)

(b)

Figure 6. The effect of material removal effect on machining process. (**a**) Displacement, (**b**) Process-damping coefficients.

Figure 6a shows that under the same cutting parameters, the vibration amplitude of the machining system in the Y directions increases with the removal of workpiece materials. The change of vibration in the X direction is just the opposite to that in the Y direction. Meanwhile, it is found that the change rate of vibration amplitude in Y direction is greater than that in X direction.

It can be seen from Figure 6b that the process-damping coefficient in the Y directions increases with the removal of workpiece materials. The process-damping coefficient considering the dynamic characteristics of the workpiece (A) is smaller than that without considering the dynamic characteristics of the workpiece (B). The process-damping coefficient in the X directions decreases with the removal of workpiece materials.

6. Milling Stability

The frequency-domain method [24], discrete method and numerical method [25] are often used to solve the dynamic equation in machining process, and then then the stability of the cutting process is obtained. The frequency-domain method has the highest calculation efficiency, but the prediction accuracy is the lowest, especially it is not applicable to small radial depth of cut. The numerical method has the highest prediction accuracy, but its computational efficiency is the lowest because many equations need to be solved directly. The discrete method can be divided into the semi-discretization method [26], the full-discretization method [27] and the time FEM [28]. The semi-discretization method and the full-discretization method are most widely used. The matrix exponential function involved in the full-discretization method is only dependent on the spindle speed and independent of the axial depth of cut, so it is more efficient than the semi-discretization method. Therefore, we use the full discrete method to obtain the stability of the machining process.

The research results in Section 5.3 show that the process-damping coefficient has a non-linear relationship with the dynamic characteristics of the process system. Taking A15 machining position as an example, the influence of coupling characteristics between process-damping and dynamic characteristics of process system on milling stability is investigated.

The stability lobe diagrams with the coupled and the uncoupled process-damping, and the dynamic characteristics of the process system are shown in Figure 7a. A series of experiments (ae = 0.5 mm, f = 0.1 mm/tooth) were carried out to verify the accuracy of the stable lobe diagram. The remaining cutting parameters are detailed in Table 3. Two points A and B are selected for experimental verification to verify the correctness of the predicted results. Figure 7b,c are acceleration signals in time domain. Figure 7d,e are Fourier transform results of acceleration signals.

Table 3. Machining state with different conditions.

	Cutting Parameters		Machining State		Cutting Parameters		Machining State
No.	N (r/min)	Ap (mm)		No.	N (r/min)	Ap (mm)	
1	1080	5.3	Stable	10	1200	4.3	Stable
2	1080	5.6	Stable	11	1200	4.7	Stable
3	1080	6.2	Chatter	12	1200	5.4	Chatter
4	1080	6.7	Chatter	13	1200	5.8	Chatter
5	1140	5.0	Stable	14	1200	6.5	Chatter
6	1140	5.5	Stable	15	1260	5.7	Stable
7	1140	6.0	Stable	16	1260	6.1	Stable
8	1140	6.6	Chatter	17	1260	6.3	Chatter
9	1140	7.2	Chatter	18	1260	6.8	Chatter

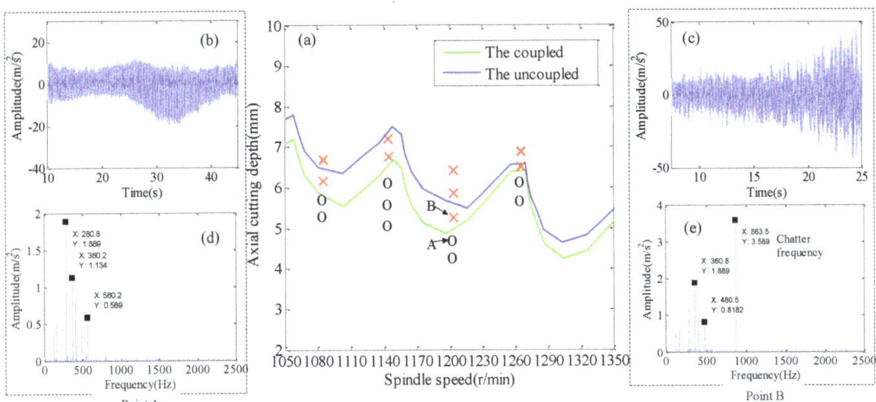

Figure 7. Validation of milling process stability. (**a**) Stability lobes diagram, (**b**) The acceleration signal at A point, (**c**) The acceleration signal at B point, (**d**) Frequency spectrum diagram at A point acceleration signal (**e**) Frequency spectrum diagram at B point acceleration signal.

The maximum amplitude of point A acceleration signal is 18 m/s^2, and its amplitude variation has better convergence, as shown in Figure 7b. Fourier transform is applied to the acceleration signal in time domain, and the result is shown in Figure 7d. It is found that the spectrum energy mainly concentrates on the cutter teeth passing frequency and its higher harmonics. The maximum amplitude of point B acceleration signal is 48 m/s^2, and its amplitude change law is from small to large, which does not have convergence, as shown in Figure 7c. Fourier transform is applied to the acceleration signal in time domain, and the result is shown in Figure 7e. It is found that the spectrum energy is mainly concentrated near the first natural frequency of the workpiece.

Meanwhile, the white light interferometer is used to measure the micromorphology of machined surface with A and B parameters and the results are shown in Figure 8a,b. Two-dimensional Fourier transform (2DFFT) is applied to the image of micro-surface topography, and the results are shown in Figure 8c,d.

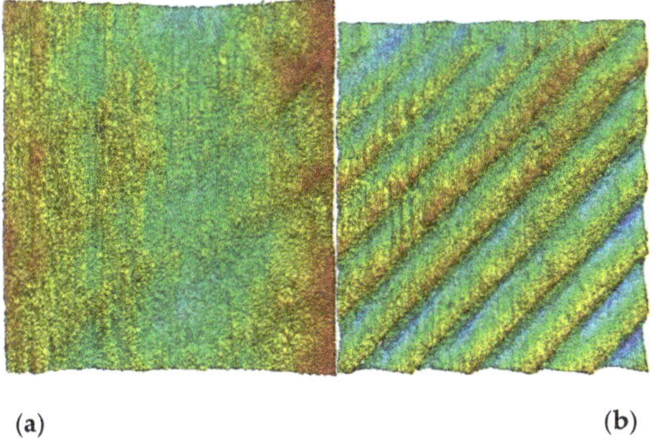

(**a**)   (**b**)

Figure 8. *Cont.*

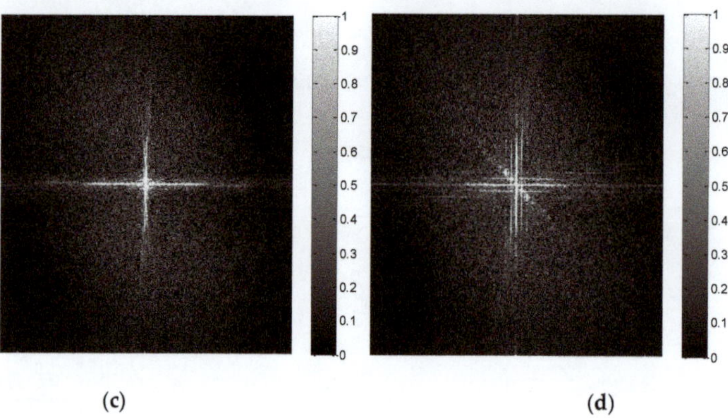

Figure 8. Microscopic morphology and frequency-domain analysis with A and B parameters. (**a**) Micromorphology at A point, (**b**) Micromorphology at B point, (**c**) 2DFFT at A point, (**d**) 2DFFT at B point.

The machined surface obtained at point A has lower level of surface roughness (Sa = 0.85 um, Ra = 0.47 um) and surface waviness, as shown in Figure 8a. The only significant spectral property of the machined surface at point A is related to the marking belonging to feed-rate, as shown in Figure 8c. The machined surface obtained at point A has the higher level of surface roughness (Sa = 1.63 um, Ra = 1.03 um) and surface waviness, as shown in Figure 8b. The spectral characteristics of the surface topography generated by point B are inclined modes related to the chatter frequency, as shown in Figure 8d. In conclusion, A is the cutting stability point and B is the cutting chatter point. It is found that the stability obtained by considering the coupling characteristics of dynamic characteristics of workpiece and process-damping has higher prediction accuracy.

Milling stability of the coupled process-damping and dynamic characteristics of process system is lower than that of the uncoupled, as shown in Figure 7a. The reason is that the process-damping coefficient obtained from the coupling process-damping and the dynamic characteristics of the process system is small. Meanwhile, the difference between them decreases with the increase of spindle speed. The reason is that the damping coefficient descends with the increase of spindle speed.

The stability lobe diagrams with and without process-damping are shown in Figure 9. A series of experiments (ae = 0.5 mm, f = 0.1 mm/tooth) were carried out to verify the accuracy of the stable lobe diagram. The remaining cutting parameters are detailed in Table 4. It is found that the milling stability considering process-damping is higher than that without process-damping in titanium alloy milling process, especially low spindle speed. The milling stability decreases with the increase of spindle speed. The reason is that the process-damping coefficient is inversely proportional to the spindle speed.

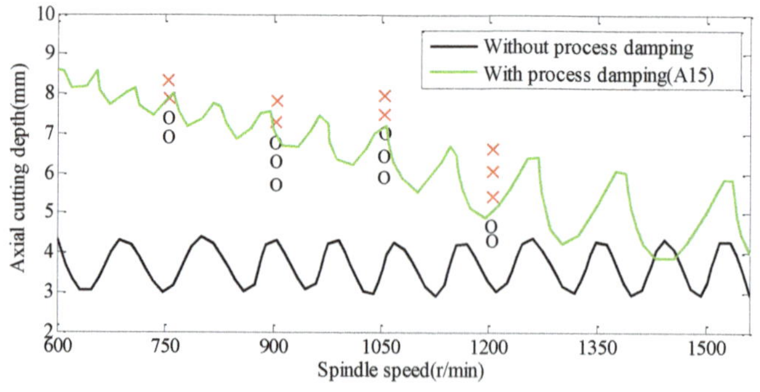

Figure 9. SLD, X-experimental stable, O-experimental unstable.

Table 4. Machining state with different conditions.

	Cutting Parameters		Machining State		Cutting Parameters		Machining State
No.	N (r/min)	Ap (mm)		No.	N (r/min)	Ap (mm)	
1	750	6.8	Stable	11	1050	6.4	Stable
2	750	7.2	Stable	12	1050	7.1	Stable
3	750	7.9	Chatter	13	1050	7.5	Chatter
4	750	8.2	Chatter	14	1050	8.0	Chatter
5	900	5.9	Stable	15	1200	4.3	Stable
6	900	6.3	Stable	16	1200	4.7	Stable
7	900	6.9	Stable	17	1200	5.4	Chatter
8	900	7.1	Chatter	18	1200	5.8	Chatter
9	900	7.5	Chatter	19	1200	6.5	Chatter
10	1050	6.0	Stable				

The 3D stability lobe diagram in milling titanium alloy thin-walled parts is shown in Figure 10. It is found that the ultimate axial cutting depth decreases with material removal. The reason is that with the removal of the material, the natural frequency of the workpiece increased, and the stiffness and damping ratio decreased gradually. The vibration resistance of workpiece is weakened.

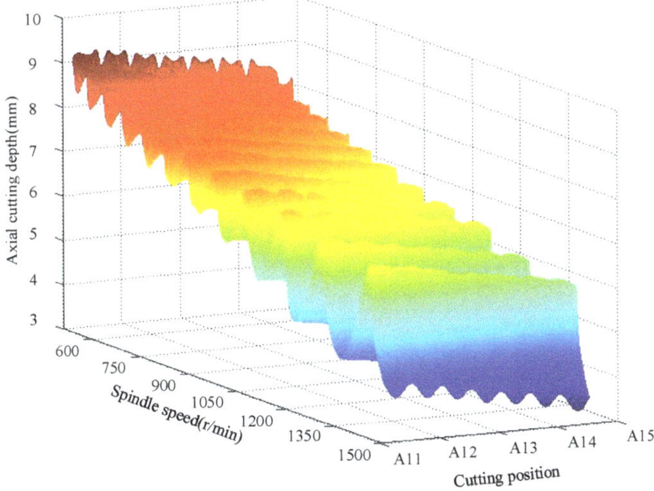

Figure 10. Three-dimensional stable lobe diagram.

7. Conclusions

In this paper, two prominent problems in the processing of titanium alloy thin-walled parts—the non-linear changes of dynamic characteristics of the parts caused by material removal effect and the damping process effect is difficult to measure—are investigated in depth. Based on the frequency-domain decomposition of the measured vibration signal and the principle of energy balance, a process-damping prediction model is obtained. Based on the structural dynamic modification method, a prediction model of workpiece dynamic characteristics is established. The effect of material removal on process-damping coefficient is studied. The full discrete method is used to solve the stability of the milling process. The correctness of the model and stability prediction is verified by experiments with different working conditions. The research results can provide theoretical support for accurate characterization and process optimization of titanium alloy thin-walled workpiece milling. The specific conclusions are as follows:

(1) Under the same cutting parameters, the vibration amplitude of the workpiece increases with the material removal, which leads to the corresponding increase of the process-damping coefficient. The variation of vibration amplitude and damping coefficient of cutting system is just the opposite.

(2) The process-damping coefficient obtained by considering the dynamic characteristics of the workpiece is less than that without considering the dynamic characteristics of the workpiece, and the gap between them increases with the removal of materials.

(3) Milling stability of the coupled process-damping and dynamic characteristics of process system is lower than that of the uncoupled. The reason is that the process-damping coefficient is relatively small. Meanwhile, the difference between them decreases with the increase of spindle speed. The reason is that the damping coefficient decreases with the increase of spindle speed.

(4) The milling stability is gradually reduced with the material is removed and the spindle speed is increased. Meanwhile, it is found that the effect of material removal on milling stability in the low-speed region is less than that in high speed region. The reason is that with the removal of materials, the vibration amplitude of the workpiece increases so that the damping coefficient of the process increases in the low-speed region, which further weakens the effect of material removal.

Author Contributions: The idea of this project was conceived by H.G. H.G. consulted the related papers on the milling stability of thin-walled parts, then sorted out the existing problems in this field, and finally wrote this paper. X.L. reviewed this project and proposed constructive guidance to make the article more complete.

Funding: This project is supported by the Projects of International Cooperation and Exchanges NSFC (51720105009).

Conflicts of Interest: The authors declare no conflict of interest.

References

1. Yue, C.; Gao, H.; Liu, X.; Liang, S.Y. A review of chatter vibration research in milling. *Chin. J. Aeronaut* **2019**, *32*, 215–242. [CrossRef]
2. Liu, X.; Gao, H.; Yue, C.; Li, R.; Jiang, N. Investigation of the milling stability based on modified variable cutting force coefficients. *Int. J. Adv. Manuf. Technol.* **2018**, *96*, 2991–3002. [CrossRef]
3. Sisson, T.R.; Kegg, R.L. An explanation of low-speed chatter effects. *J. Eng. Ind.* **1969**, *91*, 951–958. [CrossRef]
4. Peters, J.; Panherck, V.; Brussel, H.V. The measurement of the dynamic cutting coefficient. *CIRP Ann.-Manuf. Technol.* **1971**, *1*, 129–136.
5. Wu, W.D. A new approach of formulating the transfer function of dynamic cutting processes. *J. Eng. Ind. (Trans. ASME)* **1989**, *111*, 37–47. [CrossRef]
6. Lee, B.Y.; Tarng, Y.S.; Ma, S.C. Modeling of the process damping force in chatter vibration. *Int. J. Mach. Tool Manuf.* **1995**, *35*, 951–962. [CrossRef]
7. Endres, W.J.; DeVor, R.E.; Kapoor, S.G. A dual-mechanism approach to the prediction of machining forces, part 1: Model development. *J. Eng. Ind.* **1995**, *117*, 526–533. [CrossRef]
8. Cao, C.; Zhang, X.M.; Ding, H. An improved algorithm for cutting stability estimation considering process damping. *Int. J. Adv. Manuf. Technol.* **2017**, *88*, 2029–2038. [CrossRef]
9. Shawky, A.M.; Elbestawi, M.A. An enhanced dynamic model in turning including the effect of ploughing forces. *J. Manuf. Sci. E-T ASME* **1997**, *119*, 10–20. [CrossRef]
10. TunL, T.; Budak, E. Identification and modeling of process damping in milling. *J. Manuf. Sci. E-T ASME* **2013**, *135*, 021001.
11. Ahmadi, K.; Ismail, F. Stability lobes in milling including process damping and utilizing Multi-Frequency and Semi-Discretization Method. *Int. J. Mach. Tool Manuf.* **2012**, *54*, 46–54. [CrossRef]
12. Wan, M.; Feng, J.; Ma, Y.C.; Zhang, W. Identification of milling process damping using operational modal analysis. *Int. J. Mach. Tool Manuf.* **2017**, *122*, 120–131. [CrossRef]
13. Jin, X.; Altintas, Y. Chatter stability model of micro-milling with process damping. *J. Manuf. Sci. E-T ASME* **2013**, *135*, 031011. [CrossRef]
14. Feng, J.; Wan, M.; Gao, T.Q.; Zhang, W. Mechanism of process damping in milling of thin-walled workpiece. *Int. J. Mach. Tool Manuf.* **2018**, *134*, 1–19. [CrossRef]
15. Adetoro, O.B.; Sim, W.M.; Wen, P.H. An improved prediction of stability lobes using nonlinear thin wall dynamics. *J. Mater. Process Technol.* **2010**, *210*, 969–979. [CrossRef]

16. Song, Q.; Ai, X.; Tang, W. Prediction of simultaneous dynamic stability limit of time-variable parameters system in thin-walled workpiece high-speed milling processes. *Int. J. Adv. Manuf. Technol.* **2011**, *55*, 883–889. [CrossRef]

17. Campa, F.J.; De Lacalle, L.N.L.; Celaya, A. Chatter avoidance in the milling of thin floors with bull-nose end mills: model and stability diagrams. *Int. J. Mach. Tool Manuf.* **2011**, *51*, 43–53. [CrossRef]

18. Ding, Y.; Zhu, L. Investigation on chatter stability of thin-walled parts considering its flexibility based on finite element analysis. *Int. J. Adv. Manuf. Technol.* **2018**, *94*, 3173–3187. [CrossRef]

19. Zgüven, H.N. Structural modifications using frequency response functions. *Mech. Syst. Signal Process* **1990**, *4*, 53–63.

20. Alan, S.; Budak, E.; Zgüven, H.N. Analytical prediction of part dynamics for machining stability analysis. *Int. J. Autom. Technol.* **2010**, *4*, 259–267. [CrossRef]

21. Budak, E.; TunL, T.; Alan, S.; Zgüven, H.N. Prediction of workpiece dynamics and its effects on chatter stability in milling. *CIRP Annals-Manuf. Technol.* **2012**, *61*, 339–342. [CrossRef]

22. Song, Q.; Liu, Z.; Wan, Y.; Ju, G.; Shi, J. Application of Sherman-Morrison-Woodbury formulas in instantaneous dynamic of peripheral milling for thin-walled component. *Int. J. Mech. Sci.* **2015**, *96*, 79–90. [CrossRef]

23. Tunc, L.T.; Budak, E. Effect of cutting conditions and tool geometry on process damping in machining. *Int. J. Mach. Tool Manuf.* **2012**, *57*, 10–19. [CrossRef]

24. Altintas, Y.; Budak, E. Analytical prediction of stability lobes in milling. *CIRP Annals* **1995**, *44*, 357–362. [CrossRef]

25. Smith, S.; Tlusty, J. Efficient simulation programs for chatter in milling. *CIRP Annals* **1993**, *42*, 463–466. [CrossRef]

26. Insperger, T.; Stépán, G. Updated semi-discretization method for periodic delay-differential equations with discrete delay. *Int. J. Numer. Method Eng.* **2004**, *61*, 117–141. [CrossRef]

27. Bayly, P.V.; Halley, J.E.; Mann, B.P.; Davies, M. Stability of interrupted cutting by temporal finite element analysis. *J. Manuf. Sci. E-T ASM* **2003**, *125*, 220–225. [CrossRef]

28. Ding, Y.; Zhu, L.M.; Zhang, X.J.; Ding, H. A full-discretization method for prediction of milling stability. *Int. J. Mach. Tool Manuf.* **2010**, *50*, 502–509. [CrossRef]

Article

Effects of Inoculation on the Pearlitic Gray Cast Iron with High Thermal Conductivity and Tensile Strength

Guiquan Wang [1], Xiang Chen [1,2,*], Yanxiang Li [1,2] and Zhongli Liu [3]

[1] School of Materials Science and Engineering, Tsinghua University, Beijing 100084, China; wgq15@mails.tsinghua.edu.cn (G.W.); yanxiang@tsinghua.edu.cn (Y.L.)
[2] Key Laboratory for Advanced Materials Processing Technology, Ministry of Education, Beijing 100084, China
[3] School of Nuclear Equipment and Nuclear Engineering, Yantai University, Yantai 264005, China; liuzhonglimanoir@163.com
* Correspondence: xchen@tsinghua.edu.cn; Tel.: +86-10-6278-6355

Received: 5 September 2018; Accepted: 28 September 2018; Published: 1 October 2018

Abstract: With the aim of improving the thermal conductivity and tensile strength of pearlitic gray cast iron, the influence of inoculation on structure and properties was experimentally investigated. Three group of irons with similar compositions were inoculated by Zr-FeSi, Sr-FeSi, and SiC inoculants, respectively. The metallographic analysis was used to measure the maximum graphite length, primary dendrites amount and eutectic colonies counts. For a certain carbon equivalent, it was confirmed that the thermal conductivity of pearlitic gray cast iron has a direct correlation with the maximum graphite length while the tensile strength was influenced mainly by the primary dendrites amount. The optimal structure and highest thermal conductivity and tensile strength were obtained by Sr-FeSi inoculant. MnS particles act a pivotal part in modifying the structure of gray cast iron. It was found that providing nucleation sites both for graphite and primary austenite is important to promote the thermal conductivity and strength. However, excessive nuclei (MnS particles) results in shorter graphite flakes and thus the depressive growth of primary dendrites.

Keywords: thermal conductivity; tensile strength; inoculation; gray cast iron

1. Introduction

Although gray cast iron (GCI) has been considered as the primary choice to produce vital engine components for many decades, the service life is not satisfied yet. Due to the alternating firing loads and rapid thermal cycles, the failure modes in these parts are high cycle fatigue and thermal-mechanical fatigue [1]. The resistances to the fracture of these GCIs are both closely related to tensile strength and thermal conductivity [2,3]. It is, thus, necessary to develop high performance cast irons (HPCI) with high thermal conductivity while maintaining the high tensile strength.

As presented by Riposan [4], pearlitic gray cast iron, consisting of a fully pearlitic matrix and evenly distributed A-type graphite, is preferred due to its optimum comprehensive properties among all gray cast irons. However, there was a competitive relationship between the thermal conductivity and tensile strength of pearlitic GCI. Generally, the principal methods of approving tensile strength are to decrease the fraction of graphite flakes as well as to reduce their length [5,6]. On the contrary, the thermal conductivity can benefit from increasing of the graphite amount and graphite size [7]. It was reported that high tensile strength could be favored by developed primary dendrites [8], while a negative effect of the decrease of eutectic colonies size on the thermal conductivity was suggested due to a larger number of matrix discontinuities between graphite skeletons [9]. However, very few investigations have directly carried out to simultaneous promote both the thermal conductivity and tensile strength.

Inoculation has been considered as an effective method to improve the performance of GCI at low cost. It has been showed that Al, Ce, and Zr can increase the dendrites amount by directly influencing the nucleation of austenite [10]. Riposan found that Sr-FeSi was efficient in refining the eutectic colonies and nucleating the primary austenite [11]. As reported by Edalati, a homogeneous structure consisting of the uniform distribution of A-type graphite and increased eutectic colonies count could be obtained by SiC inoculation [12]. Nevertheless, limited work has been performed to study the effects of inoculation both on the microstructure and properties, especially thermal conductivity and tensile strength.

This work was undertaken to find out the possibility of the development of HPCI with a combination of high thermal conductivity and tensile strength through inoculation. Samples with various microstructure were produced using different inoculants, including Zr-FeSi, Sr-FeSi, and SiC inoculants. The influence of microstructural characteristics, mainly including of graphite length and primary dendrite percentage, was investigated to clarify the role of structural features in affecting both of thermal conductivity and strength. A detailed analysis of electronic microscopy was provided with particular attention given to the mechanism of the evolution of the microstructure.

2. Experimental Details

Nine gray cast iron ingots were melted in a 500 kg, medium-frequency induction furnace. The charge consisted of 70 wt% steel scrap and 30 wt% pig iron. Ferromolybdenum, ferromanganese, ferrosilicon and carburizer were used to meet the requirements of the composition. After superheating to 1530 °C, the liquid iron was transferred into a ladle and then poured into an EN-1561 Type II mould. Inoculants were deposited on the bottom of the ladle before pouring. The nominal composition of the samples and inoculants is given in Table 1. Carbon equivalent (CE) is calculated by the formula CE = C% + 0.31 Si% + 0.33 P%.

Table 1. The additives and chemical composition (wt%) of samples and inoculants.

Number	C	Si	Mn	P	S	Mo	Cu	Sn	CE	Additive
S1-Z	3.39	1.64	0.46	0.027	0.025	0.35	0.55	0.057	3.91	0.4 wt% Ino_1
S2-Z	3.53	1.59	0.51	0.029	0.028	0.34	0.54	0.060	4.03	0.4 wt% Ino_1
S3-Z	3.67	1.51	0.51	0.028	0.029	0.34	0.54	0.059	4.15	0.4 wt% Ino_1
S1-S	3.42	1.63	0.49	0.028	0.027	0.35	0.58	0.061	3.94	0.4 wt% Ino_2
S2-S	3.54	1.62	0.51	0.025	0.028	0.35	0.58	0.060	4.05	0.4 wt% Ino_2
S3-S	3.69	1.59	0.51	0.027	0.030	0.35	0.58	0.061	4.19	0.4 wt% Ino_2
S1-C	3.34	1.80	0.50	0.028	0.025	0.36	0.58	0.061	3.91	0.8 wt% Ino_3
S2-C	3.53	1.70	0.52	0.028	0.026	0.36	0.58	0.061	4.07	0.8 wt% Ino_3
S3-C	3.64	1.61	0.51	0.028	0.028	0.36	0.58	0.060	4.15	0.8 wt% Ino_3
Ino_1					Zr-FeSi (2.6 wt% Zr)					
Ino_2					Sr-FeSi (2.0 wt% Sr)					
Ino_3					SiC					

Tensile strength was measured using dog-bone shaped bars with 20 mm diameter in the gauge section, 60 mm gauge length and 3.2 μm surface finish according to Chinese Standard GB/T T228.1-2010. Three tests were performed for each composition and the average value was taken. Disk specimens with a diameter of 12.5 mm and a thickness of 2.5 mm were then cut from the grip section and used to determine the thermal diffusivity (α) and heat capacity (c_p) at room temperature using a NETZSCH LFA 457 laser flash apparatus (NETZSCH GABO instruments GmbH, Ahlden, Sachsen, Germany). The volume of the specimen was measured by the Archimedes method based on the fact that an object placed in a liquid displaces a volume of liquid equals to the volume of the object. The density (ρ) was obtained by weighting the specimen and dividing by the volume. And the thermal conductivity (λ) is calculated by:

$$\lambda = \alpha\, c_p \tag{1}$$

Optical microscopy and field-emission scanning electron microscopy (SEM, JEOL, Akishima, Tokyo, Japan) equipped with an energy-dispersive X-ray spectroscopy (WDS) were used to analyze the microstructure. The length and area fraction of graphite were then evaluated in an unetched condition by quantitative metallography with the software Image J Pro (Version 6.0, Media Cybernetics, Rockville, MD, USA). The samples were then etched by 4% nital to expose the matrix phases. The eutectic colonies were revealed by the Stead Le Chatelier etchant (4 g $MgCl_2$, 1 g $CuCl_2$, 2 mL HCl, 100 mL alcohol). Color etching was also performed to evaluate the area fraction of primary austenite. The color etchant was 50 g NaOH and 4 g picric acid dissolved in 100 mL distilled water. The etching procedure was carried out at 98 °C for six minutes. The volume fraction of a phase is simply assumed as the area fraction occupied by the phase on the metallographic specimen. A total of eight fields were measured on each specimen's cross-section.

3. Results

3.1. Metallographic Analysis

The typical images showing the matrix of irons inoculated by different inoculants are provided in Figure 1. According to GB/T 7216-2009, homogeneous structure consisting of a fully pearlitic matrix and evenly distributed A-type graphite was observed in all irons. Similar lamellar spaces of pearlite can also be found among different inoculations (as shown in Figure 1c,d). Other structural information of all investigated samples was shown in Figures 2–4 according to similar CE. Corresponding microstructural characteristics were summarized in Table 2. The average of the three longest flakes in the field of view was taken as the maximum length since most of the graphite flakes are incomplete in a random 2D section. With the increase of CE, the increase of graphite content and decrease of primary dendrite amount were found. Additionally, no influence of inoculation on the graphite content was observed. The important differences between various inoculation mainly appear in the maximum flake length, primary dendrite percentage and eutectic colonies size. For a similar CE, the Zr-FeSi inoculant resulted in the shortest graphite flakes, moderate primary dendrite and moderate size of the eutectic colonies. The longest graphite length, the highest primary dendrite percentage and the smallest eutectic colonies size were found in the samples inoculated by Sr-FeSi. The maximum graphite length in SiC inoculated samples is similar to that in Sr-FeSi inoculated ones. Moreover, SiC inoculated the lowest dendrite amount and the biggest eutectic colonies. As shown in Figure 5, for similar CE (graphite percentage), a clear linear relationship between primary dendrite percentage and eutectic colonies was found in the present work. Therefore, the primary dendrite percentage and graphite size are mainly concerned in the subsequent analysis.

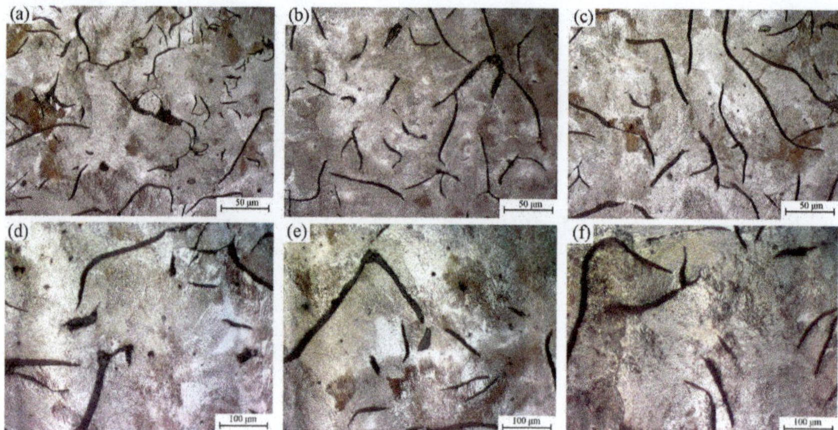

Figure 1. Optical images showing the matrix of samples inoculated by Zr-FeSi (**a**,**d**), Sr-FeSi (**b**,**e**), and SiC (**c**,**f**).

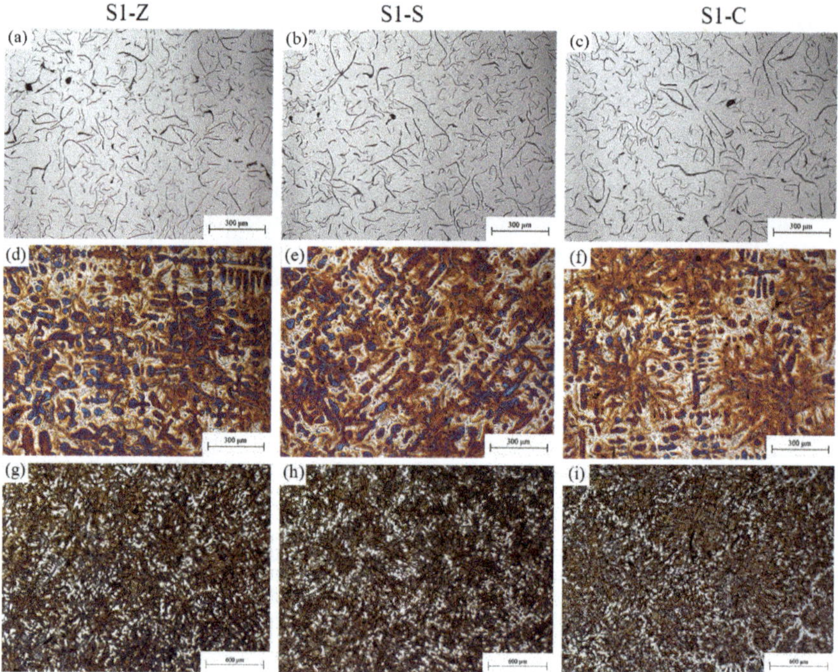

Figure 2. Metallographic images of S1-Z (**a**,**d**,**g**), S1-S (**b**,**e**,**h**) and S1-C (**c**,**f**,**i**), showing graphite (**a**–**c**), primary dendrite (**d**–**f**), and eutectic colonies (**g**–**i**).

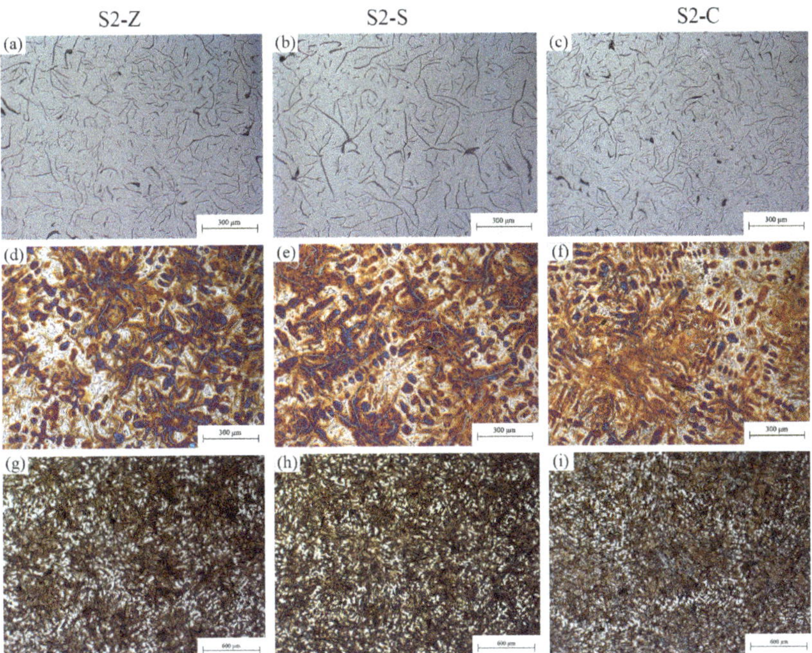

Figure 3. Metallographic images of S2-Z (**a,d,g**), S2-S (**b,e,h**) and S2-C (**c,f,i**), showing graphite (**a–c**), primary dendrite (**d–f**), and eutectic colonies (**g–i**).

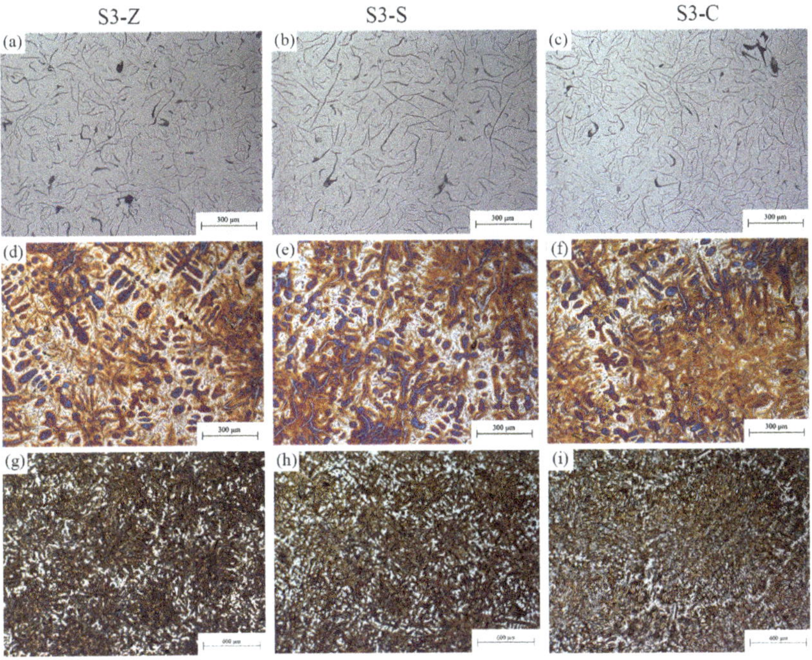

Figure 4. Metallographic images of S3-Z (**a,d,g**), S3-S (**b,e,h**) and S3-C (**c,f,i**), showing graphite (**a–c**), primary dendrite (**d–f**), and eutectic colonies (**g–i**).

Table 2. Microstructural characteristics of samples. Errors are ± one standard deviation.

Sample No.	Graphite Type	Graphite Pct. (%)	Max. Length of Graphite (μm)	Primary Dendrite Pct. (%)	Eutectic Colonies Count (/cm^2)
S1-Z	A	8.6 ± 0.5	214 ± 9	16.7 ± 0.7	258 ± 17
S2-Z	A	8.9 ± 0.4	239 ± 23	13.3 ± 0.6	242 ± 29
S3-Z	A	10.1 ± 0.6	259 ± 26	11.5 ± 0.6	263 ± 35
S1-S	A	8.5 ± 0.4	233 ± 13	18.2 ± 0.5	423 ± 29
S2-S	A	9.0 ± 0.2	273 ± 19	15.6 ± 0.9	371 ± 19
S3-S	A	9.9 ± 0.3	288 ± 25	13.1 ± 0.5	283 ± 22
S1-C	A	8.9 ± 0.3	242 ± 16	14.8 ± 0.6	123 ± 34
S2-C	A	9.4 ± 0.5	244 ± 20	10.6 ± 0.4	57 ± 19
S3-C	A	10.0 ± 0.2	260 ± 17	8.6 ± 0.5	42 ± 0

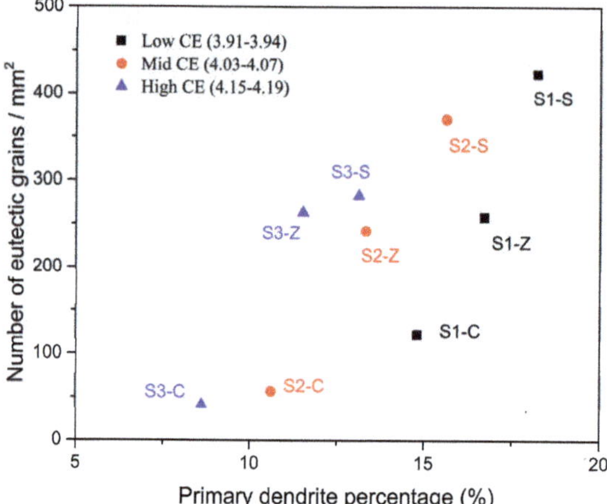

Figure 5. The primary dendrite percentage versus the number of eutectic colonies.

3.2. SEM Analysis

Typical SEM microstructures treated by different inoculants were shown in Figure 6. MnS particles were found to be embedded in the matrix and in superficial contact with graphite in all the samples. Differences mainly appear in the morphology and the distribution of particles. X oxides (X = Zr or Sr) were found in the core of most MnS particles in X-FeSi inoculated samples, while no visible inclusions in MnS were observed in SiC inoculated ones. In SiC inoculated samples, large size, and clustered MnS particles were observed as shown in Figure 6c,d. On the contrary, a significant small and evenly distributed MnS were found in X-FeSi inoculated samples.

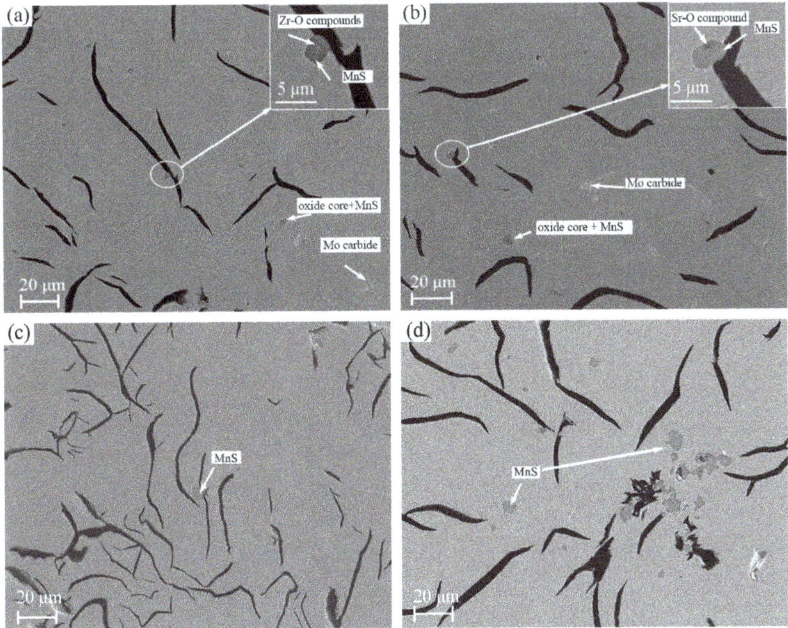

Figure 6. SEM images showing the morphology of MnS particles in S1-Z (**a**), S1-S (**b**), and S1-C (**c**,**d**).

A statistical analysis was performed on the same samples to count the MnS particles. The average value of 8 SEM images at a random location of the polished surface of samples was recorded as the count. The results are provided in Figure 7. Remarkable differences were observed in the count of MnS particles depending on the selected inoculant. The Zr-FeSi inoculated samples have the largest number of MnS particles while the least number was found in SiC inoculated irons.

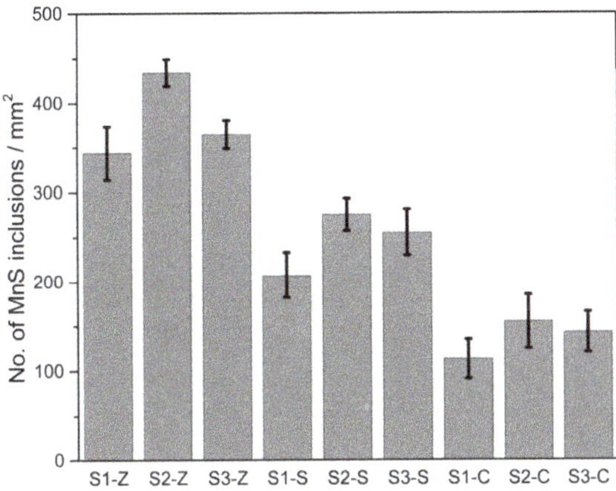

Figure 7. Statistical analysis of the quantity of MnS particles. Error bars are ± one standard deviation.

3.3. Tensile and Thermal Properties

The tensile strength and thermal conductivity of all compositions are provided in Figure 8. For a certain inoculation process, there is a clear negative correlation between tensile strength and thermal conductivity. However, for various inoculation, this relationship is untenable. The highest tensile strength and the highest thermal conductivity were achieved by the samples inoculated by Sr-FeSi. The Zr-FeSi inoculated irons have higher strength but lower thermal conductivity than that inoculated by SiC. It is suggested that improved tensile and thermal properties can be obtained simultaneously by good inoculation.

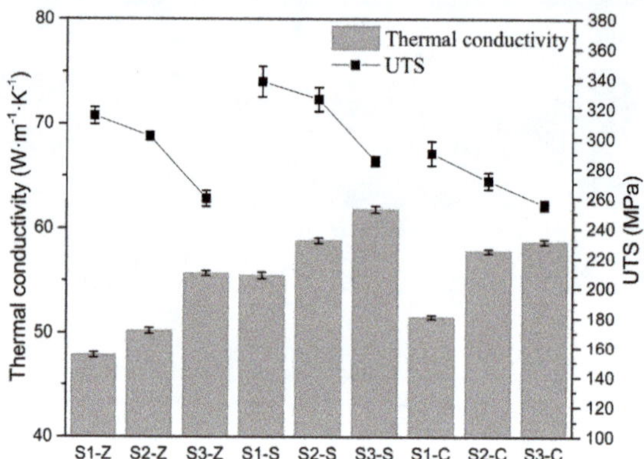

Figure 8. Tensile strength and thermal conductivity of all compositions. Error bars are ± one standard deviation.

4. Discussion

It is well known that the properties of as-cast GCI are affected by chemical composition and inoculation. The graphite amount is mainly depended on the CE, which transforms the effect of elements on the graphite precipitation into relative content of carbon. As shown in Figure 8 and Table 2, the increasing of graphite amount has a clear positive effect on the thermal conductivity but a negative effect on the strength. It can be explained by the double-edge of graphite: improving the heat conduction but dissevering the matrix. However, for a wide range of CE, the inoculation in the current work clearly modifies the microstructure and thus change the thermal and tensile properties of GCI, as shown in Table 2 and Figure 8.

The high tensile and thermal properties of Sr-FeSi inoculated GCI may result from more developed primary dendrites and longer graphite flakes. As presented in Figure 9a, the heat conductivity is found to increases with increasing maximum flake length. Although it was expected that an increased primary dendrite amount will reduce the thermal conductivity for the increasing matrix bridges over which the heat has to pass, such an effect was not established in the present work. The observed harmful effects of increasing primary dendrite amount on thermal conductivity for the same inoculation could be the results of decreasing CE and thus graphite amount. Contrary to thermal conductivity, the strength was mainly determined by the primary dendrite amount, as shown in Figure 9b. The more amount of primary dendrite, the higher strength. The impacts of graphite length are weakly, indecisive at least. The fact that tensile strength was largely determined by primary dendrite amount supports the theory that the eutectic colonies and the graphite flakes can extend over the primary arms without affecting the material strength.

As presented by Riposan, for FeSi containing deoxidizing elements X (X = Sr or Zr), X promotes the formation of small oxide micro-inclusions at high superheating temperature [13]. The precipitated oxides provide the substrate on which MnS can nucleate and grow. The so-called (Mn, X)S compound consisting of oxide and MnS provides lots of nuclei sites for primary austenite and eutectic graphite. For SiC [14], the inoculant dissolves into the melt through the reaction:

$$SiC + Fe \rightarrow FeSi + C \text{ (dissociative)} \tag{2}$$

The generated dissociative carbon then provides the nucleation sites for graphite because of a high activity and zero mismatch. The observations as shown in Table 2 and Figure 6 support the theories mentioned above, which can explain why the fewest dendrites and the smallest number of MnS particles were observed in SiC inoculated alloys.

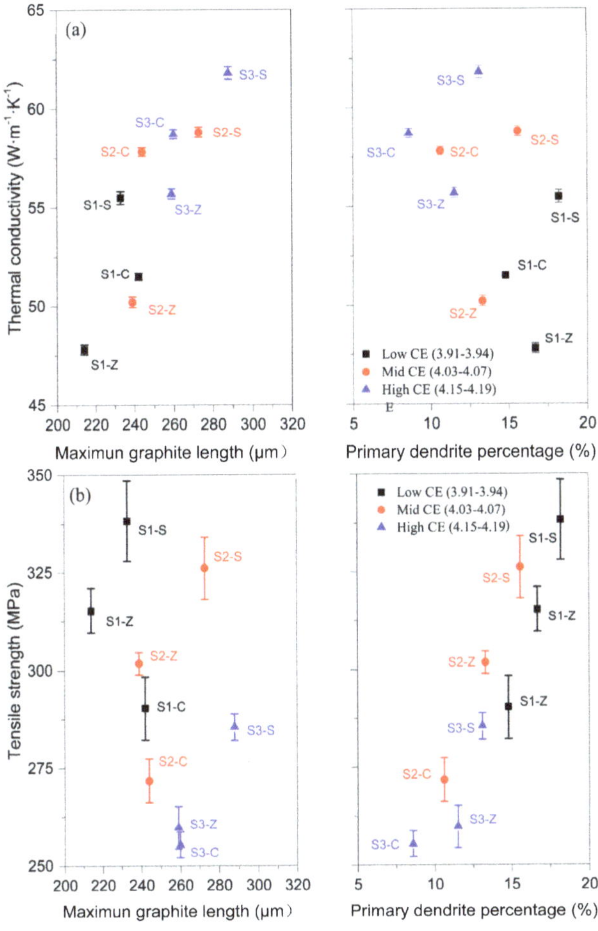

Figure 9. Relationship between the structural characteristics and properties: thermal conductivity (**a**) and tensile strength (**b**). Errors are ± one standard deviation.

Even though the Zr-FeSi inoculated samples have the largest number of MnS particles, which were considered as the effective nuclei for primary austenite and graphite, the most developed primary dendrites were found in Sr-FeSi inoculated GCI. The probable explanation can be provided by the arguments involving nucleation and growth kinetics of the phases. As reported by Rivera [15], in inoculated hypoeutectic melt solidification starts with the independent nucleation of austenite dendrites and graphite. When the dendrites grow and come into contact with the graphite as temperature drops, the units of lamellar graphite and austenite grow cooperatively and finally form the eutectic colonies. The larger amount of (Mn, Zr)S compounds in Zr-FeSi inoculated irons means more nuclei site for the austenite and graphite at the beginning of solidification and, thus, more opportunity to the interaction of primary austenite and graphite. As a result, the growth of primary dendrite is suppressed because of the earlier growth of the eutectic colonies. The fact that shorter graphite flakes and larger eutectic colonies were found in Zr-FeSi inoculated irons supports this assumption. It is suggested that the reason for the optimal structure of Sr-FeSi inoculated GCI is that Sr leads to moderate nucleation site density both for primary austenite and graphite.

5. Conclusions

For a similar CE, it was confirmed that the dominant structural factors in increasing thermal conductivity and tensile strength of pearlitic GCI are different. The thermal conductivity is determined by maximum graphite length while the tensile strength is mainly affected by primary dendrites amount. Long graphite flakes and developed dendrites are preferred for high thermal conductivity and strength.

In practice, the optimal structure can be obtained by good inoculation. While SiC additions inoculated long A-type graphite flakes, it did not appear to provide nucleation sites for primary austenite. Zr-FeSi and Sr-FeSi inoculated both primary austenite and graphite by promoting the nucleation of MnS at high temperature. However, the optimal structure and properties were found in Sr-FeSi inoculated irons. It is probably that Sr-FeSi inoculant provided the appropriate size and number of MnS particles. More dispersive MnS particles in Zr-FeSi inoculated irons resulted in shorter graphite and fewer dendrites amount because of excessively nucleation site.

Author Contributions: G.W. designed and performed the experiments; X.C. proposed the topic of this work; Y.L. and Z.L. supervised the metallographic analysis. G.W. and X.C. wrote the initial manuscript in discussions with all co-authors.

Funding: This research received no external funding.

Acknowledgments: The authors wish to express sincere thanks to Yuan Liu and Huawei Zhang for helpful discussion and encouragement during this investigate.

Conflicts of Interest: The authors declare no conflict of interest.

References

1. Langmayr, F.; Zieher, F.; Lampic, M. The thermo-mechanics of cast iron for cylinder heads. *MTZ Worldw.* **2004**, *65*, 17–20. [CrossRef]

2. Seifert, T.; Riedel, H. Mechanism-based thermomechanical fatigue life prediction of cast iron. Part I: Models. *Int. J. Fatigue* **2010**, *32*, 1358–1367. [CrossRef]

3. Holmgren, D.; Svensson, I.L. Thermal conductivity-structure relationships in grey cast iron. *Int. J. Cast Met. Res.* **2005**, *18*, 321–330. [CrossRef]

4. Riposan, I.; Chisamera, M.; Stan, S. New developments in high quality grey cast irons. *China Foundry* **2014**, *11*, 351–364.

5. Collini, L.; Nicoletto, G.; Konecna, R. Microstructure and mechanical properties of pearlitic gray cast iron. *Mater. Sci. Eng. A* **2008**, *488*, 529–539. [CrossRef]

6. Xu, W.; Ferry, M.; Wang, Y. Influence of alloying elements on as-cast microstructure and strength of gray iron. *Mater. Sci. Eng. A* **2005**, *390*, 326–333. [CrossRef]

7. Helsing, J.; Grimvall, G. Thermal-Conductivity of Cast-Iron—Models and Analysis of Experiments. *J. Appl. Phys.* **1991**, *70*, 1198–1206. [CrossRef]

8. Fourlakidis, V.; Dioszegi, A. A generic model to predict the ultimate tensile strength in pearlitic lamellar graphite iron. *Mater. Sci. Eng. A* **2014**, *618*, 161–167. [CrossRef]

9. Ormerod, J.; Taylor, R.; Edwards, R.J. Thermal-Diffusivity of Cast Irons. *Met. Technol.* **1978**, *5*, 109–113. [CrossRef]

10. Chisamera, M.; Riposan, I.; Stan, S.; Militaru, C.; Anton, I.; Barstow, M. Inoculated Slightly Hypereutectic Gray Cast Irons. *J. Mater. Eng. Perform.* **2012**, *21*, 331–338. [CrossRef]

11. Riposan, I.; Chisamera, M.; Stan, S.; Skaland, T. Graphite nucleant (microinclusion) characterization in Ca/Sr inoculated grey irons. *Int. J. Cast Met. Res.* **2003**, *16*, 105–111. [CrossRef]

12. Edalati, K.; Akhlaghi, F.; Nih-Ahmadabadi, A. Influence of SiC and FeSi addition on the characteristics of gray cast iron melts poured at different temperatures. *J. Mater. Process. Technol.* **2005**, *160*, 183–187. [CrossRef]

13. Riposan, I.; Chisamera, M.; Stan, S.; Hartung, C.; White, D. Three-stage model for nucleation of graphite in grey cast iron. *Mater. Sci. Technol.* **2010**, *26*, 1439–1447. [CrossRef]

14. Xue, W.D.; Li, Y. Pretreatments of gray cast iron with different inoculants. *J. Alloy Compd.* **2016**, *689*, 408–415. [CrossRef]

15. Rivera, G.; Calvillo, P.R.; Boeri, R.; Houbaert, Y.; Sikora, J. Examination of the solidification macrostructure of spheroidal and flake graphite cast irons using DAAS and ESBD. *Mater. Charact.* **2008**, *59*, 1342–1348. [CrossRef]

Article

Research on a New Localized Induction Heating Process for Hot Stamping Steel Blanks

Li Bao [1,2], Jingqi Chen [1], Qi Li [1], Yu Gu [1], Jian Wu [1] and Weijie Liu [1,*]

[1] School of Material Science and Engineering, Northeastern University, Shenyang 110006, Liaoning, China; lwjxs2012@163.com (L.B.); chenjq@163.com (J.C.); liqi.hnxc@gmail.com (Q.L.); cicielf1212@hotmail.com (Y.G.); wujianneu@163.com (J.W.)

[2] School of Mechanical and Electrical Engineering, Qiqihar University, Qiqihar 161006, Heilongjiang, China

* Correspondence: lwjxs2019@163.com

Received: 8 March 2019; Accepted: 25 March 2019; Published: 28 March 2019

Abstract: Localized induction heating with one magnetizer was experimentally analyzed in order to investigate the altering effect of the magnetizer on the magnetic field. A 22MnB5 blank for tailored property was locally heated to produce the parts of a car body in white, such as the B-pillars. A lower-temperature region with a temperature in the two-phase zone and a full-austenitic high-temperature region were formed on the steel blank after 30 s. After water-quenching, the mixture microstructure (F + M) and 100% fine-grained lath martensite were obtained from the lower- and high-temperature regions, respectively. Moreover, the ultimate tensile stress (UTS) of the parts from the lower- and high-temperature regions was 977 and 1698 MPa, respectively, whereas the total elongations were 17.5% and 14.5%, respectively. Compared with the parts obtained by conventional furnace heating–water quenching (UTS: 1554 MPa, total elongation: 12%), the as-quenched phase developed a tensile strength over 100 MPa greater and a higher ductility. Thus, the new heating process can be a good foundation in subsequent experiments to arbitrarily tailor the designable low-strength zone with a higher ductility by using magnetizers.

Keywords: localized inductive heating; hot stamping steel blanks; tailored properties; magnetizer

1. Introduction

One of the significant technical achievements during the last few decades in the automobile industry was the use of lightweight design to reduce the fuel consumption and greenhouse gas emission of vehicles. When the weight of a vehicle is reduced by 10%, its fuel consumption and emission will decline by 8–10% and 4–6%, respectively [1]. Although low-density materials, such as aluminum, magnesium, and/or carbon fiber-reinforced polymers, can be used to make lightweight automobile parts, their costs are considerably higher than steels [2]. As a result, steels, especially the advanced high strength steel blanks (AHSS), remain the most important materials for stamping automobile parts. Hot stamping, which transforms the microstructure of steel blanks into the stronger martensite phase while being formed, has become a famous industrial process for manufacturing lightweight parts in recent years. The applications of the above lightweight parts are important technical advances in the automobile industry, but the general invariant mechanical properties limit the maximization of the overall performances, which is a limitation that should be overcome. Obviously, it is a tailorable rather than a constant distribution of mechanical properties that facilitates the optimization of the overall performance of a stamped part, including its weight, efficiency, passenger safety, and cost [3]. For example, a B-pillar with higher strength in the upper region than in the bottom region may better protect passengers and maximize energy absorption during a vehicle crash. Such blanks with tailored mechanical properties are usually fabricated using laser welding that joins two or more tailored metal blanks, which all have different strength, thickness, or material. For example, thickness-difference blanks with tailored properties can also be produced by special rolling mails with sectional roll gaps, each of which has a different gauge [4]. Furthermore, equal-thickness and homogenous steel blanks with tailored properties can also be obtained if their microstructure can be changed locally during hot stamping.

In detail, three strategies including specific austenitization, specific cooling, and specific annealing will be introduced [5]. An absorption component which was brought into contact with the blank to suppress its austenitization was proposed [6], and then the ductile zones were formed. Two methods including spray cools and masked austenitization to generate the parts with tailored properties were presented [3]. The heated and cooled tool by varying cooling rates to produce the parts with tailored properties was designed [7]. As for specific annealing, localized tempering of the transformed martensite phase to generate ductile properties was put forward [8]. Although all the proposed techniques are capable of creating a tailored property effect by forming a mixed microstructure of martensite (hard phase) and ferrite or tempered martensite (soft phase), they are either too complicated or too time consuming to be operated and too expensive, so none of them have been industrially applied. In addition, all known parts with tailored property-related technologies, either applied in industries or under development in the laboratory, lack flexibility in designing and tailoring the property distributions on blanks, which may weaken the effort of overall performance optimization for the lightweight parts.

To find a practical process to produce hot stamped parts with finely designable tailored properties, our solution is based on induction heating, which is not a new technology and has been widely used in parts such as crankshafts, sprockets, steel tubes, and gears. However, it is currently suitable for the overall heating of small parts and the surface heating of large parts. There are many theoretical, experimental, and simulation studies on induction heating. As for the blanks for hot stamping, induction heating was first applied on 22MnB5 by a combination of a longitudinal and a transverse magnetic field and then quenched to generate the uniform martensite microstructure [9]. As an important part, the magnetizer played an important role in the induction heating. For instance, a working coil with magnetic flux concentrators to enhance the uniform heating effect on the barrel was applied [10]. Moreover, the magnetizer was used to shield the magnetic flux line to develop the uniform effect on the nonplanar mold surface [11]. With the same purpose, the effect of different magnetizer dimensions on the homogeneous heating of workpieces was investigated [12]. These induction heating methods developed the temperature distribution of the heated workpiece more uniformly with or without the magnetizer, however, they have not been used for a stamped part with tailored properties.

In this article, a new localized induction heating method was proposed. Specifically, the uniform alternating electromagnetic field generated by a flattened solenoid induction coil can be locally altered by a magnetizer piece, so that the eddy current strength in the steel blank placed in the coil can be locally weakened, leading to the formation of ferrite ponds surrounded by austenite phases. Thus, after stamping by a water-cooled mold, the austenite in the blank transformed into hard martensite and the ferrite remained. Finally, a stamped part with tailored properties was produced.

2. Experimental Method

2.1. Material

The 22MnB5 blanks with a nominal thickness of 3.5 mm (Baogang Group) were used here. The chemical composition of the material is shown in Table 1. The steel has a Curie temperature of 760 °C (T_C), below which it is ferromagnetic and above which the loss of ferromagnetic properties will be revealed. Moreover, the basic information of the steel blanks, the inductor, and the magnetizer used in localized induction heating experiments is shown in Table 2.

Table 1. Chemical composition of 22MnB5 steel blanks (wt. %).

Chemical Composition	C	Si	Mn	P	S	Cr	Ti	B	Al	Nb
wt. %	0.22	0.24	1.28	0.01	0.003	0.15	0.03	0.003	0.05	0.002

2.2. Localized Induction Heating Method

An induction coil of a high-frequency inductor was composed of six parallel copper tubes with a spacing of 10 mm to form a flat solenoid. The steel blanks were placed in an internal rectangular space, which was formed by the upper and lower copper tubes and was about 130 mm in length, 200 mm in width, and 15 mm in height. Two kinds of induction heating experiments were carried out through the inductor. The first one was uniform induction heating, and the second experiment was localized induction heating, which was based on the first one. In detail, the dimensions are shown in Table 2. Because the dimension and shape of the magnetizer may have different effects on the temperature field of the steel blank, they will be investigated in a following study. Therefore, this paper only studies the localized heating with a certain size of magnetizer piece as an example.

This solenoid induction coil can produce a uniform and stable alternating electromagnetic field, which can inductively heat up the entire body of a steel blank to a uniform temperature. It was proved that its entire body was heated up quickly and evenly from room temperature to 915–925 °C in 30 s. When a magnetizer piece was placed between the lower layer coils and the center of the steel blank, the inductive heating device became ready for localized heating. The physical model of localized induction heating consisted of a solenoid coil (yellow), a magnetizer (black), and a steel blank (gray, Figure 1b). After 30 s of localized induction heating, the steel blank was quickly transferred for water quenching. Since the quenching time should last at least 60 s [13] and in order to ensure complete quenching, we set the quenching time at 120 s.

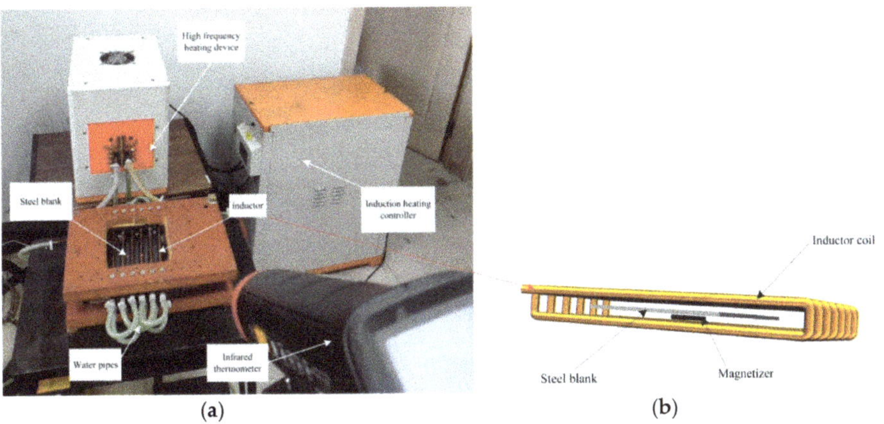

Figure 1. Localized induction heating system. (**a**) Experimental setup; and (**b**) schematic diagram of localized induction heating model.

The induction heating system, which consisted of an inductive coil, a steel blank, a magnetizer, an induction heating device, a controller, and a temperature measuring device, is also depicted in Figure 1. Since the temperature of the workpiece under the induction heating is mostly measured by the infrared measurement method, it can be noticed that a square-shaped window was made in the center of the top insulation plate where the upper coils were mounted, through which the heating up history of the blank could be recorded using an infrared thermometer ($\varepsilon = 0.7$, precision: ± 2 °C, 3i Plus, Raynger, Santa Cruz, CA, USA). Meanwhile, the temperature distribution of the blank was captured by the infrared ray thermal camera ($\varepsilon = 0.7$, precision: ± 2 °C, SC620, FLIR, Wilsonville, OR, USA). The infrared thermal camera was handheld by an operator and is not shown in Figure 1a.

Table 2. Physical dimensions.

Item	Parameter	Units
Blank material	22MnB5	-
Blank dimension	140 (L) × 130 (W) × 3.5 (H)	mm
Magnetizer material	Mn–Zn Ferrite	-
Initial relative permeability of magnetizer	400	
Magnetizer	50 (L) × 10 (W) × 4 (H)	mm
Coil material	Copper	-
Coil dimension	8.5 (outer and inner diameter)	mm
Frequency	100	kHz
Maximum power	100	kW
Total heating time	30	s
Initial and environmental temperature	25	°C
Distance between coil and blank	1	mm
Power supply	SF-C-100	-

2.3. Microstructure Characterization

After a test sample was inductively heated to austenite, it was water quenched to simulate the cooling process during stamping. The microstructures at different locations of the quenched specimen corresponding to two temperature regions were characterized by an optical microscope (OM, LEICA Q550 IW, Wetzlar, Germany) after preparation by standard mechanical grinding procedures and etching in 4% Nital solution. The Rockwell hardness on the upper surface of the quenched specimen was recorded by a Rockwell hardness tester (HR-150A, LLT, Laizhou, China) at a load of 150 N and an interval of 5 mm between the testing points. The data of three samples at each position was recorded and the arithmetic mean was selected as the basis of hardness distribution. The mechanical behavior was characterized using uniaxial tensile tests with three repetitions per microstructure. The blanks were machined into dog-bone specimens with a gauge length equal to 10 mm and a width equal to 3 mm. The tensile tests were performed at room temperature at a nominal strain rate of 0.005/s. The locations of specimens for these tests are shown in Figure 2.

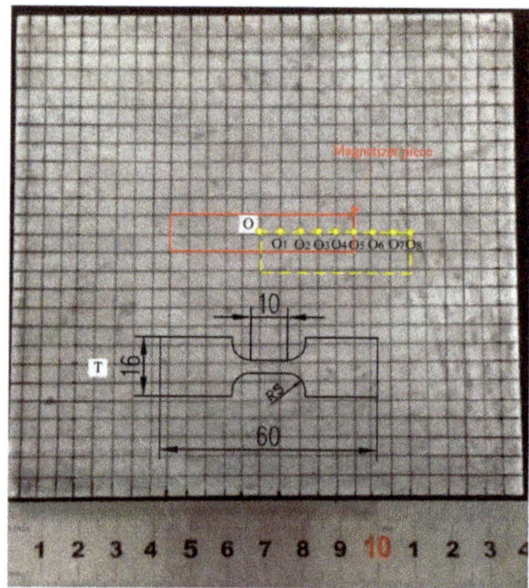

Figure 2. Sample position in the quenched blank. (The red frame: The projection position of the magnetizer piece; O_1–O_8 on yellow frame: Metallographic sampling points location; O and T: Tensile specimen size and location).

3. Results

3.1. The Temperature Field of the Steel Blank after Uniform Induction Heating

Since uniform induction heating is the basis of localized induction heating, the temperature field of the steel blank after 30 s of uniform induction heating and 2 s of transfer is illustrated in Figure 3. Moreover, the heating history profiles of the point P of temperature measurement related to time under uniform heating were compared with the curve under localized induction heating (Figure 3b). Point P was located at the intersection between the place 30 mm from the right edge of the blank and the center line of the blank width, or namely the center of the distance between the second and third copper pipes from the right side. Some representative results from three samples under the same heating process are shown in Figure 3b. Since the samples were treated at the same process conditions, they exhibited low deviations.

A uniform temperature field of the steel blank was induced and the temperature reached 925 °C (Figure 3). The two curves in Figure 3b feature two heating intervals at different heating rates. The inflexion point corresponds to the Curie temperature (T_C = 760 °C), which will rise with the increasing heating rate [13]. Moreover, the rate of the localized heating is higher than that of the uniform heating, arrives at T_C earlier, and has a higher ultimate temperature (950 °C). The reason is attributed to the effect of the magnetizer piece, which will be clearly explained later. After reaching Tc, the heating rate decreases and is nearly constant for two experiments with values ranging from 7.5 to 8.3 K·s^{-1}. This is because, due to the magnetic properties of steel, the heating rate cannot be further raised once the material is paramagnetic [14]. However, the oxide is unevenly distributed on the surface of the uncoated steel blank and the small oxide points are sporadically distributed on the blank, which suggest the oxide may noticeably affect the judgment of temperature distribution of the blank.

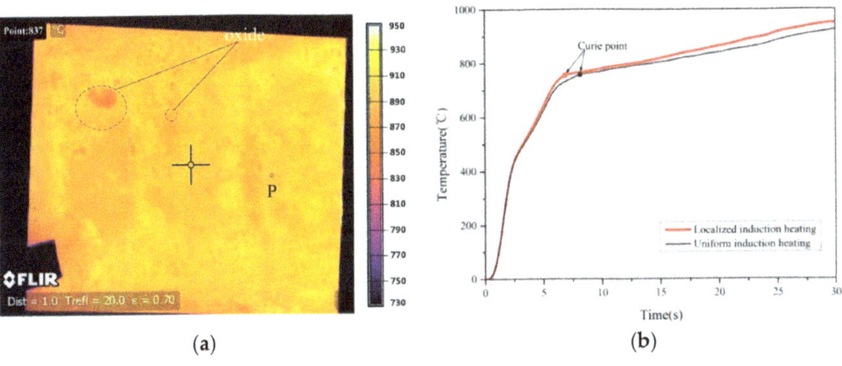

(a)   (b)

Figure 3. The temperature field and heating history curves of the steel blank due to different heating processes. (**a**) The temperature field with the measured point P and typical oxide location under uniform heating; and (**b**) the comparisons of heating history curves under two heating methods.

When the steel blank was removed from the inductor and transferred to the quenching device, its temperature greatly decreased at a rate of ~40 °C/s. This phenomenon little affected the judgment of the temperature distribution of the uniformly heated steel blank because the distribution was uniform, and most areas of the blank were uniformly reduced except the corners. However, this rapid cooling of the blank greatly influenced its temperature distribution under localized heating. The temperature regions on the steel blank after local heating may undergo different cooling rates and mutually integrate due to heat conduction, so the influence of local heating cannot be well reflected and identified. Therefore, temperature rise and distribution of the steel blank in the inductor would be reasonably observed during the localized induction heating.

3.2. Temperature Field of Steel Blank during Localized Induction Heating

The temperature fields of steel blanks investigated during 30 s of localized induction heating are shown in Figure 4, where the blue frame indicates the projection dimension and position of the magnetizer (which is right under the blank at a 1 mm distance) and the black fence indicates the location of the copper tubes. Moreover, the blank upper surface was heated. In detail, since the blank (140 mm long) was slightly longer than the internal space formed by the solenoid, the left edge of the blank was outside the solenoid, which explains the lower temperature at the left edge. Under the influence of the magnetizer piece, a dark area formed in the center of the steel blank representing the lower-temperature region.

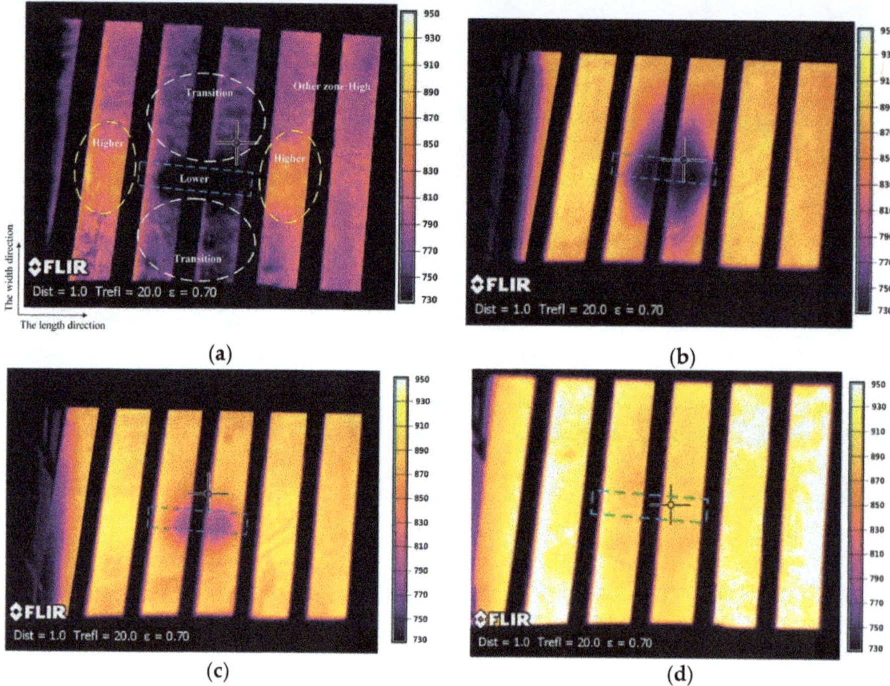

Figure 4. The thermal image of the steel blank during localized induction heating. (**a**) t = 20 s; (**b**) t = 22 s; (**c**) t = 24 s; and (**d**) t = 30 s.

The reason for this phenomenon is attributed to the shielding effect of the magnetizer on the eddy currents induced in the steel blank. The magnetic flux preferentially passed through the magnetizer, due to its higher magnetic permeability, instead of the steel blank which was right above the magnetizer. As a result, the eddy currents induced in this zone were reduced, forming a lower-temperature region. Although the lower-temperature region was obvious, its shape did not exactly match the dimension of the magnetizer piece. The purple dark area partially exceeded the magnetizer width and was named the transition region. Meanwhile, the neighborhood close to both ends of the magnetizer along its length direction was under noticeably higher temperature than at the body of the steel blank and can be named the higher-temperature region. Another region occupying a large proportion on the steel blank was unaffected by the magnetizer and called the high-temperature region. In detail, the positions of different temperature regions are schematically shown in Figure 4a as an example.

As the temperature exceeded T_C, the heating rate greatly decreased (Figure 3b), which was favorable for the steel blank under the localized induction heating and was beneficial to avoid overheating in the higher-temperature region. The heat conduction in the blank reduced the transition region. Furthermore, the lower-temperature region was shortened by the higher-temperature region in the length direction and was shorter than the projection length of the magnetizer piece. Figure 4c depicts the approximate outline of the magnetizer piece on the temperature field of the blank at t = 24 s. The higher-temperature region and other high-temperature regions seemed to have merged to reach 950 °C (Figure 4d). However, the oxides started to be gradually formed after 25 s (Figure 4d). Unfortunately, the ranges of only the high-temperature region and the lower-temperature region can be roughly judged, which means the oxides affected to distinguish between different temperature regions in the heated uncoated blank based on the thermal image at t = 30 s. In summary, after 30 s of heating, the temperature obviously minimized to 820 °C in the lower-temperature region and maximized to 950 °C in the high-temperature region and the transition region was generated on the steel blank. It is confirmed that different regions can be generated on the blank by the magnetizer under the localized induction heating. Nevertheless, the history of the time-related temperature distribution and the hardness distribution should be combined in order to clarify the boundaries of different temperature regions eventually formed in the blank.

3.3. The Tailored Properties Obtained by Water Quenching the Locally Heated Blank

The quenched blank with tailored properties was obtained by localized induction heating and water quenching. The bottom surface of the steel blank was closer to the magnetizer, which more affected different temperature regions, but the temperature distributions at the upper surface and the bottom surface should be similar. Moreover, the hardness on the two surfaces of the quenched blank was measured. It was found that the hardness of the bottom surface was similar to that of the upper surface, and the hardness was lower at the lower-temperature zones and higher at the higher-temperature zones. In order to correspond to the temperature field of the upper surface for temperature measuring, the equal hardness distribution of the upper surface was investigated and is shown in Figure 5b, in which the black frame represents the projection of the magnetizer piece. Moreover, the metallographic microstructures from different temperature regions were taken from the section of the blank every 5 mm along the two centerlines. Typical metallographic microstructures of the quenched blank are depicted in Figure 5a,c, respectively.

The hardness distribution of the quenched blank (Figure 5b) agrees well with the temperature pattern of the blank in Figure 4c. While the temperature at the center of the lower-temperature region (820 °C) corresponds to the Rockwell hardness of 33HRC, and the hardness is more consistent with the high-temperature region occupying the majority of the steel blank than 45HRC. Moreover, the hardness distribution significantly reflects the differential distribution of the temperature field. Obviously, the lower-strength region with higher ductility according to the lower-temperature region is apparently shorter than the magnetizer length. In detail, the lower-temperature region (~35 mm long) is 30% shorter than the projection of the magnetizer. Moreover, the area with the smallest hardness is about 15×5 (unit: mm^2), which will be used to make tensile specimens (Figure 5b). The transition region is about 25 mm wide, compared with the transition region (40 mm) obtained by the mold with differential temperature after the conventional heating [7] or the transition zone (80 mm) of the tailor rolled blanks [15], the transition region (25 mm) obtained in this study is shorter, and it is a gradual transition of the microstructure, which should be applicable for most automotive parts. Furthermore, there are two higher-hardness round zones which may accord with the higher-temperature regions. Fortunately, the hardness is only slightly higher than that according to other high-temperature regions, and the area is smaller.

The typical microstructure image from the lower-temperature regions shows the mixture microstructure consists of ferrite and martensite, whereas the high-temperature region (over 900 °C) is converted to lath martensite after water quenching. The martensite average grain size is less than 10 μm, which is favorable for mechanical properties. Furthermore, on the basis of sampled OM images and the blank center as the origin, the volume fraction of martensite related to distance in the horizontal positive direction as an example was determined on ImageJ. As the distance was prolonged, the temperature gradually rose from the minimum of 820 °C at the center of the lower-temperature region to the maximum of 950 °C (Figure 4d), which means the proportion of austenite and the volume fraction of martensite after quenching also increased with the rise of temperature (Figure 5e).

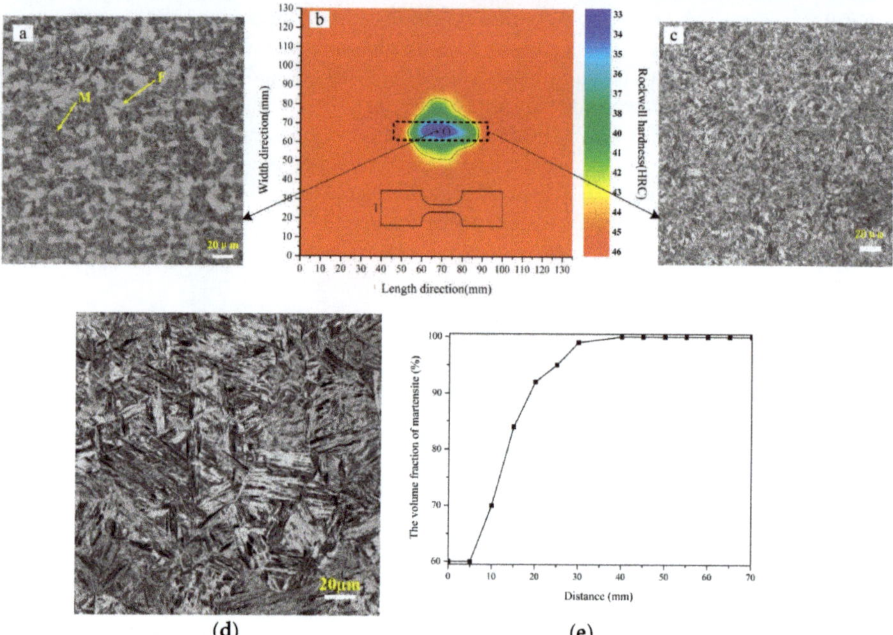

Figure 5. The hardness distribution, optical microscope (OM) images, tensile specimen size and location, the martensite volume fraction. (**a**) Mixture microstructure in the lower-temperature region; (**b**) the hardness distribution and tensile location (O and T); (**c**) martensite in the higher-temperature region; (**d**) martensite obtained by conventional heating–water quenching; and (**e**) martensite volume fraction along the horizontal positive direction from the blank center.

After taking tensile specimens in the lower-temperature region and at different locations of the high-temperature region, the data from two typical locations were selected (Figure 4b). Therefore, different mechanical properties from the two zones were determined (Table 3). Noticeably, the tensile tests were carried out at various positions in the high-hardness region, and the results were similar to the data in Table 3.

Table 3. Mechanical properties of hardened samples from different processes. UTS—ultimate tensile stress, YS—yield strength/$R_{p0.2}$.

Process	Heating/ Quenching Temperature	YS	UTS	Total Elongation (%)	Martensite Volume Fraction (%)
(time)	(°C)	(MPa)	(MPa)	-	-
Localized induction heating–water quenching	820	515	977	17.5	60
(heating: 30 s, quenching: 2 min)	950	1320	1698	14.5	100
Conventional heating–water quenching (heating: 5 min, quenching: 2 min)	950	1050	1554	12.0	100

The average ultimate tensile stress (UTS) of the quenched samples related to the lower-temperature region (820 °C) was 977 MPa with high ductility (total elongation 17.5%), but the UTS of the hardened samples reaches 1698 MPa, which is at least 100 MPa higher than the quenched samples after conventional furnace heating and water quenching (Table 3). Due to grain refinement, the strength and ductility of the quenched part can be improved simultaneously. In conclusion, the quenched part with tailored properties can be fabricated by localized induction heating and water quenching.

4. Discussion

After 30 s of localized induction heating, different temperature regions were formed in the steel blank. The variation in quenching temperature led to the difference in strength and ductility in the as-quenched blank for hot stamping.

4.1. Reasons for Different Temperature Regions on the Heated Blank

Before revealing the designable temperature distribution generated by localized induction heating, the uniform heating of the steel blank within a solenoid should be examined first. The electromagnetic field inside a solenoid is generally uniform. However, as soon as a steel blank enters, the uniformity will be altered to have a denser magnetic flux in the central region of the blank along its length direction. Fortunately, the central concentration is actually balanced by the skin effect at the edge of the workpiece and eventually leads to a uniform inductive heating throughout the entire steel blank. On the basis of uniform induction heating, the localized induction heating will be put forward. A schematic diagram of the magnetic flux path is shown in Figure 6 to easily illustrate the different temperature regions induced on the blank, which is constructed on the basis of the simulation result by the software ANSYS (version 14.5). In order to better understand the change of the magnetic flux field, the schematic diagram of magnetic flux field under uniform heating is also depicted in Figure 6c,d.

As a magnetizer piece was arranged under the steel blank, the magnetic flux lines preferentially passed through it due to its higher magnetic permeability (Figure 6a). The surrounding magnetic flux entered and passed through different planes of the magnetizer, so the magnetizer was attracted to the surrounding magnetic flux. According to the Gauss theorem of magnetic field, the magnetic flux is positively correlated with the product of magnetic flux density and coverage area [13]. As a consequence, the distribution of eddy current, which is induced by the variation of magnetic flux, determines the temperature distribution.

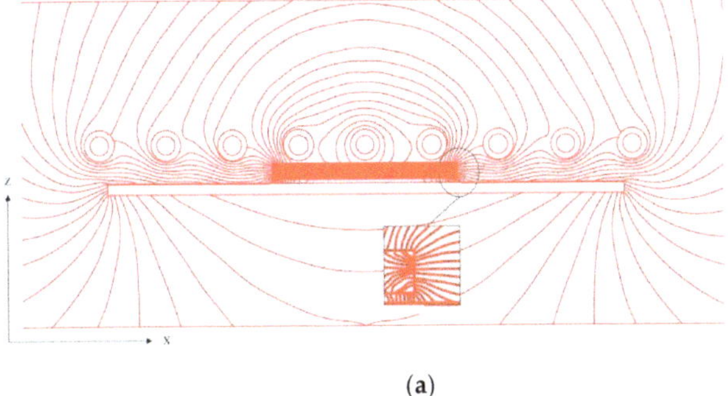

(**a**)

Figure 6. *Cont.*

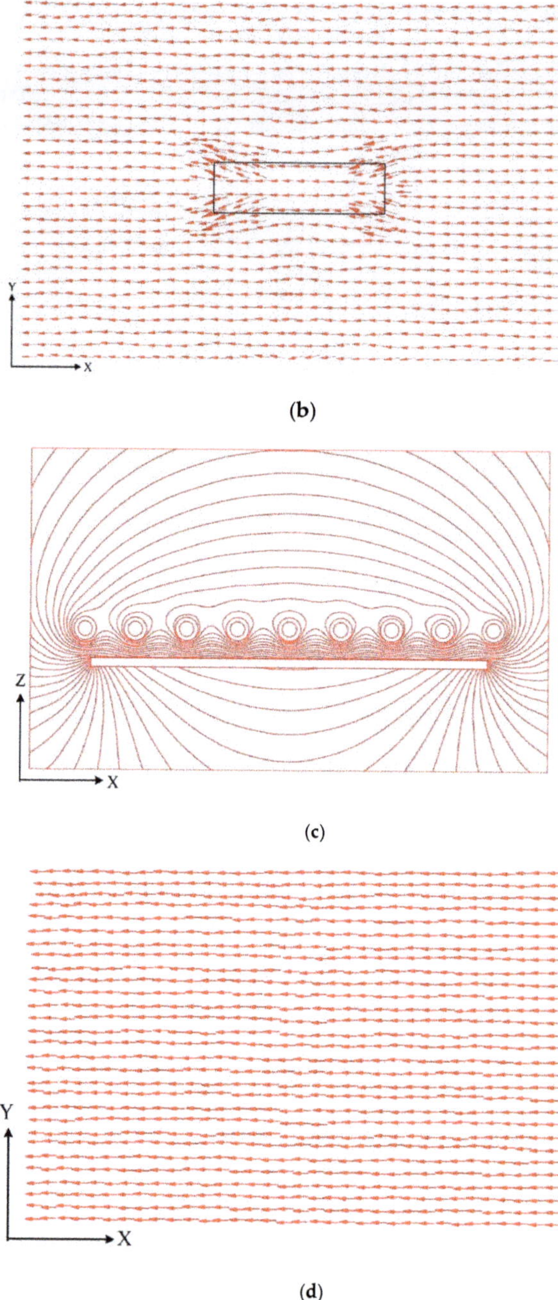

Figure 6. A schematic diagram of the magnetic flux path in two planes with/without a magnetizer piece. (**a**) X–Z plane (cross section of the coils); (**b**) X–Y plane (steel blank surface, the magnetizer depicted as the black frame); (**c**) X–Z plane without magnetizer piece; and (**d**) X–Y plane without magnetizer piece.

Due to the low intensity of the magnetic flux lines right under the magnetizer, the eddy current generated was also small and thereby a lower-temperature region was formed. Along the X direction, the magnetic flux lines concentrated near the two ends of the magnetizer and consequently the higher-temperature region was generated, compared with the other high-temperature regions which was unaffected by the magnetizer. The higher-temperature region may be a little different from other high-temperature regions (Figure 4d), which is because this region first reached the Curie temperature (T_C) and then its heating rate stabilized. The other regions that do not reach the T_C were heated up faster, and the heat in the higher-temperature region was also transferred to other temperature regions during this process. Under the comprehensive effect, the temperature difference between the higher-temperature region and other high-temperature regions decreased.

As revealed in the X–Y plane from Figure 6b, due to the sparseness of the magnetic flux lines right below the magnetizer and the attraction of the magnetizer width section in the Y direction, the nearby magnetic flux lines bent toward it, where the magnetic flux density was small. Consequently, the transition region was formed.

4.2. Reasons for Different Strength of the Quenched Part Caused by Differential Temperature

Since the different temperature regions were formed on the steel blank after localized induction heating and after the results of differential strength regions related to the temperature regions were also obtained, the reasons for the generation of these results were then explored. As reported, the start (A_{C1})–finish (A_{C3}) temperatures rise with the increasing heating rates according to the TTA-diagram of 22MnB5 steel (Time-Temperature-Austenization) [13] and the A_{C1} and A_{C3} at the heating rate of 100 K·s^{-1} correspond to 750 and 900 °C, respectively [14]. In detail, the heating rate 100 K·s^{-1} corresponds before reaching T_C, and the rate after T_C is 10 K·s^{-1}. Since the heating rate (109 K·s^{-1} before T_C and 8.3 K·s^{-1} after T_C) in this study is similar to that of the above research, it is reasonable to estimate that A_{C1} and A_{C3} are about 750 and 900 °C respectively. Therefore, the lower-temperature region with minimum 820 °C on the steel blank obtained here is in the range of the two-phase zone, so the mixed structure of ferrite and martensite can be acquired after water quenching.

Nevertheless, the high temperature of the lower-temperature region (820 °C) resulted in the high volume fraction of austenite transformed. As a consequence, the martensite volume fraction after quenching was 60% and the temperature of the high-temperature region reached 950 °C, which ensured the complete austenite formation. The fine lath martensite from the high-temperature region was transformed after water quenching.

4.3. Advantages of Localized Induction Heating

On basis of the results from the localized induction heating, the new local heating process has three advantages:

(1) The time of the localized induction heating (30 s) is only one-tenth that of the conventional heating method, which means the new heating process, has a greatly enhanced heating efficiency and is more energy saving.

(2) In the existing hot stamping production chain, heating is the important step. The new process does not require the change of other process steps, so the cost of improving the original production chain is low. Because of its short heating time, it shortens the overall hot stamping process. These are all beneficial to its future industrial application.

(3) Since only one magnetizer piece was used to redistribute eddy current induced on the steel blank, the lower-temperature region generated is still small and the temperature is high. On the basis of this result, the tailored temperature region which can customize its pattern, size, and temperature will be investigated in the future by arranging more magnetizers or changing the magnetizer parameters such as size, shape, or permeability. Therefore, the new process is flexible for tailoring the position and pattern of the ductility zones through temperature adjustment.

5. Discussion

To meet the lightweight requirements of B-pillar and similar parts with tailored properties, a new localized induction heating process is put forward, in which the high-temperature region on a steel blank is fully austenitized whereas the lower-temperature region is still in the two-phase zone. The water-quenched part with tailored properties is acquired.

(1) Since the presence of the magnetizer changes the original straight path of the magnetic flux lines and the magnetic flux is redistributed, the distribution of the eddy current induced changes. Therefore, it was experimentally validated and the steel blank with different temperature regions was obtained after 30 s. After water quenching, the high-temperature region of the blank was transformed into fine lath martensite, whereas the lower-temperature region was converted to the ferrite + martensite microstructure. Two enhancement mechanisms for the localized induction heating were realized, including fine-grain strengthening and phase transformation strengthening. The thermal images, hardness distribution, OM images, and mechanical properties of the quenched blanks prove the feasibility and effectiveness of the new localized induction heating process.

(2) The Curie temperature (T_C) is the inflexion point of the heating rate. The heating rate reaches $109 \text{ K} \cdot \text{s}^{-1}$ below T_C and is greatly slowed above T_C, which is beneficial to reducing the generation of the overheat zone and thereby is very meaningful for the localized induction heating process.

(3) A_{C1} and A_{C3} rise with the increase of the heating rate. Thus, the start and finish temperatures in the temperature range of the two-phase zone of the new heating process are both higher than those of the conventional heating method, which is consistent with other studies as $A_{C1} = 750\ ^\circ\text{C}$ and $A_{C3} = 900\ ^\circ\text{C}$.

In summary, the steel blanks with different temperature regions for hot stamping and tailored properties after water quenching can be obtained by the localized induction heating process. The new heating process will meet the industrial flexible and complex requirements as well as the short production cycle of manufacture.

Author Contributions: Data curation, J.C. and J.W.; methodology, W.L.; writing–original draft, L.B.; writing–review and editing, L.B; experimental auxiliary operation, Y.G. and Q.L.

Funding: This work is supported by the National Natural Science Foundation of China (Grant No. 51034009) and Heilong jiang Provincial Higher Education Fund Basic Research Project (135109310).

Acknowledgments: Experimental materials were provided by Baogang Group.

Conflicts of Interest: The authors declare no conflict of interest.

References

1. Zhang, Y.; Zhu, P.; Chen, G.L. Lightweight Design of automotive front side rail based on robust optimization. *Thin-Walled Struct.* **2007**, *45*, 670–676. [CrossRef]

2. Mori, K.; Bariani, P.F.; Behrens, B.-A. Hot stamping of ultra-high strength steel parts. *CIRP Ann. Manuf. Technol.* **2017**, *66*, 755–777. [CrossRef]

3. Behrens, B.A.; Bouguecha, A.; Gaebel, C.M.; Moritz, J.; Schrödter, J. Hot Stamping of Load Adjusted Structural Parts. *Procedia Eng.* **2014**, *81*, 1756–1761. [CrossRef]

4. Merklein, M.; Johannes, M.; Lechner, M.; Kupper, A. A review on tailored blanks—Production, applications and evaluation. *J. Mater. Process. Technol.* **2014**, *214*, 151–164. [CrossRef]

5. Landgrebe, D.; Putz, M.; Schieck, F.; Sterzing, A.; Rennau, A. Towards Efficient, interconnected and Flexible Value Chains–Examples and Innovations from Research on Production Technologies. In Proceedings of the 5th International Conference on Accuracy in Forming Technology, Chemnitz, Germany, 10–11 November 2015; pp. 61–78.

6. Kurz, T.; Rosner, M.; Manzenreiter, T.; Hartmann, D.; Sommer, A.; Kelsch, R.; Ademaj, A. Trends and Developments in the Usability and Production of Press-Hardened Components with Cathodic Corrosion Protection. In Proceedings of the 3rd International Conference on Hot Sheet Metal Forming of High-Performance Steel, Kassel, Germany, 13–17 June 2011; pp. 491–498.

7. George, R.; Bardelcik, A.; Worswick, M.J. Hot Forming of Boron Steels Using Heated and Cooled Tooling for Tailored Properties. *J. Mater. Process. Technol.* **2012**, *212*, 2386–2399. [CrossRef]

8. Zimmermann, F.; SpÖer, J.; Volk, W. Partial Tempering of Press Hardened Car Body Parts by a Premixed Oxygen-Methane Flame Jet. In Proceedings of the 4th International Conference on Hot Sheet Metal Forming of High-Performance Steel, Luleä, Sweden, 9–10 June 2013; pp. 267–274.

9. Kolleck, R.; Veit, R.; Merklein, M.; Lechler, J.; Geiger, M. Investigation on Induction Heating for Hot Stamping of Boron Alloyed Steels. *CIRP Ann. Manuf. Technol.* **2009**, *58*, 275–278. [CrossRef]

10. Bui, H.T.; Hwang, S. Modeling a working coil coupled with magnetic flux concentrators for barrel induction heating in an injection molding machine. *Int. J. Heat Mass Transf.* **2015**, *86*, 16–30. [CrossRef]

11. Shih, S.-Y.; Nian, S.-C. Nonplanar mold surface heating using external inductive coil and magnetic shielding materials. *Int. Commun. Heat Mass Transf.* **2016**, *71*, 44–55. [CrossRef]

12. Kai, G.; Peng Xun, Q.; Zhou, W. Effect of magnetizer geometry on the spot induction heating process. *J. Mater. Process. Technol.* **2016**, *231*, 125–136.

13. Illia, H.; Konrad, B.; Dmytro, R.; Florian, N. Phase transformations in a boron-alloyed steel at high heating rates. In Proceedings of the 17th International Conference on Metal Forming, Metal Forming, Palermo, Italy, 16–19 September 2018; pp. 1062–1070.

14. Guk, A.; Kunke, A.; Verena, K.; Verena, K.; Dirk, L. Influence of inductive heating on microstructure and material properties in roll forming processes. In Proceedings of the International Conference of Global Network for Innovative Technology & Awam International Conference in Civil Engineering, Bukit Jambul, Malaysia, 8–10 August 2017; pp. 271–276.

15. Kopp, R.; Wiedner, C.; Meyer, A. Flexibly Rolled Sheet Metal and Its Use in Sheet Metal Forming. *Adv. Mater. Res.* **2005**, *6–8*, 81–92. [CrossRef]

Article

Mechanism and Parameter Optimization in Grinding and Polishing of M300 Steel by an Elastic Abrasive

Xin Tong, Xiaojun Wu *, Fengyong Zhang, Guangqiang Ma, Ying Zhang, Binhua Wen and Yongtang Tian

School of Mechanical and Electrical Engineering, Xi'an University of Architecture and Technology, Xi'an 710055, China; tong24xin@163.com (X.T.); zhangfy12@126.com (F.Z.); ma15556515658@163.com (G.M.); zhangying831216@163.com (Y.Z.); wenbinhua123@163.com (B.W.); beitangwenfu@163.com (Y.T.)
* Correspondence: wuxiaojun@xauat.edu.cn

Received: 11 December 2018; Accepted: 18 January 2019; Published: 22 January 2019

Abstract: In order to achieve high quality polishing of a M300 mold steel curved surface, an elastic abrasive is introduced in this paper and its polishing parameters are optimized so that the mirror roughness can be achieved. Based on the Preston equation and Hertz Contact Theory, the theoretical material removal rate (MRR) equation for surface polishing of elastic abrasives is obtained. The effects of process parameters on MRR are analyzed and the polishing parameters to be optimized are as follows: particle size (S), rotational speed (Wt), cutting depth (Ap) and feed speed (Vf). The Taguchi method is applied to design the orthogonal experiment with four factors and three levels. The influence degree of various factors on the roughness of the polished surface and the combination of parameters to be optimized were obtained by the signal-to-noise ratio method. The particle swarm optimization algorithm optimized with the back propagation (BP) neural network algorithm (PSO-BP) is used to optimize the polishing parameters. The results show that the rotational speed has the greatest influence on the roughness, the influence degree of abrasive particle size is greater than that of feed speed, and cutting depth has the least influence. The optimum parameters are as follows: particle size (S) = #1200, rotational speed (Wt) = 4500 rpm, cutting depth (Ap) = 0.25 mm and feed speed (Vf) = 0.8 mm/min. The roughness of the surface polishing with optimum parameters is reduced to 0.021 μm.

Keywords: M300 mold steel; elastic abrasive; PSO-BP neural network algorithm; parameter optimization

---

1. Introduction

Due to its high Cr (16%) content, M300 mold steel has good corrosion resistance and wear resistance and has strong resistance to the erosion of general chemicals. It is often used in molds for various kinds of plastics, such as transparent plastics, camera lenses and so on. As one of the most important processes of mold surface disposing, mold polishing directly influences the quality of the mold surface and its performance. At present, mold polishing mainly adopts traditional manual polishing, which is time-consuming and laborious, and the polishing quality is difficult to guarantee [1,2]. Although manual polishing can meet the requirements of precision mold surface finish (Ra > 0.04–0.08 μm) [3], its time-consuming and laborious shortcomings make it difficult to meet the requirements of modern industry for low cost, short cycle and high quality.

In modern mold manufacturing, the proportions of free-form surfaces are increasing, and higher requirements of mold processing technology are required [4]. When the elastic abrasive tool is in contact with the surface of hard steel, the deformation of the surface of the abrasive tool is completely elastoplastic. The elastic abrasive can have well-profiled contact with a curved surface workpiece on account of its polymer elastic abrasive binder structure with greater flexibility, which is different from a rigid fixed abrasive grinding wheel where fretting of adjacent abrasives may happen on partial surface [5]. It is beneficial to improve the quality of curved surface polishing using an elastic abrasive. At the same time, an excellent profiling effect makes the elastic abrasive polishing suitable for free-form surfaces.

Conventional automated polishing uses free abrasive particles. This material is suitable for aspherical parts and workpiece surfaces with a small curvature. This method has low processing efficiency and high processing cost. The contact pressure between the polishing tool and the contact surface needs to be measured using a pressure sensor, since the conventional polishing tool is inelastic [6,7]. Compared with the free abrasive particle polishing process, the fixed abrasive polishing process has a large number of abrasive grains [8]. This type of material has a high material removal rate and a good self-twisting effect. The fully elastic contact characteristics allow the contact pressure to be determined by the depth of the cut. Due to the complex non-linear relationship between polishing parameters and roughness, the objective function that needs to be optimized cannot be obtained. Other optimization methods, such as particle swarm optimization or ant colony algorithm, are no longer suitable. However, the back propagation (BP) neural network has strong adaptive and self-organizing capabilities and is widely used in data prediction and numerical analysis.

The surface polishing mechanism and parameter optimization have been deeply studied. Beaucamp has studied the shape adaptive grinding process, which has been applied to finishing titanium alloy (Ti6Al4V) additively manufactured by EBM (Electron Beam Machining) and SLS (Selective Laser Sintering) [9]. However, this type of polishing method did not meet the requirements of precision mold surface finish. Zhang has studied the parameter optimization of five-axis polishing using an abrasive belt flap wheel for a blisk blade, in which RSM (Response Surface Methodology) is used to analyze the interactions of polishing factors on SR (Surface Roughness) and establish a predictive model between SR and various parameters [10]. A multi-objective particle swarm optimization algorithm (MOPSOA) is applied to optimize surface roughness of a work-piece after circular magnetic abrasive polishing by Nguyen [11]. A statistics parameters optimization method based on index atlases is presented for a novel 5-DOF (5-Degree of Freedom) gasbag polishing machine tool by Yan [12]· However, for elastic abrasive polishing of M300 mold steel, there is still no complete study about parameter optimization.

In order to realize high quality polishing for a M300 mold steel curved surface, based on the Preston equation and Hertz Contact Theory, the polishing mechanism of the elastic abrasive is studied in this paper [13]. The automatic polishing experiment of M300 steel was carried out using elastic abrasive tools of varying particle size. The influence of abrasive particle size, abrasive rotational speed, cutting depth and feed speed on surface roughness was analyzed.

Traditional BP neural networks use error back propagation to adjust the connection weight of the network. The BP neural network is quick to fall into the local optimal solution, the convergence speed is slow, and the network training is unstable. Therefore, the particle swarm optimization (PSO) algorithm is used to optimize the network weight and threshold to improve the network accuracy and convergence speed.

The BP Neural Network algorithm, which is optimized by the Particle Swarm Optimization algorithm (PSO-BP), is then used to achieve the optimal parameter combination. Finally, the surface quality, which is polished under the conditions of the optimal parameter combination, is verified by experiments.

2. Experiment Design

2.1. Experiment Devices

The experiments were carried out on 4-axis precision Computer Numerical Control (CNC) machine tool Mikoni430P, which is produced by LuoYang Mikoni Precision Machinery Co., Ltd., LuoYang, China. The device includes three moving axes X, Y and Z (the repeated positioning accuracy is 1 μm), and a rotation axis, A. The polishing experimental platform is shown in Figure 1a.

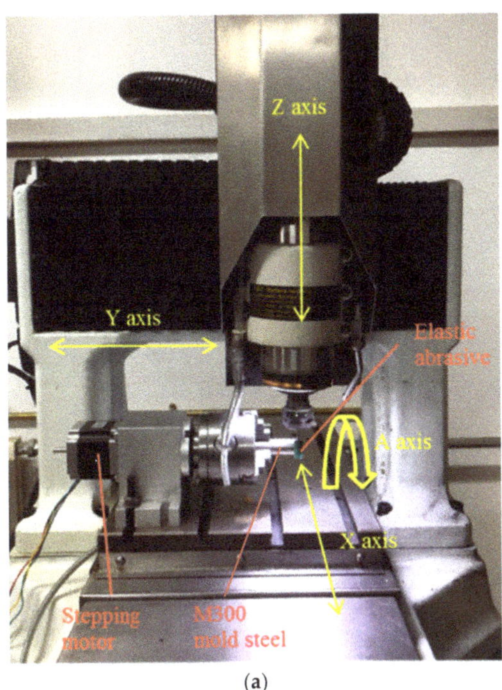

(**a**)

Figure 1. *Cont.*

(b)

Figure 1. Experimental platform of polishing. (**a**) Grinding and polishing process of elastic abrasive tool. (**b**) Measurement of workpiece surface roughness.

The surface quality of the workpiece and abrasive is measured by an Alicona INFINITE Focus three-dimensional (3D) profilometer, as shown in Figure 1b. This equipment is produced by Alicona, Austria. The magnification and the size view of the lens, which is applied for specimen measuring, are $100\times/0.6$ and 285.0027 μm $\times$ 216.2089 μm. The parameter settings for abrasive measuring are $20\times/0.4$ and 713.7553 μm $\times$ 541.4695 μm, respectively. The high pass filter is Gaussian filtering. The experiments are carried out three times to reduce the influence of random factors. When the workpiece is rotated in the YOZ plane, the polishing trajectory is strictly center symmetrical [14]. However, the workpiece in this experiment only rotates in the XOZ plane, which causes the cutting of abrasive grains mainly in the horizontal direction. Thus, the roughness (Ra) is measured along with the axial direction of the workpiece.

2.2. Experiment Conditions

The specimen size is $\phi18 \times 55$ mm and the material is M300 steel. The chemical composition of the material is shown in Table 1. The surface roughness (Ra) of the workpiece has decreased to 0.8–1 μm after preliminary semi-finishing. The workpiece is fixed on the rotation work table, which has a rotational speed of 300 r/min.

Table 1. Chemical composition of M300 steel (%).

Name	C	Cr	Mo	Mn	Si
Content	0.38	16.00	1.00	0.4	0.40

The polishing experiments are carried out using elastic abrasives of various grain size under different process parameter combinations. The experimental conditions have been presented in Table 2.

Table 2. Experimental conditions for grinding and polishing.

Name	Conditions
Specification of abrasive tools	ϕ10 mm silicone rubber based elastic abrasives
Abrasive and particle size	Silicon carbide (carborundum), #320 (42 μm), #600 (23 μm), #1000 (13 μm)
Cooling-down methods	Dry polishing

3. The Polishing Mechanism Using Elastic Abrasive

3.1. Factors of the Material Removal Rate

The mechanism of elastic abrasive tools in the polishing process is complex. The elastic–plastic deformation of abrasive surfaces and the continuous wear of the contact area lead to the decrease and fluctuation of contact surface pressure [15]. The model of material removal for the polishing process can be established according to the Preston equation for the surface polishing by axial feed abrasive on a self-rotating workpiece. The Preston equation is a commonly used empirical formula for material removal rate, which reveals that the material removal depth by a single abrasive grain is proportional to the relative pressure and line speed of the abrasive. The material removal rate (MRR) of grain in unit length of track can be expressed by Formula (1) [16]:

$$\frac{dh}{dl} = K_P \frac{V_s + V_f}{V_f} P \tag{1}$$

where K_p is the correction factor, which is related to the hardness of the workpiece and the abrasive grain, as well as the abrasive grain size; V_s is the tangential line speed of abrasive; V_f is the axial feed speed along the workpiece; and P is the pressure on the contact zone.

According to the Hertz Contact Theory, the polishing process can be simplified as the contact situation between the rigid body (workpiece) and the elastic body (abrasive). The contact surface between the workpiece and abrasive tool is ellipse as shown in Figure 2. The contact pressure submits to Elliptical Hertz distribution [17]:

$$P(y, z) = -P_0 \sqrt{1 - \left(\frac{z}{a}\right)^2 - \left(\frac{y}{b}\right)^2} \tag{2}$$

where $P_0 = \frac{3F_n}{2\pi ab}$ is the center pressure in the contact zone and F_n is the contact force when polishing, as shown in Figure 3. The elastic abrasive tool is deformed when the workpiece is pressed. Thus, the contact force is perpendicular to the tangent of the workpiece surface.

The material removal amount of the infinitesimal M along the Y direction in contact region AB is:

$$h(x) = \int_{-b'}^{b'} \frac{dh}{dl} dy = \int_{-b'}^{b'} K_p P \frac{V_s \pm V_f}{V_f} dy \tag{3}$$

in the formula $b' = b\sqrt{1 - \left(\frac{x}{c}\right)^2}$. The theoretical equation of MRR on the workpiece surface can be taken from Formula (2) and Formula (3):

$$h(x) = -K_p \frac{V_s \pm V_f}{V_f} \frac{3F_n}{\pi a} \frac{x}{c} \sqrt{1 - \left(\frac{x}{c}\right)^2} \tag{4}$$

Formula (4) shows that the MRR can be controlled by V_s, V_f and F_n. The elastic abrasive can be considered hyperelastic, and the contact pressure (F_n) of the workpiece surface is approximately proportional to the cutting depth of the abrasive tool [18]. Since the elastic abrasive steel is elastoplastically deformed when in contact with steel, the contact pressure is proportional to the cutting depth; the contact is replaced with cutting depth [19]. During the surface polishing experiment,

Vs reflects the grinding tool speed Wt, Vf reflects the feed rate along the axis and Ap stands for the set cut depth of the abrasive tool.

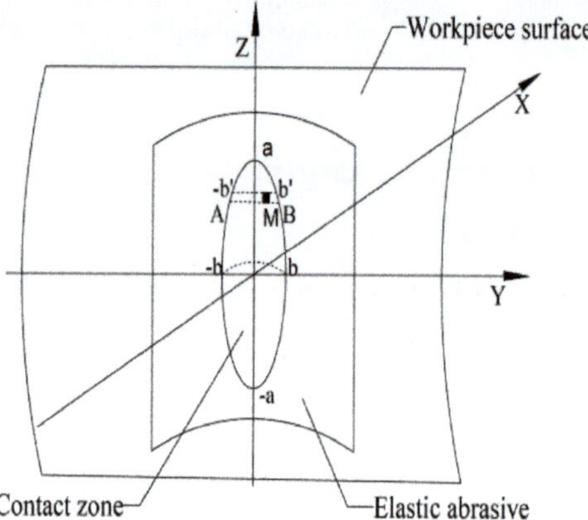

Figure 2. Contact institution of an elastic abrasive and the workpiece surface.

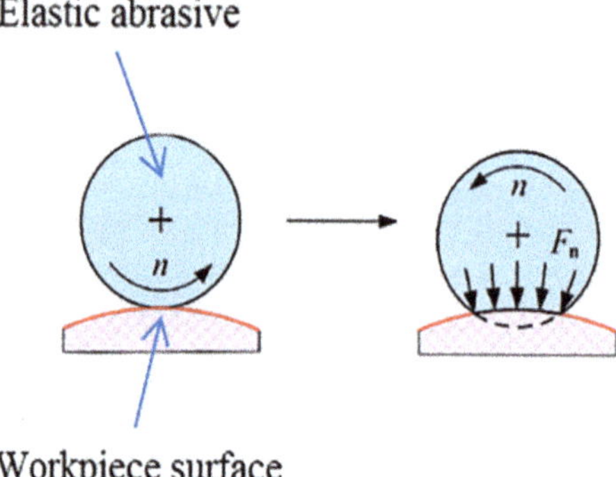

Figure 3. Contact force of an elastic abrasive and the workpiece surface.

3.2. Research on Parameters Affected by MRR

The abrasive cutting process attributed to the contact characteristics of the elastic abrasive can be roughly divided into three stages: sliding, ploughing and the cutting process [20].

Figure 4 shows the effects of processing parameters on MRR, in which the error bars indicate the range of values after three repetitions. The MRR is approximately proportional to Wt and inversely proportional to Vf in general, which is consistent with the theoretical model.

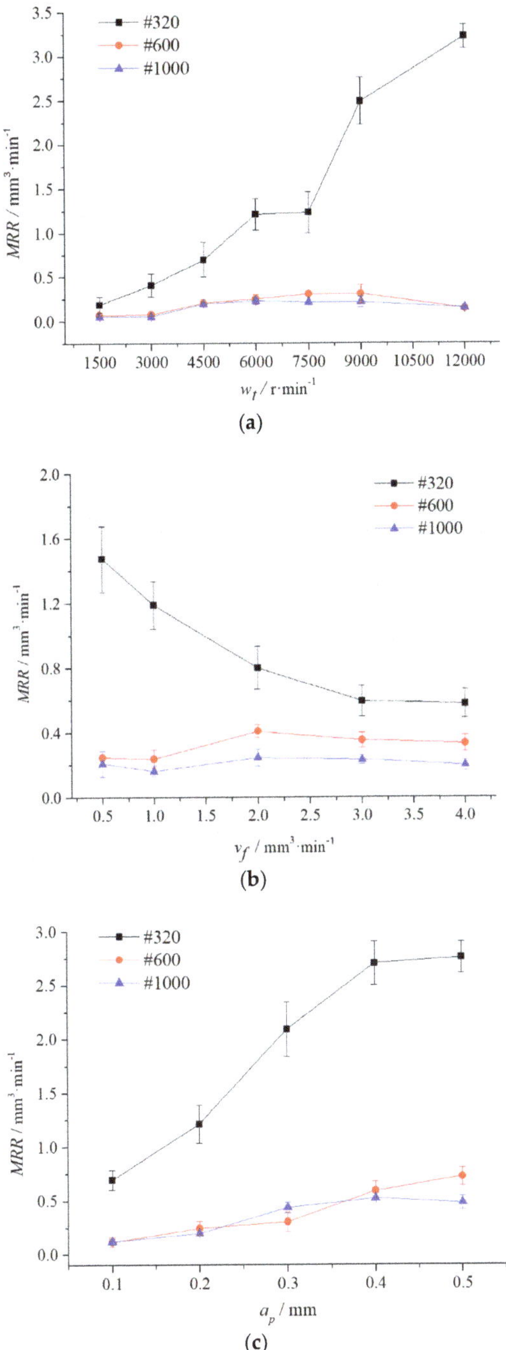

Figure 4. Plots of parameter effect on material removal rate (MRR) (**a**) Effect of the grinding tool speed (Wt) (**b**) effect of the feed rate (Vf) (**c**) effect of the set cut depth (Ap).

The smaller the particle size, the higher the MRR. Figure 5 shows that the larger the particle size, the more the abrasive grains have in per unit area and the smaller the size of every abrasive grain. Thus, a larger particle size leads to a higher number of abrasive grains in the unit contact zone and lower average pressure on a single grain. There will be more particles in the sliding friction and ploughing processes compared with a small particle size [21], while in the stage of cutting grain, the number decreased, so the material removal capacity is low.

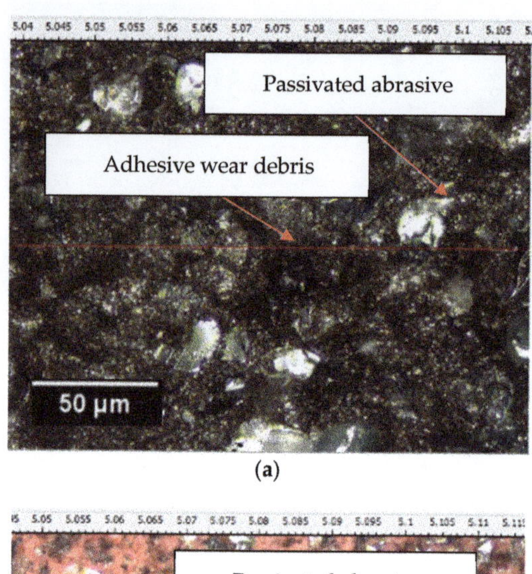

(a)

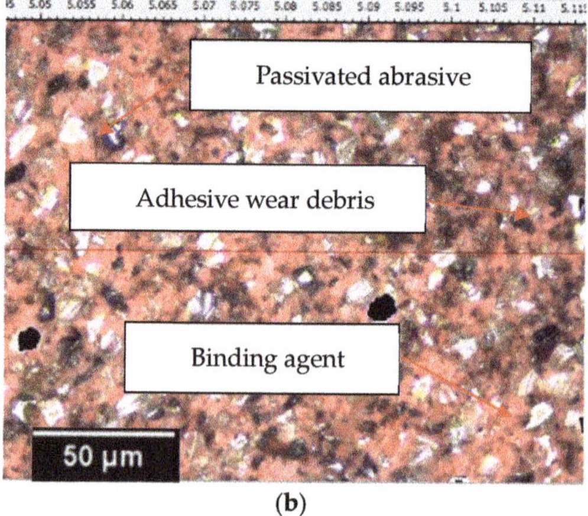

(b)

Figure 5. *Cont.*

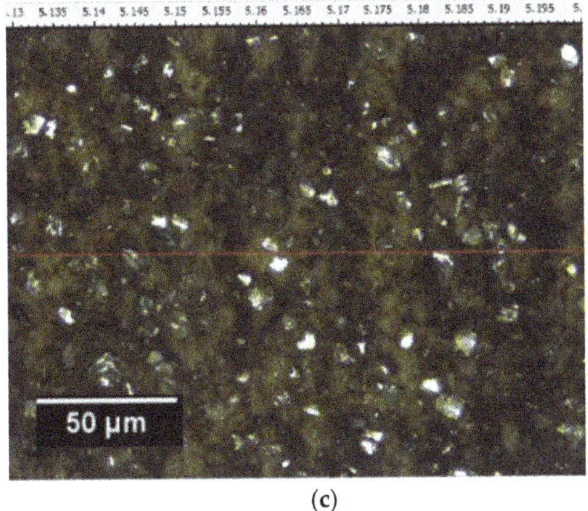

(c)

(d)

Figure 5. Images of all the three abrasives. (**a**) Morphological features of #320 (**b**) morphological features of #600 (**c**) morphological features of #1000 (**d**) elastic abrasive product.

The appearance of the elastic grinding head after wear is shown in Figure 5. It can be seen that most of the abrasive grains have passivated facets on the top and the surface is relatively smooth. In addition, the surface of each size abrasive exhibits different degrees of adhesive wear, and the #320 abrasive is covered with more fine grinding debris than the #600 abrasive. This indicates that the smaller the particle size of the abrasive tool, the higher the removal rate of the workpiece.

In Figure 4a, the MRR of the #320 grinding tool is approximately proportional to the grinding speed (Wt), which is consistent with the material removal model, and the rate of increase is gradually slowing down at 9000 r/min. The MRR of #600 and #1000 reached the peak at 6000 r/min and 9000 r/min, respectively. Then the MRR begins to decrease. Due to the contact time between the workpiece and the larger abrasive being too short with the increase of Wt and the sliding and ploughing effect being more important than the cutting effect, the removal ability of the abrasive decreased.

In Figure 4b, for a fixed cutting depth and speed of the grinding tool, the increase of Vf increases the line spacing between the polishing tracks, and the number of passes of the grinding tool decreases per unit length. Thus, the residual height of the workpiece surface increases and the material removal rate decreases [20]. After Vf is greater than 2 mm/min, the MRR gradually decreases to a stable value.

In Figure 4c, the total MRR is approximately proportional to the set depth (Ap). By the elastic contact theory, the contact area increases with the increase of Ap. The increase in the number of abrasive particles involved in grinding and polishing increases the MRR. After Ap is greater than 0.4 mm, the MRR increase of #320 and #1000 grinding tools slows down. While the #600 grinding tool reaches the peak and then decreases when the Ap is 0.4 mm, the critical value of the depth of cut is estimated to be 0.4–0.5 mm. When the Ap is larger than the critical value, the change of the effective working area is small. At the same time, the increase of contact pressure makes the abrasive blunt off worse. Eventually, the upward trend of material removal rate slows down or even declines.

4. Parameter Optimization

4.1. Experimental Design and Results

Taking into account the interaction among the factors, an orthogonal experiment with four factors and three levels [22] was designed based on the Taguchi method [23], which is shown in Table 3. The processing time of each group of experiments is 180 s. In order to reduce the processing error, each group of experiments is processed three times. The result is taken as its average value, which is as shown in Table 4.

Table 3. Experimental conditions for grinding and polishing.

Processing Parameters	Level		
	1	2	3
Particle size, S (#)	320	600	1000
Abrasive tool speed, Wt (r/min)	4500	6000	7500
Setting cut depth, Ap (mm)	0.1	0.2	0.3
Feed rate, Vf (mm/min)	0.5	1	2

In the table above: Mean 1 is the mean of the normal variance of the surface roughness of the influencing factor in the level 1 combination.

Mean 2 is the mean of the normal variance of the surface roughness of the influencing factor in the level 2 combination.

Mean 3 is the mean of the normal variance of the surface roughness of the influencing factor in the level 3 combination.

The combination of grinding parameters for a single optimized target can be achieved by the signal-to-noise ratio (SNR) analysis of the experimental data. Since the optimization target is surface roughness (Ra), the design parameters of small characters are adopted, such as Formula (5)

$$SNR = -10\lg \sum_{i=1}^{n} R_i^2 \tag{5}$$

Table 5 is the average response of SNR to Ra in each parameter level. The larger the SNR, the higher the parameter influence on Ra will be. It can be seen that the abrasive grain size and the abrasive speed have a high influence on Ra.

In order to express the influence trend of each factor level on the surface roughness more intuitively, the main effect diagram of polished roughness is obtained.

Table 4. Experimental conditions for grinding and polishing. Ra = surface roughness.

Number	S (A)	Wt (B)	Ap (C)	Vf (D)	Ra
1	320	4500	0.1	0.5	0.037
2	320	4500	0.2	1	0.074
3	320	4500	0.3	2	0.088
4	320	6000	0.3	0.5	0.069
5	320	7500	0.3	2	0.030
6	600	4500	0.1	1	0.079
7	600	4500	0.2	2	0.059
8	600	7500	0.3	1	0.072
9	600	4500	0.2	2	0.055
10	600	4500	0.3	0.5	0.108
11	600	6000	0.1	2	0.117
12	600	6000	0.2	0.5	0.145
13	600	7500	0.2	1	0.139
14	600	7500	0.3	2	0.106
15	1000	4500	0.2	0.5	0.037
16	1000	4500	0.3	2	0.047
17	1000	4500	0.1	0.5	0.072
18	1000	4500	0.2	1	0.046
19	1000	7500	0.2	1	0.035
20	1000	7500	0.1	1	0.045
21	1000	7500	0.2	2	0.111
22	1000	7500	0.3	0.5	0.105
23	320	7500	0.2	0.5	0.294
24	600	4500	0.1	1	0.083
25	600	6000	0.3	1	0.094
26	600	7500	0.1	0.5	0.074
27	1000	4500	0.1	2	0.107
28	600	6000	0.1	1	0.115
29	600	4500	0.3	1	0.082
30	320	4500	0.3	1	0.064
Mean 1	0.094	0.069	0.081	0.104	
Mean 2	0.095	0.108	0.095	0.077	
Mean 3	0.067	0.101	0.078	0.080	

Table 5. Signal-to-noise ratio (SNR) (dB) to surface roughness.

Parameter		S	Wt	Ap	Vf
	1	22.94	23.66	22.40	21.36
Level	2	20.81	19.59	22.05	22.83
	3	24.33	21.79	22.70	22.73

When the particle size is too small, the residual peak on the surface of the workpiece is not sufficiently cut, so that when the particle size increases, the roughness decreases. When the grinding speed is too fast, it results in incomplete cutting. However, if the speed is too slow, it will result in a decrease in the number of abrasive grains involved in cutting per unit time. When the depth of the cut increases, the roughness decreases due to over-cutting of the abrasive grains. As the depth of cut further increases, the deformation of the abrasive increases and the contact area with the workpiece increases. Finally, the time of cutting of the abrasive is increased, and the roughness decreases. Excessive feed rates and low feed rates will result in undercutting and overcutting, respectively.

As shown in Figure 6, particle size (S), grinding speed (Wt), cutting depth (Ap) and feed velocity (Vf) are at minimum roughness at levels 3, 1, 3, 2, respectively. The minRa parameter combinations are A3 B1 C3 D2, since the roughness is negative indicator.

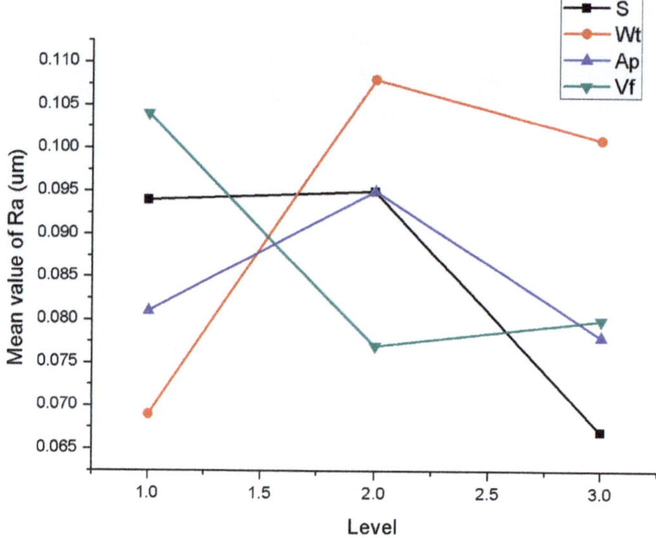

Figure 6. Main effect diagram of Ra.

4.2. PSO-BP Neural Network Model

The surface roughness polished with an elastic abrasive is affected by many factors, and the complex non-linear relationship between roughness and influencing factors is difficult to fit to a linear model or common non-linear model. The BP neural network has a high mapping ability and can realize any non-linear mapping from input to output. By using this high mapping ability and generalization ability of the BP neural network, the mapping model between particle size (S), rotational speed (Wt), cutting depth (Ap), feed speed (Vf) and polished surface roughness can be established to solve the problem of parameter optimization. However, the BP neural network easily falls into the local extremum [24].

Particle swarm optimization (PSO) is a swarm intelligence optimization algorithm which finds out the optimal region in a complex search space by the interactions among particles [25]. The learning of the BP neural network is mainly reflected in the adjustment process of the weight value and the threshold. The optimization operation of particle swarm optimization corresponds to the weight value and the threshold of the BP neural network algorithm, and then the PSO-BP neural network model is established.

Particle size (S), rotational speed (Wt), cutting depth (Ap) and feed speed (Vf) are input factors. The polishing surface roughness is used as the output factor. The BP neural network model with one hidden layer is established, as shown in Figure 7. The number of neurons in the hidden layer is 11. The transfer function of the hidden layer is "tansig". The transfer function of the output layer is "pureline". The training function is "trainlm". The training accuracy, learning rate and cycle times are 0.0001, 0.05 and 3000, respectively.

When the weight is optimized by particle swarm optimization, the connection weights of each layer of the neural network are encoded into particles and the fitness is the mean square error of the network output. The goal is to search for the optimal network weights within the default number of iterations.

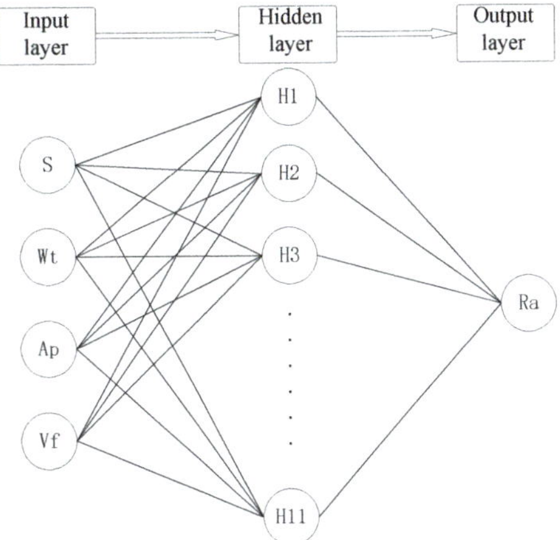

Figure 7. Schematic diagram of back propagation (BP) neural network structure.

The PSO algorithm functions to find the optimal solution in a group of particles by iterating. The particle is updated by the Pbest values and the Gbest values. The Pbest is the best location which is searched by particles. The Gbest is the best location which is searched by the whole particle swarm.

Supposing $z_i = (z_{i1}, z_{i2}, \ldots, z_{id}, \ldots z_{iD})$ is the D-dimensional position vector of the No.i particle, the position of the particle can be measured by the fitness function. The $v_i = (v_{i1}, v_{i2}, \ldots, v_{id}, \ldots, v_{iD})$ is the fly velocity of particle i. The $p_i = (p_{i1}, p_{i2}, \ldots, p_{id}, \ldots, p_{iD})$ is the optimal position of particle i so far. The $p_g = (p_{g1}, p_{g2}, \ldots, p_{gd}, \ldots, p_{gD})$ is the optimal position found so far by the particle swarm. The fly velocity and position are updated according to Formula (6).

$$v_{id}^{k+1} = wv_{id}^{k} + c_1 r_1 \left(p_{id} - z_{id}^{k} \right) + c_2 r_2 \left(p_{gd} - z_{id}^{k} \right) i = 1, 2, \ldots, md = 1, 2, \ldots, D \tag{6}$$

where k is the current number of iterations; r_1, r_2 is the random number (0, 1); c_1, c_2 are the learning factors; and W is inertia weight.

In order to maintain the equilibrium of particle swarm convergence speed and convergence efficiency, the initial algorithm should have a large global search capability and the latter algorithm should have strong local search capability. Therefore, the linear variations of Formulas (7)–(9) are used to improve the global optimization ability of the particle group at the initial stage and improve the local optimization ability of the particle group in the later stage.

$$c_1 = (c_{1f} - c_{1i}) \times (k \div k_{max}) + c_{1i} \tag{7}$$

$$c_2 = (c_{2f} - c_{2i}) \times (k \div k_{max}) + c_{2i} \tag{8}$$

$$w = w_{max} - \frac{w_{max} - w_{min}}{k_{max}} \times k \tag{9}$$

In general, when $C_1 + C_2 < 4$, the optimization ability of the example group is best [16], so c_{1f} and c_{1i} are 0.5 and 2.5, respectively; c_{2f} and c_{2i} are 2.5 and 0.5, respectively. wmax and wmin are 0.9 and 0.4, respectively.

Setting the maximum speed as 0.8, the number of particles as 40, and the minimum error as 0.001, a PSO-BP network model was built (Figure 7) to train the data for rows 1–25 in Table 4. The data from rows 26–30 is used to examine the trained network model. The comparison between the PSO-BP neural network and the BP neural network is shown in Figure 8.

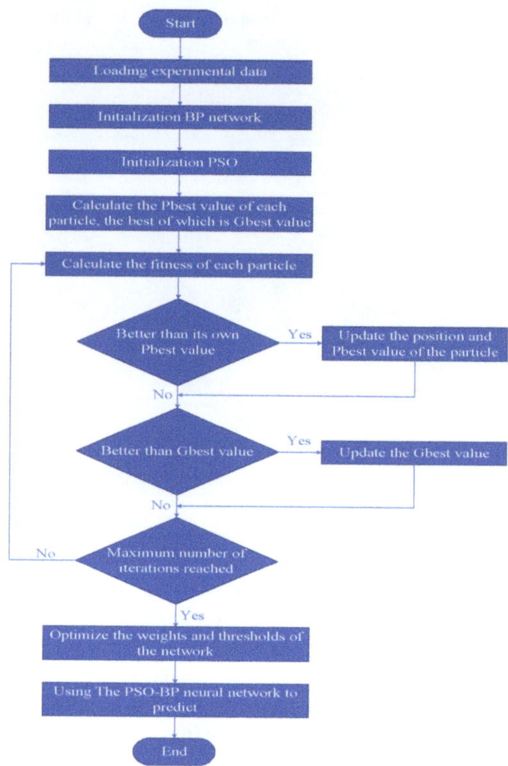

Figure 8. Flow-process diagram of the particle swarm optimization algorithm optimized with the back propagation neural network algorithm (PSO-BP).

Compared with Figures 9a and 9b,e, it can be seen that the PSO-BP neural network converges to the preset precision in only six steps, and the efficiency of the PSO-BP neural network is obviously improved compared with the basic BP neural network. By comparing Figure 9c with Figure 9d, the predicted value of the former is very close to the experimental value, but the latter has a large deviation.

As shown in Table 6, the prediction error of the PSO-BP network model is within 0.3%, so the PSO-BP network model has a high accuracy and can be used as a prediction model.

Table 6. Predicted error.

Number	26	27	28	29	30
Actual value (µm)	0.074	0.107	0.115	0.082	0.064
Predicted value (µm)	0.0741	0.1067	0.1148	0.0818	0.0643
Error(%)	0.14	0.28	0.17	0.24	0.47

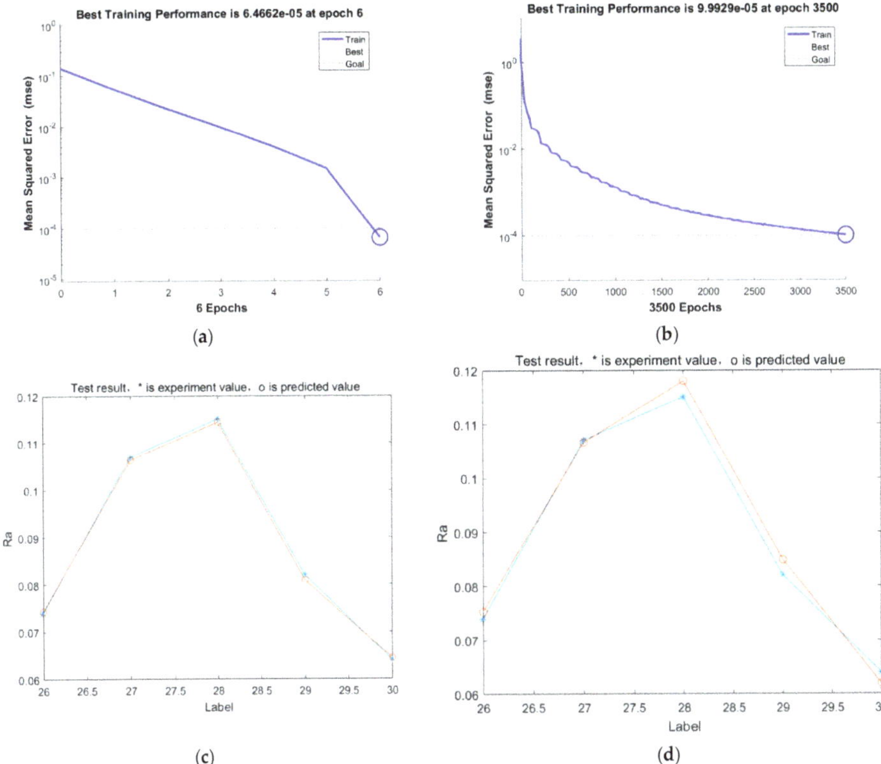

Figure 9. The comparison of PSO-BP and BP: (**a**) Training process of the PSO-BP neural network algorithm; (**b**) training process of the basic BP neural network algorithm; (**c**) Prediction error of the PSO-BP neural network algorithm; (**d**) Prediction error of the basic BP neural network algorithm.

4.3. Optimization Results

Based on the minRa parameter combination of each factor, each factor is set to be five levels, and the distribution is shown in Table 7. The orthogonal test is designed by using the Taguchi method and the data is input into the trained PSO-BP neural network model for prediction. The results are shown in Table 8.

Table 7. The distribution of each factor.

Processing Parameters	Level				
	1	**2**	**3**	**4**	**5**
Particle size, S (#)	700	800	1000	1100	1200
Abrasive tool speed, Wt (r/min)	4300	4400	4500	4600	4700
Set cut depth, Ap (mm)	0.2	0.25	0.3	0.35	0.4
Feed rate, Vf (mm/min)	0.8	0.9	1	1.1	1.2

Table 8. Predicted results.

Number	A (#)	B (r/min)	C (mm)	D (mm/min)	Ra (μm)
1	700	4300	0.2	0.8	0.0850
2	700	4400	0.25	0.9	0.0971
3	700	4500	0.3	1	0.0761
4	700	4600	0.35	1.1	0.0637
5	700	4700	0.4	1.2	0.0891
6	800	4300	0.25	1	0.0896
7	800	4400	0.3	1.1	0.0702
8	800	4500	0.35	1.2	0.0705
9	800	4600	0.4	0.8	0.0621
10	800	4700	0.2	0.9	0.0611
11	1000	4300	0.3	1.2	0.0491
12	1000	4400	0.35	0.8	0.0600
13	1000	4500	0.4	0.9	0.0729
14	1000	4600	0.2	1	0.0456
15	1000	4700	0.25	1.1	0.0438
16	1100	4300	0.35	0.9	0.0548
17	1100	4400	0.4	1	0.0634
18	1100	4500	0.2	1.1	0.0393
19	1100	4600	0.25	1.2	0.0360
20	1100	4700	0.3	0.8	0.0414
21	1200	4300	0.4	1.1	0.0460
22	1200	4400	0.2	1.2	0.0339
23	1200	4500	0.25	0.8	0.0211
24	1200	4600	0.3	0.9	0.0392
25	1200	4700	0.35	1	0.0536

4.4. Experimental Verification

In Table 8, the optimized polishing parameter combination is obtained as follows: A5 B3 C2 D1 (S: #1200, Wt: 4500 rpm, Ap: 0.25 mm, Vf: 0.8 mm/min). The confirmatory experiments of the minRa parameter combination A3 B1 C3 D2 and optimized parameter combination A5 B3 C2 D1 are carried out respectively.

The surface morphology of the M300 workpiece polished under the conditions of the optimized parameter combination A5 B3 C2 D1 is shown in Figure 10. It can be seen that the polishing pattern is obviously reduced, and the surface damage is greatly improved. The surface roughness (Ra) is reduced to 0.021 μm after machining. Compared with the minRa parameter combination (as shown in Figure 11), the roughness is reduced significantly, and the surface quality is improved, which means that the parameter optimization method used is feasible.

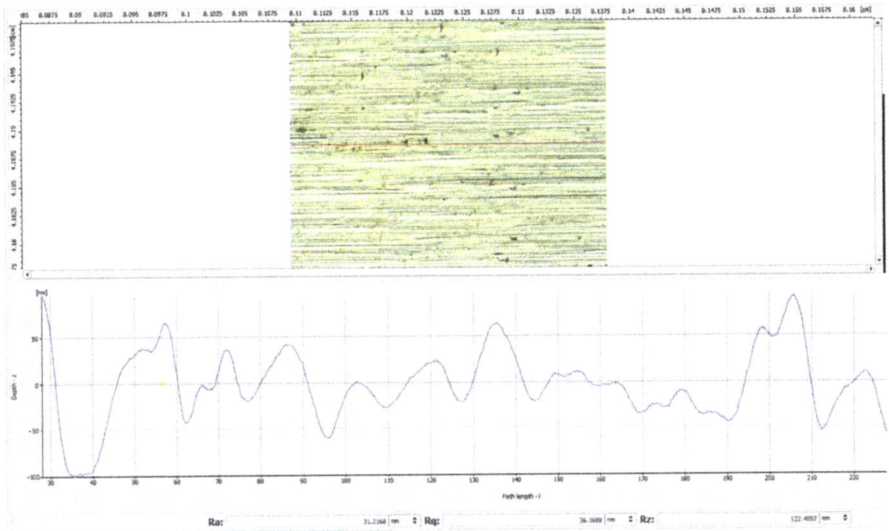

Figure 10. Surface topographies of the workpiece polished by optimized parameters.

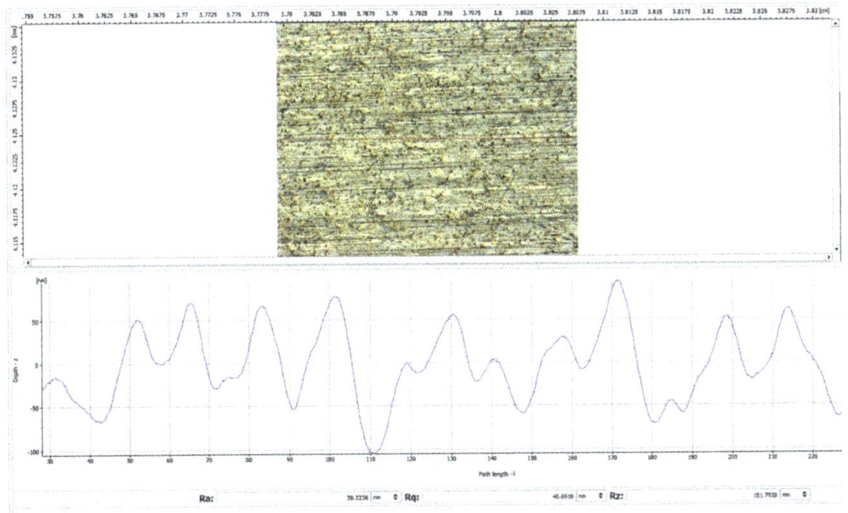

Figure 11. Surface topographies of the workpiece polished by the minRa parameter combination (S = #1000, Wt = 4500 r/min, Ap = 0.3 mm, Vf = 1 mm/min).

5. Conclusions

The silicon carbide abrasive and silicone–rubber based elastic abrasive is cheaper and has a better profile when polishing the curved surface of M300 mold steel. It is easy to obtain high surface quality and provide a feasible method for high efficiency and high-quality polishing of M300 mold steel.

1. The elastic abrasive has high material removal ability in the initial stage of processing. With the increase in processing time, the material removal rate decreases rapidly and tends to be stable. Generally, the abrasive with large particle size (S) has low removal ability, and it is easy to obtain a stable polished surface quality. This is because the larger the particle size (S), the more the

    abrasive grains in per unit area and the smaller the size of every abrasive grain. A larger particle size (S) leads to a higher number of abrasive grains in the unit contact zone and a lower average pressure on a single grain.

2. Based on the parameter combinations of particle size, grinding speed, cutting depth and feed speed, an orthogonal experiment is carried out and the range analysis of the experimental results is performed. The results show that the speed of the grinding tool has the greatest influence on roughness, and the influence of particle size and feed speed on roughness is close. The degree of cutting depth is the least influential. The minRa parameters of each level are as follows: S = #1000, Wt = 4500 rpm, Ap = 0.3 mm and Vf = 1 mm/min.

3. The experimental parameters are trained and examined by the PSO-BP neural network algorithm. The results show that the prediction roughness error is less than 0.3%, which means that the network structure has high precision.

4. The surface roughness is taken as the optimization index. Based on the combination of minRa parameters, the polishing parameters are optimized by using the trained PSO-BP neural network structure. The optimization results show that the optimal parameter combination is S = #1200, Wt = 4500 rpm, Ap = 0.25 mm and Vf = 0.8 mm/min. The verified experiment shows that the roughness of the polished surface is reduced to 0.021 μm under the optimal parameter combination conditions, which is consistent with the predicted optimization results. The parameter optimization method based on the PSO-BP neural network algorithm is feasible to optimize the polishing parameters.

Author Contributions: Conceptualization, X.T. and F.Z.; methodology, X.T.; software, X.T.; validation, X.T., G.M. and F.Z.; formal analysis, Y.Z.; investigation, X.W.; resources, X.W.; data curation, X.T.; writing—original draft preparation, X.T.; writing—review and editing, X.T.; visualization, Y.T.; supervision, X.W.; project administration, X.W.; funding acquisition, X.W.

Funding: National Natural Science Foundation of China (Grant No. 51375361 and No. 51475353).

Acknowledgments: This project is financially supported by the National Natural Science Foundation of China (Grant No. 51375361 and No. 51475353). The authors are also grateful for the Industrial Research Projects of Shaanxi Province (2012K09-15).

Conflicts of Interest: The authors declare no conflict of interests.

References

1. Davim, J.P. (Ed.) *Machining of Complex Sculptured Surfaces*; Springer-Verlag London: London, UK, 2012.
2. Sun, J.J.; Taylor, E.J.; Srinivasan, R. MREF-ECM process for hard passive materials surface finishing. *J. Mater. Proc. Technol.* **2001**, *108*, 356–368. [CrossRef]
3. Ma, Y.Q.; Qi, L.H.; Zhou, J.M.; Zhang, T.; Li, H.J. Effects of Process Parameters on Fabrication of 2D-Cf/Al Composite Parts by Liquid–Solid Extrusion Following the Vacuum Infiltration Technique. *Metall. Mater. Trans. B* **2016**, *48*, 582–590. [CrossRef]
4. Tsai, M.J.; Huang, J.F. Efficient automatic polishing process with a new compliant abrasive tool. *Int. J. Adv. Manuf. Technol.* **2006**, *30*, 817–827. [CrossRef]
5. Wei, W. *Research on Progressive Soft-Bonded Abrasive Grain Pneumatic Wheel Finishing*; Zhejiang University of Technology: Hangzhou, China, 2016.
6. Brecher, C.; Tuecks, R.; Zunke, R.; Wenzel, C. Development of a force controlled orbital polishing head for free form surface finishing. *Prod. Eng.* **2010**, *4*, 269–277. [CrossRef]
7. Moumen, M.; Chaves-Jacob, J.; Bouaziz, M.; Linares, J.-M. Optimization of pre-polishing parameters on a 5-axis milling machine. *Int. J. Adv. Manuf. Technol.* **2016**, *85*, 443–454. [CrossRef]
8. Qi, L.H.; Ma, Y.Q.; Zhou, J.M.; Hou, X.H.; Li, H.J. Effect of fiber orientation on mechanical properties of 2D-Cf/Al composites by liquid–solid extrusion following Vacuum infiltration technique. *Mater. Sci. Eng. A* **2015**, *625*, 343–349. [CrossRef]
9. Beaucamp, A.T.; Namba, Y.; Charlton, P.; Jain, S.; Graziano, A.A. Finishing of additively manufactured titanium alloy by shape adaptive grinding (SAG). *Surf. Topogr. Metrol. Prop.* **2015**, *3*, 024001. [CrossRef]

10. Zhang, J.; Shi, Y.; Lin, X. Parameter optimization of five-axis polishing using abrasive belt flap wheel for blisk blade. *J. Mech. Sci. Technol.* **2017**, *31*, 4805–4812. [CrossRef]
11. Nguyen, N.T.; Yin, S.H.; Chen, F.J. Multi-objective optimization of circular magnetic abrasive polishing of SUS304 and Cu materials. *J. Mech. Sci. Technol.* **2016**, *30*, 2643–2650. [CrossRef]
12. Li, Y.; Tan, D.; Wen, D. Parameters optimization of a novel 5-DOF gasbag polishing machine tool. *Chin. J. Mech. Eng.* **2013**, *26*, 680–688. [CrossRef]
13. Ji, S.M.; Li, W.; Tan, D.P. Processing Characteristics of Soft Abrasive Flow Based on Preston Equation. *Chin. J. Mech. Eng.* **2011**, *47*, 156–163. [CrossRef]
14. Wu, X.J.; Tong, X. Study of trajectory and experiment on steel polishing with elastic polishing wheel device. *Int. J. Adv. Manuf. Technol.* **2018**, *97*, 199–208. [CrossRef]
15. Wu, C.L.; Wang, W.; Li, Q. Analysis of Polishing Processing Parameters of Stainless Steel Curved Surface Based on Preston Equation. *Mech. Sci. Technol.* **2012**, *41*, 102–105. (In Chinese)
16. Ji, S.M.; Li, C.; Tan, D.P. Study on Mach inability of Softness Abrasive Flow Based on Preston Equation. *J. Mech. Eng.* **2011**, *47*, 156–163. (In Chinese) [CrossRef]
17. Ji, S.M.; Zeng, X.; Jin, M.S. New Finishing Method of Softness Consolidation Abrasives and Material Removal Characteristic Analysis. *J. Mech. Eng.* **2013**, *49*, 173–181. (In Chinese) [CrossRef]
18. Lu, Y.; Huang, Y. Experimental Investigation in the Grinding and Wear Performance of Abrasive Belt Grinding. *Mech. Sci. Technol.* **2014**, *33*, 1865–1867.
19. Wu, X.J.; Chen, Z.; Zhou, T.Z.; Ma, C.J.; Shu, X.; Dong, J.Y. Study on the contact characteristics of high-strength grinding and polishing M300 steel curved surface. *J. Mech. Eng.* **2018**, *54*, 171–177. [CrossRef]
20. Wang, Y.J. *Precision Abrasive Belt Grinding Research Based on Contact Theory*; Chongqing University Press: Chongqing, China, 2015. (In Chinese)
21. Zhang, J.F.; Shi, Y.Y.; Lin, X.J. Parameters optimization in belt polishing process of blade based on Grey Relational Analysis. *Comput. Integr. Manuf. Syst.* **2017**, *23*, 806–812. (In Chinese)
22. Zhang, L.; Yuan, C.M. Modeling and Experimental Research on Surface Removal of Die Surface Polishing. *Chin. J. Mech. Eng.* **2002**, *38*, 98–102. [CrossRef]
23. Xu, X.B.; Shao, H. Multi-objective Optimization of Turning ParametersBased on Taguchi Algorithm and Grey Relational Theory. *Tool Technol.* **2015**, *49*, 15–18.
24. Li, M. *Research on Tool Wear Monitoring Technology Based on Particle Swarm Optimization Neural Network*; Southwest Jiaotong University: Chengdu, China, 2012.
25. Eberhart, R.; Kennedy, J. A new optimizer using particle swarm theory. In Proceedings of the International Symposium on MICRO Machine and Human Science, Nagoya, Japan, 4–6 October 2002; pp. 39–43.

 *materials* 

Article

Suppression of the Growth of Intermetallic Compound Layers with the Addition of Graphene Nano-Sheets to an Epoxy Sn–Ag–Cu Solder on a Cu Substrate

Min-Soo Kang, Do-Seok Kim and Young-Eui Shin *

School of Mechanical Engineering, Chung-Ang University, 84, Heukseok-ro, Dongjak-gu, Seoul 06974, Korea; kang10101@cau.ac.kr (M.-S.K.); kdoseok@naver.com (D.-S.K.)
* Correspondence: Shinyoun@cau.ac.kr

Received: 15 February 2019; Accepted: 14 March 2019; Published: 21 March 2019

Abstract: This study investigated the suppression of the growth of the intermetallic compound (IMC) layer that forms between epoxy solder joints and the substrate in electronic packaging by adding graphene nano-sheets (GNSs) to 96.5Sn–3.0Ag–0.5Cu (wt %, SAC305) solder whose bonding characteristics had been strengthened with a polymer. IMC growth was induced in isothermal aging tests at 150 °C, 125 °C and 85 °C for 504 h (21 days). Activation energies were calculated based on the IMC layer thickness, temperature, and time. The activation energy required for the formation of IMCs was 45.5 KJ/mol for the plain epoxy solder, 52.8 KJ/mol for the 0.01%-GNS solder, 62.5 KJ/mol for the 0.05%-GNS solder, and 68.7 KJ/mol for the 0.1%-GNS solder. Thus, the preventive effects were higher for increasing concentrations of GNS in the epoxy solder. In addition, shear tests were employed on the solder joints to analyze the relationship between the addition of GNSs and the bonding characteristics of the solder joints. It was found that the addition of GNSs to epoxy solder weakened the bonding characteristics of the solder, but not critically so because the shear force was higher than for normal solder (i.e., without the addition of epoxy). Thus, the addition of a small amount of GNSs to epoxy solder can suppress the formation of an IMC layer during isothermal aging without significantly weakening the bonding characteristics of the epoxy solder paste.

Keywords: graphene nano-sheets (GNSs); epoxy solder; intermetallic compound (IMC)

1. Introduction

The use of lead (Pb) in the electronic packaging industry has been greatly reduced because it poses a threat to the environment and to public health. As such, worldwide environmental regulations such as the "Restriction of Hazardous Substances" (RoHS) from the European Union ban the use of dangerous and harmful substances in electronic products [1–4]. As a response to this, most electronic appliance manufacturers employ lead-free solder. Several environmentally friendly Sn–based alloys such as Sn–3.0Ag–0.5Cu [5], Sn–14Bi-5In [6], Sn–0.7Cu [7], Sn–9Zn [8], Sn–8Zn-3Bi [9], and Sn–58Bi [10] are considered the most promising candidates to replace toxic Sn–Pb alloys in electronic packaging systems. Of these, Sn–Ag–Cu alloys have drawn particular attention for their advantageous properties, such as their relatively low melting point compared with Sn–Ag binary eutectic alloys, their generally superior mechanical properties, and relatively good solderability [10–14].

However, Sn–Ag–Cu solder joints are prone to cracking due to the difference in the coefficients of thermal expansion (CTEs) between the copper (Cu) substrate and the solder. Another serious drawback is the formation of an intermetallic compound (IMCs) layer within the solder joint. Past research has looked to solve the issue of crack propagation in solder joints using a polymer underfill system.

Recently, underfill bonding mechanisms have become particularly common in ball grid array (BGA) chip solder joints. In addition, solders containing epoxy have been developed to suppress crack propagation in solder joints. Generally, epoxy solder joints enhance the bonding strength of solder joints through mechanical locking around the solder, ensuring firm bonding between the solder and the substrate [15–17]. However, the growth of IMC layers at high temperatures remains a problem. A dual Cu–Sn IMCs layer consisting of Cu_6Sn_5 and Cu_3Sn can form within solder joints between the Cu substrate and the solder interface. Generally, IMC growth in Sn–Ag–Cu solder joints is more rapid than in eutectic Sn–Pb solder joints due to their higher reflow temperature and significant proportion of Sn [18–20]. Thick intermetallic growth degrades the bonding strength due to the brittle nature of the IMC layer and the mismatch in physical properties such as the CTE and elastic modulus. Excessive thickness may also decrease the ductility and strength of the solder joint [21] and increase the electrical resistance [22]. As such, it is important to suppress the growth of the IMC layer in all kinds of solder joint.

In this study, the effect of the addition of graphene nano-sheets (GNSs) to SAC305 epoxy solder on the growth of the Cu–Sn IMC layer was investigated. The influence of the GNSs on the bonding characteristics of the solder joints was also investigated in shear tests.

2. Experimental

Plain epoxy solder paste was produced using SAC305 solder powder with a particle size of 10–25 μm, a flux, an amine-containing epoxy as a binder, a small amount of rosin, a surfactant, and a carboxyl-group activator. GNS epoxy solder pastes were prepared by mixing GNSs (at 0.01, 0.05, 0.1 wt %) with plain epoxy solder, with the mixture blended mechanically for 2 h at 200 rpm. Scanning electron microscopy (SEM, Hitachi S-3400N, Tokyo, Japan) images of the GNSs used in this study are shown in Figure 1b. GNSs are typically produced from natural graphite through chemical exfoliation, thermal shock and shear, or the use of a plasma reactor. In this paper, the GNSs had an average thickness of approximately 6–8 nanometers and an average particle size of 10 microns. The test specimens consisted of a chip resistor (R3216; L: 3.2 mm, W: 1.6 mm, H: 0.55 mm) soldered to a printed circuit board (PCB) with an organic solderability preservative (OSP) finish. The solder paste (plain epoxy or GNS epoxy) was screen-printed onto a Cu pad using the squeeze method, and the R3216 chip was placed on the solder paste. The solder paste was reflowed at a temperature profile of 240 °C for 100 s. An example specimen is presented in Figure 1a. It has been found that the melting point of conventional solder increases slightly if GNSs are added [23,24], while SAC305 epoxy solder also requires higher temperatures for the reflow process than does conventional SAC305 solder. Thus, in this study, the increase in the melting point with the addition of GNSs did not have a significant influence on the reflow process because the epoxy solder needed to be reflowed at a much higher temperature regardless.

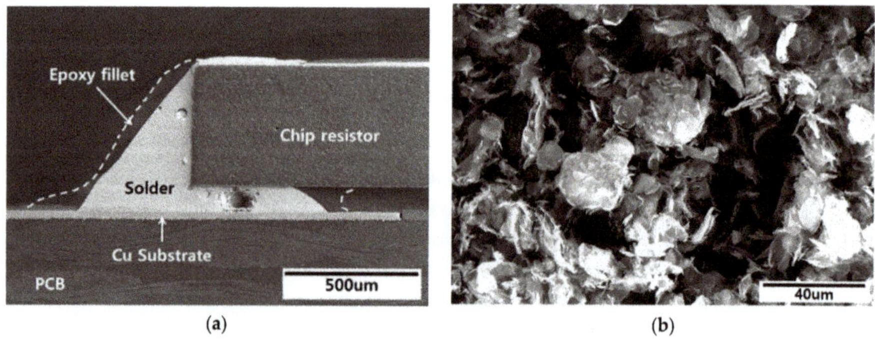

Figure 1. SEM images of (**a**) the structure of the specimens and (**b**) graphene nano-sheets (GNSs).

Following fabrication of the specimens, the solder joints were aged in a thermal chamber at 85 °C, 125 °C and 150 °C for 21 days (504 h) following the JESD22-A103C standard [25] (i.e., high-temperature storage life). Shear forces were then measured to determine the effect of the GNSs on the bonding characteristics of the solder joints. The shear forces were tested using a JIS Z 3198-7, which operates at a shear speed of 5–30 mm/min and a height that is 25% or less of the component thickness (Figure 2). We measured shear forces at a shear speed of 10 mm/min and a height of 100 μm (0.1 mm) from the Cu substrate. Finally, the solder joints were cross-sectioned, polished, and etched for 30 s with 10 vol % hydrochloric acid in ethanol in order to observe the IMC layer in the solder joints. The thickness of the IMC layer was measured based on SEM images, with the average of the maximum and minimum thickness for each specimen used in the analysis.

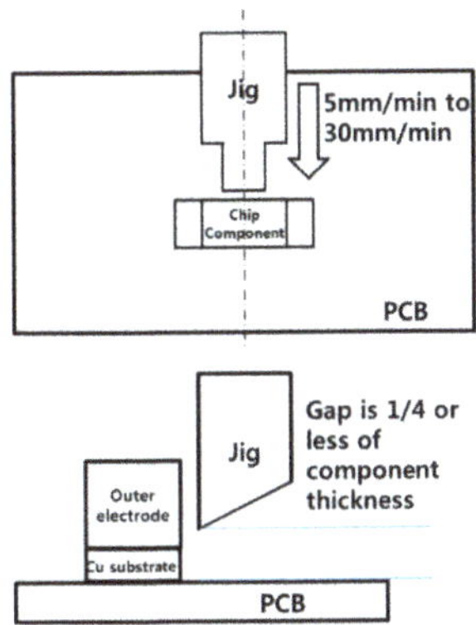

Figure 2. Schematic diagram of the shear testing.

3. Results

The SEM image in Figure 3 confirms that the GNSs were distributed throughout the solder joint, while the SEM images in Figure 4 depict the growth of the IMC layer in the solder joint. Generally, IMC layers form at the interface between the solder and a substrate, i.e., between the solder and the chip electrode and/or between the solder and the Cu substrate. In this study, the IMC layer formed at the solder/Cu substrate interface only because the chip electrode was coated with a nickel (Ni) layer. The Ni layer suppresses Sn diffusion in the solder joint, protecting the chip from damage. However, the solder/Cu substrate interface does not have a barrier that prevents Sn diffusion given that it was finished with only OSP and flux for soldering. As can be observed in Figure 4, the IMC layer grew consistently over time. Two sub-layers were observed in the solder: Cu_6Sn_5 adjacent to the solder matrix, which appeared from the start of the aging process, and Cu_3Sn adjacent to the Cu substrate, which appeared after 336 h of aging. After aging at 150 °C, the thickness of the IMC layer was 7.9 μm in the plain epoxy solder joint, 6.3 μm in the 0.01%-GNS solder joint, 5.3 μm in the 0.05%-GNS solder joint, and 3.6 μm in the 0.1%-GNS solder joint. Figure 5 presents a summary of the thickness of the IMC layer over time for the three test temperatures and four solder paste formulations.

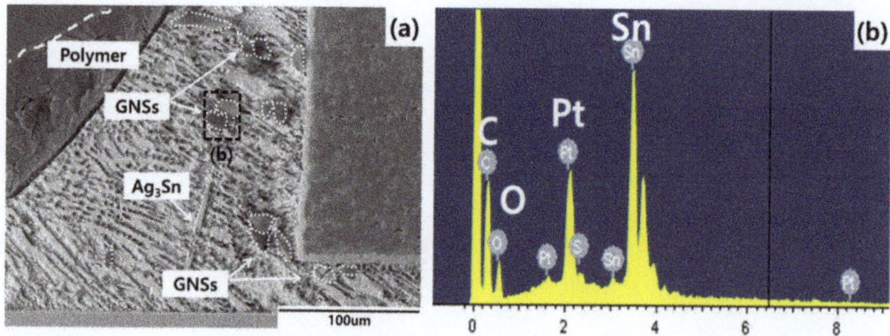

Figure 3. Dispersion of graphene nano-sheets(GNSs) in the solder joints: (**a**) cross-section image of a GNS epoxy solder joint and (**b**) EDS analysis results.

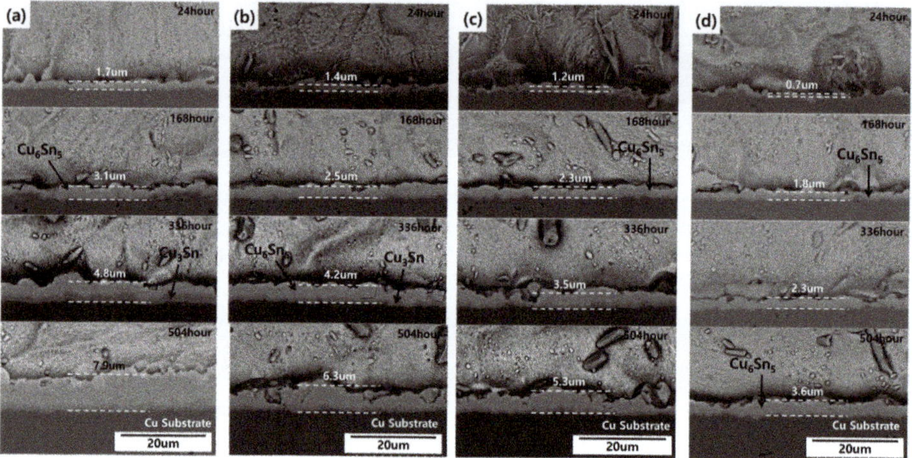

Figure 4. SEM images showing the growth of the IMC layer during the 150°C aging test: (**a**) plain epoxy solder joints, (**b**) 0.01%-GNS epoxy solder joints, (**c**) 0.05%-GNS epoxy solder joints, (**d**) 0.1%-GNS epoxy solder joints.

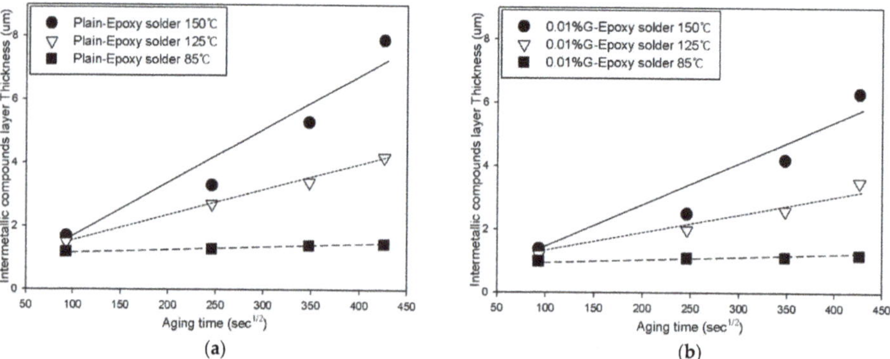

Figure 5. *Cont.*

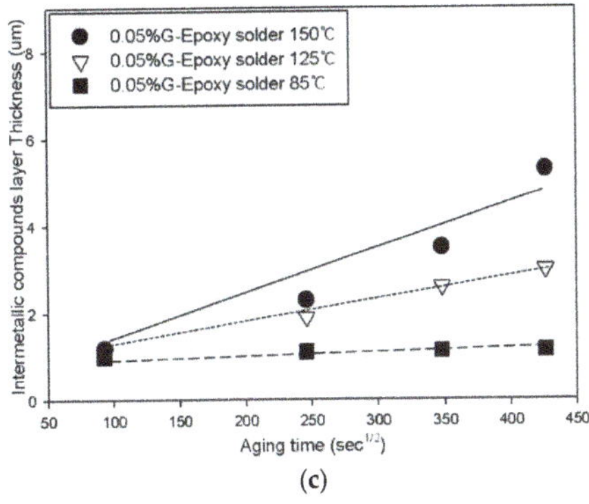
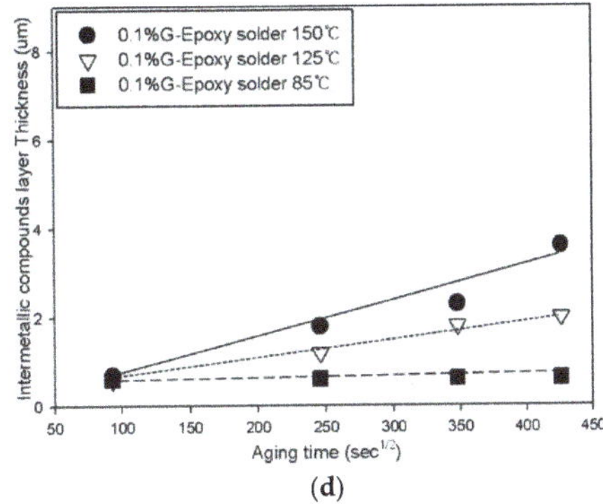

Figure 5. IMC layer growth over time in the solder joints: (**a**) plain solder joint; (**b**) 0.01%-GNS solder joint; (**c**) 0.05%-GNS solder joint; (**d**) IMC thickness in the 0.1%-GNS solder joint.

The kinetics of IMC layer formation is diffusion-controlled, depending on Sn and Cu diffusion at the Sn/Cu substrate interface as a function of time and temperature [26–29]. In general, the growth of the IMC layer should follow the square root time law, with the thickness of the layer in a diffusion couple expressed as the simple parabolic equation shown in Equation (1):

$$d_t = d_0 + (Dt)^{1/2} \tag{1}$$

where d_t is the thickness of the IMC layer (μm), d_0 is the initial thickness of the IMC layer (μm), D is the growth coefficient, and t is the reaction time. Equation (1) can be re-written as Equation (2):

$$d_t - d_0 = (Dt)^{1/2} \tag{2}$$

D is related to the diffusion coefficient of the atomic elements in the IMC layer and can be obtained from a linear regression line. The growth of an IMC layer is a diffusion-dominant process, thus, the Arrhenius relationship is applicable. The activation energy for IMC growth can be calculated using this relationship:

$$D = D_0 \exp\left(-\frac{Q}{RT}\right) \tag{3}$$

where D_0 is the diffusion constant, Q is the activation energy, R is the gas constant, and T is the absolute temperature.

$$\ln D = \ln D_0 - Q/RT \tag{4}$$

The calculated Arrhenius plot is shown in Figure 6, and the activation energies were estimated from the slope of the Arrhenius plot [30–32]. The activation energy required to form the IMC layer in the plain epoxy solder, 0.01%-GNS solder, 0.05%-GNS solder, and 0.1%-GNS solder was 45.5 KJ/mol, 52.8 KJ/mol, 62.5 KJ/mol, and 68.7 KJ/mol, respectively. Generally, the activation energy for SAC305 solder joint/Cu substrate interfaces has been calculated at around 50.6–62.8 kJ/mol in past research [33,34]. The epoxy does not influence the activation energy in solder joints; it merely provides support for the solder joints. Thus, the activation energy of the plain epoxy solder joint in the present research was similar to conventional solder joints. In this study, the 0.1%-GNS solder joint exhibited the highest activation energy. This means that the presence of GNSs increases the amount of energy required for Sn/Cu diffusion and the subsequent formation of IMCs in GNS solder joints. The GNSs can, thus, be regarded as an effective diffusion barrier in solder joints. GNSs have a large surface area compared to their thickness, and this high specific surface area can suppress the diffusion of atoms [23,27,35]. Importantly, the GNSs do not react with or coarsen the solder matrix.

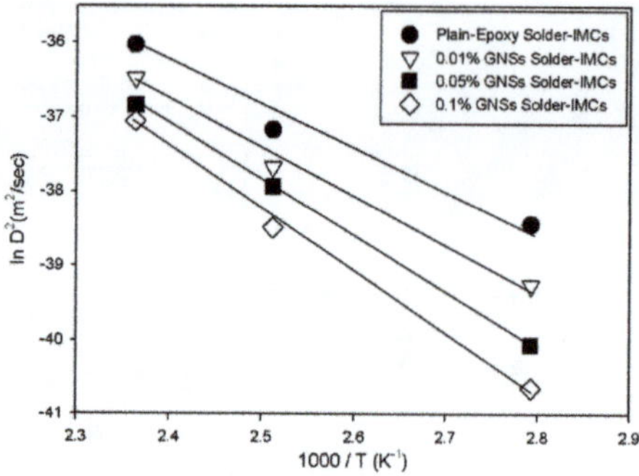

Figure 6. Arrhenius plot of the IMC layer for the four solder types.

Shear forces were assessed to investigate the effect of GNSs on the bonding characteristics of the solder joints (Figure 7). The initial shear force was 95.6 N, 92.6 N, 91.8 N, and 90.2 N for the plain epoxy solder, 0.01%-GNS solder, 0.05%-GNS solder, and 0.1%-GNS solder, respectively. The conventional SAC305 (i.e., without epoxy) solder joint had a shear force of 50.1 N. The GNS-containing solders had a lower shear force than the plain epoxy solder, with the 0.1%-GNS formulation exhibiting the lowest force. However, the shear forces for all of the epoxy solder formulations in this study (i.e., plain, 0.01%-GNS, 0.05%-GNS, and 0.1%-GNS) were almost twice as high as conventional SAC305 solder joints.

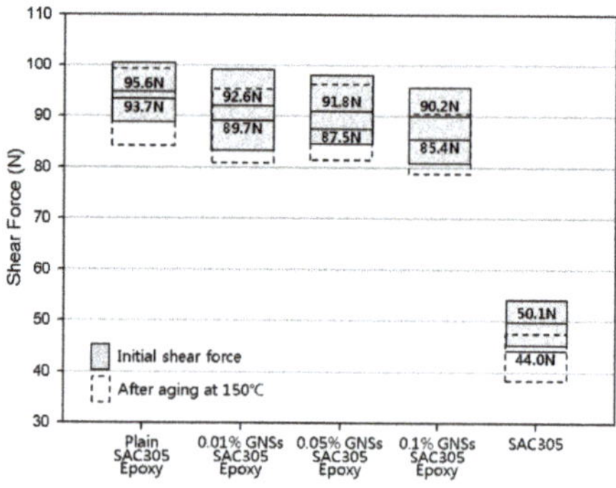

Figure 7. The weakening of the shear force after aging tests at 150 °C.

The shear force was then compared with specimens that had been aged at 150 °C because the thickness of the IMC layers was greatest at this temperature and because the effect of GNSs in the epoxy solder was similar in every solder joint. After the aging at 150 °C, the decrease in the shear force was 2.0% for plain epoxy solder joint, 3.1% for the 0.01%-GNS solder joint, 4.7% for the 0.05%-GNS solder joint, and 5.3% for the 0.1%-GNS solder joint, while that for the conventional SAC305 solder joint was 12.2%. The addition of GNSs to the solder matrix, thus, weakens the bonding characteristics, but this is not a critical problem in solder joints because the epoxy reinforces the bonding and mechanical properties of the solder. Rather, the thickness of the IMC layer was lower in GNS solder joints because the diffusion of atoms was blocked due to the high specific surface area of the GNSs, as shown in Figure 8 [31,33]. Thus, the addition of small amounts of GNSs to epoxy solder can prevent atomic diffusion and continuously suppress IMC layer formation during isothermal aging.

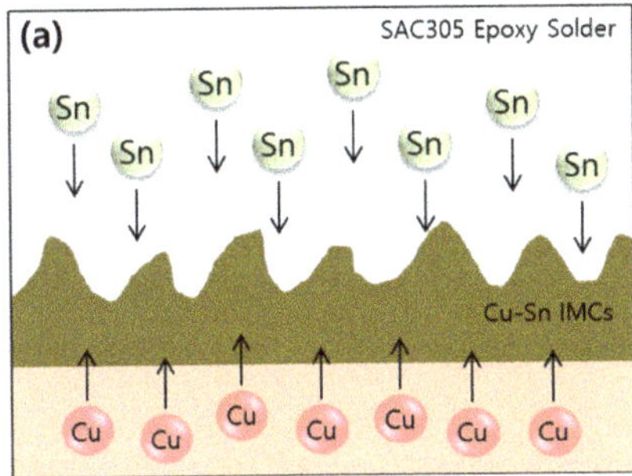

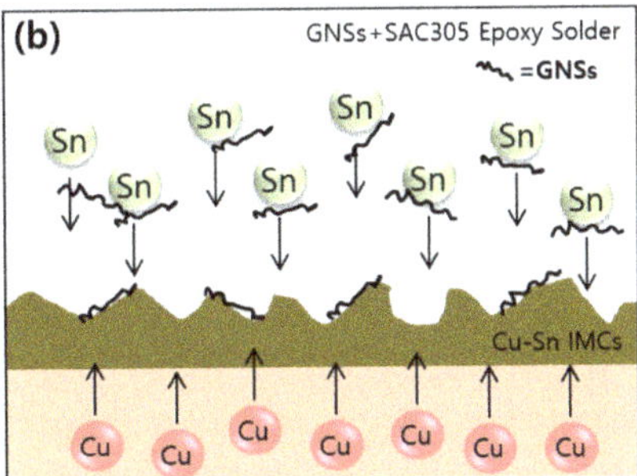

Figure 8. Schematic diagram of IMC growth in (**a**) SAC305 epoxy solder joints and (**b**) GNS/SAC305 epoxy solder joints.

4. Conclusions

The effect of GNSs on the growth of the IMC layer at the solder/Cu substrate interface in epoxy solder (SAC305) joints was investigated in this study. The addition of GNSs effectively suppressed this IMC layer without greatly weakening the bonding characteristics of the solder. In detail, the key results of this study were as follows:

(1) Growth of the Cu–Sn IMC layer at the solder/Cu substrate interface due to Sn and Cu atomic diffusion was confirmed with aging tests at 150 °C, 125 °C, and 85 °C. It was found that the IMC layer was suppressed by the addition GNSs to the solder. Under the harshest conditions (isothermal 150 aging for 21 days), the thickness of the IMC layer was 7.9 μm in plain epoxy solder, 6.3 μm for the 0.01%-GNS solder, 5.3 um for the 0.05%-GNS solder, and 3.6 μm for the 0.1%-GNS solder.

(2) The suppression of IMC growth was stronger with increasing GNS levels in the solder. The activation energy for the plain epoxy solder, 0.01%-GNS solder, 0.05%-GNS solder, and 0.1%-GNS solder was 45.5 KJ/mol, 52.8 KJ/mol, 62.5 KJ/mol, and 68.7 KJ/mol, respectively.

(3) The presence of GNSs in the epoxy solder paste had a slight negative effect on the bonding characteristics of the solder, but this was not critical to the function of the solder joint because the epoxy reinforced the bonding and mechanical characteristics of solder joints compared to conventional SAC305 solder joints.

Author Contributions: Writing—original draft, M.-S.K. and D.-S.K.; Writing—review & editing, Y.-E.S.

Funding: This research was supported by the Basic Science Research Program through the National Research Foundation of Korea (NRF) funded by the Ministry of Education (NRF-2017R1D1A1B03034423).

Conflicts of Interest: The authors declare no conflict of interest.

References

1. Chen, G.; Wu, F.; Liu, C.; Silberschmidt, V.V.; Chan, Y.C. Microstructures and properties of new Sn–Ag–Cu lead-free solder reinforced with Ni-coated graphene nanosheets. *J. Alloys Compd.* **2016**, *656*, 500–509. [CrossRef]

2. Kang, M.S.; Jeon, Y.J.; Kim, D.S.; Shin, Y.E. Degradation characteristics and Ni_3Sn_4 IMC growth by a thermal shock test in SAC305 solder joints of MLCCs applied in automotive electronics. *Int. J. Precis. Eng. Manuf.* **2016**, *17*, 445–452. [CrossRef]

3. Jeon, Y.J.; Kim, D.S.; Shin, Y.E. Study of characteristics of solar cells through thermal shock and high-temperature and high-humidity testing. *Int. J. Precis. Eng. Manuf.* **2015**, *15*, 355–360. [CrossRef]

4. Yang, Z.J.; Yang, S.M.; Yu, H.S.; Kang, S.J.; Song, J.H.; Kim, K.J. IMC and creep behavior in lead-free solder joints of Sn-Ag and Sn-Ag-Cu alloy system by SP method. *Int. J. Autom. Technol.* **2014**, *15*, 1137–1142. [CrossRef]

5. Gain, A.K.; Zhang, L.; Chan, Y.C. Microstructure, elastic modulus and shear strength of alumina (Al_2O_3) nanoparticles-doped tin–silver–copper (Sn–Ag–Cu) solders on copper (Cu) and gold/nickel (Au/Ni)-plated Cu substrates. *J. Mater. Sci. Mater. Electron.* **2015**, *26*, 7039–7048. [CrossRef]

6. Gnecco, F.; Ricci, E.; Amore, S.; Giuranno, D.; Borzone, G.; Zanicchi, G.; Novakovic, R. Wetting behaviour and reactivity of lead free Au–In–Sn and Bi–In–Sn alloys on copper substrates. *Int. J. Adhes. Adhes.* **2007**, *27*, 409–416. [CrossRef]

7. Gain, A.K.; Zhang, L.; Quadir, M.Z. Thermal aging effects on microstructures and mechanical properties of an environmentally friendly eutectic tin-copper solder alloy. *Mater. Des.* **2016**, *110*, 275–283. [CrossRef]

8. Kotadia, H.R.; Howes, P.D.; Mannan, S.H. A review: On the development of low melting temperature Pb-free solders. *Microelectron. Reliab.* **2014**, *54*, 1253–1273. [CrossRef]

9. Gain, A.K.; Zhang, L. Microstructure, thermal analysis and damping properties of Ag and Ni nano-particles doped Sn–8Zn–3Bi solder on OSP–Cu substrate. *J. Alloys Compd.* **2014**, *617*, 779–786. [CrossRef]

10. Gain, A.K.; Zhang, L. Growth mechanism of intermetallic compound and mechanical properties of nickel (Ni) nanoparticle doped low melting temperature tin–bismuth (Sn–Bi) solder. *J. Mater. Sci. Mater. Electron.* **2016**, *27*, 781–794. [CrossRef]

11. Kumar, K.M.; Kripesh, V.; Tay, A.A. Single-wall carbon nanotube (SWCNT) functionalized Sn–Ag–Cu lead-free composite solders. *J. Alloys Compd.* **2008**, *450*, 229–237. [CrossRef]

12. Hu, X.; Chan, Y.C.; Zhang, K.; Yung, K.C. Effect of graphene doping on microstructural and mechanical properties of Sn–8Zn–3Bi solder joints together with electromigration analysis. *J. Alloys Compd.* **2013**, *580*, 162–171. [CrossRef]

13. Gao, F.; Nishikawa, H.; Takemoto, T.; Qu, J. Mechanical properties versus temperature relation of individual phases in Sn–3.0Ag–0.5Cu lead-free solder alloy. *Microelectron. Reliab.* **2009**, *49*, 296–302. [CrossRef]

14. Cheng, F.; Nishikawa, H.; Takemoto, T. Microstructural and mechanical properties of Sn–Ag–Cu lead-free solders with minor addition of Ni and/or Co. *J. Mater. Sci.* **2008**, *43*, 3643–3648. [CrossRef]

15. Kim, K.S.; Huh, S.H.; Suganuma, K. Effects of intermetallic compounds on properties of Sn–Ag–Cu lead-free soldered joints. *J. Alloys Compd.* **2003**, *352*, 226–236. [CrossRef]

16. Myung, W.-R.; Kim, Y.; Jung, S.-B. Mechanical property of the epoxy-contained Sn–58Bi solder with OSP surface finish. *J. Alloys Compd.* **2014**, *615*, S411–S417. [CrossRef]

17. Kim, J.; Myung, W.R.; Jung, S.B. Effect of aging treatment and epoxy on bonding strength of Sn–58Bi solder and OSP-finished PCB. *J. Microelectron. Packag. Soc.* **2014**, *21*, 97–103. [CrossRef]

18. Cho, I.; Ahn, J.H.; Yoon, J.W.; Shin, Y.E.; Jung, S.B. Mechanical strength and fracture mode transition of Sn-58Bi epoxy solder joints under high-speed shear test. *J. Mater. Sci. Mater. Electron.* **2012**, *23*, 1515–1520. [CrossRef]

19. Chen, B.L.; Li, G.Y. Influence of Sb on IMC growth in Sn–Ag–Cu–Sb Pb-free solder joints in reflow process. *Thin Solid Films* **2004**, *462*, 395–401. [CrossRef]

20. Li, G.-Y.; Shi, X.Q. Effects of bismuth on growth of intermetallic compounds in Sn-Ag-Cu Pb-free solder joints. *Trans. Nonferrous Met. Soc. China* **2006**, *16*, s739–s743. [CrossRef]

21. Lee, T.Y.; Choi, W.J.; Tu, K.N.; Jang, J.W.; Kuo, S.M.; Lin, J.K.; Frear, D.R.; Zeng, K.; Kivilahti, J.K. Morphology, kinetics, and thermodynamics of solid-state aging of eutectic SnPb and Pb-free solders (Sn–3.5 Ag, Sn–3.8 Ag–0.7 Cu and Sn–0.7 Cu) on Cu. *J. Mater. Res.* **2002**, *17*, 291–301. [CrossRef]

22. Tsao, L.C. Suppressing effect of 0.5 wt.% nano-TiO2 addition into Sn–3.5 Ag–0.5 Cu solder alloy on the intermetallic growth with Cu substrate during isothermal aging. *J. Alloys Compd.* **2011**, *509*, 8441–8448. [CrossRef]

23. Noh, B.I.; Koo, J.M.; Kim, J.W.; Kim, D.G.; Nam, J.D.; Joo, J.; Jung, S.B. Effects of number of reflows on the mechanical and electrical properties of BGA package. *Intermetallics* **2006**, *14*, 1375–1378. [CrossRef]

24. Huang, Y.; Xiu, Z.; Wu, G.; Tian, Y.; He, P.; Gu, X.; Long, W. Improving shear strength of Sn-3.0 Ag-0.5 Cu/Cu joints and suppressing intermetallic compounds layer growth by adding graphene nanosheets. *Mater. Lett.* **2016**, *169*, 262–264. [CrossRef]

25. *High Temperature Storage Life, JESD22A-103C*; JEDEC Solid State Technology Association: Arlington, VA, USA, 2004.

26. Mayappan, R.; Salleh, A. Intermetallic growth retardation with the addition of graphene in the Sn-3.5 Ag lead-free solder. *IOP Conf. Ser. Mater. Sci. Eng.* **2017**, *205*, 012017. [CrossRef]

27. Flanders, D.R.; Jacobs, E.G.; Pinizzotto, R.F. Activation energies of intermetallic growth of Sn-Ag eutectic solder on copper substrates. *J. Electron. Mater.* **1997**, *26*, 883–887. [CrossRef]

28. Liu, B.; Tian, Y.; Wang, C.; An, R.; Liu, Y. Extremely fast formation of CuSn intermetallic compounds in Cu/Sn/Cu system via a micro-resistance spot welding process. *J. Alloys Compd.* **2016**, *687*, 667–673. [CrossRef]

29. Vianco, P.T.; Rejent, J.A.; Hlava, P.F. Solid-state intermetallic compound layer growth between copper and 95.5 Sn-3.9 Ag-0.6 Cu solder. *J. Electron. Mater.* **2004**, *33*, 991–1004. [CrossRef]

30. Madeni, J.C.; Liu, S. Effect of thermal aging on the interfacial reactions of tin-based solder alloys and copper substrates and kinetics of formation and growth of intermetallic compounds. *Soldagem Inspeção* **2011**, *16*, 86–95. [CrossRef]

31. Mayappan, R.; Ahmad, Z.A. Effect of Bi addition on the activation energy for the growth of Cu_5Zn_8 intermetallic in the Sn–Zn lead-free solder. *Intermetallics* **2010**, *18*, 730–735. [CrossRef]

32. Ghosh, G. Interfacial microstructure and the kinetics of interfacial reaction in diffusion couples between Sn–Pb solder and Cu/Ni/Pd metallization. *Acta Mater.* **2000**, *48*, 3719–3738. [CrossRef]

33. El-Daly, A.A.; El-Taher, A.M.; Dalloul, T.R. Enhanced ductility and mechanical strength of Ni-doped Sn–3.0 Ag–0.5 Cu lead-free solders. *Mater. Des.* **2014**, *55*, 309–318. [CrossRef]

34. Kim, S.H.; Tabuchi, H.; Kakegawa, C.; Park, H. IMC growth mechanisms of SAC305 lead-free solder on different surface finishes and their effects on solder joint reliability of FCBGA packages. *J. Microelectron. Electron. Packag.* **2009**, *6*, 119–124. [CrossRef]

35. Ma, D.; Wu, P. Improved microstructure and mechanical properties for Sn58Bi0. 7Zn solder joint by addition of graphene nanosheets. *J. Alloys Compd.* **2016**, *671*, 127–136. [CrossRef]

Article

Concrete Properties Comparison When Substituting a 25% Cement with Slag from Different Provenances

María Eugenia Parron-Rubio [1], Francisca Perez-García [2],*, Antonio Gonzalez-Herrera [2] and María Dolores Rubio-Cintas [1]

1. Departamento de Ingeniería Industrial y Civil, Universidad de Cádiz, 11205 Algeciras, Spain; m.eugenia.parron@uca.es (M.E.P.-R.); mariadolores.rubio@uca.es (M.D.R.-C.)
2. Departamento de Ingeniería Civil, Materiales y Fabricación, Universidad de Málaga, 29071 Málaga, Spain; agh@uma.es (A.G.-H.)
* Correspondence: perez@uma.es

Received: 2 May 2018; Accepted: 13 June 2018; Published: 17 June 2018

Abstract: Concrete consumption greatly exceeds the use of any other material in engineering. This is due to its good properties as a construction material and the availability of its components. Nevertheless, the present worldwide construction increases and the high-energy consumption for cement production means a high environmental impact. On the other hand, one of the main problems in the iron and steel industry is waste generation and byproducts that must be properly processed or reused to promote environmental sustainability. One of these byproducts is steel slag. The cement substitution with slag strategy achieves two goals: raw materials consumption reduction and waste management. In the present work, four different concrete mixtures are evaluated. The 25% cement substitution is carried out with different types of slag. Tests were made to evaluate the advantages and drawbacks of each mixture. Depending on the origin, characteristics, and treatment of the slag, the concrete properties changed. Certain mixtures provided proper concrete properties. Stainless steel slag produced a fluent mortar that reduced water consumption with a slight mechanical strength loss. Mixtures with ground granulated blast furnace slag properties are better than the reference concrete (without slag).

Keywords: concrete; slag; valorization; cement; circular economy

---

1. Introduction

Nowadays, the increasing growth of waste generated as a result of industrial activity is unavoidable, and dealing with this complex problem has become a difficult issue in part because of increasingly strict environmental regulations and policies. The analysis of industrial and construction waste and its transformation into raw materials in order to be introduced again into the production chain is part of the circular economy, a trend in economics initiated by Kenneth Boulding in 1965 [1].

According to the World Steel Association, 1600 millions of metric tons of steel were produced in the world in 2016, and around 162 million were produced in the European Union alone. Steel slag is a byproduct of steelmaking, produced during the separation of the molten steel from impurities in steelmaking furnaces. Making efforts to valorize this type of industrial waste can help achieve a more sustainable environment. Moreover, storing slags in landfills has negative effects on the environment, due to not only the extension of land occupied by this kind of waste, but also the leaching problems that it generates. According to the Nippon Slag Association, the pH of steel slag increases when reacting with water, which results in a highly alkaline fluid containing heavy metals that is harmful to the environment. This requires important and costly safety measures in landfills in order to verify the integrity of surrounding land and aquifers [2].

The properties of the slag produced by steelmaking depend on many factors, mainly the manufacturing process. According to Setie et al. [3], four types of steel slag can be distinguished: electric arc furnace (EAF) slag, blast furnace slag (BFS), basic oxygen furnace slag (BOFS), and ladle furnace slag (LFS).

Electric arc furnace slag is generated when iron scrap is melted and refined during the process of steelmaking. In a previous paper, Black Spanish EAF was characterized and was found to be composed of minerals such as anhydrous calcium silicates and silicoaluminates, magnetite and magnesioferrite, and manganese oxides (Luxán et al. [4]). Many researchers have observed the suitability of EAF as an aggregate substitution in concrete, showing similar compressive strength and high workability in concrete mixtures [5–8].

Regarding the specific case of steel slag waste valorization, in the European Union, ground granulated blast furnace slag (GGBF) has been used as an addition to the mix of cement manufactured in plant [9–13] to make up the cements CEM II and CEM III. This type of slag is a derivative of blast furnace slag, which is formed when iron ore or iron pellets, coke, and a flux (either limestone or dolomite) are melted together in a blast furnace. It has been used in various applications, mainly in aggressive environments (maritime and hydraulic facilities) because of its resistance against salt water and sulphates. Since the 80s, much research [14–19] has been conducted in order to use this type of slag as an additive or as fine and coarse aggregate substitutes [20–29].

Ladle furnace slag is produced during the secondary refining of steel through the addition of lime (CaO) and dolomitic limestone (CaO·MgO) in the electric arc furnace. The resulting byproduct, ladle furnace slag, has been used as a sand and cement substitute in masonry mortars, showing improvements in workability and mechanical strength in the medium term [30]. However, the use of LFS as a substitute in civil construction is less suitable due to its high expansiveness, especially when used in roads as an aggregate substitute [31–34].

During the manufacturing process of steel, 110–130 kg per metric ton of EAF and 20–30 kg per metric ton of LFS are generated [35].

In this paper, the suitability of three types of slag (EAF, GGBFS, and LFS) produced in still mills from different locations in Spain will be put to the test as cement substitutes in concrete mixtures.

A comparative analysis of different concrete mixtures will be carried out, substituting the cement with the aforementioned slags. The results of this paper are part of a wider program intended to elaborate a standard to guide the use of steel slag as a cement substitute in concrete mixtures in infrastructure building based on its chemical composition and mechanical behavior.

This investigation is an extension of the patent "Method for producing cinder concrete" [36], in which a manufacturing process for the production of concrete was developed with a stainless steel addition or cement substitution, obtaining a concrete mixture proportioning especially indicated for the construction of retaining walls and precast voussoirs, reducing energy and resources consumption.

2. Materials

Four different concrete mixes have been designed by substituting 25% of the weight of the cement with slag obtained from four different ladle furnaces in Spain. Because this investigation is part of a project studying the use of concrete with landfill slag substitution in precast voussoir manufacturing, the concrete dosage has been designed in order to meet certain consistency requirements (dry). The percent of the cement/slag substitution (25%) is based on the results of the patent "Method for producing cinder concrete" in order to make a performance comparison between the four types of slags. The chemical composition is shown in Table 1. The mix proportion shown in Table 2 was used as reference. The following materials were used to make the mixtures:

- Cement: Portland Cement CEM I 52.5 R with the composition given in Table 1. This cement was selected due to the absence of any kind of additive that could mask the results. It was used as reference pattern. Density: 2.5 g/cm^3. Specific surface area: >2800 cm^2/gr.

- Sand: crushed limestone sand was used. Size ratio: fine aggregate 0/2, medium aggregate (sand) 0/4, and gravel 4/16.
- Water: domestic tap water.
- Additive: Superplasticizer. Concrete additive: UNE EN 934-2.

The different slag used for the test were the following:

- Ground granulated blast furnace slag (GGBFS) with mechanical processing (M2): Initial aggregates are sand-like type 0/3 with a high humidity content (around 8–10%). They are dried and ground in origin. This is made by means of vertical roller mills specific to this material, which dries during grinding. This results in a maximum grain size of 0.063 mm; thus, it doesn't require sieving. Density: 2.91 g/cm^3. Specific surface area: 4620 cm^2/gr.
- Unprocessed ladle furnace slag (LFS). Two different materials (with different compositions) were tested, coming from two different steel mills (used in M3 and M4, respectively). The only process they were subjected to was sieving in the lab with a 0.063 mm sieve. The fraction obtained through sieving was 23% and 15%, respectively.
- Unprocessed electric furnace slag from stainless steelmaking (M5), except sieving in the lab with a 0.063 mm sieve. The fraction obtained through sieving is 82%.

The chemical composition of the slags used in the concrete mixtures (M2, M3, M4, and M5) is shown in Table 1. This chemical composition is proportionated by the slag supplier companies, and it refers to the slags before the sieving process when needed in the laboratory. This provides more representative values for the different slags, because they do not depend on a particular batch of slag. However, chemical analysis of the slags that were introduced into the different concrete mixtures (after sieving when needed) were carried out in our laboratory and showed similar values to those given by the suppliers, without any remarkable deviation.

The different concrete mixtures were named as follow:

Mix 1 (M1): Ordinary concrete without slag.
Mix 2 (M2): Concrete with a 25% cement replaced with processed slag.
Mix 3 (M3): Concrete with a 25% cement replaced with unprocessed slag.
Mix 4 (M4): Concrete with a 25% cement replaced with unprocessed slag.
Mix 5 (M5): Concrete with a 25% cement replaced with stainless steel slag.

Table 1. Cement and slag chemical composition (data provided by the supplying company).

Slag Origin/Chemical Composition	Type of Slag	SiO$_2$	AL$_2$O$_3$	Fe$_2$O$_3$	CaO	MgO	Na$_2$O	K$_2$O	S	TiO$_2$	Cl	Limestone	P$_2$O$_5$	Cr$_2$O$_3$	MnO	Fe
		%	%	%	%	%	%	%	%	%	%	%	%	%		
Cement (M1)	-	20–22	4–10	4	55–62	2	0.3	0.3	-	-	-	-	-	-	-	-
Processed slag (M2)	GGBFS	35.9	11.2	0.3	40	7.7	0.2	0.4	0.8	0.6	<0.1	0.5	-	-	-	-
Unprocessed slag type 1 (M3)	LFS	15.85	16.53	0.83	57	7.7	-	-	1.46	-	-	-	<0.1	<0.1	0.53	-
Unprocessed slag type 2 (M4)	LFS	22.28	9.37	0.84	56.94	7.37	0	-	-	0.46	-	-	0	0	0.44	0.58
Stainless steel slag (M5)	EAF	23	5.27	1.41	56.9	6.23	-	-	-	1.5	-	-	<0.1	2.96	1.68	-

Table 2. Concrete mixture proportion.

Mix	Water (*w/b* Ratio)	Binder					Aggregates		
		Dosage	Cement	Slag	Additive	Dosage	Fine Sand 0–2	Sand 0–4	Gravel 4–16
M1	0.5	300 kg/m^3	100%	0%	3.9 kg/m^3	2033.8 kg/m^3	15%	35%	50%
M2-M3-M4-M5	0.5	300 kg/m^3	75%	25%	3.9 kg/m^3	2033.8 kg/m^3	15%	35%	50%

3. Tests Description

The different mixes described in the previous section were subject to different standard tests. The goal of these tests was to evaluate how the cement–slag substitution may affect the main properties of consistency and workability (slump test) and mechanical capabilities (compressive strength). Additionally, as one of the main uses of this kind of concrete is in the marine environment, the depth of penetration of water became a key characteristic to study.

Concrete was made with the mixture proportions shown in Table 2, where a 25% of the cement was substituted by the different slag according to Table 1, providing the five different mixes previously described.

The concrete mixture proportion was made according to the norm EN 12390-2 [37] for testing hardened concrete, where the making and curing of specimens for strength tests is described. It covers the preparation and filling of molds, the compaction of the concrete, the levelling of the surface, the curing of test specimens, and the transporting of the test specimens.

The compressive strength test specimen was a cube with a 10 cm size. Four concrete mixtures were made for each mix type. From each mixture, ten specimens were made; two were tested at 7, 28, and 90 days, leaving the others as reserve. The depth of penetration of water test was made with a cylindrical specimen with a 15 cm diameter and 30 cm height. Two different specimens were tested for every mix obtaining the average mean value.

3.1. Slump Test

The consistency of fresh concrete was determined by the slump test according to the norm EN 12350-2 [38]. A truncated conical mold was used where the fresh concrete was poured and compacted. The mold was placed over a base and was raised upwards. The concrete cone slumped and the distance slumped provided a measure of the consistency of the concrete. Table 3 presents the results of these tests.

Table 3. Slump test results.

Mix	Slump	Standard Deviation	Consistency
M1	2.0	0.3	Dry
M2	2.0	0.2	Dry
M3	1.0	0.3	Dry
M4	0.1	0.1	Dry
M5	8.0	0.3	Soft

This test evaluated the workability of the concrete. It was intended to evaluate how the cement–slag substitution altered the consistency, leading indirectly to change the water cement ratio or suggesting different uses.

3.2. Compressive Strength Test

Compressive strength tests were performed according to norms EN 12390-3 [39] and EN 12390-4 [40]. The test was carried out with a servo-controlled compact compression testing machine (Proeti, Madrid, Spain) with a maximum capacity of 2000 kN (ETIMATIC-Proetisa H0224). The load control system emulates a servo-valve using a pump that accurately controls oil flow to the piston, controlling the rotation speed of the pump motor and the load gradient. The superior compression plate is supported by a spherical bearing ring that accommodates the alignment inaccuracy and avoids any lateral force. Once the specimen is placed, the display continuously shows the load, the failure load, and the strength calculation in real time (Figure 1).

(**a**)   (**b**)

Figure 1. Compressive strength (**a**) test and (**b**) specimens.

These tests provided the results shown in Table 4. The tests were performed at 7, 28, and 90 days to evaluate the behavior of slag in time. It was performed with a cubic specimen (10 cm size), and the load was applied at a constant speed of 0.5 MPa/s.

Table 4. Compressive strength test results and standard deviation.

Days/Mixes	7	Standard Deviation (7)	28	Standard Deviation (28)	90	Standard Deviation (90)	% Strength Gain
M1	52.12	5.23	59.34	3.87	66.05	2.38	0%
M2	54.73	2.52	63.69	3.08	71.51	5.15	8%
M3	34.22	1.65	37.07	0.98	45.01	5.69	−32%
M4	44.48	1.02	48.42	0.34	51.54	0.54	−22%
M5	37.04	4.54	44.38	4.66	48.94	5.64	−26%

3.3. Depth of Penetration of Water Under Pressure Test

This test was made according to norm EN 12390-8 [41], depth of penetration of water under pressure, with a cylindrical specimen with a 15 cm diameter and 30 cm height.

The test was performed with a 28-days cured specimen. As a first step, the specimen was placed in a drying oven for 24–48 h to be completely dried.

Then, the specimen was placed in the apparatus and a water pressure of (500 $\pm$ 50) kPa was applied for (72 $\pm$ 2) h. The appearance of the surfaces was controlled during the test in order to detect any water leakage.

At the end of the test, the specimen was taken from the system and any excess of water was removed and wiped. Then, the specimen was broken in two halves perpendicularly to the face where the water pressure was applied. The water penetration front could be clearly seen and marked on the specimen (as seen in Figure 2). The maximum depth of penetration was measured in mm.

(a)   (b)

Figure 2. Depth of penetration of water under pressure (**a**) test and (**b**) specimens.

4. Results and Discussion

The results obtained with the different mixes shown in Section 2 are analyzed and discussed in this section.

4.1. Consistency

The slump test results are shown in Table 3. The main observation is that most of the mixtures appear unaltered compared to the reference M1, except mixture M5 in which a higher fluency and workability could be observed (Figure 3c). It corresponded to the stainless-steel slag (M5), which was the only mix with a soft consistency. Consequently, the other mixtures showed a dry consistency, meeting the requirements for precast voussoir molds. A slump below half a centimeter was observed in the different tests. The material became very thick with a high presence of internal and external pores (Figure 3d). This behavior can be compared in Figure 3 with the other mixtures.

This means that the stainless-steel slag provoked a low level of friction in the material causing high fluency, as well as keeping the necessary viscosity to ensure the proper cohesion of the particles, thus avoiding segregation. This is the effect produced in self-compacting concrete [42], where fluency and viscosity is achieved by means of additives [43]. The opposite effect is observed in mix M4, where the material becomes denser and thicker, with a reduction in fluency and cohesion, promoting internal hollows and penalizing its permeability (as will be shown in the following subsections).

Figure 4 shows the standard deviation values of the experiment. Average values are considered in the range from 0.5 to 2 cm. Three mixtures are present within this range, while M4 is below and M5 is very above.

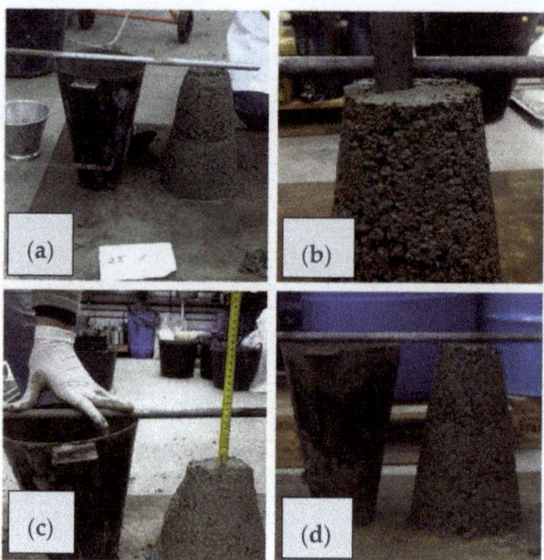

Figure 3. Concrete consistency. Slump test. (**a**) M1; (**b**) M2; (**c**) M5; and (**d**) M4.

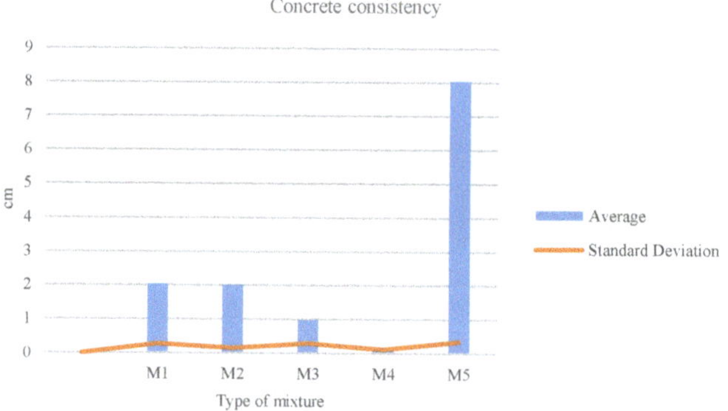

Figure 4. Consistency standard deviation.

4.2. Compressive Strength

The compressive strength test results are shown in Table 4 and plotted in Figure 5, comparatively.

This experimental data shows how the M2 mixture provides a good compressive strength, even slightly above the reference sample, and 8% higher than that of M1. In Figure 5b, it can be observed how this relative compressive strength improvement is sustained over time. The other mixtures present certain strength loss, ranging from 32% for M3 to 22% for M4. This broad range—40% considering the four mixtures—proves that the origin, characteristics, and treatment given provide very different mechanical properties.

Regarding chemical composition, the proportion of SiO_2 is the main factor to consider. Mix M4, with similar SiO_2 content to the reference M1 (cement), shows the lower strength loss, and in the case of M2, where there is a strength gain, the SiO_2 percentage is higher (Table 1). On the other hand, M2 and M5 with higher strength loss, have lower SiO_2 content. In Figure 6, a linear regression is displayed showing the relationship between pozzolanic capacity and compressive strength. The correlation is strong with a R^2 of 0.967.

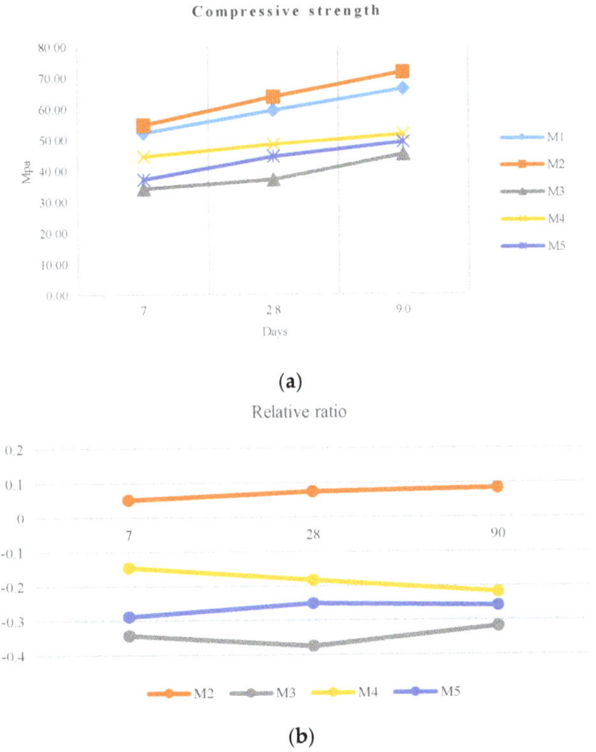

Figure 5. Compressive strength test results comparison: (**a**) evolution of different mixes over time and (**b**) relative ratio respect M1.

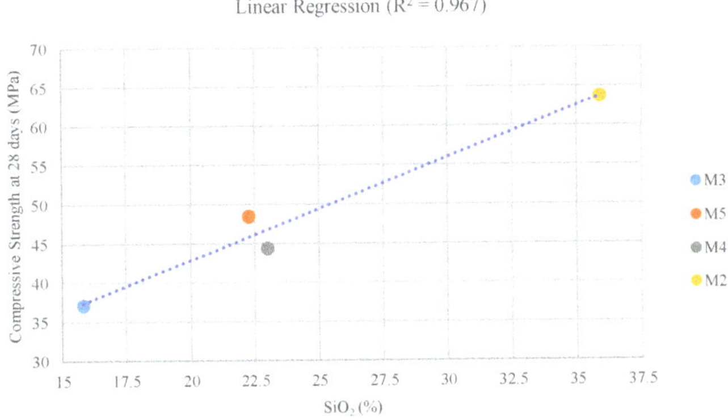

Figure 6. Pozzolanic capacity and compression strength correlation.

According to Canovas et al. [44], pozzolanic materials with a high content of SiO_2 present a high capacity to yield tobermorite (calcium hydrosilicates (C-S-H)) by reacting with portlandite (a product of concrete mineral hydration). This is the case of the M2 slag, which presents the highest SiO_2 content, obtaining a superior compressive strength performance [45]. The slag used in M3, which provided the minimum compressive strength result at 28 days, presents the lowest percentage of SiO_2. Thus, for the extreme cases (the highest and lowest SiO_2 content) we can observe a direct effect of the pozzolanic capacity of the slag used in compressive strength performance. However, M4 slag and M5 slag, despite presenting similar SiO_2 content (21–23%) to the cement, showed uneven results (48.42 and 44.38 MPa for M4 and M5, respectively, and 59.34 MPa for the 100% cement mix).

4.3. Depth of Penetration of Water

Depth of penetration of water under pressure test results are shown in Table 5 and plotted in Figure 7. Maximum and average penetration values are provided.

Table 5. Depth of penetration of water under pressure test results.

Mix	Maximum Penetration (mm)	Standard Deviation (mm)	Average Penetration (mm^2)	Standard Deviation (mm^2)
M1	28.3	12.09	15.8	6.42
M2	24.5	7.83	13.0	4.43
M3	25.0	3	15.3	2.91
M4	101.5	2.5	57.8	1.5
M5	21.5	3.5	9.8	0.015

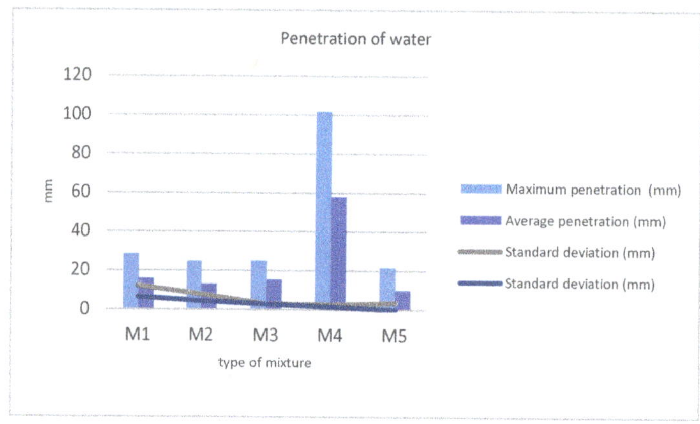

Figure 7. Average and standard deviation of the depth of penetration of water under pressure tests.

Excepting M4, all of the mixtures are in accordance with the norm EN 12390-8 [41]. The maximum deeper penetration is below the 30 mm limit established, and the average values do not exceed the 20 mm limit. M4 doubles both limits (Table 5). As seen in Section 4.1, it is the consequence of the appearance of internal cavities due to the very dry consistency of the concrete making a disperse aggregate mix, which cannot be eliminated by vibration. Water penetration is favored by this circumstance.

On the other extreme, we find the results obtained for the M5 mixture. In this case, penetration is below the average of the other samples. It is provoked by a reduction in the number and size of the cavities due to the good fluency and workability of this concrete.

The average values and the standard deviation of the experimental results are shown in Figure 7. M4 was not considered in the statistic calculation to avoid distortion.

Average values are considered in the range from 0.5 to 2 cm. Three mixtures are present within this range while M5 is below and M4 is very above.

5. Conclusions

According to the results described, we can outline the following conclusions:

(1)   Regarding concrete consistency, stainless-steel slag furnaces (M5) provided excellent workability properties with a higher fluency, keeping a 0.5 water–cement ratio. On the other extreme, the M4 mixture provided a consistency that was extremely dry with cavities that undermined its properties. The other mixtures provided dry consistencies similar to that obtained with the reference mix (formulated with a low water–cement ratio).

(2)   The ground granulated blast furnace slag (M2), with the highest content of SiO_2, showed a compressive strength gain of 8% relative to concrete with no slag substitution. On the other hand, the slag with the lowest content of SiO_2 performed the worst in the compressive strength tests, obtaining a strength loss of 32%. Thus, as has been pointed out by other authors [44,45], the content of SiO_2 affects directly the resultant compressive strength of the concrete.

(3)   Excepting the M4 mixture, the water penetration tests provided similar results for the different mixtures. In some cases, penetration was lower (M2 and M5) than the reference, and the reduction observed with M5 due to its exceptional fluency and workability was remarkable. M4 is the great exception and presents an extremely high maximum water penetration. This can be explained due to a possible lack of cohesion between particles during the curing process, because the consistency was too dry.

Summarizing, we can conclude that stainless-steel slag (M5) provides a good concrete formula, with excellent workability and a low water consumption. The negative point is a reduction in compressive strength. Regarding the type of slag, no correlation with compressive strength or workability has been found. It has been found that the results of the slag treated in a similar way to the cement (M2) are better than those obtained with the reference concrete (without slag), especially regarding the compressive strength, with the consequent savings in cement consumption.

As a final conclusion, we can say that cement substitution with slag can be considered a good strategy to reduce cement consumption while alleviating the waste management problem. Promising results have been obtained for certain mixtures, where the properties of the resultant concrete are even improved compared with the reference mix. This conclusion has been proven with 25% of cement substitution. Except for the pozzolanic capacity of the slag (SiO_2 content), no strong correlation has been found to explain the resultant concrete properties. Ladle furnace slags (M3 and M4) showed dissimilar results for compressive strength and workability; thus, specific analysis must be carried out in order to assess the suitability of this type of steel waste to be valorized.

Author Contributions: All the authors conceived and designed the experiments; M.E.P.-R. and F.P.-G. performed the experiments and analyzed the data; and M.E.P.-R. and A.G.-H. wrote the paper.

Funding: The authors acknowledge the financial support provided to this work by the Center of Industrial Technological Development (CDTI) of the Ministry of Economy and Competitiveness as a part of the research project IDI-20160509 through the companies DRACE and GEOCISA.

Conflicts of Interest: The authors declare no conflict of interest.

References

1.   Boulding, K.E. *Earth as a Space Ship*; Washington State University: Pullman, WA, USA, 1965; pp. 1–2.
2.   Chemical Characteristics of Iron and Steel Slag: NIPPON SLAG ASSOCIATION. Available online: http://www.slg.jp/e/slag/character.html (accessed on 17 May 2018).
3.   Setién, J.; Hernández, D.; González, J.J. Characterization of ladle furnace basic slag for use as a construction material. *Constr. Build. Mater.* **2009**, *23*, 1788–1794. [CrossRef]
4.   Luxán, M.P.; Sotolongo, R.; Dorrego, F.; Herrero, E. Characteristics of the slags produced in the fusion of scrap steel by electric arc furnace. *Cem. Concr. Res.* **2000**, *30*, 517–519. [CrossRef]
5.   Abu-Eishah, S.I.; El-Dieb, A.S.; Bedir, M.S. Performance of concrete mixtures made with electric arc furnace (EAF) steel slag aggregate produced in the Arabian Gulf region. *Constr. Build. Mater.* **2012**, *34*, 249–256. [CrossRef]
6.   Gokce, A.; Beyaz, C.; Ozkan, H. Influence of fines content on durability of slag cement concrete produced with limestone sand. *Constr. Build. Mater.* **2016**, *111*, 419–428. [CrossRef]

7. Pellegrino, C.; Cavagnis, P.; Faleschini, F.; Brunelli, K. Properties of concretes with Black/Oxidizing Electric Arc Furnace slag aggregate. *Cem. Concr. Compos.* **2013**, *37*, 232–240. [CrossRef]

8. Pepe, M.; Koenders, E.A.B.; Faella, C.; Martinelli, E. Structural concrete made with recycled aggregates: Hydration process and compressive strength models. *Mech. Res. Commun.* **2014**, *58*, 139–145. [CrossRef]

9. Manso, J.M.; Polanco, J.A.; Losañez, M.; González, J.J. Durability of concrete made with EAF slag as aggregate. *Cem. Concr. Compos.* **2006**, *28*, 528–534. [CrossRef]

10. Cabrera-Madrid, J.A.; Escalante-García, J.I.; Castro-Borges, P. Resistencia a la compresión de concretos con escoria de alto horno. Estado del arte re-visitado. *Rev. ALCONPAT* **2016**, *6*, 64–83. [CrossRef]

11. Ganjian, E.; Pouya, H.S. Effect of magnesium and sulfate ions on durability of silica fume blended mixes exposed to the seawater tidal zone. *Cem. Concr. Res.* **2005**, *35*, 1332–1343. [CrossRef]

12. Colangelo, F.; Cioffi, R. Use of cement kiln dust, blast furnace slag and marble sludge in the manufacture of sustainable artificial aggregates by means of cold bonding pelletization. *Materials* **2013**, *6*, 3139–3159. [CrossRef] [PubMed]

13. Andini, S.; Cioffi, R.; Colangelo, F.; Montagnaro, F.; Santoro, L. Effect of Mechanochemical Processing on Adsorptive Properties of Blast Furnace Slag. *J. Environ. Eng.* **2013**, *139*, 1446–1453. [CrossRef]

14. Ortega, J.M.; Esteban, M.D.; Sánchez, I.; Climent, M.A. Performance of sustainable fly ash and slag cement mortars exposed to simulated and real in situ Mediterranean conditions along 90 warm season days. *Materials* **2017**, *10*, 1254. [CrossRef] [PubMed]

15. Rubio Cintas, M.D.; Parrón Vera, M.A.; Contreras de Villa, F. Nuevos usos de las escorias y polvos de humo provocados por la siderurgia. *An. Ing. Mec.* **2008**, *16*, 1233–1238.

16. Ortega, J.M.; Esteban, M.D.; Rodríguez, R.R.; Pastor, J.L.; Sánchez, I. Microstructural Effects of Sulphate Attack in Sustainable Grouts for Micropiles. *Materials* **2016**, *9*, 905. [CrossRef] [PubMed]

17. Rubio, M.D.; Parrón, M.A.; Contreras, F. Resistencia mecánica de hormigones con sustitución de un porcentaje de cemento por polvos de humo de sílice y escoria de horno de arco eléctrico. In *Comunicaciones V Congreso ACHE*; Asociación Científico-Técnica del Hormigón Estructural: Madrid, Spain, 2011; pp. 1–10.

18. Shi, C.; Zheng, K. A review on the use of waste glasses in the production of cement and concrete. *Resour. Conserv. Recycl.* **2007**, *52*, 234–247. [CrossRef]

19. Wang, H.; Sun, X.; Wang, J.; Monteiro, P.J.M. Permeability of concrete with recycled concrete aggregate and pozzolanic materials under stress. *Materials* **2016**, *9*, 252. [CrossRef] [PubMed]

20. Gambhir, M.L. *Concrete Technology: Theory and Practice*; Education, M.G.H., Ed.; Tata Mcgraw Hill Education Private Limited: New York, NY, USA, 2013; ISBN 8121900034.

21. Hadjsadok, A.; Kenai, S.; Courard, L.; Michel, F.; Khatib, J. Durability of mortar and concretes containing slag with low hydraulic activity. *Cem. Concr. Compos.* **2012**, *34*, 671–677. [CrossRef]

22. Gökalp, İ.; Uz, V.E.; Saltan, M.; Tutumluer, E. Technical and environmental evaluation of metallurgical slags as aggregate for sustainable pavement layer applications. *Transp. Geotech.* **2018**, *14*, 61–69. [CrossRef]

23. Behiry, A.E.A.E.M. Evaluation of steel slag and crushed limestone mixtures as subbase material in flexible pavement. *Ain Shams Eng. J.* **2013**, *4*, 43–53. [CrossRef]

24. Mahmoud, E.; Ibrahim, A.; El-Chabib, H.; Patibandla, V.C. Self-Consolidating Concrete Incorporating High Volume of Fly Ash, Slag, and Recycled Asphalt Pavement. *Int. J. Concr. Struct. Mater.* **2013**, *7*, 155–163. [CrossRef]

25. Sas, W.; Gluchowski, A.; Radziemska, M.; Dzieciol, J.; Szymanski, A. Environmental and geotechnical assessment of the steel slags as a material for road structure. *Materials* **2015**, *8*, 4857–4875. [CrossRef] [PubMed]

26. Xue, Y.; Wu, S.; Hou, H.; Zha, J. Experimental investigation of basic oxygen furnace slag used as aggregate in asphalt mixture. *J. Hazard. Mater.* **2006**, *138*, 261–268. [CrossRef] [PubMed]

27. Shahbazi, M.; Rowshanzamir, M.; Abtahi, S.M.; Hejazi, S.M. Optimization of carpet waste fibers and steel slag particles to reinforce expansive soil using response surface methodology. *Appl. Clay Sci.* **2017**, *142*, 185–192. [CrossRef]

28. Goodarzi, A.R.; Salimi, M. Stabilization treatment of a dispersive clayey soil using granulated blast furnace slag and basic oxygen furnace slag. *Appl. Clay Sci.* **2015**, *108*, 61–69. [CrossRef]

29. Manso, J.M.; Ortega-López, V.; Polanco, J.A.; Setién, J. The use of ladle furnace slag in soil stabilization. *Constr. Build. Mater.* **2013**, *40*, 126–134. [CrossRef]

30. Rodriguez, Á.; Manso, J.M.; Aragón, Á.; Gonzalez, J.J. Strength and workability of masonry mortars manufactured with ladle furnace slag. *Resour. Conserv. Recycl.* **2009**, *53*, 645–651. [CrossRef]

31. Hybská, H.; Hroncová, E.; Ladomerský, J.; Balco, K.; Mitterpach, J. Ecotoxicity of concretes with granulated slag from gray iron pilot production as filler. *Materials* **2017**, *10*, 505. [CrossRef] [PubMed]

32. Juckes, L.M. The volume stability of modern steelmaking slags. *Miner. Process. Extr. Metall. Trans. Inst. Min. Metall. Sect. C* **2003**, *112*, 177–197. [CrossRef]

33. Monkman, S.; Shao, Y.; Shi, C. Carbonated Ladle Slag Fines for Carbon Uptake and Sand Substitute. *J. Mater. Civ. Eng.* **2009**, *21*, 657–665. [CrossRef]

34. Wang, G.; Wang, Y.; Gao, Z. Use of steel slag as a granular material: Volume expansion prediction and usability criteria. *J. Hazard. Mater.* **2010**, *184*, 555–560. [CrossRef] [PubMed]

35. Nishigaki, M. Producing permeable blocks and pavement bricks from molten slag. *Stud. Environ. Sci.* **1997**, *71*, 31–40. [CrossRef]

36. Rubio, M.D.; Parrón, M.A.; Contreras, F. Method for producing cinder concrete. ES20130000758 20130803. 3 June 2015.

37. *European Committee for Standardization EN 12390-2:2009—Testing Hardened Concrete—Part 2: Making and Curing Specimens for Strength Tests*; European Comittee for Standardization: Brussels, Belgium, 2009.

38. *European Committee for Standardization EN 12350-2:2009 Testing Fresh Concrete. Slump-Test*; European Comittee for Standardization: Brussels, Belgium, 2009.

39. *European Committee for Standardization EN 12390-3: 2001.Testing Hardenes Concrete. Part 3: Compressive Strength of Test Specimens*; European Comittee for Standardization: Brussels, Belgium, 2001.

40. *European Comittee for Standardization EN 12390-4: 2009 Testing of Hardened Concrete. Part 4: Compressive Strength Specification for Testing Machines*; European Comittee for Standardization: Brussels, Belgium, 2009.

41. *European Comittee for Standardization EN 12390-8:2009 Testing Hardened Concrete. Part 8: Depth of Penetration of Water under Pressure*; European Comittee for Standardization: Brussels, Belgium, 2009.

42. Flatt, R.J. Towards a prediction of superplasticized concrete rheology. *Mater. Struct. Constr.* **2004**, *37*, 289–300. [CrossRef]

43. Ferraris, C.F.; Obla, K.H.; Hill, R. The influence of mineral admixtures on the rheology of cement paste and concrete. *Cem. Concr. Res.* **2001**, *31*, 245–255. [CrossRef]

44. Cánovas, M.F.; Gaitan, V.H. Comportamiento de hormigones de alta resistencia reforzados con fibras de acero frente al impacto de proyectiles. *Mater. Constr.* **2012**, *62*, 381–396. [CrossRef]

45. Sinica, M.; Sezeman, G.A.; Mikulskis, D.; Kligys, M.; Česnauskas, V. Impact of complex additive consisting of continuous basalt fibres and SiO_2 microdust on strength and heat resistance properties of autoclaved aerated concrete. *Constr. Build. Mater.* **2014**, *50*, 718–726. [CrossRef]

MDPI

St. Alban-Anlage 66

4052 Basel

Switzerland

Tel. +41 61 683 77 34

Fax +41 61 302 89 18

www.mdpi.com

Materials Editorial Office

E-mail: materials@mdpi.com

www.mdpi.com/journal/materials